CIM Series in Mathematical Sciences

The CIM Series in Mathematical Sciences is published on behalf of and in collaboration with the Centro Internacional de Matemática (CIM) in Portugal. Proceedings, lecture course material from summer schools and research monographs will be included in the new series.

More information about this series at http://www.springer.com/series/11745

Michael Hintermüller • José Francisco Rodrigues
Editors

Topics in Applied Analysis and Optimisation

Partial Differential Equations, Stochastic and Numerical Analysis

Joint CIM-WIAS Workshop, TAAO 2017,
Lisbon, Portugal, December 6-8, 2017

Springer

Editors
Michael Hintermüller
Weierstrass Institute for Applied Analysis
and Stochastics
Berlin, Germany

José Francisco Rodrigues
CMAF&IO, Faculdade de Ciências
Universidade de Lisboa
Lisboa, Portugal

ISSN 2364-950X ISSN 2364-9518 (electronic)
CIM Series in Mathematical Sciences
ISBN 978-3-030-33115-3 ISBN 978-3-030-33116-0 (eBook)
https://doi.org/10.1007/978-3-030-33116-0

Mathematics Subject Classification (2010): 35-06, 35Qxx, 35Rxx, 49-06, 49Nxx, 60-06, 65-06, 65Kxx, 65Mxx, 65Zxx

This Springer imprint is published by the registered company Springer Nature Switzerland AG
The registered company address is: Gewerbestrasse 11, 6330 Cham, Switzerland

Preface

This volume of the Springer-CIM Series is related to a workshop on "Topics in Applied Analysis and Optimisation (Stochastic, Partial Differential Equations and Numerical Analysis)" which was held in Lisbon from December 6–8, 2017. It brought together twenty four speakers in applied mathematics invited from both, the Portuguese International Center for Mathematics (CIM) and the Weierstrass-Institute for Applied Analysis and Stochastics (WIAS) Berlin, Germany. Both institutions are members of ERCOM (European Research Centres on Mathematics). CIM is a not-for-profit, privately-run association that aims at developing and promoting research in Mathematics. At present CIM has 20 associates, including three Portuguese Universities, thirteen Research Centres in the Mathematical Sciences, the Institute of Telecommunications, and three national scientific societies, respectively of Mathematics, of Statistics and of Mechanics. WIAS is a member of the Leibniz Association and conducts project oriented research in applied mathematics with the aim of solving complex problems in technology, science and the economy as well as biomedicine.

We also feel that a brief word on the birth of the idea of having the aforementioned workshop in Lisbon and on editing the present special volume within the Springer-CIM Series is in order: In fact, during a Workshop on "Emerging Developments in Interfaces and Free Boundaries" at the Mathematical Research Institute in Oberwolfach, yet another member of the Leibniz Association in Germany and ERCOM, the two editors of this volume came together in order to develop a format to foster joint research between ERCOM members. As a result, a scientific event was created with the aim to present and discuss current scientific interests among the research groups of the Weierstrass Institute in Berlin and mathematics centres in Portugal, in particular the Centro de Matemática, Aplicações Fundamentais e Investigação Operacional (CMAFcIO), of the University of Lisbon, and the Centro de Matemática of the University of Coimbra (CMUC).

This CIM -WIAS workshop finally brought together a selection of experts in Europe, and launched and strengthened further scientific collaborations in applied mathematics. Topics of particular interest included partial differential equations with applications to material sciences, thermodynamics and laser dynamics, scientific computing, nonlinear optimisation and stochastic analysis. In the outcome of the workshop, this collective book gathers fourteen contributions from four specific topical areas that were addressed in the meeting.

The three surveys on nonsmooth optimisation start with a chapter by A. Alphonse, M. Hintermüller and C. N. Rautenberg on elliptic stationary quasi-variational inequalities (QVIs), which is a class of quasi-equilibrium non-convex and non-smooth problems that has recently received increasing interest. Latest progress in that field includes the development of numerical solutions algorithms. In particular for the so-called gradient-constrained case, a Moreau-Yosida technique combined with a variable splitting approach can be addressed efficiently by alternating minimization schemes. The latter are the subject of the contribution by M. A. T. Figueiredo on the Alternating Direction Method of Multipliers (ADMM). The scope of that section, however, is much wider than just addressing (discretized) QVIs. Rather general classes of minimization problems are considered. The third survey, which is by J. F. Rodrigues and L. Santos, presents a general framework for a class of problems with gradient type constraints that can be formulated as stationary and evolution variational and quasi-variational inequalities and is illustrated with several physical applications.

The four contributions that related to stochastic methods involve a statistical approach to constructing non-asymptotic confidence sets in 2-Wasserstein space by J. Ebert, V. Spokoiny, and A. Suvorikova; recent modeling and developments from stochastic geometry to analyse spatial multiop communication systems by B. Jahnel and W. König; a model to deal with uncertain friction for the transport of natural gas through passive networks of pipelines, involving a simplification of the Euler equations and the use of a Markov chain Monte Carlo method, is considered by H. Heitsch and N. Strogies; and the use of Malliavin calculus to study invariant and quasi-invariant measures for the two dimensional Euler equation is surveyed by A. B. Cruzeiro and A. Symeonides.

Partial Differential Equations for dissipative and conservative models are ubiquitous in mathematical-physics problems, like in the contribution by M. Kantner, A. Mielke, M. Mittnenzweig and N. Rotundo on the modeling of semiconductors, from quantum mechanics to devices, respecting fundamental principles of non-equilibrium thermodynamics. These models are also of importance in describing gradient structures for two-phase flows of concentration suspensions, as in the chapter by D. Peschka, M. Thomas, T. Ahnert, A. Münch and B. Wagner. The numerical solutions for electrolyte flows modeled by coupling the Nernst-Planck-Poisson drift diffusion and Navier-Stokes equations is considered by J. Fuhrmann, C. Guhlke, A. Linke, C. Merdon and R. Müller, while models of dynamic damage and phase-field fracture, and their various time discretisations are presented by T. Roubíček.

Analytic and geometrical insights into modeling phase change problems are presented in the remaining three papers. In this respect, S. Amstutz and N. van Goethem present a survey and motivation of the incompatibility operator, a recent geometrical object introduced for a novel approach to elasto-plasticity problems, including a model for continua with dislocations; P. Colli, G. Gilardi and J. Sprekels show the well-posedness and stability for a classical nonlocal phase-field system of viscous Cahn-Hilliard type; and S. Dipierro and E. Valdinoci present recent progress on the fractional Allen-Cahn equation for long-range phase coexistence models.

Finally, we would like to acknowledge financial support from CMAFcIO and CMUC for the event held at the Faculdade de Ciências da Universidade de Lisboa and the Weierstrass Institute Berlin. For technical help in putting together the LaTeX collection of articles, our sincere thanks go to Anja Schröter (WIAS) and to Assis Azevedo (UMinho).

February 7, 2019.

Michael Hintermüller (Berlin)

José Francisco Rodrigues (Lisboa)

Contents

Recent Trends and Views on Elliptic Quasi-Variational Inequalities

Amal Alphonse, Michael Hintermüller, and Carlos N. Rautenberg

Abstract We consider state-of-the-art methods, theoretical limitations, and open problems in elliptic Quasi-Variational Inequalities (QVIs). This involves the development of solution algorithms in function space, existence theory, and the study of optimization problems with QVI constraints. We address the range of applicability and theoretical limitations of fixed point and other popular solution algorithms, also based on the nature of the constraint, e.g., obstacle and gradient-type. For optimization problems with QVI constraints, we study novel formulations that capture the multivalued nature of the solution mapping to the QVI, and generalized differentiability concepts appropriate for such problems.

1 Introduction

Quasi-Variational Inequalities represent a specific subclass of quasi-equilibrium problems in which non-convexity and non-smoothness are present. They play an important role in the modelling of complex phenomena in applied sciences, engineering, and economy, where compliancy or other state dependent bound constraints have to be taken care of. The nonlinear nature of the constraint set challenges the derivation of existence results and the design and analysis of associated solution algorithms.

A. Alphonse[1], M. Hintermüller[2], and C. N. Rautenberg[3]
Weierstrass Institute for Applied Analysis and Stochastics, Mohrenstraße 39, 10117 Berlin, Germany
e-mail: alphonse@wias-berlin.de, hintermueller@wias-berlin.de, rautenberg@wias-berlin.de

[2,3]Department of Mathematics, Humboldt-University of Berlin, Unter den Linden 6, 10099 Berlin, Germany.
e-mail: hint@math.hu-berlin.de, carlos.rautenberg@math.hu-berlin.de

M. Hintermüller, J. F. Rodrigues (eds.), *Topics in Applied Analysis and Optimisation*, CIM Series in Mathematical Sciences, https://doi.org/10.1007/978-3-030-33116-0_1

In the majority of available models, the state dependent constraint is of the form

$$\psi(Gy) \leq \Phi(y),$$

where ψ is a real-valued nonlinear function, G a linear operator, and Φ a nonlinear operator that is of superposition type or it is defined by the solution mapping of a nonlinear partial differential equation (PDE). For example, in the case of unilateral constraints, $\psi(x) = x$ and $G = \mathrm{id}$, and for gradient constraints, $\psi(x) = |x|$ and $G = \nabla$ is the weak gradient. Applications involving these restrictions include, but are not limited to, the magnetization of superconductors, Maxwell systems, thermohydraulics, image processing, game theory, surface growth of granular (cohensionless) materials, hydrology, and solid and continuum mechanics. For more details, we refer the reader to [21,25,34,52,54,56,64,65] and the monographs [14,53].

The goal of this paper is to present state-of-the-art results including mathematical limitations and open questions that arise in the treatment of QVIs. Specific focus topics involve existence of solutions, development of appropriate solvers together with some problematic issues found in the literature, optimal control, and directional differentiability of the QVI solution map.

Due to our aim of keeping the paper compact, we have not been able to include certain important approaches. In particular, in the case of gradient constraints, the QVI can be rewritten as a generalized equation. It then follows that these QVIs become a particular instance of a more general problem class; see, e.g., [48,51]. The latter approach was pioneered by Kenmochi and collaborators, and further work can be found in [24,26,46,47]. Also, we have not included the L^∞ contraction results from Hanouzet and Joly which are well documented in [31,32] and [15]. As we focus on the infinite dimensional setting in this paper, we have not included recent finite dimensional solvers associated with KKT-type and augmented Lagrangian methods; see [22,23,35,49,50]. In a similar vein, we have avoided discretization issues of closed convex sets which are required for consistency of numerical schemes and are deeply related to the density of smooth functions on the aforementioned sets; see [41,43].

The paper is organized as follows. The class of problems under consideration is described in section 1.1, where the basic functional analytic framework is established, and solutions to the QVIs of interest are equivalently described as fixed points of a specific nonlinear map T. In section 2, we consider some existence results involving compactness or increasing properties of the map T. Furthermore, we provide sufficient conditions for both properties and mention open questions concerning both approaches. Section 3 concerns iterative methods for solving QVIs. We state some results for obstacle, gradient, and more general constraints. Also, we focus on an unfortunate trend of the QVI literature that intends to extend the technique of the Lions–Stampacchia existence result to the QVI setting. We show that in general the approach is rather restrictive and that the assumption of the Lipschitz continuity of the projection map $\mathbf{K} \mapsto \mathrm{P}_{\mathbf{K}}$, frequently made, does not hold in general. Subsequently, we consider iterations that converge in case of multiple solutions and regularization approaches of Moreau–Yosida and Gerhardt type. We finalize the section by addressing drawbacks associated with the simple fixed point iteration $y_n = T(y_{n-1})$. In section 4, we state optimal control problems with QVI constraints that take into account the multivalued nature of the solution set. In particular, utilizing a control reduced form of the problem leads to a formulation in terms of minimum and maximum solutions to the QVI. A newly established directional differentiability result for the QVI map is provided in section 5, where the classical result of Mignot is extended accordingly to the QVI framework.

1.1 The basic setting and problem formulation

We consider V to be a reflexive real Banach space of (equivalence) classes of maps of the type $v : \Omega \to \mathbb{R}$ for some Lipschitz domain $\Omega \subset \mathbb{R}^N$ with $N \in \mathbb{N}$ and norm denoted by $\|\cdot\|_V$. Its topological dual is denoted by V' and the pairing between V' and V is given by $\langle\cdot,\cdot\rangle$. If V is a Hilbert space, then $(\cdot,\cdot)$ denotes its inner product. For a sequence $\{v_n\}$ in V, strong and weak convergence to $v \in V$ are written as "$v_n \to v$" and "$v_n \rightharpoonup v$", respectively.

For a map $K\colon V \to W$, where W is a Banach space, we say that K is *completely continuous* if $v_n \rightharpoonup v$ in V implies $K(v_n) \to K(v)$ in W. Since V is reflexive, a completely continuous map is compact; see [71, Chapter II, Lemma 1.1].

Throughout the paper we consider a (possibly nonlinear) operator $A\colon V \to V'$ that is Lipschitz continuous and uniformly monotone, i.e., there exist constants $c > 0$ and $C > 0$ such that for all $u, v \in V$,

$$\|A(u) - A(v)\|_{V'} \leq C\|u - v\|_V, \tag{A1}$$

and

$$\langle A(u) - A(v), u - v\rangle \geq c\|u - v\|_V^r, \tag{A2}$$

for some constant $r > 1$. If V is a Hilbert space, then $r = 2$. In addition, we assume that $A(0) = 0$.

The typical setting that we consider here is with $V := W_0^{1,p}(\Omega)$, with $\Omega \subset \mathbb{R}^N$ a bounded Lipschitz domain, $2 \leq p < +\infty$, and $A := -\Delta_p$, the p–Laplacian, given by

$$\langle -\Delta_p(u), v\rangle := \int_\Omega |\nabla u|^{p-2}\nabla u \cdot \nabla v \, dx, \quad \text{for } u, v \in W_0^{1,p}(\Omega).$$

In this case, $c = 1$ and $r = p$.

The general problem class under consideration is given as follows.

Problem $(\mathrm{P_{QVI}})$: Given $f \in V'$,

$$\text{find } y \in \mathbf{K}(y) : \langle A(y) - f, v - y\rangle \geq 0, \quad \forall v \in \mathbf{K}(y). \tag{$\mathrm{P_{QVI}}$}$$

The general structure of $v \mapsto \mathbf{K}(v)$ is given by

$$\mathbf{K}(v) := \{w \in V : \psi(Gw) \le \Phi(v)\}, \tag{1.1}$$

where $\Phi(v)\colon \Omega \to \mathbb{R}$ is a measurable function for each v and "$v \le w$" means that $v(x) \le w(x)$ for almost all (f.a.a.) $x \in \Omega$, unless stated otherwise. We assume that $G \in \mathscr{L}(V, L^p(\Omega)^d)$ for some $1 < p < +\infty$ and $d \in \mathbb{N}$, that is, $G\colon V \to L^p(\Omega)^d$ is linear and bounded. Additionally, we suppose that $\psi : \mathbb{R}^d \to \mathbb{R}$ is convex. For the sake of simplicity, we assume that $\mathbf{K}(v)$ is always non-empty for each v. The closedness and convexity of $\mathbf{K}(v)$ follow from the assumptions invoked here. Additionally, we assume that $v \mapsto \max(0, v)$, and $v \mapsto \min(0, v)$ are continuous with respect to the weak and strong topologies of V.

We distinguish at least two notable cases both for $V = W_0^{1,p}(\Omega)$. If $\psi(Gw) = w$, we refer to the problem as the *obstacle* case. If $\psi(x) = |x|$ and $G = \nabla$ is the weak gradient so that $G\colon W_0^{1,p}(\Omega) \to L^p(\Omega)$, we refer to the problem as the *gradient* case.

We denote the solution set to ($\mathrm{P_{QVI}}$) for a given $f \in V'$ by $\mathbf{Q}(f)$, and note that in general $\mathbf{Q}(f)$ contains more than one element. It is convenient to characterize $\mathbf{Q}(f)$ as the set of fixed points of a certain map. In this light, consider $\mathbf{K} \subset V$ non-empty, closed and convex. Then for any $f \in V'$, we define $S(f, \mathbf{K})$ as the unique solution to:

$$\text{Find } y \in \mathbf{K} : \langle A(y) - f, v - y \rangle \ge 0, \quad \forall v \in \mathbf{K}. \tag{1.2}$$

Also, for the map $v \mapsto \mathbf{K}(v)$ given as above, we consider

$$T(v) := S(f, \mathbf{K}(v)). \tag{1.3}$$

It then follows that solutions to ($\mathrm{P_{QVI}}$) are equivalently defined as solutions to

$$T(v) = v.$$

In general for an operator R, we denote the set of fixed points by $\mathrm{Fix}(R)$.

2 Some existence theory

In this section we provide an overview of techniques available to prove existence of solutions to QVIs and the limitations and caveats associated with the utilized techniques. We focus on compactness results and ordering approaches. Contraction methods, however, are left for the section on solution algorithms.

2.1 Compactness and Mosco convergence

One approach to prove existence of a fixed point of T is based on compactness of the map T. In particular, since V is reflexive, it is enough to consider the complete continuity of T, i.e., given $v_n \rightharpoonup v$, then $T(v_n) \to T(v)$ in V; see [71, Chapter II,

Lemma 1.1.]. Then, a suitable fixed point theorem yields existence. Note, however, that this is directly associated with a notion of set convergence for $\{\mathbf{K}(v_n)\}$, as introduced by Mosco; see [62].

Definition 1 (Mosco convergence) Let $\mathbf{K}$ and $\mathbf{K}_n$, for each $n \in \mathbb{N}$, be non-empty, closed and convex subsets of V. Then the sequence $\{\mathbf{K}_n\}$ *is said to converge to* $\mathbf{K}$ *in the sense of Mosco* as $n \to \infty$, denoted by $\mathbf{K}_n \xrightarrow{\text{M}} \mathbf{K}$, if the following two conditions are fulfilled:

(i) For each $w \in \mathbf{K}$, there exists $\{w_{n'}\}$ such that $w_{n'} \in \mathbf{K}_{n'}$ for $n' \in \mathbb{N}' \subset \mathbb{N}$ and $w_{n'} \to w$ in V.
(ii) If $w_n \in \mathbf{K}_n$ and $w_n \rightharpoonup w$ in V along a subsequence, then $w \in \mathbf{K}$.

The importance of Mosco convergence lies in the following continuity result: Let $f_n \to f$ in V', then

$$\mathbf{K}_n \xrightarrow{M} \mathbf{K} \quad \text{implies} \quad S(f_n, \mathbf{K}_n) \to S(f, \mathbf{K}) \text{ in } V.$$

The proof can be found in [66]. The above fact implies that if $v_n \rightharpoonup v$ in V yields $\mathbf{K}(v_n) \xrightarrow{M} \mathbf{K}(v)$, then $T\colon V \to V$ is compact. Using $v = 0$ in (1.2), we observe that $T(V) \subset \overline{B}_{c^{-1}\|f\|_{V'}}(0; V)$, the closed ball in V of radius $c^{-1}\|f\|_{V'}$ and center at 0. Hence by Schauder's fixed point theorem, the equation $T(y) = y$ has solutions in V.

The full characterization of Mosco convergence of $\{\mathbf{K}(v_n)\}$ based on properties of Φ, ψ, and G, is a complex task and to this day, only partial answers are available. Specifically, condition (i) in Definition 1, commonly referred to as the *recovery sequence condition*, is delicate to check in applications, while (ii) admits the following simple and general characterization.

Proposition 2.1 *Suppose that* $\Phi\colon V \to L^q(\Omega)$, *for some* $1 \le q \le +\infty$, *is completely continuous, and* $v_n \rightharpoonup v$ *in* V. *Then* (ii) *in Definition 1 holds true for* $\mathbf{K}_n = \mathbf{K}(v_n)$ *and* $\mathbf{K} = \mathbf{K}(v)$.

Proof For $w_n \in \mathbf{K}(v_n)$, we have $\psi(Gw_n) \le \Phi(v_n)$, and if $w_n \rightharpoonup w$ in V, it follows that $Gw_n \rightharpoonup Gw$ in $L^p(\Omega)^d$. By Mazur's lemma, there exists $z_n = \sum_{k=n}^{N(n)} \alpha(n)_k Gw_k$ where $\sum_{k=n}^{N(n)} \alpha(n)_k = 1$ and $\alpha(n)_k \ge 0$ such that $z_n \to Gw$ in $L^p(\Omega)^d$. Since $\psi : \mathbb{R}^d \to \mathbb{R}$ is convex,

$$\psi(z_n) \le \sum_{k=n}^{N(n)} \alpha(n)_k \psi(Gw_k) \le \Phi(v) + \underbrace{\sum_{k=n}^{N(n)} \alpha(n)_k |\Phi(v_k) - \Phi(v)|}_{\epsilon_n}.$$

As $v_n \rightharpoonup v$ in V, we have $\Phi(v_n) \to \Phi(v)$ in $L^q(\Omega)$, and hence $\epsilon_n \to 0$ in $L^q(\Omega)$. Therefore, we obtain $w \in \mathbf{K}(v)$ by taking the limit above (over some subsequence converging in the pointwise almost everywhere sense). □

Perhaps the simplest situation in which (i) holds is the obstacle case with $\Phi\colon V \to V$ completely continuous. Let $w \le \Phi(v)$ be arbitrary and $v_n \rightharpoonup v$ in V, and define

$w_n := \min(w, \Phi(v_n))$ so that $w_n \le \Phi(w_n)$. Since $\Phi(v_n) \to \Phi(v)$ in V, it follows that $w_n \to w$ in V. Consequently (i) holds true. Note that we assume that $V \ni z \mapsto \min(0, z) \in V$ is continuous. The relaxation of the complete continuity assumption for Φ is an arduous task that we consider in what follows.

2.1.1 The result of Boccardo and Murat

A typical function space setting for our focus problem ($\mathrm{P_{QVI}}$) is given by $V = W_0^{1,p}(\Omega)$, for $1 < p < +\infty$, and obstacle-type constraints. As seen above, if $\Phi\colon W^{1,p}(\Omega) \to W^{1,p}(\Omega)$ is completely continuous, then the map T is compact. This can be relaxed substantially by means of the compactness result of Murat in [63]. It states that if $F_n \rightharpoonup F$ in $H^{-1}(\Omega)$ with $F_n \ge 0$ for all $n \in \mathbb{N}$, then $F_n \to F$ in $W^{-1,q}(\Omega)$ with $q < 2$. Here, $F_n \ge 0$ refers to $\langle F_n, \sigma \rangle \ge 0$ for all $\sigma \in V$ with $\sigma \ge 0$. Moreover, the regularity of $\partial\Omega$ can be dropped and the result still remains intact [19]. In our setting, this result leads to the following useful assertion; see [17, 18].

Theorem 2.1 (Boccardo-Murat) *Suppose that $v_n \rightharpoonup v$ in $W_0^{1,p}(\Omega)$ implies $\Phi(v_n) \rightharpoonup \Phi(v)$ in $W^{1,q}(\Omega)$ or $W_0^{1,q}(\Omega)$ for some $q > p$. Then $\mathbf{K}(v_n) \xrightarrow{M} \mathbf{K}(v)$.*

We note that counterexamples can be constructed for $q = p$. In words, the above result relies on the fact that Φ realizes an increase in regularity and preserves weak continuity.

Open problems. For QVIs with similar constraint types as considered here but with fractional order operators A, a result analogous to the one in Theorem 2.1 appears unavailable. For this kind of operators, the QVI can be equivalently formulated in weighted Sobolev spaces, see [4]. In this context, it is an open question whether it is possible to extend the above result of Boccardo and Murat to weighted Sobolev spaces $W_0^{1,p}(\Omega; w)$ for some w in a Muckenhoupt class. □

2.1.2 Gradient and further cases

The cases other than the obstacle one are significantly more difficult, mainly due to the possible nonlinearity $\psi : \mathbb{R}^d \to \mathbb{R}$. Here, we consider the setting where $V = W_0^{1,p}(\Omega)$ with $1 < p < +\infty$. The following result is based on [12, 40, 54]

Proposition 2.2 *Let $G \in \mathscr{L}(W_0^{1,p}(\Omega), L^p(\Omega)^d)$ for some $d \in \mathbb{N}$, and let $\psi : \mathbb{R}^d \to \mathbb{R}$ be (positive) homogeneous of degree one, i.e., $\psi(tx) = t\psi(x)$ for any $x \in \mathbb{R}^d$ and $t > 0$. Suppose that $\Phi\colon W_0^{1,p}(\Omega) \to L_\eta^\infty(\Omega) \subset L^\infty(\Omega)$ is completely continuous, where $L_\eta^\infty(\Omega) := \{v \in L^\infty(\Omega) : v \ge \eta > 0 \text{ a.e.}\}$. Then, we have that*

$$v_n \rightharpoonup v \text{ in } W_0^{1,p}(\Omega) \quad \text{implies} \quad \mathbf{K}(v_n) \xrightarrow{M} \mathbf{K}(v).$$

Proof First note that by assumption, $\Phi\colon W_0^{1,p}(\Omega) \longrightarrow L^p(\Omega)$ is also completely continuous. Thus, by Proposition 2.1 we only need to prove the recovery sequence part for Mosco convergence. For this purpose and for $v_n \rightharpoonup v$ in $W_0^{1,p}(\Omega)$, define

$$\beta_n := \left(1 + \frac{\|\Phi(v_n) - \Phi(v)\|_{L^\infty}}{\eta}\right)^{-1}.$$

If $\psi(Gw) \leq \Phi(v)$, then it follows for $w_n := \beta_n w$ that $w_n \longrightarrow w$ in $W_0^{1,p}(\Omega)$ and $\psi(Gw_n) \leq \Phi(v_n)$ (see [40]) which finishes the proof. □

We note that the previous result only provides *sufficient* conditions for Mosco convergence; this leads to another open problem.

Open problems. Find sufficient and necessary conditions on ϕ_n, ϕ such that

$$\{w \in W_0^{1,p}(\Omega) : \psi(Gw) \leq \phi_n\} \xrightarrow{M} \{w \in W_0^{1,p}(\Omega) : \psi(Gw) \leq \phi\}.$$

Similarly, it is an open question whether $\phi_n \rightharpoonup \phi$ in $W^{1,q}(\Omega)$ for some q suffices to guarantee the above Mosco convergence in the gradient case by other means than embeddings. □

2.2 Order approaches

We consider now an approach based on order that was pioneered by Tartar; see [72] and also [8, Chapter 15, §15.2]. Let (V, H, V') be a Gelfand triple of Hilbert spaces, that is, we have $V \hookrightarrow H \hookrightarrow V'$, where the embedding $V \hookrightarrow H$ is dense and continuous, and H is identified with its topological dual H' so that the embedding $H \hookrightarrow V'$ is also dense and continuous. Within this section, $(\cdot, \cdot)$ denotes the inner product in H.

We assume that $H_+ \subset H$ is a convex cone with

$$H_+ = \{v \in H : (v, y) \geq 0 \text{ for all } y \in H_+\}.$$

Based on this, we use the following ordering denoted by "$\leq$":

$$x \leq y \quad \text{if and only if} \quad y - x \in H_+.$$

For $x \in H$, we have the decomposition $x = x^+ - x^- \in H_+ - H_+$ with $(x^+, x^-) = 0$ such that x^+ denotes the orthogonal projection onto H_+ and $x^- = x - x^+$ the one onto $H_- = -H_+$. The infimum and supremum of two elements $x, y \in H$ are defined as $\sup(x, y) := x + (y - x)^+$ and $\inf(x, y) := x - (x - y)^+$ respectively.

The supremum of an arbitrary subset of H that is bounded (in the order) above is also correctly defined since H is Dedekind complete: A set $\{x_i\}_{i \in J}$ where J is completely ordered and bounded from above implies that $\{x_i\}_{i \in J}$ is a generalized Cauchy sequence in H (see [9, Chapter 15, §15.2, Proposition 1]). From this Dedekind completeness follows (see [2, Chapter 4, Theorem 4.9 and Corollary 4.10]). This additionally implies that norm convergence preserves order. Indeed, if $z_n \leq y_n$ for each $n \in \mathbb{N}$ and $z_n \to z$ and $y_n \to y$ both in H, then $z \leq y$.

Finally, we assume that

$$y \in V \Rightarrow y^+ \in V \qquad \text{and} \qquad \exists \mu > 0 : \|y^+\|_V \leq \mu \|y\|_V, \forall y \in V.$$

Then the order in H induces one in V', as well. In fact, for $f, g \in V'$, we write $f \leq g$ if $\langle f, \phi\rangle \leq \langle g, \phi\rangle$ for all $\phi \in V_+ := V \cap H_+$ and define $V'_+ := \{f \in V' : f \geq 0\}$.

The typical example in this framework is given by the Gelfand triple $(V, H, V') = (H_0^1(\Omega), L^2(\Omega), H^{-1}(\Omega))$. Here, $H_+ = L^2(\Omega)^+$, the set of almost everywhere (a.e.) non-negative functions, and $v \leq w$ denotes that $v(x) \leq w(x)$ for almost all (f.a.a.) $x \in \Omega$.

In this section, we assume that the operator $A\colon V \to V'$ is strictly T-monotone, i.e.,

$$\langle A(y) - A(z), (y - z)^+\rangle > 0, \qquad \forall y, z \in V : (y - z)^+ \neq 0. \tag{A3}$$

In particular, if A is linear, then the above is equivalent to $\langle Ay^-, y^+\rangle \leq 0$ for all $y \in V$, and we have maximum principles available for A. In addition, consider the following definition.

Definition 2 A map $R\colon V \to V$ is said to be increasing if for $y, z \in V$ we have that

$$y \leq z \quad \text{implies} \quad R(y) \leq R(z).$$

The following general result concerning existence of fixed points for increasing maps is the fundamental tool to prove existence of solutions to problem $(\mathrm{P_{QVI}})$.

Theorem 2.2 (Tartar–Birkhoff) *Let $R\colon V \to V$ be increasing, and suppose that there exist $\underline{y}, \overline{y} \in V$ such that*

$$\underline{y} \leq \overline{y}, \qquad \underline{y} \leq R(\underline{y}), \quad \textit{and} \quad R(\overline{y}) \leq \overline{y}.$$

Then the set $\mathrm{Fix}(R) \cap [\underline{y}, \overline{y}]$ is non-empty. Furthermore, there exist $y_1, y_2 \in \mathrm{Fix}(R) \cap [\underline{y}, \overline{y}]$ such that

$$y \in \mathrm{Fix}(R) \cap [\underline{y}, \overline{y}] \quad \Rightarrow \quad y \in \mathrm{Fix}(R) \cap [y_1, y_2].$$

The above theorem mainly states that if a map is increasing, has a subsolution y_1 and a supersolution y_2, then it has a fixed point between (with respect to the order induced in H) y_1 and y_2. Moreover, there are minimal and maximal fixed points in $[y_1, y_2]$.

For the map $T\colon V \to V$ to be increasing, some assumptions are required on the structure of $\mathbf{K}$. For this purpose consider the obstacle case and assume that $\Phi\colon V \to H$ is increasing. Also, suppose that $f_{\min} \leq f \leq f_{\max}$ for some $f_{\min}, f_{\max} \in V'$, and that $\Phi(A^{-1} f_{\min}) \geq A^{-1} f_{\min}$. Then, it follows that

$$\underline{y} = A^{-1} f_{\min} \qquad \text{and} \qquad \overline{y} = A^{-1} f_{\max}$$

are sub- and supersolutions, respectively, of T, and all assumptions of the previous theorem are satisfied. Hence, defining $\mathbf{A}_{\text{ad}} = \{g \in V' : f_{\min} \le g \le f_{\max}\}$, we have the operators

$$\mathsf{m}\colon \mathbf{A}_{\text{ad}} \to V \qquad \text{and} \qquad \mathsf{M}\colon \mathbf{A}_{\text{ad}} \to V$$

that take elements of $\mathbf{A}_{\text{ad}}$ to minimal and maximal solutions to ($\mathrm{P}_{\mathrm{QVI}}$) in the interval $[\underline{y}, \overline{y}] = [A^{-1} f_{\min}, A^{-1} f_{\max}]$.

Open problems. Characterize the stability of the maps $f \mapsto \mathsf{m}(f)$ and $f \mapsto \mathsf{M}(f)$. Specifically, if $\{f_n\}$ is in $\mathbf{A}_{\text{ad}}$, identify conditions on the sequence $\{f_n\}$ so that

$$\mathsf{m}(f_n) \to \mathsf{m}(f) \quad \text{and} \quad \mathsf{M}(f_n) \to \mathsf{M}(f)$$

in H and in V. □

3 Solution methods and algorithms

Next we concentrate on solution methods for problem ($\mathrm{P}_{\mathrm{QVI}}$) which are constructive in the sense that they can also be used to show existence of solutions. We focus first on contraction results without the aid of T-monotonicity properties of A, i.e., assumption (A3). In section 3.2, we focus on some problematic tendencies in the literature that attempt to generalize the Lions–Stampacchia existence result on VIs [57] to QVIs. We show that in general, such approaches provide worse results than a simple change of variables and the direct use of (A1). In section 3.3, we exploit ordering properties and consider iterations that converge to $\mathsf{m}(f)$ and $\mathsf{M}(f)$ under appropriate assumptions. Additionally, we consider regularization methods for the constraint $y \in \mathbf{K}(y)$ of the Moreau–Yosida and Gerhardt-type in section 3.4. In the former case, we show how the approach is suitable for Newton-type solvers. We end this section with considerations of the iteration $y_{n+1} = T(y_n)$ when only compactness of T is available.

3.1 Contraction results for *T*

Uniqueness of solutions to ($\mathrm{P}_{\mathrm{QVI}}$) is rarely available. However, in some cases it is possible to obtain that $v \mapsto S(f, \mathbf{K}(v))$ is contractive for a sufficiently small f and with Φ Lipschitz with sufficiently small Lipschitz constant. The interpretation of these prerequisites is as follows: If the Lipschitz constant of Φ satisfies $L_\Phi \ll 1$, then $\Phi(\cdot) \simeq$ constant, and hence it is expected that ($\mathrm{P}_{\mathrm{QVI}}$) is close to a variational inequality and admits a unique solution under such assumptions.

3.1.1 Obstacle case

We provide first a simple example associated to the obstacle case that arises when Φ preserves the regularity of the state space (the reason to describe such a simple case is related to the digression in section 3.2).

In the obstacle case, provided that $\Phi\colon V \to V$ is Lipschitz, we can consider the change of variable $z = y - \Phi(v)$. Hence, it is straightforward to prove, via the monotonicity of A, that T satisfies

$$\|T(v_1) - T(v_2)\|_V \leq \frac{1}{c}\|A\Phi(v_1) - A\Phi(v_2)\|_{V'} \leq \frac{C}{c}L_\Phi\|v_1 - v_2\|_V.$$

Consequently, for

$$\frac{C}{c}L_\Phi < 1,$$

the map T has a unique fixed point and the iteration $y_{n+1} = T(y_n)$ converges to this fixed point for any initial $y_0 \in V$. The extent of the usage of this technique is limited to the very case described here. Note also that if $V = H_0^1(\Omega)$, then the assumptions here also imply that $\Phi(v) = 0$ on $\partial\Omega$ in the sense of the trace.

The case $L_\Phi = 1$ may lead to a degenerate situation: Consider $\Phi(y) = y$. Then $y \in \mathbf{K}(y)$ is always satisfied and $v \leq \mathbf{K}(y)$ implies $v - y \leq 0$, so that ($\mathrm{P_{QVI}}$) is equivalent to the problem: Find $y \in V$ such that $Ay \leq f$ in V'. This implies that $A^{-1}g$ is a solution to this problem for every $g \leq f$ in V'.

3.1.2 Gradient and further cases

In other than the obstacle case, contraction results are far more elusive and when available, the contraction rates depend heavily on the regularity and magnitude of the data as we see next. The result is a slight generalization of [40, 42].

We consider the case $V = W_0^{1,p}(\Omega)$ with $A\colon W_0^{1,p}(\Omega) \to W^{-1,p'}(\Omega)$ not necessarily linear, but homogeneous with degree $\beta \geq 1$, i.e., $A(ty) = t^\beta A(y)$ for $t > 0$ and $y \in W_0^{1,p}(\Omega)$, and with monotonicity exponent $r \leq \min(2, p)$ in (A2). We consider $f \in L^{r'}(\Omega) \subset W^{-1,p'}(\Omega)$ where $1/r + 1/r' = 1$ and $1/p + 1/p' = 1$.

Let $G \in \mathscr{L}(W_0^{1,p}(\Omega), L^p(\Omega)^d)$ for some $d \in \mathbb{N}$, and $\psi\colon \mathbb{R}^d \to \mathbb{R}$ such that $\psi(tx) = t\psi(x)$ for $t > 0$. Many examples fit this setting. For instance $G := \nabla$, the weak gradient, or $G :=$ div, the weak divergence, together with $\psi(x) = |x|$ corresponding to the Euclidian norm in $\mathbb{R}^N$ or the absolute value respectively. Consider the map $\Phi\colon W_0^{1,p}(\Omega) \to L_\nu^\infty(\Omega)$ defined as $\Phi(u) = \lambda(u)\phi$ where λ is a nonlinear Lipschitz continuous functional and $\phi \in L^\infty(\Omega)$.

Theorem 3.3 (**[40]**) *In the above described setting, we have*

$$\|T(v_1) - T(v_2)\|_{W_0^{1,p}} \leq L(f)\|v_1 - v_2\|_{W_0^{1,p}},$$

where $L(f) \to 0$ *as* $\|f\|_{L^{r'}} \to 0$.

Hence, for small data we observe existence of a unique fixed point of T, and thus a unique solution to ($\mathrm{P}_{\mathrm{QVI}}$). A proof is given in [40, Theorem B.1].

Relaxing the hypothesis on the structure of Φ typically rules out contraction or even Lipschitz continuity. In order to see this, note that if $\phi_1, \phi_2 \in L^\infty(\Omega)$ and $\mathbf{K}_i := \{v \in W_0^{1,p}(\Omega) : \|\nabla v\|_{\mathbb{R}^N} \leq \phi_i \text{ a.e.}\}$ then

$$\|S(f, \mathbf{K}_1) - S(f, \mathbf{K}_2)\|_{W_0^{1,p}} \leq M(f)\|\phi_1 - \phi_2\|_{L^\infty}^{1/r} \tag{3.1}$$

where r is the constant in (A2). That is, the map is only Hölder continuous in general; see [40, 70]

Open problems. The extension of the result of Theorem 3.3 from the rank one case, $\Phi(y) = \lambda(y)\phi$, to the finite rank case, $\Phi(y) = \lambda_1(y)\phi_1 + \lambda_2(y)\phi_2 + \cdots + \lambda_m(y)\phi_m$, is still an open task.

Additionally, improvements (if possible) on the exponent $1/r$ in (3.1) have yet to be found, although the Lipschitz continuity result seems unattainable; see also section 3.2. □

3.2 The map $\mathbf{K} \mapsto \mathrm{P}_{\mathbf{K}}$ and extensions to Lions–Stampacchia

We restrict ourselves in this section to the Hilbert space setting and describe now a common misleading approach found in the literature. This unfortunate technique is based on aiming to extend the theorem of Lions and Stampacchia in [57] to the QVI framework.

Let $i : V \to V'$ denote the duality operator, that is, the canonical isomorphism defined as $\langle iu, v\rangle := (u, v)$, and its inverse $i^{-1} := j$ is the Riesz map for V. Here, problem ($\mathrm{P}_{\mathrm{QVI}}$) can be equivalently written as

$$\text{Find } y \in \mathbf{K}(y) : (y - jH_\rho(y), v - y) \geq 0, \quad \forall v \in \mathbf{K}(y)$$

for $H_\rho(w) = iw - \rho(A(w) - f)$ with $w \in V$, and any $\rho > 0$. Then, the existence of a solution to ($\mathrm{P}_{\mathrm{QVI}}$) can be transferred to finding $y \in V$ satisfying $y = B_\rho(y)$ with

$$B_\rho(y) := \mathrm{P}_{\mathbf{K}(y)}(y - \rho j(A(y) - f))$$

for some $\rho > 0$. Here $\mathrm{P}_{\mathbf{K}(y)} : V \to V \subset \mathbf{K}(y)$ is the projection map, i.e., for any $v \in V$, $\mathrm{P}_{\mathbf{K}(y)}(v)$ is the unique element in $\mathbf{K}(y)$ such that

$$\|\mathrm{P}_{\mathbf{K}(y)}(v) - v\|_V = \inf_{w \in \mathbf{K}(y)} \|w - v\|.$$

In the case where $\Phi(y) = \phi$ for all y, it follows that B_ρ is a contraction provided that $0 < \rho < 2c/C^2$, where c, C are the monotonicity and Lipschitz constant of A, respectively, given in (A1) and (A2). In fact, we have

$$\|B_\rho(v) - B_\rho(w)\|_V \leq \sqrt{1 - 2\rho c + \rho^2 C^2}\|v - w\|_V.$$

A significant amount of literature on QVIs is based on trying to extend this result to the quasi-variational setting. This approach relies on the hard assumption

$$\|\mathrm{P}_{\mathbf{K}(y)}(w) - \mathrm{P}_{\mathbf{K}(z)}(w)\|_V \leq \eta\|y - z\|_V \tag{3.2}$$

for some $0 < \eta < 1$ and all y, z, w in a bounded set in V. This should not be confused with the non-expansiveness of the map $z \mapsto \mathrm{P}_{\mathbf{K}(y)}(z)$, i.e., we have that $\|\mathrm{P}_{\mathbf{K}(y)}(z_1) - \mathrm{P}_{\mathbf{K}(y)}(z_2)\|_V \leq \|z_1 - z_2\|_V$, for all $y, z_1, z_2 \in V$. In general, (3.2) is *not* valid, and the only framework (in our setting) where it seems to work is in the obstacle type case with $\Phi\colon V \to V$. Indeed, in the latter case we see that the projection map can be rewritten in simpler terms as

$$\mathrm{P}_{\mathbf{K}(y)}(w) = \Phi(y) + \mathrm{P}_{\{z \in V : z \leq 0\}}(w - \Phi(y)). \tag{3.3}$$

Note that it is necessary for this representation that Φ preserves the V regularity. For example if $V = H_0^1(\Omega)$ and Φ maps V into $L^2(\Omega)$ but not into $H_0^1(\Omega)$, this V-regularity requirement is no longer valid.

In case (3.3) holds, a solution to the QVI is equivalently a fixed point of the map B_ρ now defined as

$$B_\rho(y) := \Phi(y) + \mathrm{P}_{\{z \in V : z \leq 0\}}((y - \rho j(A(y) - f) - \Phi(y)),$$

which satisfies

$$\|B_\rho(v) - B_\rho(w)\|_V \leq (2L_\Phi + \sqrt{1 - 2\rho c + \rho^2 C^2})\|v - w\|_V.$$

In order for B_ρ to be contractive, a first observation is that we need

$$2L_\Phi + \sqrt{1 - \left(\frac{c}{C}\right)^2} < 1,$$

which implies that

$$\frac{C}{c} L_\Phi < \frac{1}{2}.$$

This is a much more restrictive and convoluted approach than the one described in section 3.1.1, where only $\frac{C}{c} L_\Phi < 1$ is required! Furthermore, the linear convergence rate (in case of a contraction) in this case is worse than the one in section 3.1.1, given by $\frac{C}{c} L_\Phi$.

There is a deep and interesting reason why condition (3.2) fails in a general setting. The result in question was described by Attouch and Wets in [5–7], and it involves continuity properties of $\mathbf{K} \mapsto \mathrm{P}_{\mathbf{K}}$. This is given in the following section.

3.2.1 The map $\mathbf{K} \mapsto \mathrm{P}_{\mathbf{K}}$

For any closed, non-empty and convex set $\mathbf{K}$ in V, we define the distance function of an element $y \in V$ to the set $\mathbf{K}$ as

$$\mathrm{d}(y, \mathbf{K}) := \inf_{z \in \mathbf{K}} \|z - y\|_V,$$

and for two closed, non-empty, and convex sets $\mathbf{K}_1, \mathbf{K}_2$ we define the excess function e as

$$\mathrm{e}(\mathbf{K}_1, \mathbf{K}_2) := \sup_{z \in \mathbf{K}_1} \mathrm{d}(z, \mathbf{K}_2).$$

For any $\rho \geq 0$, the ρ-Hausdorff distance between $\mathbf{K}_1$ and $\mathbf{K}_2$ is given by

$$\mathrm{haus}_\rho(\mathbf{K}_1, \mathbf{K}_2) := \sup(\mathrm{e}(\mathbf{K}_1^\rho, \mathbf{K}_2), (\mathrm{e}(\mathbf{K}_2^\rho, \mathbf{K}_1)),$$

where $\mathbf{K}_i^\rho := \mathbf{K}_i \cap \rho B$, $i = 1, 2$, and B is the open unit ball centered at zero. Then, we have (see [6, Proposition 5.3]) the following.

Theorem 3.4 (Attouch–Wets) *Let V be a Hilbert space and $\mathbf{K}_1, \mathbf{K}_2$ any two closed, convex, non-empty subsets of V. For $y_0 \in V$, we have that*

$$\|\mathrm{P}_{\mathbf{K}_1}(y_0) - \mathrm{P}_{\mathbf{K}_2}(y_0)\|_V \leq \rho^{1/2} \mathrm{haus}_\rho(\mathbf{K}_1, \mathbf{K}_2)^{1/2}$$

for $\rho := \|y_0\| + \mathrm{d}(y_0, \mathbf{K}_1) + \mathrm{d}(y_0, \mathbf{K}_2)$.

The $1/2$ exponent in the right hand side expression is optimal, and examples (even in finite dimensions) can be found where equality holds. Additionally, in Banach spaces like $L^p(\Omega)$ or $\ell^p(\mathbb{N})$, the exponent degrades even further: it is $1/p$ if $2 < p < +\infty$ and $1/p'$ if $1 < p < 2$ where p' is the Hölder conjugate of p.

In order to understand how this result fully translates into our class of maps $y \mapsto \mathbf{K}(y)$, consider the following example. Let $\Omega = (0, 1)$ and $V = \{v \in H^1(\Omega) : v(0) = 0\}$ with norm $\|v\|_V^2 := \int_\Omega |v'|^2 \,\mathrm{d}x$, where v' stands for the weak derivative of $v : \Omega \to \mathbb{R}$.

Suppose that, for $i = 1, 2$, $\mathbf{K}_i := \{v \in V : |v'| \leq \phi_i\}$ with $\phi_2 > \phi_1 > 0$ constants. Then, if $v_i \in \mathbf{K}_i$ for $i = 1, 2$, we have

$$\int_\Omega |v_2' - v_1'|^2 \,\mathrm{d}x \geq \int_{\{v_2' \geq \phi_1\}} |v_2' - \phi_1|^2 \,\mathrm{d}x + \int_{\{v_2' \leq -\phi_1\}} |v_2' + \phi_1|^2 \,\mathrm{d}x. \tag{3.4}$$

Define $\tilde{v}_1(x) = \int_0^x F(\phi_1, v_2'(s)) \,\mathrm{d}s$, where

$$F(\phi_1, t) := \begin{cases} \min(\phi_1, t), & t \geq 0, \\ \max(-\phi_1, t), & t < 0. \end{cases}$$

This implies that $\tilde{v}_1$ is bounded and $\tilde{v}_1' = F(\phi_1, v_2')$ in the sense of distributions, so that $\tilde{v}_1 \in H^1(\Omega)$, and in particular $\tilde{v}_1 \in V$; note that $\tilde{v}_1(0) = \lim_{x \downarrow 0} \tilde{v}_1(x) = 0$. Additionally,

$$\int_\Omega |v_2' - \tilde{v}_1'|^2\,\mathrm{d}x = \int_{\{v_2' \geq \phi_1\}} |v_2' - \phi_1|^2\,\mathrm{d}x + \int_{\{v_2' \leq -\phi_1\}} |v_2' + \phi_1|^2\,\mathrm{d}x,$$

so by (3.4), we have that

$$\mathrm{d}(v_2, \mathbf{K}_1)^2 = \inf_{v_1 \in \mathbf{K}_1} \int_\Omega |v_2' - v_1'|^2\,\mathrm{d}x = \int_{\{v_2' \geq \phi_1\}} |v_2' - \phi_1|^2\,\mathrm{d}x + \int_{\{v_2' \leq -\phi_1\}} |v_2' + \phi_1|^2\,\mathrm{d}x.$$

Since $-\phi_2 \leq v_2' \leq \phi_2$, for any $v_2 \in \mathbf{K}_2$ we have the bound

$$\mathrm{d}(v_2, \mathbf{K}_1)^2 \leq \int_\Omega |\phi_2 - \phi_1|^2\,\mathrm{d}x.$$

Further, if we choose $\tilde{v}_2(x) := \phi_2 x$, we have $\mathrm{d}(\tilde{v}_2, \mathbf{K}_1)^2 = \int_\Omega |\phi_2 - \phi_1|^2\,\mathrm{d}x$. Therefore

$$\mathrm{e}(\mathbf{K}_2, \mathbf{K}_1) = \sup_{v_2 \in \mathbf{K}_2} \mathrm{d}(v_2, \mathbf{K}_1) = \left(\int_\Omega |\phi_2 - \phi_1|^2\,\mathrm{d}x \right)^{1/2} = |\phi_2 - \phi_1|.$$

Also, since $\mathbf{K}_1 \subset \mathbf{K}_2$, $\mathrm{d}(v_1, \mathbf{K}_2) = 0$ for any $v_1 \in \mathbf{K}_1$ and hence $\mathrm{e}(\mathbf{K}_1, \mathbf{K}_2) = 0$. Thus, for sufficiently large $\rho > 0$, we have $\mathrm{haus}_\rho(\mathbf{K}_1, \mathbf{K}_2) = |\phi_2 - \phi_1|$. This establishes that if $\Phi \colon V \to \mathbb{R}$ is Lipschitz, then

$$\|\mathrm{P}_{\mathbf{K}(y)}(y_0) - \mathrm{P}_{\mathbf{K}(w)}(y_0)\|_V \leq \eta \|y - w\|_V^{1/2},$$

for some $\eta > 0$. Note however, that in this setting it is indeed possible to obtain a contraction for the map T; see section 3.1.2.

3.3 Order approaches: solution methods for $\mathsf{m}(f)$ and $\mathsf{M}(f)$

We consider the Gelfand triple (V, H, V') and the framework of section 2.2 including the assumptions on $A \in \mathscr{L}(V, V')$ and Φ. Then the map T is increasing and on the interval of sub- and supersolutions $[\underline{y}, \overline{y}] = [A^{-1} f_{\min}, A^{-1} f_{\max}]$, there exists a minimal and a maximal solution to $(\mathrm{P}_{\mathrm{QVI}})$, denoted $\mathsf{m}(f)$ and $\mathsf{M}(f)$, respectively. We follow a similar approach as in [15].

Consider the iterations

$$\begin{aligned} m_{n+1} &:= T(m_n), \quad m_0 := \underline{y}, \quad \text{and} \\ M_{n+1} &:= T(M_n), \quad M_0 := \overline{y}, \quad \text{for } n = 0, 1, \ldots \end{aligned}$$

Since $\underline{y} \leq T(\underline{y})$, $T(\overline{y}) \leq \overline{y}$, and $\underline{y} \leq \overline{y}$, the fact that T is increasing implies that $m_n \leq m_{n+1}$ and $M_{n+1} \leq M_n$, and additionally $m_n, M_n \in [\underline{y}, \overline{y}]$.

It can be proven than $\{m_n\}$ and $\{M_n\}$ are Cauchy sequences in H, and since they are also bounded in V, we obtain

$$m_n \uparrow m^*, \quad M_n \downarrow M^*, \qquad \text{in } H, \qquad \text{and} \qquad m_n \rightharpoonup m^*, \quad M_n \rightharpoonup M^*, \qquad \text{in } V.$$

Note that $M^* \leq \Phi(M_{n-1})$ for all $n \in \mathbb{N}$ so that

$$c\|M_n - M^*\|_V^2 \leq \langle AM_n - AM^*, M_n - M^*\rangle \leq \langle f - AM^*, M_n - M^*\rangle$$

by the fact that $M_n = S(f, \mathbf{K}(M_{n-1}))$, and hence $M_n \to M^*$ in V. Further, provided that $\Phi\colon H \to H$ is continuous, it is not hard to prove that M^* is a solution to ($\mathrm{P_{QVI}}$): from $M^* \leq \Phi(M_{n-1})$, we have that $M^* \leq \Phi(M^*)$, and for any $v \leq \Phi(M^*)$, we have $v \leq \Phi(M_{n-1})$ for any $n \in \mathbb{N}$. Hence,

$$\langle AM^* - f, v - M^*\rangle = \lim_{n\to\infty} \langle AM_n - f, v - M_n\rangle \geq 0,$$

i.e., $M^* = S(f, \mathbf{K}(M^*))$. Since $\mathsf{M}(f)$ is the maximum solution to ($\mathrm{P_{QVI}}$) on $[\underline{y}, \overline{y}]$, $M^* \leq \mathsf{M}(f)$. Further, since $\mathsf{M}(f) \leq \overline{y}$, by repeated iteration of T on the previous inequality we have that $\mathsf{M}(f) \leq M^*$, i.e., $\mathsf{M}(f) = M^*$.

In order to prove that $m^* = \mathsf{m}(f)$, additional assumptions are required. Let $\Phi\colon V \to V$ be completely continuous. Then $v_n := \min(m^*, \Phi(m_{n-1}))$ satisfies $v_n \to m^*$ in V and $v_n \leq \Phi(m_{n-1})$. Hence,

$$c\|m_n - v_n\|_V^2 \leq \langle Am_n - Av_n, m_n - v_n\rangle \leq \langle f - Av_n, m_n - v_n\rangle,$$

where we have used that $m_n = S(f, \mathbf{K}(m_{n-1}))$. Thus, $m_n \to m^*$ in V. From $m_n \leq \Phi(m_{n-1})$, and since strong convergence in H preserves order, we have $m^* \leq \Phi(m^*)$. Choose $v \leq \Phi(m^*)$ arbitrary and define $v_n := \min(v, \Phi(m_{n-1}))$, so that $v_n \to m^*$ in V and $v_n \leq \Phi(m_{n-1})$. Then

$$\langle Am^* - f, v - m^*\rangle = \lim_{n\to\infty} \langle Am_n - f, v_n - m_n\rangle \geq 0.$$

That is, m^* is a solution to ($\mathrm{P_{QVI}}$) within $[\underline{y}, \overline{y}]$. Hence, by definition of $\mathsf{m}(f)$, we have $\mathsf{m}(f) \leq m^*$, and from $\underline{y} \leq \mathsf{m}(f)$ and the consecutive iteration of T on the previous inequality, we have $m^* \leq \mathsf{m}(f)$, i.e., $m^* = \mathsf{m}(f)$. Overall, we have the following result.

Proposition 3.3 *In addition to the assumptions for Φ in section 2.2, suppose that $\Phi\colon V \to V$ is completely continuous. Then $m_n \uparrow \mathsf{m}(f)$ and $M_n \downarrow \mathsf{M}(f)$ in H and $m_n \to \mathsf{m}(f)$ and $M_n \to \mathsf{M}(f)$ in V.*

Open problems. The speed of convergence of $\{m_n\}$ and $\{M_n\}$ is, in general, slower than linear. This hinders their applicability when addressing large scale problems, or when considering optimization problems involving $\mathsf{m}(f)$ and $\mathsf{M}(f)$, as in section 4. It is an open question whether it is possible to accelerate such iterations by combining them with intermediate steps. Additionally, it is open wether linearly convergent methods can be designed in general when the solution is non-unique. □

3.4 Regularization methods

3.4.1 Extended Moreau–Yosida and Semismooth Newton

It is convenient to consider regularizations of QVIs by smoothing. The type of regularization or smoothing that we consider in this section consists of approximating the QVI in question by a sequence of parameter-dependent PDEs. Regularization methods are useful for numerical purposes as well as for theoretical efforts. For example, they can be used to prove fundamental results such as existence of solutions as well as to derive stationarity conditions for optimal control problems with QVI constraints[1], which is a subject of work under preparation by the authors. Moreover, even for VIs, obtaining mesh independence requires regularization.

Obstacle case

For simplicity, we consider $V = H_0^1(\Omega)$ and $H = L^2(\Omega)$. In this section we present some results on the Moreau–Yosida regularization of the obstacle type $(\mathrm{P}_{\mathrm{QVI}})$ given by the nonlinear PDE

$$F(y) := Ay - f + \frac{1}{\beta}(y - \Phi(y))^+ = 0 \tag{3.5}$$

for $\beta > 0$. Under suitable assumptions it is expected that as $\beta \downarrow 0$, the sequence of solutions y_β^* converges to the solution of $(\mathrm{P}_{\mathrm{QVI}})$. In fact, if $\Phi\colon V \to V$ is increasing and completely continuous with $\Phi(0) \geq 0$ and $f \in V'_+$, then $\{y_{\beta_n}^*\}$ has a subsequence that converges in V to a solution of $(\mathrm{P}_{\mathrm{QVI}})$, for any $\beta_n \downarrow 0$.

Focusing on (3.5), we consider $y_0 \in \tilde{V} \subset V$, and the Newton iteration

$$y_{k+1} = y_k - G_F(y_k)^{-1}F(y_k), \qquad k = 0, 1, 2, \ldots \tag{3.6}$$

where $G_F(y) \in \mathcal{L}(V, V')$ is a (presumably invertible) Newton derivative of F [36], which is defined to satisfy

$$\lim_{h \to 0} \frac{\|F(y+h) - F(y) - G_F(y+h)h\|_{V'}}{\|h\|_V} = 0.$$

It is know that $(\cdot)^+\colon L^p(\Omega) \to L^2(\Omega)$ is Newton differentiable for any $p > 2$ with Newton derivative $G_{\max}(y) = \mathrm{Heaviside}(y)$. Suppose that $\Phi\colon V \to L^q(\Omega)$ is Fréchet differentiable for some $q \geq p$, then we have (see [45, Lemma 8.15]) that $G_F(y) \in \mathcal{L}(V, V')$ is given by

$$G_F(y)h = Ah + \frac{1}{\beta}G_{\max}(y - \Phi(y))(I - \Phi'(y))h.$$

[1] Naturally optimality conditions obtained through regularization will not be as strong as those potentially obtained through using the directional differentiability of the QVI solution mapping, see section 5.

Suppose that $(\chi_{\Omega_0}\Phi'(y)h, h) \leq (\chi_{\Omega_0}h, h)$ for any $\Omega_0 \subset \Omega$ and for all $y, h \in V$. Then $\langle G_F(y)h, h\rangle \geq \tilde{c}\|h\|_V^2$ for some $\tilde{c} > 0$ so that $\|G_F(y)^{-1}\|_{\mathcal{L}(V',V)} \leq 1/\tilde{c}$ and hence (3.6) converges superlinearly to the solution y_β^* of (3.5), provided that $\|y_0 - y_\beta^*\|_V$ is sufficiently small; see [36–38,45].

Example on thermoforming. The production of plastic parts is in general done by thermoforming. In this procedure, a plastic sheet is heated to its pliable temperature and then forced via air pressure (positive or negative) towards a mold, commonly made of metal, and involving some cooling mechanism. Such a manufacturing process involves several scales: it is used for microfluidic structures, plastic cups, and large parts in the automotive industry.

We consider the following time-asymptotic behaviour of the thermoforming process leading to an elliptic problem. We let a plastic membrane y lie over the domain Ω, and let the temperature of the membrane be constant (this simplification frees us from considering changing rheological properties of the heated membrane).

The mathematical problem is then given by: Find $(y, \Phi, T) \in V \times V \times W$ such that

$$y \leq \Phi, \quad \langle Ay - f, y - v\rangle \leq 0, \qquad \forall v \in V : v \leq \Phi, \tag{3.7}$$

$$\langle kT - \Delta T, w\rangle = (g(\Phi - u), w) \qquad \forall w \in W, \tag{3.8}$$

$$\Phi = \Phi_0 + LT \qquad \text{in } V, \tag{3.9}$$

where $f \in H_+$, $k > 0$ is a constant, $\Phi_0 \in V$ is the desired mold, and $L\colon W \to V$ is a bounded linear operator such that

$$\text{for every } \Omega_0 \subset \Omega, \text{ if } u \leq v \text{ a.e. on } \Omega_0 \text{ then } Lu \leq Lv \text{ a.e. on } \Omega_0,$$

and $g : \mathbb{R} \to \mathbb{R}$ is decreasing with $g(0) = M > 0$ a constant, $0 \leq g \leq M$ and g' bounded.

The above problem can be equivalently formulated as problem ($\mathrm{P}_{\mathrm{QVI}}$) where $\Phi\colon W \to V$ is defined as follows. Let $v \in W$ and consider the problem: Find $\phi \in V$ such that

$$\langle kT - \Delta T, w\rangle = (g(\phi - v), w) \qquad \forall w \in W, \tag{3.10}$$

$$\phi = \Phi_0 + LT \qquad \text{in } V. \tag{3.11}$$

We define $\Phi(v) = \phi$.

In Figure 1, we see the membrane y, the obstacle $\Phi(y)$, the coincidence set, and the difference $\Phi(y) - \Phi_0$, all computed with the semismooth Newton method described above for β sufficiently large (full details of the analysis and numerical implementation of the models presented here can be found in section 6 of [3]).

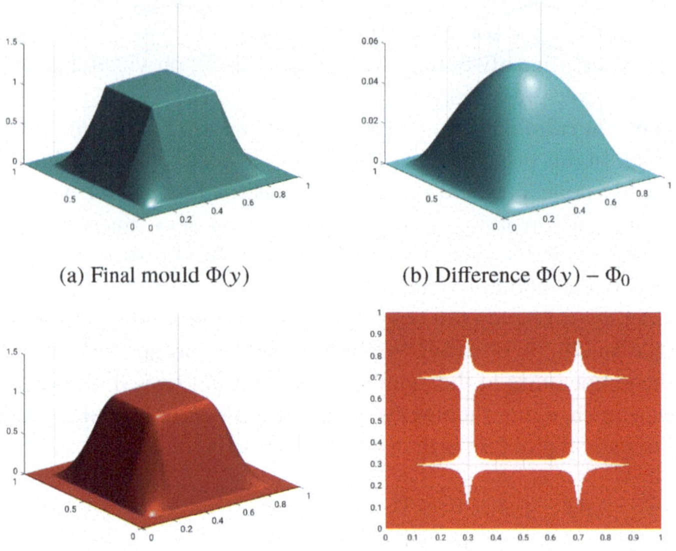

(a) Final mould $\Phi(y)$ (b) Difference $\Phi(y) - \Phi_0$

(c) Membrane y (d) Coincidence set $\{y = \Phi(y)\}$ (in red)

Fig. 1: Results for the thermoforming example

Gradient case

We consider here $V = W_0^{1,p}(\Omega)$ and $H = L^2(\Omega)$. The type of regularization used in (3.5) is not amenable for direct application in the gradient case. In fact, provided A is symmetric, one can consider the minimization problem

$$\min_{y \in V} \frac{1}{2}\langle Ay, y\rangle - \langle f, y\rangle + \frac{1}{\beta}\left\|(|\nabla y| - \Phi(y))^+\right\|_H^2 \tag{3.12}$$

associated to the QVI with the gradient constraint. In connection with (3.12), it was proven in [40, Theorem 3.2] that there is a sequence of β such that the associated solutions to the penalized minimization problems converge to the solution of the minimization problem $\min_{y \in V} \frac{1}{2}\langle Ay, y\rangle - \langle f, y\rangle$ subject to $|\nabla y| \leq \Phi(y)$ a.e in Ω, which is not in general a solution of $(\mathrm{P_{QVI}})$. This fact is in sharp contrast to the VI setting: in fact, if $\Phi(y)$ is replaced by $\Phi(w)$ in (3.12) for some $w \in V$, then the problem is suitable for a semismooth Newton approach and the sequence of solutions $\{y_\beta(w)\}_\beta$ converges, as $\beta \downarrow 0$, in V to $y^* = S(f, \mathbf{K}(w))$. In this case, we have that $y_\beta(w) \in V$ satisfies

$$\left.\begin{aligned} &F(y) := \langle A(y), v\rangle - \langle f, v\rangle + \tfrac{1}{\beta}((|\nabla y| - \Phi(w))^+, q(\cdot)\nabla v) = 0 \\ &q(x) \in \begin{cases} \frac{\nabla y}{|\nabla y|}(x), \text{ if } |\nabla y(x)| > 0 \\ \bar{B}_1(0)^N, \text{ otherwise,} \end{cases} \\ &\text{for all } v \in V, \end{aligned}\right\} \tag{3.13}$$

where $\bar{B}_1(0)^N$ denotes the closed unit ball in $\mathbb{R}^N$.

The application of the semismooth Newton method for the resolution of $F(y) = 0$ in this case has several subtleties. Specifically, the existence of a Newton derivative of the map $y \mapsto \mathcal{P}(y) := -\mathrm{div} q(\cdot)^{\mathrm{T}}((|\nabla y| - \Phi(w))^+$ requires a delicate interplay of the domain and image spaces. In contrast to $(\cdot)^+ : L^p(\Omega) \to L^2(\Omega)$, which is Newton differentiable for any $p > 2$, the aforementioned map is Newton differentiable when considered as $\mathcal{P} : W_0^{1,p}(\Omega) \to W^{-1,s}(\Omega)$, with $3 \leq 3s \leq p < \infty$; see [40].

3.5 Gerhardt-type regularization for the gradient case

For simplicity we consider the gradient case where $A = -\Delta$ is simply the Laplacian. We briefly discuss here an extension of a technique introduced by Gerhardt [30] which was developed by Rodrigues, Santos and collaborators in a series of papers; see [10, 11, 61, 68].

One way to regularize problem ($\mathrm{P_{QVI}}$) in the case described above is through the PDE

$$-\nabla \cdot (g_\epsilon(|\nabla y|^2 - \Phi^2(y))\nabla y) - f = 0 \tag{3.14}$$

where $g_\epsilon : \mathbb{R} \to \mathbb{R}$ is a bounded non-decreasing function which is twice continuously differentiable with

$$g_\epsilon(t) = \begin{cases} 1 & : t \leq 0, \\ e^{t/\epsilon} & : \epsilon \leq t \leq \frac{1}{\epsilon} - \epsilon, \\ e^{1/\epsilon^2} & : t \geq \frac{1}{\epsilon}, \end{cases} \tag{3.15}$$

for $\epsilon > 0$. Formally, it can be thought of as an approximation to

$$g_0(t) = \begin{cases} 1 & : t \leq 0, \\ \infty & : t > 0. \end{cases}$$

This suggests that in the limiting process (as $\epsilon \to 0$) for the nonlinear term not to blow up, the argument inside the regularization function needs to be non-positive, which of course then retrieves the gradient constraint. This type of regularization was first introduced by Gerhardt [30] with the aim of approximating the solution to an elliptic minimization problem, and the specific form (3.15) was used in [61, 67] to tackle parabolic variational inequality problems. See also [13, 69]. The function g_ϵ satisfies the useful monotonicity property [13]

$$(g_\epsilon(|x|^2 - a)x - g_\epsilon(|y|^2 - a)y)(x - y) \geq 0$$

which allows one to pass to the limit in the weak formulation of (3.14) after having obtained uniform estimates. Rigorous details of this can be found in the cited works.

This type of regularization, though powerful in the theoretical setting, has not been proven useful yet in the development of solution algorithms. In fact, if we formulate (3.14) as $F(y) = 0$ and try to identify a Newton derivative (as done in the previous section), we face differentiating the highest order terms of the associated nonlinear differential operator, a complex task in its own right. Furthermore, the Newton-type iterations would require, in the case of discretization by finite elements, a time consuming reassembling of the stiffness matrix in each iteration.

3.6 Drawbacks of the iteration $y_{n+1} = T(y_n)$

Since problem (1.2) is suitable for numerical resolution via diverse methods, a first approach for computing fixed points of T is to consider the iteration

$$y_{n+1} = T(y_n), \qquad n = 0, 1, \ldots,$$

with $y_0 \in V$ given.

The properties of A determine that the sequence $\{u_n\}$ is bounded in V and hence it contains weakly convergent subsequences. Additionally, suppose that sufficient properties of Φ are available so that $\mathbf{K}(v_n) \to \mathbf{K}(v)$ in the sense of Mosco if $v_n \rightharpoonup v$ in V. Then it follows that $T\colon V \to V$ is completely continuous: if $v_n \rightharpoonup v$, then $T(v_n) \to T(v)$ in V.

This seemingly amenable circumstance described above leads to the following erroneous argument that is common in the literature: "*Denote also by* $\{y_n\}$ *a weakly convergent subsequence of* $\{y_n\}$ *with limit* y^*. *Then taking the limit on both sides of* $y_{n+1} = T(y_n)$, *we observe that* y^* *is a fixed point of* T". The mistake clearly lies in assuming that if $y_{n_k} \rightharpoonup y^*$, then $\{y_{n_k+1}\}$ has the same weak limit. In particular what this attemps to show is that the compactness properties of T determine that the sets of weak and strong accumulation points of $\{y_n\}$ (denoted as $\mathcal{A}$) are identical, and if $y \in \mathcal{A}$, then $T(y) \in \mathcal{A}$.

Since $T(\mathcal{A}) \subset \mathcal{A}$, we can try to extend the digression further and study the possibility of finding a fixed point since we have now a T-invariant set. If $\mathcal{A}$ can be proven to be convex (it is usually not), then T has a fixed point in $\mathcal{A}$ via Schauder's fixed point. The alternative is to consider the search of a fixed point in the closed convex hull of $\mathcal{A}$ denoted by $\overline{\text{co}}\,\mathcal{A}$. If $T(\overline{\text{co}}\,\mathcal{A}) \subset \overline{\text{co}}\,T(\mathcal{A})$ holds true, then $T(\overline{\text{co}}\,\mathcal{A}) \subset \overline{\text{co}}\,\mathcal{A}$ and Schauder's fixed point theorem can be used to deduce that T has a fixed point in $\overline{\text{co}}\,\mathcal{A}$. However, for obstacle type problems, if Φ is concave, the map T is too (see [15]), so that $T(\text{co}\,\mathcal{A}) \geq \text{co}\,T(\mathcal{A})$.

4 Optimal control problems

There exist several applications for optimization problems where the QVI is a constraint. Then, often solutions of the QVI are controlled in such a way that they are close to some desired state. These types of problems have been almost completely neglected in the literature. An instance of such a problem is

Problem $(\mathbb{P})$:

$$\begin{aligned} &\text{minimize} \quad J(y,f) \text{ over } (y,f) \in V \times U,\\ &\text{subject to} \quad f \in U_{\text{ad}} \subset U \subset V', \ \text{ and } \ y \text{ solves } (\mathrm{P}_{\mathrm{QVI}}), \end{aligned} \tag{$\mathbb{P}$}$$

where $J\colon V \times U \longrightarrow \mathbb{R}$ is weakly lower semicontinuous and U_{ad} is compact in V'. Note that if $\mathbf{K}(v_n) \xrightarrow{M} \mathbf{K}(v)$ whenever $v_n \rightharpoonup v$ in V, then problem $(\mathbb{P})$ has a solution: Indeed, let $\{(y_n, f_n)\}$ be an infimizing sequence. Then, there exists a subsequence, denoted also as $\{(y_n, f_n)\}$, such that $y_n \rightharpoonup y^*$ in V, $f_n \rightharpoonup f^*$ in U and $f_n \longrightarrow f^*$ in V'. We have that $y_n = S(f_n, \mathbf{K}(y_n))$ and $y^* = S(f^*, \mathbf{K}(y^*))$ by taking limits on both sides, and hence $\lim J(y_n, f_n) = J(y^*, f^*)$ so that (y^*, f^*) is a minimizer of the problem.

The literature on such problems is scarce; see [1, 20] for exceptions. Further, it falls short in tackling the real problems in the QVI setting. The solution set of the QVI is in general not a singleton, and in case of industrial applications it is of interest to control the entire solution set. In view of this, we have the following open questions.

Open problems. In the QVI context, it is sometimes important to control the full solution set $\mathbf{Q}(f)$ on a certain interval of interest $[\underline{y}, \overline{y}]$. We consider the Gelfand triple setting of section 2.2. A possible formulation for such control problems is as follows:

Problem $(\tilde{\mathbb{P}})$:

$$\begin{aligned} &\text{minimize } J(\mathbf{O}, f) := J_1(T_{\sup}(\mathbf{O}), T_{\inf}(\mathbf{O}), f)\\ &\text{over} \quad (\mathbf{O}, f) \in 2^H \times U,\\ &\text{subject to } f \in U_{\text{ad}},\\ &\qquad\qquad y \in \mathbf{O}, \quad \mathbf{O} = \{z \in V : z \text{ solves } (\mathrm{P}_{\mathrm{QVI}}) \cap [\underline{y}, \overline{y}]\}. \end{aligned} \tag{$\tilde{\mathbb{P}}$}$$

In the above problem we consider $J_1 : H \times H \times U \longrightarrow \mathbb{R}$ and for $\underline{y}, \overline{y} \in H$ we define the set map $T_{\sup}$

$$T_{\sup}(\mathbf{O}) := \begin{cases} \sup_{z \in \mathbf{O} \cap [\underline{y}, \overline{y}]} z, & \mathbf{O} \cap [\underline{y}, \overline{y}] \neq \emptyset\ ;\\ \underline{y}, & \text{otherwise.} \end{cases}$$

The map $T_{\inf}$ defined analogously as

$$T_{\inf}(\mathbf{O}) := \begin{cases} \inf_{z \in \mathbf{O} \cap [\underline{y}, \overline{y}]} z, & \mathbf{O} \cap [\underline{y}, \overline{y}] \neq \emptyset\,; \\ \overline{y}, & \text{otherwise.} \end{cases}$$

As explained in section 2.2, the supremum of an arbitrary subset of H that is bounded above (in the order) is also correctly defined since H is Dedekind complete, which shows that $T_{\inf}$ and $T_{\sup}$ are well defined in our setting.

Recall the framework of section 2.2 where $\underline{y}$ and $\overline{y}$ are respectively sub- and supersolutions of the map $T(\cdot) = S(f, \mathbf{K}(\cdot))$. Then the reduced version of problem $(\tilde{\mathbb{P}})$ is formulated in terms of the operators m and M as

$$\begin{aligned} &\text{minimize } J_1(\mathsf{M}(f), \mathsf{m}(f), f) \\ &\text{subject to } f \in U_{\text{ad}}. \end{aligned} \tag{$\tilde{\mathbb{P}}_{\text{red}}$}$$

An important example is when it is required to force the solution set to be a singleton and the element in question to be close to some desired state y_d. Here, a possible choice for J_1 is given by

$$J_1(\mathsf{M}(f), \mathsf{m}(f), f) = \frac{1}{2} \int_\Omega |\mathsf{M}(f) - \mathsf{m}(f)|^2 \,\mathrm{d}x + \frac{\sigma}{2} \int_\Omega |y_d - \mathsf{m}(f)|^2 \,\mathrm{d}x.$$

To the best of our knowledge, problem $(\tilde{\mathbb{P}})$ (and its reduced version) has not been considered in the literature, and it is a topic of active research by the present authors. Important (and currently still open) subtasks for analyzing the above control problem are (i) the study of stability properties of the maps $f \mapsto \mathsf{M}(f), \mathsf{m}(f)$ and (ii) their (generalized) differentiability properties. While (i) typically helps to establish existence of a solution to the optimization problem, (ii) allows for suitable stationarity conditions characterizing solutions.

5 Differentiability

We consider in this section the differential stability of the solution map associated to $(\mathrm{P}_{\mathrm{QVI}})$, in particular, the mapping taking the source term into the set of the solutions. Showing that this map is differentiable (in some sense) is not only an interesting analytical task in its own right but is also of use for optimal control, numerics and applications.

The corresponding differentiability study for variational inequalities has been thoroughly investigated [33, 59, 74]. Let us set the scene and outline this theory first before moving on to QVIs.

Let X be a locally compact topological space which is countable at infinity with ξ a Radon measure on X. Suppose $V \subset L^2(X; \xi) =: H$ is a Hilbert space with the embedding continuous and dense and such that $|u| \in V$ whenever $u \in V$, and let $A\colon V \to V'$ be now a linear operator satisfying the boundedness, coercivity and T-monotonicity properties from before, i.e., (A1), (A2), and (A3). The pair (V, A) falls into the class of *positivity preserving coercive forms* with respect to $L^2(X;\xi)$ [16,58]. We further assume that

$$V \cap C_c(X) \subset C_c(X) \quad \text{and} \quad V \cap C_c(X) \subset V \quad \text{are dense embeddings,} \tag{5.1}$$

thus (V, a) is a *regular* form [27, §1.1] [16, §2][2]. This framework allows us to define the notions of capacity, quasi-continuity and related objects, see [59, §3] and [33, §3]. Several concrete examples of V and A are given in [59, §3] and [3, §1.2].

Given an obstacle $\phi \in V_+$, we define the set

$$\mathbf{K} := \{w \in V : w \leq \phi\},$$

and given a source term $f \in V'$, we make an abuse of notation here and define by $S\colon V' \to V$ the mapping $S(f) := S(f, \mathbf{K})$ with the latter defined in (1.2). It is useful to introduce the well known notions of the *tangent cone* and the *critical cone* associated to $\mathbf{K}$, given respectively by

$$T_{\mathbf{K}}(y) := \{\varphi \in V : \varphi \leq 0 \text{ q.e. on } \{y = \phi\}\} \text{ and } \mathcal{K}_{\mathbf{K}}(y) := T_{\mathbf{K}}(y) \cap [f - Ay]^{\perp}. \tag{5.2}$$

The coincidence set appearing in the tangent cone is of course calculated over X. This is worth emphasis since for example if V is chosen to be the Sobolev space $H^1(\Omega)$ on a bounded Lipschitz domain Ω, then X should be $\bar{\Omega}$, the closure of the domain, and not Ω itself; see [3, §1.2].

The following result of Mignot tells us that the mapping S is directionally differentiable.

Theorem 5.5 (Theorem 3.3 of [59]) *Given $f \in V'$ and $d \in V'$, there exists a function $S'(f)(d) \in V$ such that*

$$S(f + td) = S(f) + tS'(f)(d) + o(t) \quad \forall t > 0$$

holds where $t^{-1}o(t) \to 0$ as $t \to 0^+$ in V and $\delta := S'(f)(d)$ satisfies the VI

$$\delta \in \mathcal{K}_{\mathbf{K}}(y) : \langle A\delta - d, v - \delta \rangle \geq 0 \quad \forall v \in \mathcal{K}_{\mathbf{K}}(y), \text{ where } y = S(f).$$

The directional derivative $\delta = \delta(d)$ is positively homogeneous in d.

In [44], the authors essentially extended the results of Mignot to a more general setting and turned the question of directional differentiability for VIs with more general constraint sets (than those of obstacle type) into a geometric question of the *polyhedricity* of the underlying constraint set, and more details and background can be found in the cited text.

One says that *strict complementarity* holds if the critical cone simplifies to the linear subspace

---

[2] A space V under all of the previous assumptions except the second density assumption in (5.1) is referred to by Mignot in [59] as a 'Dirichlet space' — this is rather inconsistent with the modern literature [27] where Dirichlet spaces and Dirichlet forms are defined differently (see [27, §1.1]).

$$\mathcal{K}_{\mathbf{K}}(y) = \mathcal{S}_{\mathbf{K}}(y) := \{\varphi \in V : \varphi = 0 \text{ q.e. on } \{y = \phi\}\}. \tag{5.3}$$

In this case, the VI satisfied by δ simplifies to a variational equality due to the relaxation of constraints on the test functions for the inequality. It is not hard to see that, at least formally, strict complementarity arises when the biactive set $\{Ay - f = 0\} \cap \{y - \phi = 0\}$ is empty; see [28,29] for some technical details regarding biactivity that include its proper definition under low regularity of y and f. Under strict complementarity, the derivative in Theorem 5.5 is in fact a Gâteaux derivative as the next result shows.

Theorem 5.6 (Theorem 3.4 of [59]) *In the context of Theorem 5.5, if strict complementarity holds, then the derivative δ satisfies*

$$\delta \in \mathcal{S}_{\mathbf{K}}(y) : \langle A\delta - d, v - \delta \rangle = 0 \quad \forall v \in \mathcal{S}_{\mathbf{K}}(y).$$

In this case, $\delta = \delta(d)$ is linear in d.

5.1 Directional differentiability for QVIs

To formulate the QVI case, let $\Phi\colon V \to V$ be increasing with $\Phi(0) \geq 0$. Given $f \in V'$, consider ($\mathrm{P}_{\mathrm{QVI}}$) in the obstacle case (i.e., $\psi \circ G \equiv \mathrm{id}$):

$$y \in \mathbf{K}(y) : \quad \langle Ay - f, v - y \rangle \geq 0 \quad \forall v \in \mathbf{K}(y). \tag{5.4}$$

We consider $\mathbf{Q}\colon V'_+ \rightrightarrows V$, the multi-valued solution mapping taking $f \mapsto y$. To show that this map is directionally differentiable (in some sense), the obvious idea that springs to mind is to rewrite (5.4) by transforming the obstacle onto the source term and then to apply Mignot's theory. Indeed, the inequality implies that the quantity $\hat{y} := (\mathrm{id} - \Phi)y$ solves

$$\hat{y} \in \mathbf{K}_0 : \langle A(\mathrm{id} - \Phi)^{-1}\hat{y} - f, \phi - \hat{y} \rangle \geq 0 \quad \forall \phi \in \mathbf{K}_0,$$

with $\mathbf{K}_0 := \{w \in V : w \leq 0\}$; however, in general, the elliptic operator $A(\mathrm{id} - \Phi)^{-1}$ is not linear, coercive nor T-monotone, so the VI theory is not applicable and a different approach is needed.

The idea in [3] is the following: approximate the QVI solution $q(t) \in \mathbf{Q}(f + td)$ by a sequence $q_n(t)$ of solutions of VIs (each of which by definition has a explicit obstacle), obtain suitable differential formulae for those VIs and then pass to the limit to (hopefully) obtain an expansion formula relating elements of $\mathbf{Q}(f + td)$ to $\mathbf{Q}(f)$. There are some delicacies in this procedure:

1. **derivation of the expansion formulae** for the above-mentioned VI iterates $q_n(t)$; they must relate $q(t)$ to a solution $y \in \mathbf{Q}(f)$, and recursion plays a highly nonlinear role in the relationship between one iterate and the preceding iterates;

2. **obtaining uniform bounds on the directional derivatives**; even though the derivatives satisfy a VI, it requires the handling of a recurrence inequality unless some regularity is available (see [3, §4.3]);
3. **identifying the limit of the higher-order terms as a higher-order term**; this procedure involves two limits: one as $t \to 0^+$ and one as $n \to \infty$, and commutation of limits in general requires an additional uniform convergence.

The main difficulty is indeed the final point above. Although the directional derivatives and higher-order terms of the VI iterates do possess some monotonicity properties, this information unfortunately does not help as much as one may hope.

The iteration scheme alluded to above requires some further restrictions on the data f and the direction d that the derivative is taken in, and we shall outline these in the following. We assume that $f \in V'_+$ and define $\bar{y} \in V$ as the (non-negative) weak solution of the unconstrained problem $A\bar{y} = f$. In a similar fashion to $\bar{u}$, define $\bar{q}(t) \in V$ as the solution of the unconstrained problem with right hand side $f + td$: $A\bar{q}(t) = f + td$.

Since we are considering the issue of sensitivity of QVIs with (by definition) implicit obstacles defined through the mapping Φ, it is clear that further regularity is required of Φ. We introduce these further assumptions below where we state the main theorem of [3], but first let us define

$$\mathcal{K}_{\mathbf{K}(y)}(y, \alpha) := \Phi'(y)(\alpha) + \mathcal{K}_{\mathbf{K}(y)}(y)$$

which can be thought as a *translated* critical cone.[3]

Theorem 5.7 (Theorem 1.6 of [3]) *Let $f, d \in V'_+$. Given $y \in \mathbf{Q}(f) \cap [0, \bar{y}]$, assume the following:*

(H1) *the map $\Phi\colon V \to V$ is Hadamard directionally differentiable*[4]
(H2) *either*

- a. *$\Phi\colon V \to V$ is completely continuous, or*
- b. *$V = H^1(\Omega)$, $X = \overline{\Omega}$ where Ω is a bounded Lipschitz domain, $\Phi\colon L^\infty_+(\Omega) \to L^\infty_+(\Omega)$ and is concave with $\Phi(0) \geq c > 0$, and $f, d \in L^\infty_+(\Omega)$*[5]

(H3) *the map $\Phi'(v)\colon V \to V$ is completely continuous (for fixed $v \in V$)*
(H4) *for any $b \in V$, $h\colon (0,T) \to V$ and $\lambda \in [0,1]$,*

$$\frac{\|\Phi'(y + tb + \lambda h(t))h(t)\|_V}{t} \to 0 \text{ as } t \to 0^+ \text{ if } \frac{h(t)}{t} \to 0 \text{ as } t \to 0^+$$

(H5) *given $T_0 \in (0,T)$ small, if $z\colon (0,T_0) \to V$ satisfies $z(t) \to y$ as $t \to 0^+$, then*

---

[3] Explicitly this set is $\{\varphi \in V : \varphi \leq \Phi'(y)(w)$ q.e. on $\{y = \Phi(y)\}$ and $\langle Ay - f, \varphi - \Phi'(y)(w)\rangle = 0\}$.

[4] In fact, (H1) can be weakened significantly by requiring Hadamard differentiability of Φ only at the point y, i.e., locally, as in assumptions (H4) and (H5).

[5] In this case, solutions of the QVI (5.4) are unique [55].

$$\|\Phi'(z(t))b\|_V \leq C_\Phi \|b\|_V \quad \text{where } C_\Phi < \frac{1}{1 + c^{-1}C}$$

for all $t \in (0, T_0)$*, where* C *and* c *are from* (A1) *and* (A2).

Then there exists $q(t) \in \mathbf{Q}(f + td) \cap [y, \bar{q}(t)]$ *and* $\alpha = \alpha(d) \in V_+$ *such that*

$$q(t) = y + t\alpha + o(t) \quad \forall t > 0$$

holds where $t^{-1}o(t) \to 0$ *as* $t \to 0^+$ *in* V *and* α *satisfies the QVI*

$$\alpha \in \mathcal{K}_{\mathbf{K}(y)}(y, \alpha) : \langle A\alpha - d, v - \alpha \rangle \geq 0 \quad \forall v \in \mathcal{K}_{\mathbf{K}(y)}(y, \alpha)$$

The directional derivative $\alpha = \alpha(d)$ *is positively homogeneous in* d.

It should be emphasized that the assumptions (H4) and (H5) depend on the specific function y, i.e., these are *local* conditions. The assumption (H5) implies certain restrictions: in the case that Φ is linear, it imposes a smallness condition on the operator norm of Φ which enforces uniqueness of solutions of the QVI. However, it does not necessarily rule out the multivalued setting in the case of nonlinear Φ.

Open problems. The result in the general multi-valued setting given in Theorem 5.7 is a differentiability result for a specific selection mechanism that associates to a function $y \in \mathbf{Q}(f)$ a function $q(t) \in \mathbf{Q}(f + td)$ (the precise mechanism is expounded in [3, §3.2.1]). A useful variant of the theorem would be to obtain the result for the mapping that selects the minimal or maximal solution to the QVI, i.e., if $\mathsf{M}(f) \in \mathbf{Q}(f)$ is the maximal solution of the QVI with source term f, is M directionally differentiable? A difficulty lies in the approximation scheme we use; in the proof of Theorem 5.7 we chose $q_0 = y$; instead we could choose $q_0 = y_0$ where $0 \leq y_0 \leq \bar{y}$ which leads to the equality

$$q_n(t) = y_n(t) + t\hat{\alpha}_n + \hat{o}_n(t)$$

where $y_n = S(f, \mathbf{K}(y_{n-1}))$. The main problem is in dealing with the limiting behaviour of the higher-order terms $\hat{o}_n(t)$, which now depends on the base point y_n which depends on n. This fact constrains us in this direction. For more details see [3, Remark 3.9]. □

It is worth restating Theorem 5.7 in the case when $\mathbf{Q}\colon V'_+ \rightrightarrows V$ is single-valued (i.e., the QVI problem has a unique solution).

Theorem 5.8 *Suppose* $\mathbf{Q}$ *is single-valued and let the hypotheses of Theorem 5.7 hold given* $f, d \in V'_+$*. There exists a function* $\mathbf{Q}'(f)(d) \in V_+$ *such that*

$$\mathbf{Q}(f + td) = \mathbf{Q}(f) + t\mathbf{Q}'(f)(d) + o(t) \quad \forall t > 0$$

holds where $t^{-1}o(t) \to 0$ *as* $t \to 0^+$ *in* V *and* $\mathbf{Q}'(f)(d)$ *satisfies the QVI given in Theorem 5.7.*

Similarly to Theorem 5.6, under a modification of the notion of strict complementarity, we obtain a regularity result on the directional derivative. In this setting, *strict complementarity* holds if the set $\mathcal{K}_{\mathbf{K}(y)}(y, w)$ simplifies to

$$\begin{aligned}\mathcal{K}_{\mathbf{K}(y)}(y, w) &= \mathcal{S}_{\mathbf{K}(y)}(y, w) \\ &:= \{\varphi \in V : \varphi = \Phi'(y)(w) \text{ q.e. on } \{y = \Phi(y)\}\}.\end{aligned}$$

Theorem 5.9 (Theorem 1.7 of [3]) *In the context of Theorem 5.7, if strict complementarity holds, then the derivative α satisfies*

$$\alpha \in \mathcal{S}_{\mathbf{K}(y)}(y, \alpha) : \langle A\alpha - d, \alpha - v\rangle = 0 \quad \forall v \in \mathcal{S}_{\mathbf{K}(y)}(y, \alpha).$$

In this case, if $h \mapsto \Phi'(v)(h)$ is linear, $\alpha = \alpha(d)$ satisfies $\alpha(c_1 d_1 + c_2 d_2) = c_1\alpha(d_1) + c_2\alpha(d_2)$ for constants $c_1, c_2 > 0$ and directions $d_1, d_2 \in V'_+$.

Naturally, we recover the results of [59] in the case where Φ is a constant mapping.

Open problems. A focus of ongoing work by the authors is the study of optimal control problems with QVI constraints of the following type:

Problem $(\mathbb{P}')$:

$$\begin{aligned} &\text{minimize} \quad \frac{1}{2}\|y - y_d\|_H^2 + \frac{\lambda}{2}\|f\|_U^2 \ \text{ over } (y, f) \in V \times U, \\ &\text{subject to} \quad f \in U_{\text{ad}} \subset U \subset V', \ \text{ and } \ y \text{ solves (5.4)}, \end{aligned} \tag{5.5}$$

Here, the data y_d is a desired state and $\lambda > 0$ is a constant. Under certain assumptions on the mapping Φ and the spaces featured above, existence of an optimal control and state can be shown using relatively standard methods. Obtaining stationarity conditions that explicitly characterize the optimal control and optimal state (which would, in particular, allow for a feasible numerical resolution of the problem) is of prime importance in optimization. Typically, *strong* stationarity conditions are sought and such conditions in the VI case have been obtained [60] by making use of the differentiability of the VI solution mapping, and we would like to extend this result also to the QVI case. A challenge lies in the fact that, in Theorem 5.7, differentiability (in the QVI setting) is only obtained for non-negative directions. Hence, problem (5.5) would contain pointwise a.e. bounds on the control. From [73] it is however known that obtaining strong stationarity is impossible in the VI case with such pointwise a.e. control bounds (without further restrictions on the bounds themselves). This represents a major issue. However, there are other notions of stationarity (see [39]) that could be obtained. □

6 Conclusion

We have considered a variety of key topics and we have highlighted limitations and open questions associated to QVIs of elliptic type. For the existence results we focused on compactness approaches and the lack of necessity and sufficiency results for Mosco convergence in cases other than constraint sets of obstacle type, and we also tackled some order approaches. For the simple fixed point arguments, we provided some positive results, and showed that the popular extension approaches to Lions–Stampacchia are in the best case scenario unnecessary. Additionally, we have provided some second-order solution algorithms of the semismooth Newton type. Finally, we have established some novel optimization problems that take into account the multivalued nature of the solution set of the QVI and gave an account of the newly established directional differentiability for the QVI solution map.

Acknowledgements This research was carried out in the framework of MATHEON supported by the Einstein Foundation Berlin within the ECMath projects OT1, SE5, CH12 and SE15/SE19 as well as project A-AP24. The authors further acknowledge the support of the DFG through the DFG-SPP 1962: Priority Programme "Non-smooth and Complementarity-based Distributed Parameter Systems: Simulation and Hierarchical Optimization" within Projects 10, 11, and 13, through grant no. HI 1466/7-1 Free Boundary Problems and Level Set Methods, and SFB/TRR154.

References

1. S. Adly, M. t. Bergounioux, and M. Ait Mansour. Optimal control of a quasi-variational obstacle problem. *J. Global Optim.*, 47(3):421–435, 2010.
2. C. D. Aliprantis and O. Burkinshaw. *Positive operators*, volume 119. Springer Science & Business Media, 2006.
3. A. Alphonse, M. Hintermüller, and C. N. Rautenberg. Directional differentiability for elliptic quasi-variational inequalities of obstacle type. *Calc. Var. Partial Differential Equations*, 58(1):58:39, 2019.
4. H. Antil and C. N. Rautenberg. Fractional elliptic quasi-variational inequalities: theory and numerics. *Interfaces Free Bound.*, 20(1):1–24, 2018.
5. H. Attouch and R. J.-B. Wets. Quantitative stability of variational systems: I. the epigraphical distance. *Transactions of the American Mathematical Society*, pages 695–729, 1991.
6. H. Attouch and R. J.-B. Wets. Quantitative stability of variational systems ii. a framework for nonlinear conditioning. *SIAM Journal on Optimization*, 3(2):359–381, 1993.
7. H. Attouch and R. J.-B. Wets. Quantitative stability of variational systems: Iii. ε-approximate solutions. *Mathematical Programming*, 61(1-3):197–214, 1993.
8. J.-P. Aubin. *Mathematical methods of game and economic theory*, volume 7 of *Studies in Mathematics and its Applications*. North-Holland Publishing Co., Amsterdam-New York, 1979.
9. J.-P. Aubin. *Mathematical methods of game and economic theory*. North-Holland, 1979.
10. A. Azevedo, F. Miranda, and L. Santos. Variational and quasivariational inequalities with first order constraints. *J. Math. Anal. Appl.*, 397(2):738–756, 2013.
11. A. Azevedo, F. Miranda, and L. Santos. *Stationary Quasivariational Inequalities with Gradient Constraint and Nonhomogeneous Boundary Conditions*, pages 95–112. Springer Berlin Heidelberg, Berlin, Heidelberg, 2014.

12. A. Azevedo and L. Santos. Convergence of convex sets with gradient constraint. *J. Convex Anal.*, 11(2):285–301, 2004.
13. A. Azevedo and L. Santos. A diffusion problem with gradient constraint depending on the temperature. *Adv. Math. Sci. Appl.*, 20(1):153–168, 2010.
14. C. Baiocchi and A. Capelo. *Variational and Quasivariational Inequalities*. Wiley-Interscience, 1984.
15. A. Bensoussan and J.-L. Lions. *Contrôle impulsionnel et inéquations quasi variationnelles*, volume 11 of *Méthodes Mathématiques de l'Informatique [Mathematical Methods of Information Science]*. Gauthier-Villars, Paris, 1982.
16. J. Bliedtner. Dirichlet forms on regular functional spaces. pages 15–62. Lecture Notes in Math., Vol. 226, 1971.
17. L. Boccardo and F. Murat. Nouveaux résultats de convergence dans des problèmes unilatéraux. In *Nonlinear partial differential equations and their applications. Collège de France Seminar, Vol. II (Paris, 1979/1980)*, volume 60 of *Res. Notes in Math.*, pages 64–85, 387–388. Pitman, Boston, Mass.-London, 1982.
18. L. Boccardo and F. Murat. Homogenization of nonlinear unilateral problems. In *Composite media and homogenization theory (Trieste, 1990)*, volume 5 of *Progr. Nonlinear Differential Equations Appl.*, pages 81–105. Birkhäuser Boston, Boston, MA, 1991.
19. H. Brezis. Remarque sur l'article précédent de f. murat. *J. Math. Pures Appl.*, 60:321–322., 1981.
20. H. Dietrich. Optimal control problems for certain quasivariational inequalities. *Optimization*, 49(1-2):67–93, 2001. In celebration of Prof. Dr. Alfred Göpfert 65th birthday.
21. G. Duvaut and J.-P. Lions. *Les Inéquations en Mécanique et en Physique*. Dunod, Paris, 1972.
22. F. Facchinei, C. Kanzow, S. Karl, and S. Sagratella. The semismooth Newton method for the solution of quasi-variational inequalities. *Comput. Optim. Appl.*, 62(1):85–109, 2015.
23. F. Facchinei, C. Kanzow, and S. Sagratella. Solving quasi-variational inequalities via their KKT conditions. *Math. Program.*, 144(1-2, Ser. A):369–412, 2014.
24. T. Fukao and N. Kenmochi. Abstract theory of variational inequalities with Lagrange multipliers and application to nonlinear PDEs. *Math. Bohem.*, 139(2):391–399, 2014.
25. T. Fukao and N. Kenmochi. A thermohydraulics model with temperature dependent constraint on velocity fields. *Discrete Contin. Dyn. Syst. Ser. S*, 7(1):17–34, 2014.
26. T. Fukao and N. Kenmochi. Quasi-variational inequality approach to heat convection problems with temperature dependent velocity constraint. *Discrete Contin. Dyn. Syst.*, 35(6):2523–2538, 2015.
27. M. Fukushima, Y. Oshima, and M. Takeda. *Dirichlet forms and symmetric Markov processes*, volume 19 of *De Gruyter Studies in Mathematics*. Walter de Gruyter & Co., Berlin, extended edition, 2011.
28. A. Gaevskaya. *Adaptive finite elements for optimally controlled elliptic variational inequalities of obstacle type*. PhD Thesis, Universität Augsburg, 2013.
29. A. Gaevskaya, M. Hintermüller, R. H. W. Hoppe, and C. Löbhard. Adaptive finite elements for optimally controlled elliptic variational inequalities of obstacle type. In *Optimization with PDE constraints*, volume 101 of *Lect. Notes Comput. Sci. Eng.*, pages 95–150. Springer, Cham, 2014.
30. C. Gerhardt. On the existence and uniqueness of a warpening function in the elastic-plastic torsion of a cylindrical bar with multiply connected cross-section. pages 328–342. Lecture Notes in Math., 503, 1976.
31. B. Hanouzet and J.-L. Joly. Convergence uniforme des itérés définissant la solution d'une inéquation quasi variationnelle abstraite. *C. R. Acad. Sci. Paris Sér. A-B*, 286(17):A735–A738, 1978.
32. B. Hanouzet and J. L. Joly. Convergece uniforme des itérés définissant la solution d'inéquations quasi-variationnelles et application à la régularité. *Num. Funct. Anal. and Optimiz.*, 4:399–414, 1979.
33. A. Haraux. How to differentiate the projection on a convex set in Hilbert space. Some applications to variational inequalities. *J. Math. Soc. Japan*, 29(4):615–631, 1977.

34. P. T. Harker. Generalized Nash games and quasi-variational inequalities. *European Journal of Operational Research*, 54:81–94, 1991.
35. N. Harms, T. Hoheisel, and C. Kanzow. On a smooth dual gap function for a class of quasi-variational inequalities. *J. Optim. Theory Appl.*, 163(2):413–438, 2014.
36. M. Hintermüller, K. Ito, and K. Kunisch. The primal-dual active set strategy as a semismooth Newton method. *SIAM J. Optim.*, 13(3):865–888 (2003), 2002.
37. M. Hintermüller, V. Kovtunenko, and K. Kunisch. Semismooth Newton methods for a class of unilaterally constrained variational problems. *Adv. Math. Sci. Appl.*, 14(2):513–535, 2004.
38. M. Hintermüller and K. Kunisch. PDE-constrained optimization subject to pointwise constraints on the control, the state, and its derivative. *SIAM J. Optim.*, 20(3):1133–1156, 2009.
39. M. Hintermüller, B. S. Mordukhovich, and T. M. Surowiec. Several approaches for the derivation of stationarity conditions for elliptic MPECs with upper-level control constraints. *Math. Program.*, 146(1-2, Ser. A):555–582, 2014.
40. M. Hintermüller and C. N. Rautenberg. A sequential minimization technique for elliptic quasi-variational inequalities with gradient constraints. *SIAM J. Optim.*, 22(4):1224–1257, 2012.
41. M. Hintermüller and C. N. Rautenberg. On the density of classes of closed convex sets with pointwise constraints in Sobolev spaces. *J. Math. Anal. Appl.*, 426(1):585–593, 2015.
42. M. Hintermüller and C. N. Rautenberg. On the uniqueness and numerical approximation of solutions to certain parabolic quasi-variational inequalities. *Port. Math.*, 74(1):1–35, 2017.
43. M. Hintermüller, C. N. Rautenberg, and S. Rösel. Density of convex intersections and applications. *Proc. A.*, 473(2205):20160919, 28, 2017.
44. M. Hintermüller and T. Surowiec. First-order optimality conditions for elliptic mathematical programs with equilibrium constraints via variational analysis. *SIAM J. Optim.*, 21(4):1561–1593, 2011.
45. K. Ito and K. Kunisch. *Lagrange Multiplier Approach to Variational Problems and Applications*. SIAM, 2008.
46. A. Kadoya, N. Kenmochi, and M. Niezgódka. Quasi-variational inequalities in economic growth models with technological development. *Adv. Math. Sci. Appl.*, 24(1):185–214, 2014.
47. R. Kano, N. Kenmochi, and Y. Murase. Parabolic quasi-variational inequalities with non-local constraints. *Adv. Math. Sci. Appl.*, 19(2):565–583, 2009.
48. R. Kano, Y. Murase, and N. Kenmochi. Nonlinear evolution equations generated by subdifferentials with nonlocal constraints. In *Nonlocal and abstract parabolic equations and their applications*, volume 86 of *Banach Center Publ.*, pages 175–194. Polish Acad. Sci. Inst. Math., Warsaw, 2009.
49. C. Kanzow. On the multiplier-penalty-approach for quasi-variational inequalities. *Math. Program.*, 160(1-2, Ser. A):33–63, 2016.
50. C. Kanzow and D. Steck. Augmented Lagrangian and exact penalty methods for quasi-variational inequalities. *Comput. Optim. Appl.*, 69(3):801–824, 2018.
51. N. Kenmochi. Parabolic quasi-variational diffusion problems with gradient constraints. *Discrete Contin. Dyn. Syst. Ser. S*, 6(2):423–438, 2013.
52. N. Kenmochi and U. Stefanelli. Existence for a class of nonlocal quasivariational evolution problems. In *Nonlinear phenomena with energy dissipation*, volume 29 of *GAKUTO Internat. Ser. Math. Sci. Appl.*, pages 253–264. Gakkōtosho, Tokyo, 2008.
53. A. S. Kravchuk and P. J. Neittaanmäki. *Variational and quasi-variational inequalities in mechanics*, volume 147. Springer Science & Business Media, 2007.
54. M. Kunze and J. Rodrigues. An elliptic quasi-variational inequality with gradient constraints and some of its applications. *Mathematical Methods in the Applied Sciences*, 23:897–908, 2000.
55. T. H. Laetsch. A uniqueness theorem for elliptic q. v. i. *J. Functional Analysis*, 18:286–288, 1975.
56. J.-L. Lions. Asymptotic behaviour of solutions of variational inequalitites with highly oscillating coefficients. *Applications of Methods of Functional Analysis to Problems in Mechanics, Proc. Joint Symp. IUTAM/IMU. Lecture Notes in Mathematics, Springer, Berlin*, 503, 1975.

57. J.-P. Lions and G. Stampacchia. Variational inequalities. *Commun. Pure Appl. Math.*, 20:493–519, 1967.
58. Z. M. Ma and M. Röckner. Markov processes associated with positivity preserving coercive forms. *Canad. J. Math.*, 47(4):817–840, 1995.
59. F. Mignot. Contrôle dans les inéquations variationelles elliptiques. *J. Functional Analysis*, 22(2):130–185, 1976.
60. F. Mignot and J. P. Puel. Inéquations variationelles et quasi variationelles hyperboliques du premier ordre. *J. Math Pures Appl.*, 55:353–378, 1976.
61. F. Miranda, J.-F. Rodrigues, and L. Santos. On a p-curl system arising in electromagnetism. *Discrete Contin. Dyn. Syst. Ser. S*, 5(3):605–629, 2012.
62. U. Mosco. Convergence of convex sets and of solutions of variational inequalities. *Advances in Mathematics*, 3(4):510–585, 1969.
63. F. Murat. L'injection du cône positif de H^{-1} dans $W^{-1,q}$ est compacte pour tout $q < 2$. *J. Math. Pures Appl. (9)*, 60(3):309–322, 1981.
64. J.-S. Pang and M. Fukushima. Quasi-variational inequalities, generalized Nash equilibria, and multi-leader-follower games. *Computational Management Science*, 3:373–375, 2009.
65. L. Prigozhin. On the Bean critical-state model in superconductivity. *European Journal of Applied Mathematics*, 7:237–247, 1996.
66. J. F. Rodrigues. *Obstacle Problems in Mathematical Physics*. North-Holland, 1987.
67. J.-F. Rodrigues. On the mathematical analysis of thick fluids. *J. Math. Sci. (N.Y.)*, 210(6):835–848, 2015.
68. J. F. Rodrigues and L. Santos. A parabolic quasi-variational inequality arising in a superconductivity model. *Ann. Scuola Norm. Sup. Pisa Cl. Sci.*, XXIX:153–169, 2000.
69. J. F. Rodrigues and L. Santos. Quasivariational solutions for first order quasilinear equations with gradient constraint. *Arch. Ration. Mech. Anal.*, 205(2):493–514, 2012.
70. L. Santos. Variational problems with non-constant gradient constraints. *Portugaliae Mathematica*, 59:205–248, 2002.
71. R. E. Showalter. *Monotone Operators in Banach Space and Nonlinear Partial Differential Equations*. American Mathematical Society, 1997.
72. L. Tartar. Inéquations quasi variationnelles abstraites. *CR Acad. Sci. Paris Sér. A*, 278:1193–1196, 1974.
73. G. Wachsmuth. Strong stationarity for optimal control of the obstacle problem with control constraints. *SIAM Journal on Optimization*, 24(4):1914–1932, 2014.
74. E. H. Zarantonello. Projections on Convex Sets in Hilbert Space and Spectral Theory. *Contributions to Nonlinear Functional Analysis*, pages 237–424, 1971.

The Incompatibility Operator: from Riemann's Intrinsic View of Geometry to a New Model of Elasto-Plasticity

Samuel Amstutz, Nicolas Van Goethem

Abstract
The mathematical modelling in mechanics has a long-standing history as related to geometry, and significant progresses have often been achieved by the invention of new geometrical tools. Also, it happened that the elucidation of practical issues led to the invention of new scientific concepts, and possibly new paradigms, with potential impact far beyond. One such example is Riemann's intrinsic view in geometry, that offered a radically new insight in the Physics of the early 20th century. On the other hand, the rather recent intrinsic approaches in elasticity and elasto-plasticity also share this philosophical standpoint of looking from inside, i.e., from the "manifold" point of view. Of course, this approach requires smoothness, and is thus incomplete for an analyst. Nevertheless, its first aim is to highlight the concepts of metric, curvature and torsion; these notions are addressed in the first part of this survey paper. In a second part, they are given a precise functional meaning and their properties are studied systematically. Further, a novel approach to elasto-plasticity constructed upon a model of incompatible elasticity is designed, carrying this intrinsic spirit. The main mathematical object in this theory is the incompatibility operator, i.e., a linearized version of Riemann's curvature tensor. So far, this route not only has led the authors to a new model with a solid functional foundation and proof of existence results, but also to a framework with a minimal amount of ad-hoc assumptions, and complying with both the basic principles of thermodynamics and invariance principles of Physics. The questions arising from this novel approach are complex and intriguing, but we believe that the model is now sufficiently well posed to be studied simultaneously as a problem of mathematics and of mechanics. Most of the

Samuel Amstutz
Université d'Avignon, Laboratoire de Mathématiques d'Avignon, 301 rue Baruch de Spinoza, 84916 Avignon, France, e-mail: samuel.amstutz@univ-avignon.fr

Nicolas Van Goethem
Universidade de Lisboa, Faculdade de Ciências, Departamento de Matemática, CMAFCIO, Alameda da Universidade, C6, 1749-016 Lisboa, Portugal, e-mail: vangoeth@fc.ul.pt

M. Hintermüller, J. F. Rodrigues (eds.), *Topics in Applied Analysis and Optimisation*,
CIM Series in Mathematical Sciences, https://doi.org/10.1007/978-3-030-33116-0_2

research programme remains to be done, and this survey paper is written to present our model, with a particular care to put this approach into a historical perspective.

1 On the origin of curvature in science and the birth of intrinsic views

A first axiomatisation of geometry can be assigned to Euclid of Alexandria (circa 300 B.C.) with *The elements* gathering the knowledge of the time in planar and solid geometry. A little later, a remarkable calculation of the perimeter of the Earth by Eratosthenes of Cyrene provided a first indirect measure of the curvature of the Earth. In the same third century B.C. Archimedes and Apollonius developed the first ideas of plane curvature in the theory of conic sections. The problem of mapping the earth on a planisphere raised the first projections by the geographers of the Antiquity, the principal one being Ptolemy (150 B.C.). Euler, in his *Recherches sur la courbure des surfaces* (1767) [22], laid the foundations of the differential geometry of surfaces. Two centuries before, in the age of discoveries arose the problem of navigation with planar representations of the Earth. A precursor in this field is the Portuguese mathematician Pedro Nunes, who distinguished the great circles from the rhumb lines, and in his *Opera* [55] of 1566 proposed the rectification of the rhumb line in the sea charts. Three years later, the Belgian mathematician and cartographer Gérard Mercator produced the first wall map of the world, that he called *New and more complete representation of the terrestrial globe properly adapted for its use in navigation*, with the conformal projection (i.e., angle-preserving) that now bears his name. In a certain sense, this can be regarded as the dawn of the co-existence of the intrinsic representation (as the map, with its intrinsic metric used by Mercator, preserves the shapes but not the areas) with the embedding view (the map can be embedded in the Euclidean space, yielding the terrestrial globe). The next crucial achievement arose 250 years later with the Gauss' *Theorema Egregium* [28], stating that the scalar curvature is indeed an intrinsic property of a surface, in the sense that it can be computed only by local measurements of distances independently of the ambient space in which the surface is embedded.

The generalization to higher dimensions of Gauss' work for surfaces was initiated by one of his students, Bernhard Riemann in 1854 [57]. Riemann was the first to introduce the notion of differentiable manifold and of a quadratic form (the so-called Riemannian metric) in order to compute the length of a generalized notion of curve. Moreover, the manifold curvature is represented by a complicated object generalizing the Gauss curvature, which is called today the Riemann curvature tensor. In some sense, Gauss' point of view of a geometry of embeddings (he was also concerned with the embedding in the Euclidean space of non-Euclidean – such as hyperbolic – geometries) was somehow bypassed with Riemann's new standpoint of forgetting embeddings for a while, and thinking instead of surfaces intrinsically, with the notion of manifold. It should also be stressed that Riemann's intrinsic approach is the essence of Einstein-Poincaré's new paradigms of physics in the early 20th century.

After Riemann (and Weyl [80]), Whitney was first in 1936 [69, 81] to provide a complete formulation of the notion of manifold. His famous result is the following: any m-dimensional differentiable manifold can be smoothly embedded in $\mathbb{R}^{2m}$. By embedding it is intended an injective immersion (an immersion is a differentiable function between manifolds where the derivative has everywhere full rank). Later, the Nash-Kuiper embedding theorem of 1954/55 [42,53], in conjunction with Whitney's result, states that any m-dimensional Riemannian manifold (M, g) can be C^1-isometrically embedded in an Euclidean space of dimension $2m + 1$ [13]. The importance of this theorem in the history of Mathematics is unvaluable, since it in particular reconciles (Gauss') embeddings- with (Riemann's) intrinsic views.

With this survey, we would like to emphasize the potential of intrinsic approaches in the mathematical modeling of elasto-plastic solids. In particular we begin by identifying the deformation of our body with a metric g and emphasize the role of curvature and torsion in the presence of line-like defects, ultimately leading to plastic effects at the macroscale. Precisely, the notion of incompatibility is at the heart of a new paradigm to describe inelastic effects, since incompatibility is indeed a linearized version of Riemannian curvature, as we will see. The starting point of the theory is to accept that, in the spirit of Riemann, we should attack the modeling problem of elasto-plasticity with an instrinsic approach, that is geometric in nature. Indeed, incompatibility is a physical notion related to non-smooth, in a certain sense singular deformations in the following acception: quoting Cartan [11][1], "the Riemannian space is for us an ensemble of small pieces of Euclidean space, lying however to a certain degree amorphously", while about crystals where dislocation form, Kondo [38] suggests that "the defective crystal is, by contrast [with respect to the above by him given definition of perfect crystal], an aggregation of an immense number of small pieces of perfect crystals (i.e. small pieces of the defective crystal brought to their natural state in which the atoms are arranged on the regular positions of the perfect crystal) that cannot be connected with one other so as to form a finite lump of perfect crystals as an organic unity". In the sequel, we develop these ideas, and in particular we set up a precise mathematical understanding of these concepts.

2 Curvature in nonlinear elasticity

Consider now the manifold to be an open subset Ω of $\mathbb{R}^3$, and consider a symmetric positive definite tensor g_{ij}. The question raised is the following: under which conditions is (Ω, g) *flat*, that is, there exists an immersion $\Theta : \Omega \to \mathbb{R}^3$ such that g is the metric tensor of the open set $\hat{\Omega} := \Theta(\Omega)$? (Note that manifold and embedding spaces have here the same dimension, as opposed to Whitney-Nash theorems). To answer this question, let us first assume that such an immersion exists. Then, by definition, it exists a local frame $\{\underline{g}_i\}_{1 \le i \le 3}$ with $\underline{g}_i := \partial_i \Theta$ such that $g_{ij} := \underline{g}_i \cdot \underline{g}_j$. Then the fol-

[1] This was in essence Riemann's definition of a manifold: each point is in a neigborhood ressembling a distorted Euclidean space.

lowing theorem holds: necessarily, the Riemann curvature tensor Riem_g associated to g_{ij} vanishes in Ω. Of particular interest is the reciprocal statement: given an open, connected and simply connected set Ω and an arbitrary metric $C := g_{ij}$, if $\mathrm{Riem}_g = 0$, then there exists $\Theta \in C^3(\Omega; \mathbb{R}^3)$ such that $C = (\nabla\Theta)^T\nabla\Theta$, i.e., $g_{ij} = \partial_i\Theta \cdot \partial_j\Theta$. Proof of these theorems and their variants can be found in [13].

In elasticity Ω is the reference body, $\hat{\Omega}$ is its deformation, $F := D\Theta = \nabla\Theta$ is the associated deformation tensor, and C the right Cauchy-Green tensor. As a matter of fact, a restatement of this theorem reads: given a reference body Ω and an intrinsic measure C of its deformation (in elasticity C is known to account for local stretch and rotations, [67]), when does it exist a tensor F such that $C = F^T F$ also satisfying $\mathrm{Curl}\, F = 0$? Indeed, it is well known (Helmholtz-Weyl-de Rahm decompositon-type results, also known, in the sense of distributions, as Poincaré's Lemma [13], [39]) that

$$\mathrm{Curl}\, F = 0 \quad \Leftrightarrow \quad F = D\phi \text{ for some } \phi, \tag{2.1}$$

with Curl and D intended in the sense of distributions. Therefore we can write that given Ω open connected and simply connected, and provided a symmetric, positive definite tensor C one has:

$$\exists F : \mathrm{Curl}\, F = 0, C = F^T F \quad \Leftrightarrow \quad \mathrm{Riem}_C = 0. \tag{2.2}$$

In the presence of line-like defects such as Volterra dislocations [34], we are typically faced to the following issue: we have an elastic body Ω with a dislocation loop $\mathcal{L}$ and we assume that we have the means for determining at any point of $\Omega \setminus \mathcal{L}$ the stretch and rotation, in other words we are given a metric tensor C. It turns out that we are only able to construct such a deformation F as in (2.1) and (2.2) in $\Omega \setminus \Pi_{\mathcal{L}}$ where $\Pi_{\mathcal{L}}$ is a surface containing $\mathcal{L}$ and dividing Ω into two subdomains Ω^+ and Ω^- [59]. Let $S_{\mathcal{L}} \subset \Pi_{\mathcal{L}}$ be the surface enclosed by the loop, i.e. $\partial S_{\mathcal{L}} = \mathcal{L}$. Therefore, above and below $\Pi_{\mathcal{L}}$ there exists ϕ^+ and ϕ^-, respectively, such that $C = F^T F$ with $F = D\phi^\pm$ in $\Omega^\pm$. However, it turns out that by the Volterra construction, there is constant jump on $S_{\mathcal{L}}$ that we denote by b, the Burgers vector of $\mathcal{L}$. Specifically, $\phi := \mathrm{Id} + u$ exists globally in Ω by means of a function of bounded variation u, i.e. whose distributional derivative satisfies $Du = \nabla u + b \otimes \nu_{S_{\mathcal{L}}} d\mathcal{H}^2_{\lfloor S_{\mathcal{L}}}$ where $\nu_{S_{\mathcal{L}}}$ is the unit normal to $S_{\mathcal{L}}$. Therefore, in $\Omega \setminus \mathcal{L}$ one may define the deformation tensor as $F = \mathbb{I} + \nabla u$ ($= D\phi^\pm$ in $\Omega \setminus S_{\mathcal{L}}$) such that $C = (\mathbb{I} + \nabla u)^T(\mathbb{I} + \nabla u)$. However in $\mathcal{L}$ this representation fails, and hence the aforementioned approach holds piecewise. Nonetheless, something can be said at $\mathcal{L}$, namely by use of Stokes theorem [59], one finds

$$-\,\mathrm{Curl}\, F = -\,\mathrm{Curl}\, \nabla u = \Lambda_{\mathcal{L}}^T := b \otimes \tau_{\mathcal{L}} d\mathcal{H}^1_{\lfloor \mathcal{L}}, \tag{2.3}$$

where $\Lambda_{\mathcal{L}}$ is called the dislocation density. This reasoning can be generalized for $\mathcal{L}$ a countable union of rectifiable dislocations. This leads us to the following conclusion: the dislocations prevent the deformation to be Euclidean and indeed no global embedding exists. We refer to [82] for a recent contribution to the topic.

The mathematical nature of the displacement field has been clarified in [60, 61]: given the set of dislocations $\mathcal{L}$, an integral current (i.e., the generalization of a closed Lipschitz loop) and a deformation tensor F satisfying (2.3), there exists $u \in W^{1,p}(\Omega, \mathbb{T}^3) \cap SBV(\Omega, \mathbb{R}^3) \cap C^\infty(\Omega \setminus \mathcal{L}, \mathbb{T}^3)$ such that the following holds in the sense of distributions:

$$\text{Div}\,\nabla u = \Delta u = \text{Div}\,F \text{ and } -\text{Curl}\,\nabla u = b \otimes \tau_{\mathcal{L}} d\mathcal{H}^1_{\lfloor \mathcal{L}} = -\text{Curl}\,F. \quad (2.4)$$

The crucial point to note is that the displacement field can be seen in various ways: either as a multi-valued (i.e., in the three-dimensional distorted flat torus $\mathbb{T}^3$ for a normalized Burgers vector, meaning that each component u_i is identified with $u_i + 2\pi\mathbb{Z}$) Sobolev vector field (note that (2.3) imposes that $1 \leq p < 2$), or as a special function of bounded variation, whereby it exhibits a jump on a surface $S_{\mathcal{L}}$ which is not unique (it must be a Lipschitz surface with $\mathcal{L}$ as boundary). Moreover away from $\mathcal{L}$ the displacement is smooth, however multivalued. We can write

$$\text{Dislocation in nonlinear elastic bodies} \Longrightarrow \text{Curl}\,F \neq 0 \text{ and } \exists u \in \mathbb{T}^3 \text{ s.t. } F = \nabla u. \quad (2.5)$$

3 Incompatibility in linearized elasticity and path integral formulae

Linearized elasticity is obtained by neglecting the quadratic terms in $C = (\mathbb{I} + \nabla u)^T(\mathbb{I} + \nabla u)$, that is, we consider a metric $g(u)$ defined as $g_{ij} = \delta_{ij} + 2e_{ij}(u)$ where $e(u) = \nabla^S u := \frac{1}{2}(\nabla^T u + \nabla u)$. Let us consider Riem_e the associated Riemann curvature tensor. It was proved in [45, Proposition 3.11] that

$$(\text{Riem}_e)_{ijkl} = \epsilon_{ijm}\epsilon_{kln}(\text{inc}\,e)_{mn} + \text{h.o.t.}, \quad (3.1)$$

where inc is the symbol standing for the incompatibility operator, writing in Cartesian coordinates as

$$\text{inc}\,e := \text{Curl Curl}^T\,e. \quad (3.2)$$

In the above, ϵ is the Levi-Civita symbol, and the Curl of a tensor is calculated row-wise, hence for a symmetric tensor inc takes the Curl column-wise then row-wise. The equation $\text{inc}\,e = 0$ is equivalent to the Saint-Venant compatibility conditions recalled and discussed thereafter. Pioneer contributions linking compatibility conditions and the Riemann curvature tensor can be found in [26, 63]. We also define the Frank tensor as

$$\mathcal{F}(e) := \text{Curl}^T\,e. \quad (3.3)$$

Let us recall now the problem of reconstructing a displacement from a given symmetric tensor[2]. In linearized elasticity, if all the functions involved are smooth

[2] The history of this construction roots in the end of the 19th century. We have identified by chronological order the following relevant contributions: Kirchhoff in 1876 [36], Beltrami in 1886 [8], Volterra in 1887 [76] (see also [77, 78]), Love in 1892 [44], Michell in 1899 [47],

enough, the displacement field u turns out to be completely defined in terms of the linearized strain tensor e by an explicit recursive integral formula. Let $e \in C^\infty(\Omega, \mathbb{M}^3)$ be a symmetric tensor field in Ω. Let us fix $x_0, x \in \Omega$, and let $\gamma \in C^1([0,1], \Omega)$ be a curve in Ω such that $\gamma(0) = x_0$ and $\gamma(1) = x$. We define the following quantities:

$$w_i(x;\gamma) := w_i(x_0) + \int_\gamma \epsilon_{ipn}\partial_p e_{mn}(y)\mathrm{d}y_m, \tag{3.4}$$

$$u_i(x;\gamma) := u_i(x_0) + \int_\gamma (e_{il}(y) - \epsilon_{ilk}w_k(y))\, dy_l. \tag{3.5}$$

Note that $\mathcal{F}_{im} = \epsilon_{ipn}\partial_p e_{mn}$, thence the name Frank tensor, since its integral on a closed curve γ making one loop around the dislocation $\mathcal{L}$ provides the jump of the rotation vector w, classically known as the Frank tensor. The quantities $w(x)$ and $u(x)$ defined in (3.5) do a-priori depend on the choice of the path from x_0 to x. In such a case the quantities w and u define two C^∞ functions on Ω that will be called the multi-valued rotation and displacement vectors associated to the strain e, respectively (see [74] for the exact meaning of multivaluedness in this context). However, if one has $\operatorname{inc} e = 0$ then u and w are single-valued fields, i.e., are unambiguously defined. Thus, in particular in (3.5), one can use the notation $\int_\gamma = \int_{x_0}^x$ to mean that the integral is path independent. In order to prove this fact, we compute the jump of w and u between two arbitrary curves with the same endpoints, namely γ and $\tilde{\gamma}$, and observe that this quantity is zero if the incompatibility tensor vanishes. These are exactly the well-known Saint-Venant compatibility relations [13]. The rotation and displacement jumps at x are defined as

$$[\![w_i]\!] = w_i(x;\gamma) - w_i(x;\tilde{\gamma}), \quad [\![u_i]\!] = u_i(x;\gamma) - u_i(x;\tilde{\gamma}), \tag{3.6}$$

respectively, and hence depend on the chosen closed path $\gamma - \tilde{\gamma}$ at x. Let $\Omega \subseteq \mathbb{R}^3$ be a simply-connected domain, let $x_0 \in \Omega$ be prescribed, and let $w, u \in C^\infty(\Omega, \mathbb{R}^3)$ be the functions defined in (3.5). Then the following formulae hold:

$$[\![w_i]\!] = \int_S (\operatorname{inc} e(y))_{im}\mathrm{d}S_m(y), \quad [\![u_i]\!] = \int_S (y_m - x_m)\epsilon_{imk}(\operatorname{inc} e(y))_{qk}\mathrm{d}S_q(y), \tag{3.7}$$

for all $x \in \Omega$, and where S is a surface enclosed by the the closed path $\gamma - \tilde{\gamma}$. In particular,

$$\operatorname{inc}\ e = 0 \quad \Rightarrow \quad [\![w_i]\!] = [\![u_i]\!] = 0 \text{ for every } x \text{ and } S\,.$$

Thus, given the tensors e and $\mathcal{F}(e) = \operatorname{Curl}^T e$, and as a consequence of $\operatorname{inc} e = 0$, the vector fields w and u are univoquely defined in (3.5). We refer to [45, Proposition 2.2, Corollary 2.4] for a proof. Moreover, the following classical quantities can be

Cesàro in 1906 [12] and the Cosserat brothers in 1909 [15]. The first rigourous proof must be assigned to Beltrami. According to Love, though, the bulk compatibility conditions should be credited to Barré de Saint-Venant in 1864 [58].

introduced: (i) $e_{ij} := \frac{1}{2}\left(\partial_j u_i + \partial_i u_j\right)$ is the linearized strain tensor (i.e., the linear part of Green St-Venant tensor $C_{ij} = e_{ij} + \partial_i u_k \partial_k u_i$); (ii) $\omega_{ij} := \frac{1}{2}\left(\partial_j u_i - \partial_i u_j\right)$ is said rotation tensor, with $w_i := \frac{1}{2}\epsilon_{ijk}\omega_{kj}$ the rotation vector.

Therefore, the linearized counterpart of (2.5) reads by (3.7),

$$\operatorname{inc} e \neq 0 \Rightarrow \exists u, w \text{ multiple-valued fields, s.t. } e = \nabla^S u. \tag{3.8}$$

In particular, this happens if, given a dislocation loop L, $\gamma - \tilde{\gamma}$ is a curve making one or more loops around L. The multiplicity is precisely the number of loops made, while the jump of u is the Burgers vector b (and of w, the Frank tensor Ω).

4 The legacy of Ekkehart Kröner: the geometry of a crystal with dislocations

Given the dislocation density $\Lambda_{\mathcal{L}}$ we have seen that there exists a map $\varphi \in W^{1,p}(\Omega, \mathbb{T}^3)$, $1 \leq p < 2$ such that $-\operatorname{Curl} \nabla\varphi = \Lambda_{\mathcal{L}}^T$ with $\operatorname{Div} \nabla\varphi = 0$ in Ω, and $(\nabla\varphi)N = 0$ on $\partial\Omega$. The first equality stems directly from the Stokes theorem and from the property $[\![\varphi]\!] = b$ on any enclosing surface $S_{\mathcal{L}}$ [60]. In the same spirit [73], one can find a displacement field $u \in W^{1,p}(\Omega, \mathbb{T}^3)$, $1 \leq p < 2$ satisfying $-\operatorname{Div}(\mathbb{A}\nabla^S u) = f$ and $-\operatorname{Curl} \nabla u = \Lambda_{\mathcal{L}}^T$ in Ω, $(\mathbb{A}\nabla^S u)N = g$ on $\partial\Omega$ and satisfying $[\![u]\!] = b$ and $[\![\mathbb{A}\nabla^S u]\!] = 0$ on $\Omega \cap S_{\mathcal{L}}$. Moreover, it is deduced in [73] that the following expression holds in the sense of distributions:

$$\operatorname{inc} e(u) = \operatorname{inc} \nabla^S u = \operatorname{Curl}\left(\Lambda_{\mathcal{L}} - \frac{\mathbb{I}}{2}\operatorname{tr}\Lambda_{\mathcal{L}}\right) \quad \text{in } \Omega. \tag{4.1}$$

This establishes at the mesoscopic scale the famous macroscopic Kröner formula relating elastic strain incompatibility and dislocation density.

4.1 The geometric approach at the macroscale

At the mesoscale we have seen that the dislocation density reads $\Lambda_{\mathcal{L}} = \tau \otimes b d\mathcal{H}^1_{\lfloor\mathcal{L}}$ and represents the quantity of Burgers vector per unit area (as the density measure $d\mathcal{H}^1_{\lfloor\mathcal{L}}$ has the dimension of an inverse area). Let $\mathcal{S}$ be a small surface in Ω with unit normal n and define the cylinder $\mathcal{V}_\epsilon = \{x + tn, x \in \mathcal{S}, -\epsilon \leq t < \epsilon\}$. Consider a family $\mathcal{L}$ of N parallel mesoscopic dislocations with Burgers vector b and orientation $\tau = n$. One has

$$\frac{1}{2\epsilon}\int_{\mathcal{V}_\epsilon} \Lambda_{\mathcal{L}}^T n dV = Nb.$$

This corresponds to the definition of dislocation density as used by practitioners and lead Kröner [40] to define the macroscopic Burgers vector of a surface $\mathcal{S} \subset \Omega$ as

$$B(\mathcal{S}) := \int_{\mathcal{S}} \Lambda^T dS, \tag{4.2}$$

a macroscopic quantity related to the number of dislocation lines crossing $\mathcal{S}$, with dS the oriented area element, and where Λ is the assumed smooth macroscopic dislocation density. Moreover, one defines the contortion tensor[3] as $\kappa = \Lambda - \frac{\mathbb{I}}{2}\operatorname{tr}\Lambda$. Recalling (3.1), there is a direct link between intrinsic curvature and dislocation density, since the macroscopic expression of Kröner's formula (see [40, 71]) reads

$$\operatorname{inc}\varepsilon = \operatorname{Curl}\kappa. \tag{4.3}$$

Further, at the macroscale we are given a smooth linearized strain ε, and consider the elastic metric

$$g_{ij} := \delta_{ij} + \varepsilon_{ij}. \tag{4.4}$$

So far, we have introduced a metric and an intrinsic curvature related to the presence of dislocations. Yet an important notion is missing: that of connection. The connection in geometry is a notion that permits a comparison between the local geometry at one point and the local geometry at another point. It is thus related to the differentiation of tensor fields and indeed it is well-known [18] that a "good notion" of gradient on a manifold is induced by the choice of the connection Γ (also known as the Christoffel symbols). Denoting this gradient by ∇_Γ, one can introduce the notion of parallel transport along a curve: it is said that a vector v is parallely transported along $\gamma(t)$ if $\dot{\gamma} \cdot \nabla_\Gamma v = 0$. It is said that a connection is compatible with the metric if $\nabla_\Gamma g = 0$: in this case two vector fields v, w parallely transported along a curve have the property that their scalar product $g(v, w)$ is constant. Further, in case of compatible symmetric connections, and in this case only, the Christoffel symbols write as (see [18, Theorem 29.3.2])

$$\Gamma_{ij}^k = \frac{1}{2} g^{kl}(\partial_i g_{lj} + \partial_j g_{il} + \partial_l g_{ij}).$$

This connection is termed Riemannian or Levi-Civita after the name of the Italian mathematician Tullio Levi-Civita (1873-1943). As a matter of fact, a manifold is said Riemannian if it is endowed with both a Riemannian metric and a Riemannian connection (see, e.g., [24]) and, for us, a non-Riemannian manifold means that the connection need not be symmetric and compatible. Note that this latter expression of Γ_{ij}^k is the unique symmetric connection compatible with the metric $g := \mathbb{I} + \varepsilon$ and is that considered to obtain (3.1) from the Riemann curvature tensor $\mathrm{Riem}_\varepsilon$ associated to it. In the sequel this metric connection will be denoted [4] by Γ^B.

Now, for non-Riemannian connections, the crucial notion is that of connection's torsion, defined as the tensor T writing component-wise as

[3] This object has a geometrical meaning in non-Riemannian manifolds, see e.g. [62], that we will not detail here.

[4] Subscript B stands for the Bravais crystal, cf. [70].

$$T_{ij}^k = \Gamma_{[ij]}^k := \frac{1}{2}(\Gamma_{ij}^k - \Gamma_{ji}^k). \tag{4.5}$$

4.2 Parallel displacement and curvature

The role of parallel displacement in a perfect crystal is emphasized by Kröner as he says [41]: "when a lattice vector is parallely displaced using Γ^B along itself, say 1000 times, then its start and goal are separated by 1000 atomic spacings, as measured by g. Because the result of the measurement by parallel displacement and by counting lattice steps is the same, we say that the space is metric with respect to the connection $[\Gamma^B]$." Let us translate Kröner's words in formulae. Consider a closed loop $C = \{\gamma(t) : t \in [0, 1]\}$, with $\dot{\gamma}(t)$ the unit tangent vector. Let us transport a lattice vector e along C. In the absence of defects one has an Euclidean connection and hence e is parallely transported along C, i.e. $\nabla_{;C} e = \dot{\gamma}\nabla e = \dot{\gamma}^k \partial_k e = 0$. Assume now the presence of dislocations along C. We know that the manifold has some curvature and consider in a first step the Levi-Civita connection $\Gamma = \Gamma^B$ associated to g. To transport the contravariant vector e we need to compute its covariant derivative along C, namely $(\nabla_{;C} e)^i := \dot{\gamma}^k(\partial_k e^i + \Gamma_{jk}^i e^j)$ (cf. [18, §28.2., Eq. (23)]). Therefore the instantaneous deviation of e due to dislocations is given by

$$(\nabla_{;C} e)^i = \Gamma_{jk}^i e^j \dot{\gamma}^k. \tag{4.6}$$

In linearized elasticity it is assumed $g_{kl} = \delta_{kl} + 2\varepsilon_{kl}$ and thus lowering and raising the indices can be considered indifferently, to the first order. So, from $\Gamma_{k;ji} := g_{kl}\Gamma_{ji}^l$, the deviation of e^i in a time interval dt writes as $\Gamma_{i;jk} e_j dx_k$+h.o.t., with $dx^k := \dot{\gamma}^k dt$. Let us compute the total amount of deviation on a closed loop C up to the first order, that is, we calculate $\int_C \Gamma_{i;jk} dx_k$ that by Stokes theorem rewrites as $\int_{S_C} \epsilon_{mqk} \partial_q \Gamma_{i;jk} dS_m$, with S_C a surface enclosed by C. The covariant derivative of $\Gamma_{i;jk}$ reads (cf. [18, Eq. (32)])

$$\nabla_q \Gamma_{i;jk} = \partial_q \Gamma_{i;jk} - (\Gamma_{i;pk}\Gamma_{p;jq} + \Gamma_{i;pq}\Gamma_{p;jk}) - \Gamma_{i;jp}\Gamma_{p;kq}. \tag{4.7}$$

By the symmetry in q and k of the term inside the parenthesis and of the last term, this yields in the absence of torsion

$$\epsilon_{mqk} \partial_q \Gamma_{i;jk} = \epsilon_{mqk} \nabla_q \Gamma_{i;jk}. \tag{4.8}$$

Now, let $\{\mathbf{a_i}\}$ denote an orthonormal basis, and recall that $\Gamma_{jk}^i = (\nabla_k \mathbf{a}_j)^i$ (cf. [18, Eq. (33)]). Thus, by a property of the Riemann curvature tensor (or a definition, see [18, Theorem 30.1.1])), one has

$$\begin{aligned}\int_C \Gamma^B_{i;jk} dx_k &= \int_{S_C} \epsilon_{mqk} \nabla_q (\nabla_k \mathbf{a}_j)_i dS_m + \text{h.o.t.}\\ &= \int_{S_C} \frac{1}{2} \epsilon_{mqk} \left(\text{Riem}^B_g\right)_{i;pkq} (\mathbf{a}_j)_p dS_m + \text{h.o.t.}\\ &= \int_{S_C} \frac{1}{2} \epsilon_{mqk} \left(\text{Riem}^B_g\right)_{i;jkq} dS_m + \text{h.o.t.}.\end{aligned}$$

By virtue of (3.1)-(3.3) this yields

$$\begin{aligned}-2\int_C \Gamma^B_{i;jk} dx_k &= \epsilon_{ijp} \int_{S_C} (\operatorname{inc}\varepsilon)_{pm} dS_m + \text{h.o.t.}\\ &= \epsilon_{ijp} \int_{S_C} \epsilon_{mql} \partial_q \left(\mathcal{F}(\varepsilon)\right)_{pl} dS_m + \text{h.o.t.},\end{aligned} \tag{4.9}$$

that is, by (3.5),

$$-2\int_C \Gamma^B_{i;jk} dx_k = \epsilon_{ijp} \int_C \left(\mathcal{F}(\varepsilon)\right)_{pl} dx_l + \text{h.o.t.} = \epsilon_{ijp} [\![w_p]\!](S_C) + \text{h.o.t.}.$$

Therefore we see that at the first order, the total deviation of e^i around C depends on the rotation jump on the surface, since

$$\int_C \nabla_k e_i dx_k = \frac{1}{2} \epsilon_{ikl} \Omega_k (S_C) e_l + \text{h.o.t.},$$

where we have introduced the Frank tensor $\Omega_k := [\![w_k]\!]$. So far we have established the following:

Point view of the external observer:

- g Riemannian metric, Γ^B Riemannian connection
  $\Rightarrow (\Omega, g, \Gamma^B)$ Riemannian manifold.
- *Observation 1:* $\nabla\Lambda \neq 0, T = T^B = 0 \Rightarrow \text{Riem}_g \neq 0$:
  non-homogeneous dislocation density

$\Rightarrow$ crystal manifold with curvature but no torsion.

- *Observation 2:* $\int_C \nabla e \cdot dL = \frac{1}{2}\Omega(S_C) \times e + \text{h.o.t.}$:
  disclinations $\Rightarrow$ crystal curvature.

In this case the anholonomy is given by the presence of disclinations, where the term "holonomy"[5] of a connection refers to the extent to which parallel transport around closed loops achieves, or fails, to preserve the geometrical data being transported: for instance under a metric connection, the orthogonality of two parallely transported vector fields is preserved.

In this first perspective, of the so-called *external* observer, the absence of isometric embedding in $\mathbb{R}^3$ is translated into the nonvanishing Riemann curvature tensor

[5] The holonomy group is defined as the set of linear transformations arising from parallel transport along closed loops.

associated with the Levi-Civita metric connection of g. Though, one can see that something is missing in this formalism, since a homogeneous (constant) density of dislocations yields a crystal manifold with vanishing curvature, that is, a flat manifold. Moreover a pure dislocation (i.e, with vanishing jump of the rotation tensor) would permit parallel transport of lattice vectors, in contradiction with the nature of the dislocations which is responsible for atomic jumps, hence inducing a geometry that is not Euclidean. Instead, we would like to find a geometry that is specific for crystals with dislocations. This will be achieved by means of a new, non-Riemannian connection, i.e. a connection with torsion. This new perspective requires from the standpoint of physics the introduction of the so-called *internal* observer, able to determine crystallographic defects. For a recent review on incompatible deformation fields and the Riemann curvature field, we refer to [65].

4.3 The non-Riemannian crystal manifold

To emphasize the perspective of the internal observer we define the following geometrical objects:

DISLOCATION TORSION: $$T_{i;kj} := \epsilon_{kjq}\Lambda_{qi} \tag{4.10}$$

CONNECTION CONTORTION: $$\Delta\Gamma_{k;ij} := -T_{j;ik} - T_{i;jk} + T_{k;ji} \tag{4.11}$$

NON SYMMETRIC CONNECTION: $$\Gamma_{k;ij} := \Gamma^B_{k;ij} + \Delta\Gamma_{k;ij}. \tag{4.12}$$

It is easily seen that Γ is a connection, since Γ^B is a connection and $\Delta\Gamma$ is a tensor (by the transformation property of a connection, cf. [18, Eq. (22)]). Moreover the symmetric part of Γ namely $\Gamma_{k;(ij)}$ is a connection and it can be proved [70, Theorem 5.2] that $\Gamma_{k;[ij]} = T_{k;ij}$. Denoting the symmetric part of the connection by $\Gamma^\circ_{k;ij} = \Gamma^B_{k;ij} + \Delta\Gamma_{k;(ij)}$ we have, by the above calculations, one has

$$\int_C \Gamma^\circ_{i;jk} e_j dx_k = \int_{S_C} \frac{1}{2}\epsilon_{mqk} \left(\mathrm{Riem}^\circ(\varepsilon)\right)_{i;jkq} e_j dS_m + \text{h.o.t.},$$

where $\mathrm{Riem}^\circ(\varepsilon)$ is the Riemann curvature tensor associated to Γ°. Let us now compute the remaining term, namely

$$\int_C \Gamma_{i;[jk]} e^j dx_k = \int_C T_{i;jk} e_j dx_k = \int_C \Lambda_{pi} dS_p,$$

where we have introduced the surface element $dS_p = \epsilon_{pjk} e_j dx_k$. The deviation of the second term is just simply the integral on C of the macrosopic Burgers vector $B_i(dS)$, that is

$$\int_C \Gamma_{i;[jk]} e^j dx_k = \int_C (n \cdot \tau) B_i(dS),$$

with $dS_p = n_p dS$.

Thus we see that provided the metric and the connection of the internal observer, the lattice vector is deviated: the first source of deviation is the Riemannian curvature, whereas the second is the connection's torsion. Note that this connection might also be not metric, i.e., if $\nabla g \neq 0$. The metric tensor is considered to model point defects in crystals, but this is beyond the scope of this survey (cf [83]).

4.4 Internal and external observers

These are physical concepts related to fictitious thought experiments. The external observer is only able to make experiments from the outside, namely it measures fiber elongations and/or rotations and hence it measures a deformation ε that provides the geometer with a metric g. For the external observer the utmost geometrical quantity available is the Riemannian curvature as derived from the metric connection Γ^B. On the contrary, the internal observer acts at another scale. It is able to recognize crystallographic direction and count atomic steps, hence can measure a density of dislocations. Thus torsion is available to the internal observer, but not to the external. It is only a combined view that provides a complete geometrical picture of the dislocated crystal, thence described as a non-Riemannian manifold, summarized as follows:

Combined views of the internal and external observers:

• g Riemannian metric, Γ non-Riemannian connection
$\Rightarrow (\Omega, g, \Gamma)$ non-Riemannian manifold.
• *Property:* $\big(\Lambda \neq 0 \Rightarrow T \neq 0\big)$ and $\big(\nabla\Lambda \neq 0 \Rightarrow \mathrm{Riem}_\Gamma \neq 0\big)$:
non-homogeneous dislocation density $\Rightarrow$ crystal manifold with curvature and torsion;
homogeneous dislocation density $\Rightarrow$ crystal manifold with torsion.
• Anholonomy by displacement and rotation jumps
(i.e., dislocations and disclinations):
$\int_C \nabla_k e_i dx_k = \int_C (n \cdot \tau) B_i(dS) + \epsilon_{ikl}\Omega_k(S_C)e_l + \cdots$.

We note that if one restricts to the first order then it holds true that $\int_C \nabla_k e_i dx_k = \int_{S_C} \frac{1}{2}\epsilon_{mqk}\,(\mathrm{Riem}_\Gamma(\varepsilon))_{i;jkq}\, e_j dS_m$ with Riem_Γ the Riemann curvature of the connection Γ defined as a sum with Γ^B in such a way that $\mathrm{Riem}_\Gamma = \mathrm{Riem}^B + \cdots$ with the remaining terms related to $\Delta\Gamma$. Note also that it is solely Riem^B that yields the term $\Omega(S_C) \times e$.

As we see, Kröner's macroscopic framework allows us to come back to the language of geometry, by stressing that the crystal geometry and the physical laws governing defects are inseparable, as is the case in the Einstein's general theory of relativity. We entirely agree with Noll when he writes [54] that "the geometry [must be] the natural outcome, not the first assumption, of the theory"[6]. Many geometrical

[6] As in the *Continuous Distribution of Dislocation* (CDD) theory of Bilby et al. [9].

tools and mathematical theory required for a rigorous description of the dislocated crystal geometry can be found in the landmark papers by Noll [54] and Wang [79], while also pointing out a recent book on continuum mechanics in this spirit [21]. We emphasize that $\Delta\Gamma_{k;[ji]}$ was also introduced by Noll [54] and called the crystal *inhomogeneity tensor*.

4.5 Inelastic effects and notion of eigenstrain

The geometric description of a dislocated body has been made so far for static dislocations, that is, at the macroscale for a constant in time dislocation density tensor. We have seen that spatial variation of Λ and hence of the contortion tensor κ induces a non vanishing Curl κ thence a nonzero elastic strain incompatibility $\operatorname{inc}\varepsilon$. A further notion introduced by Kröner is the eigenstrain $\bar{\varepsilon}$ satisfying $\operatorname{inc}\bar{\varepsilon} = -\operatorname{inc}\varepsilon$. Physically it represents the additional strain to recover compatibility, since $\operatorname{inc}(\bar{\varepsilon} + \varepsilon) = 0$ implies the existence of a vector field u such that $e(u) = \bar{\varepsilon}+\varepsilon$ as related to the so-called Beltrami decomposition of symmetric tensor fields and Saint-Venant conditions [45]. Plasticity is the macroscopic behaviour of a body whose dislocation density tensor varies in time, since dislocation motion is the physical cause of plasticity. The last sections of this survey will be dedicated to the description of a novel model of elasto-plastic bodies based on ε and $\operatorname{inc}\varepsilon$, hence the model variables are (ε, κ), two intrinsic and objective tensors, in an extended sense to be defined later. Before, we intend to recall the intrinsic approach to linearized elasticity.

5 A geometric conception of linearized elasticity: the intrinsic approach

5.1 Gauss vs. Riemann in linearized elasticity

In the conventional mathematical treatment of linearized elasticity, the basic model variable is the displacement field $u : \Omega \to \mathbb{R}^3$, with respect to which elastic problems are stated and solved. Further, the linearized strain is introduced in a second step as the symmetric gradient of the displacement field, $\varepsilon = e(u) := \nabla^S u$. However, in many computations and experiments, the strain is most naturally the "observable" field, thence becoming the main model variable. In this spirit, the stress might also be considered as a root variable, in the sense that it is a field that is observable, measurable and controllable: by a possibly fictitious thought experiment, the stress is obviously measurable by extracting from the elastic body a small enough volume element and then measuring the Newtonian forces exerted on its facets. Moreover, given the stress tensor, the strain is well defined as soon as a constitutive law is provided, here a linear homogeneous, isothermal and isotropic law: the strain-stress constitutive law reads $\varepsilon = \mathbb{C}\sigma$, with $\mathbb{C}$ the compliance tensor, i.e., fourth-rank (inverse) tensor of elasticity.

However, today most elasticity problems are treated using the displacement as basic model variable, from which the strain is defined by the kinematic relation $\varepsilon = \nabla^S u$, thence the stress by a constitutive law. This approach presumably comes from the study of elliptic boundary-value problems, where the elasticity system is most often presented as a vector-valued extension of elliptic equations in divergence form. Moreover, weak and variational formulations are most easily derived by means of the displacement, and show a convenient and elegant way of solving problems in elasticity.

There are nonetheless profound theoretical reasons to refrain from taking the displacement as main model variable. For instance, its possible multi-valuedness, which is not to avoid from a physical standpoint, since multi-valuedness may have a meaning, but which must be addressed in an adequate manner in an appropriate mathematical formalism (see above). Another example is the reference configuration, from which the displacement is defined and which by definition is arbitrary: although natural in finite elasticity, it becomes somehow artificial in linearized elasticity, since Eulerian and Lagrangian representations coincide. Moreover, in elasto-plasticity or for elastic bodies with defects, the stress and defect-free reference configurations might not exist (simultaneously). Hence, to remedy this issue, it is often appealed to "intermediate" reference configurations, from which plastic and elastic deformations are defined, but whose physical as well as mathematical meaning are far from clear. In fact, what is a plastic distortion (i.e., the "plastic" part of the displacement gradient) as long as no constitutive law exist for the rotations (i.e., the skew-symmetric part of the gradient)? Not to mention some models which introduce plastic and elastic displacements, whose physical and mathematical meanings are extremely vague. Further, we should also mention the fact that any rigorous model should in principle be proven independent of the choice the reference configuration. Lastly, in the presence of crystal defects like dislocations the very notion of displacement or velocity is not clearly defined at any scale. For instance, at the atomic scale, bonds can move while atoms remain fixed.

5.2 Ciarlet's intrinsic approach to linearized elasticity

As we have seen, it may happen that because of defects or other incompatibilities, the very notion of a displacement field does not make sense as a conventional single-valued field. Instead, one would like to state the linear-elastic problem in terms of the strain ε, which needs not a-priori be taken as a symmetric gradient. For these reasons, the intrinsic approach in linearized elasticity by Ph. Ciarlet and C. Mardare [14] constitutes a major breakthrough in mathematical elasticity, which was able to reconcile in an elegant manner the two aforementioned approaches. In their presentation, the strain is the main model variable in terms of which strong as well as variational formulations are sought. The displacement only appears in a second step if the Riemannian curvature tensor associated to the elastic metric vanishes (see above). In the approach of Ciarlet and Mardare, a differential geometry setting is chosen (see above), where the boundary under analysis is defined by means of smooth enough immersions, from which the curvilinear basis, the metric, the symmetric connection and the curvature tensors are derived. It should be emphasized that such derived curvilinear bases are indeed defined in the body itself as well as on its boundary, but are not mutually orthogonal, and have no particular physical meaning. The main difficulty is the treatment of the boundary conditions, since a condition such as $u_{|\Gamma_0} = 0$ for some $\Gamma_0 \subset \partial\Omega$ is not easily translated to a boundary condition on the deformation tensor.

Let ε be a compatible strain tensor, i.e., by Saint-Venant theorem (see below), there exists a displacement field u such that $\varepsilon = e(u) = \nabla^S u$. Let ε_T be the tangential strain (i.e. the projection of e perpendicularly to the normal to the boundary) $\partial\Omega$. Let $\gamma^\sharp := \varepsilon_{T\,|\partial\Omega}$ be the linearized change-of-metric and $\rho^\sharp$ the linearized change-of-curvature tensors as introduced by Ciarlet-Mardare in [14]. Let $\Gamma \subset \partial\Omega$ be a connected relatively open set, and let $\mathcal{R}(\Gamma)$ be the set of rigid motions (i.e., roto-translations) on Γ. Their main theorem states the following:

Theorem 5.1 (Ciarlet-Mardare [14, Theorem 6.1]) *Let $u \in H^1(\Omega)$. Let either*

(i) $u_{|\Gamma} = 0$
(ii) $\bar{\gamma}^\sharp(e) = \bar{\rho}^\sharp(e) = 0$
(iii) $u_{|\Gamma} \in \mathcal{R}(\Gamma)$.

One has (i) $\Rightarrow$ (ii) $\Rightarrow$ (iii), where $\bar{\gamma}^\sharp$ and $\bar{\rho}^\sharp$ are suitable extensions of $\gamma^\sharp$ and $\rho^\sharp$.

To give a more physical understanding of the linearized change-of-curvature tensor, we need to anticipate the specific trace operators $\mathcal{T}_0$ and $\mathcal{T}_1$ that have been introduced in [3]. They have been obtained through a Green-like formula (see below Theorem 8.2 for details) where T and η are smooth enough symmetric tensors,

$$\int_\Omega T \cdot \operatorname{inc} \eta dx = \int_\Omega \operatorname{inc} T \cdot \eta dx + \int_{\partial\Omega} \mathcal{T}_1(T) \cdot \eta \, dS(x) + \int_{\partial\Omega} \mathcal{T}_0(T) \cdot \partial_N \eta \, dS(x).$$

It has been proved in [2, Lemma 2.11] that that

$$\text{(i)} \Rightarrow \text{(ii)}' \quad \varepsilon_T = \mathcal{T}_0(\varepsilon) = 0 \text{ and } \mathcal{T}_1(\varepsilon) = 0 \text{ on } \Gamma,$$

and that, for $\Gamma = \partial\Omega$ (see [2, Proposition 2.19]),

$$\text{(ii)}' \Rightarrow \text{(iii)}.$$

Note that the rigid displacement is set to zero as soon as the normal components of ε, i.e., $\varepsilon - \varepsilon_T$ vanish. The link with the Frank tensor is the following:

$$\varepsilon = 0 \text{ on } \Gamma \Rightarrow \left(\operatorname{Curl}^T \varepsilon \times N = 0 \Leftrightarrow \mathcal{T}_1(\varepsilon) = 0 \text{ on } \Gamma \right),$$

with N the outer normal to $\partial\Omega$. In this case either conditions $\operatorname{inc} \varepsilon = 0$ in Ω together with $\varepsilon = \operatorname{Curl}^T \varepsilon \times N = 0$ or $\varepsilon = \mathcal{T}_1(\varepsilon) = 0$ on Γ implies that $\varepsilon = \nabla^S v$ with $v = 0$

on Γ (since in this case $\bar{\rho}^\sharp(\varepsilon) = 0$ [72]). Thus, the intrinsic elasticity system writes in strong form as

$$\begin{cases} -\operatorname{div}\left(\mathbb{C}^{-1}\varepsilon\right) = f & \text{in} \quad \Omega \\ \operatorname{inc}\varepsilon = 0 & \text{in} \quad \Omega \\ \mathcal{T}_0(\varepsilon) = \mathcal{T}_1(\varepsilon) = 0 & \text{on} \quad \Gamma \\ \left(\mathbb{C}^{-1}\varepsilon\right) N = g & \text{on } \partial\Omega \setminus \Gamma \end{cases},$$

where the traces are intended in a weak sense, and with f and g the volume and surface loads, respectively. The corresponding intrinsic variational formulation reads

$$\inf_{\substack{\operatorname{inc} e = 0 \\ \mathcal{T}_0(e) = \mathcal{T}_1(e) = 0 \text{ on } \Gamma}} \int_\Omega \left(\frac{1}{2}\mathbb{C}^{-1}e - \mathbb{K}\right) \cdot e dx,$$

where $\mathbb{K}$ is a tensor of external forces satisfying

$$\begin{cases} -\operatorname{div}\mathbb{K} = f & \text{in} \quad \Omega \\ \mathbb{K}N = g & \text{on } \partial\Omega \setminus \Gamma \end{cases}.$$

6 The classical route to plasticity

After the pioneer works by Coulomb (1773), the theory of plasticity finds its roots in the mid 19th and early 20th centuries by experimentalists (Tresca, 1864) and engineers/mathematicians/physicists (e.g., Saint-Venant[7], 1870, Von Mises, 1913, Prandtl, 1924), and was later developed by Drucker (1947) and Hill (1950). Importants works were achieved by Taylor and Orowan (1934) putting the spot light on the link between plasticity and dislocation motion.

6.1 The mathematical approaches: two perspectives

The mathematical literature starts with Hodge and Prager around 1950 who were first to understand plasticity in modern mathematical terms, proposing a variational formulation in terms of the stress rate and based on the principle of virtual power, while in parallel Greenberg [27] proposed a variational formulation in terms of the velocities. Thus we see that historically the two approaches of strain vs. displacement-based elasto-plasticity models have already coexisted since the beginning. This is a key aspect of our approach, which is an intrinsic model, i.e., considering the elastic strain as basic model variable, in the sense of the geometers mentioned above, of Prager, and recently (notably after decades of displacement-based approach) of

[7] A name that we have encountered already above an which is also related to the incompatibility operator, see below.

Ciarlet and coauthors [13, 14]. The philosophical standpoint can be recast in view of the aformentioned history as follows:

Riemann's view	**Gauss' view**
Intrinsic models	Embeddings models
No reference configuration	Reference configurations
Strain/strain rate-based approaches	Displacement/velocity-based approaches.

(6.1)

Let us now describe the main ingredients of the conventional, historical approach, which, we recall, has proven so far in complete agreement with the observations. It is hence an excellent model for all practical purposes. In the general framework of thermodynamics, one postulates the existence of a dissipation potential which provides the evolution laws for plastic deformation (the so-called flow rules) and possible other internal variables. This part of the theory is strongly linked with convex analysis and thus was fostered in the 60ies by the remarkable works by Rockafellar and Moreau [51]. In the end of the 70ies and in the 80ies, Strang and Temam [68] and their collaborators introduced a new functional space to describe plasticity, the space of bounded deformations, whose main feature is to allow the strain to have a regular part and a measure part, thus well-suited to modeling discontinuous phenomena, such as plastic slip on part of the boundary, or shear band formation in perfect plasticity. This theory yielded on the one hand a first rigorous mathematical formulation (i.e., with proofs of existence, etc., see the excellent textbook by Han and Reddy [31]), and on the other hand gave rise to performent numerical schemes. Summarizing, the conventional approach to elasto-plasticity is based on

(i) the velocities (or the displacement field) as first model variable,
(ii) a postulated decomposition in elastic and plastic parts of the total compatible strain, where each part is assumed to model distinct sub-scale phenomena (elastic deformation means variation in inter-atomic distance, whereas plastic deformation is a macroscopic manifestation of the modification of inter-atomic bonds, where dislocations play a role),
(iii) separate constitutive laws for the elastic and plastic strains,
(iv) a convex elastic domain (whose boundary is the so-called yield surface), which is a sufficient condition in order to satisfy the 2nd principle of thermodynamics and further permit the use of convex calculus.

First existence results for linear elasticity/perfect plasticity were provided around 1980 by Johnson [35], Suquet [66], by means of visco-plastic approximations (see also Anzelotti, Giaquinta, Luckhaus [5, 6], Hardt and Kinderlehrer [32]). About 20 years later, a quasi-static evolutionary variational formulation was sucessfully proposed by Mielke and coauthors (Mainik, Roubíček, Stefanelli, etc. [48,49]) in the early 2000s, based on a balance between dissipative and potential restoring forces. Slightly later a series of refinements were provided by Dal Maso's school [16], and also by other authors (such as, e.g., Ortiz, Francfort [17], De Simone, Neff [23]), some of them grounding their approach on the energy dissipation principle and De Giorgi's theory of minimizing movements. In the 80ies and 90ies, computational plasticity has been developed, based on strain incremental schemes, see for instance the classical textbooks of Hughes and Simo [64] and Han and Reddy [31].

6.2 Conventional (0th-order) elasto-plasticity models

Conventional models of small strain elasto-plasticity start with the following postulate. It is assumed that $\varepsilon^e = \mathbb{A}^{-1}\sigma$ is the elastic strain, with $\mathbb{A}$ the isotropic elasticity tensor and σ the Cauchy stress tensor. Then, the eigen-strain is called plastic strain, $\bar{\varepsilon} = \varepsilon^p$, and there exists a vector field u called the displacement field, satisfying

$$e(u) = \varepsilon^e + \varepsilon^p. \tag{6.2}$$

Whereas at time t, the elastic strain $\varepsilon^e(t)$ obeys the elasticity system

$$-\operatorname{div}(\mathbb{A}\varepsilon^e(t)) = f(t) \text{ in } \Omega, \quad (\mathbb{A}\varepsilon^e(t))N = g(t) \text{ on } \Gamma_N, \tag{6.3}$$

the plastic strain satisfies other laws, called the flow rules. These are based on another series of postulates. Before recalling these rules, it is important to have in mind three facts regarding (6.2).

(i) The partition is local, i.e., $\varepsilon(x) = \varepsilon^e(x) + \varepsilon^p(x)$ for any $x \in \Omega$ and is purely of physical nature, that is, there is not any sort of mathematical structure behind it.
(ii) The geometric meaning of ε as an intrinsic metric has been lost, since each part has its own definition, given by solving some equations, whereas the total deformation is defined by their sum.
(iii) By (6.2), it is conventionally postulated that the total deformation is compatible, that is, that there exists a displacement field u such that $\varepsilon = \nabla^S u = e(u)$. This statement is not justified by any mathematical argument and the adoption of this hypothesis is made for simplicity. Indeed it automatically implies that the incompatibilities of elastic and plastic parts mutually compensate, without the need to let the flow rules comply with this property.

As for the plastic part, following Moreau [51] it is assumed that:

- There exists a compact and convex subset $\mathcal{K}$ of symmetric 3×3-matrices such that the condition $\sigma \in \mathcal{K}$ is always satisfied. The yield surface is represented by the boundary $\partial\mathcal{K}$. Moreover, in the general context of plasticity with hardening, the elastic domain $\mathcal{K}(t)$ at time t depends on $\sigma(t)$, to account for the back-stress tensor, and we write $\mathcal{K}(t)$ to mean $\mathcal{K}(\sigma(t))$.
- Let $I_{\mathcal{K}}$ be the indicator function of $\mathcal{K}$, i.e. $I_{\mathcal{K}}(\eta) = 0$ if $\eta \in K$ and $I_{\mathcal{K}}(\eta) = +\infty$ if $\eta \notin \mathcal{K}$. Then the so-called associated flow rule (a special case commonly used) can be written as

$$\begin{cases} \sigma(t) \in \operatorname{int} \mathcal{K}(t) \Rightarrow & \dot{\varepsilon}^p(t) = 0 \\ \sigma(t) \in \partial\mathcal{K}(t) & \Rightarrow \dot{\varepsilon}^p(t) \in \partial I_{\mathcal{K}(t)}(\sigma(t)) \Leftrightarrow (\eta - \sigma(t)) \cdot \dot{\varepsilon}^p(t) \leq 0, \forall \eta \in \mathcal{K}(t). \end{cases} \tag{6.4}$$

Here $\partial I_{\mathcal{K}(t)}$ is the normal cone $N_{\mathcal{K}}(\sigma(t))$ and $\sigma(t) \in \operatorname{int} \mathcal{K}(t) \Leftrightarrow N_{\mathcal{K}}(\sigma(t)) = \{0\}$. This formalism is to be compared with the model of Hill and Rice [33, 56], which is summarized as follows:

- Introduce the dissipation potential $\mathcal{D}(\dot{\varepsilon}^p) := \sup\{\eta \cdot \dot{\varepsilon}^p | \eta \in \mathcal{K}(t)\}$, that is the support function of $\mathcal{K}(t)$. Convex calculus entails

$$\mathcal{D}(\dot{\varepsilon}^p) = \sigma(t) \cdot \dot{\varepsilon}^p \Leftrightarrow \sigma(t) \in \partial\mathcal{D}(\dot{\varepsilon}^p). \tag{6.5}$$

- It is easily proven that the two formalisms are equivalent [31]:

$$(6.4) \Leftrightarrow (6.5).$$

Note that the $\Leftarrow$ implication shows that Hill and Rice formalism does not require the notion of elastic domain and yield surface, which are obtained as consequences. However they postulate the existence of a potential $\mathcal{D}$ that is convex, positively homogeneous and lower semicontinuous. Thus we see that in both cases the model is strongly based on convex analysis.

7 Gradient elasto-plasticity for continua with dislocations: towards an incompatibility-driven model

7.1 The size effect

Zeroth-order models as described in Section 6 may not be sufficient with a view to several technological applications. For instance, it is essential for micro-electronics devices (such as silicon wafers for semiconductor production [52]) to accurately identify the mechanical properties of micro-structured materials, since there exist important strength differences that result from modification of the material micro-structural characteristics with changing size (where in general smaller sizes correlate with stronger responses). Indeed, the mechanical properties of micro-structured materials (e.g., yield strength, strain-hardening rate) with small-scale structures are extremely size-sensitive, and the increase in strength with decreasing scale can be related to increasing the strain gradients. For instance, industrial silicon is produced by the growth of a crystal seed, which by definition incorporates all material sizes and various types of micro-structures (individual point-defects, voids, dislocations, volumic clusters, dislocations): present in the small-size material, they will grow together with the crystal and form defect structures of various sizes [52, 75]. Further, serious issues prevent the use of classical (local) theories of plasticity and fracture (see [25] or [1]), since classical continuum mechanics cannot accurately capture size effects and highly localized deformations. On the other hand, atomistic simulations are out of reach in terms of computational cost, and therefore are restricted to small samples. In order to address the size effect problem, the so-called gradient plasticity models have gained an increasing interest in the scientific and technological communities.

7.2 Gradient models

The success of gradient theories stems from the incorporation of a micro-structural length scale parameter. Indeed, it is a general feature that gradient-theories do not assume stress as function of the sole history of strain at a point x, rather they take into account possible interactions with other material points in its neighbourhood. For example, the internal state variable that is responsible for isotropic hardening in classical plasticity theory is the effective plastic strain p, that, in a non-local media can be replaced by a weighted average $\bar{p}(x) = \int_V p(x+\xi)h(\xi)dx$, where x is the point of interest, ξ is the size of the localized plastic zone, and $h(\xi)$ is a weighting function. Then $\bar{p}(x)$ can be approximated in terms of $p(x)$, $L\nabla p(x)$ and $L^2\nabla^2 p(x)$, for a certain characteristic length L. The physical basis of the gradient plasticity theory rests on theoretical developments of geometrically necessary dislocations (GNDs), see Daya Reddy, Gurtin and Neff works for instance [7, 20, 29, 30]. For instance, the micro-mechanical modeling of the inelastic material behavior of metallic single crystals is based on the fact that resistance to glide is due to random trapping of mobile dislocations, "statistically stored" dislocations (SSDs), and acting as obstacles to further dislocation motion. On the other hand, the GND are related to the elastic strain incompatibility and are responsible for the observed macroscopic plastic behaviour, as stated by the famous Kröner's relation [40] "$\text{inc}\,\epsilon = \text{Curl}\,\kappa$" with κ related to dislocation density. In the last two decades, another class of gradient theories was introduced, assuming higher-order gradients of the plastic strain field, as proposed in our model. These theories are a particular case of the generalized continua, such as continua with micro-structure, which were inspired by the pioneering work of the Cosserat brothers[8] [15].

7.3 Our approach: a gradient model based on the strain incompatibility

We believe that the intrinsic point of view together with continuum-and gradient-based theories as proposed by our approach (see Sections 8 and 9) are needed to bridge the gap between classical continuum and micro-mechanical theories. The main feature of our model is that we focus on dislocation micro-structure in single

[8] Indeed, the Cosserat (or micro-polar) continuum enhances the kinematic description of deformation by an additional field of local rotations.

crystals with a novel continuum model making an explicit link between plasticity and dislocations, and based on a novel model paradigm. As we will explain, the model we propose is of gradient type, involving the curl and the incompatibility of the strain. Further, the deformation tensor is seen as a metric in the aforementioned geometric sense. In particular we do not distinguish between elastic or plastic deformations.

The novel approach we propose has been introduced and discussed in [4]. In our model, neither of the three above postulates are considered. Our paradigm is radically different and our approach is based on the following rationales.

1. Strain rate is prefered to strain and is given the following, primordial definition. Identify three fibers at x, denoted by a_1, a_2, a_3, which at time t are oriented along the axes of a Cartesian coordinate system and of unit lengths. Then the deformation rate is defined at x as (see, e.g., [19, 67])

$$d_{ij}(t) = \frac{1}{2}\left(\frac{d}{dt}(a_i \cdot a_j)\right)_t. \tag{7.1}$$

   Having fixed an initial time $t_0 = 0$, the time integral of the objective tensor d, called the strain or deformation tensor reads $\varepsilon(t) = \int_0^t d(s)ds$. Note that (7.1) holds for infinitesimal as well as for finite strains and hence one is not forced to specify the quantitative nature of the deformations before they take place.
2. This strain defined in this fashion is neither elastic nor plastic, it simply has a compatible and an incompatible part, that are given by a structure theorem called Beltrami decomposition [45]:

$$\varepsilon = \nabla^S u + E^0. \tag{7.2}$$

   As opposed to elastic-plastic splittings this decomposition is unique once boundary conditions for u are prescribed. Moreover, while ε is an objective field (in a general sense discussed in section 9, neither $\nabla^S u$ nor E^0 are objective. Therefore the model will be constructed upon ε and its derivatives.
3. The governing equations should generalize classical linear elasticity in the sense that it must take into account the possible strain incompatibility. The idea behind is that the model should explicitely account for the physical cause of plasticity: the presence and motion of dislocations as microstructural perturbations.

As detailed below, the key point upon which our model relies is the fact that strain incompatibility is directly related to the density of dislocations by (4.3). Moreover, our model involves a new tangent material coefficient ℓ, with the dimension of a force, representing the resistance of matter against incompatibility. In general, this scalar is space- and time-dependent, and evolves with the course of (at first quasi-static) deformation. If ℓ depends on space only, we proved in [2] that our model is a special case of the classical Mindlin gradient elasticity [50], whereas for a time-dependent ℓ, it gives rise to a drastically new approach and a novel nonlinear plasticity model in direct relation with dislocation motion.

A difficult modelling problem that we have not yet achieved is to find an appropriate evolution equation for the incompatibility modulus ℓ. In principle it should be derived based on dislocation mechanics as a function of temperature, strain, strain rate, and a set of measurable micro-structural physical parameters. Further, due to the existence of plastic strain gradient terms, higher-order boundary conditions are required on both external (free surfaces) and internal boundary (interfaces) regions where plastic deformation occurs. Also, these higher-order boundary conditions, which are motivated from the physical understanding of the dislocation mechanics, may vary with the course of plastic deformation. It is a further principle of our approach that boundary conditions should be naturally integrated in the functional framework of our equations. Hence internal transmission conditions are only byproducts of weak formulations.

7.4 Link with classical elasto-plasticity models

Recall that classical linearized elasto-plasticity models are based on the a priori decomposition $\varepsilon^{\rm tot} = \varepsilon^{\rm e} + \varepsilon^{\rm p}$, where the total strain $\varepsilon^{\rm tot}$ is compatible ($\operatorname{inc}\varepsilon^{\rm tot} = 0$), the elastic strain $\varepsilon^{\rm e}$ is derived from the Cauchy stress by Hooke's law, and the plastic strain $\varepsilon^{\rm p}$ obeys flow rules. We now compare this decomposition with the Beltrami decomposition $\varepsilon = \nabla^S u + \varepsilon^0$. Since $\operatorname{inc}\varepsilon^{\rm tot} = \operatorname{inc}\nabla^S u = 0$, there exists a vector field w (see [45]) such that $\varepsilon^{\rm tot} = \nabla^S u - \nabla^S w$ and we can write

$$\varepsilon^{\rm tot} = \nabla^S u - \nabla^S w = -\left(\varepsilon^0 + \nabla^S w\right) + \left(\nabla^S u + \varepsilon^0\right).$$

We then recognize $\nabla^S u + \varepsilon^0$ as the strain ε. The correspondence with the Beltrami decomposition $\varepsilon = \nabla^S u + \varepsilon^0$ can be made upon setting $\varepsilon^{\rm e} = \varepsilon$, $\varepsilon^{\rm p} = -(\varepsilon^0 + \nabla^S w)$.

The interpretation is the following (see Fig. 1): for us, ε represents the deformation from a reference state, say state A to a neighbour state B of the same material. It can be viewed as the composition of the incompatible deformation ε^0 from state A to an intermediate state A', and the compatible deformation $\nabla^S u$ from A' to B. In the classical approach, another configuration A'' serves as reference configuration. The total deformation $\varepsilon^{\rm tot}$ from A'' to B is the sum of the plastic deformation $\varepsilon^{\rm p} = -(\varepsilon^0 + \nabla^S w)$ from A'' to A and the elastic deformation $\varepsilon^{\rm e} = \varepsilon$ from A to B. Therefore, plastic effects play somehow the role of configurational forces [28] that account for the change of reference configuration. Of course, choosing $w = 0$ (thus $A'' = A'$) would be a choice of simplicity, but in that case $\varepsilon^{\rm p}$ would be identified with $-\varepsilon^0$, hence it would not be trace-free as assumed in some standard flow rules. We emphasize the arbirariness of A", whereas A' is uniquely determined from Beltrami decomposition. However, a significant difference between the kinematical frameworks of the two approaches is that $\varepsilon^{\rm p}$ is usually supposed trace-free whereas ε^0 is divergence-free. For us, incompressibility could be realized by an enrichment of the model, with possible additional variables and equations, but it is not prescribed a priori in the general setting.

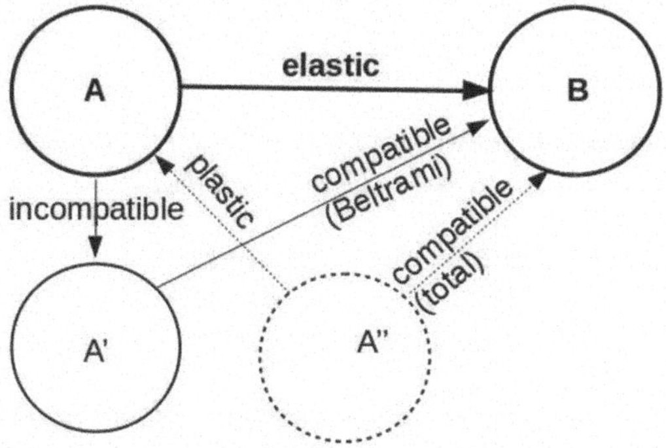

Fig. 1: The Beltrami decomposition ($A \longrightarrow A' \longrightarrow B$) vs the standard elastic/plastic decomposition ($A'' \longrightarrow A \longrightarrow B$)

8 The incompatibility operator: functional framework

We have seen so far that the notion of incompatibility is a crucial ingredient in the modeling of dislocated crystals and therefore in the understanding of the plastic behavior of solids. From the mathematical point of view, inc is a second order differential operator that acts on (usually symmetric) tensor fields. We address in this section the analysis of this operator in the framework of Sobolev spaces. The Beltrami[9] decomposition asserts that any symmetric tensor field can be split into a divergence-free part and an incompatibility-free part. It appears therefore natural to study the incompatibility operator in divergence-free spaces. As orthogonal complements, incompatibility-free spaces will also be discussed.

Let Ω be a regular (C^∞) bounded domain of $\mathbb{R}^3$. We denote by $\partial\Omega$ its boundary and by N its outward unit normal. Moreover $\mathbb{S}^3$ denotes the set of symmetric 3-matrices.

8.1 Divergence-free lifting, Green formula and applications

We begin with the divergence-free lifting of traces of symmetric tensor fields. Set

[9] Eugenio Beltrami (1835-1900) is an Italian physicist and mathematician known in particular for his works on elasticity–stating the equilibrium equations of a body in terms of the stress in place of the strain [8]– but also in non-Euclidean geometries in the wake of Gauss and Riemann. He was indeed a friend of Riemann whom he met at Pisa university where he had a chair. Moreover, his chair of mathematical physics in Rome was later transmitted to Volterra in 1900. Vito Volterra (1860-1940) is presumably the first who gave a correct definition of dislocations and disclinations in [77]. It is thus not mere coincidence that the name of Beltrami will take a crucial place in our survey on incompatibility and dislocations. Neither that the second author of this survey started his study of dislocations with a book found in the main Scuola Normale library in Pisa in 2000 [37].

$$\tilde{H}^{3/2}(\partial\Omega, \mathbb{S}^3) = \left\{ E \in H^{3/2}(\partial\Omega, \mathbb{S}^3) : \int_{\partial\Omega} ENdS(x) = 0 \right\}.$$

Theorem 8.2 (Divergence-free lifting [3]) *Let* $\mathbb{E} \in \tilde{H}^{3/2}(\partial\Omega, \mathbb{S}^3)$ *and* $\mathbb{G} \in H^{1/2}(\partial\Omega, \mathbb{S}^3)$. *There exists* $E \in H^2(\Omega, \mathbb{S}^3)$ *such that*

$$\begin{cases} E = \mathbb{E} & \text{on } \partial\Omega, \\ (\partial_N E)_T = \mathbb{G}_T & \text{on } \partial\Omega, \\ \operatorname{div} E = 0 & \text{in } \Omega, \end{cases}$$

where the subscript T stands for the tangential part. In addition, such a lifting can be obtained through a linear continuous operator

$$\mathcal{L}_{\partial\Omega} : (\mathbb{E}, \mathbb{G}) \in \tilde{H}^{3/2}(\partial\Omega, \mathbb{S}^3) \times H^{1/2}(\partial\Omega, \mathbb{S}^3) \mapsto E \in H^2(\Omega, \mathbb{S}^3).$$

Define the subset of $C^\infty(\partial\Omega, \mathbb{S}^3)$

$$\mathcal{G} = \{V \odot N, V \in \mathbb{R}^3\},$$

with the notation $U \odot V := (U \otimes V + V \otimes U)/2$.

Lemma 8.1 (Dual trace space [3]) *Every* $\mathbb{E} \in H^{-3/2}(\partial\Omega, \mathbb{S}^3)/\mathcal{G}$ *admits a unique representative* $\tilde{\mathbb{E}}$ *such that*

$$\int_{\partial\Omega} \tilde{\mathbb{E}} NdS(x) = 0. \tag{8.1}$$

Moreover, the dual space of $\tilde{H}^{3/2}(\partial\Omega, \mathbb{S}^3)$ *is canonically identified with* $H^{-3/2}(\partial\Omega, \mathbb{S}^3)/\mathcal{G}$.

We define the spaces of symmetric tensor fields

$$H^{\text{div}}(\Omega, \mathbb{S}^3) := \{E \in L^2(\Omega, \mathbb{S}^3) : \operatorname{div} E \in L^2(\Omega, \mathbb{R}^3)\},$$
$$H^{\text{inc}}(\Omega, \mathbb{S}^3) := \left\{E \in L^2(\Omega, \mathbb{S}^3) : \operatorname{inc} E \in L^2(\Omega, \mathbb{S}^3)\right\},$$

endowed with the norms defined by $\|E\|^2_{H^{\text{div}}} = \|E\|^2_{L^2} + \|\operatorname{div} E\|^2_{L^2}$, $\|E\|^2_{H^{\text{inc}}} = \|E\|^2_{L^2} + \|\operatorname{inc} E\|^2_{L^2}$, respectively.

Recall that the Green formula for the divergence allows to define, for any $T \in H^{\text{div}}(\Omega, \mathbb{S}^3)$, its normal trace $TN \in H^{-1/2}(\partial\Omega, \mathbb{R}^3)$ by

$$\int_{\partial\Omega} (TN) \cdot \varphi dS(x) := \int_{\Omega} \left(\operatorname{div} T \cdot \tilde{\varphi} + T \cdot \nabla^S \tilde{\varphi} \right) dx \qquad \forall \varphi \in H^{1/2}(\partial\Omega, \mathbb{R}^3),$$

with $\tilde{\varphi} \in H^1(\Omega, \mathbb{R}^3)$ an arbitrary lifting of φ. For the incompatibility operator one has the following counterpart.

Lemma 8.2 (Green formula for the incompatibility [3]) *Suppose that* $T \in C^2(\overline{\Omega}, \mathbb{S}^3)$ *and* $\eta \in H^2(\Omega, \mathbb{S}^3)$. *Then*

$$\int_\Omega T \cdot \operatorname{inc}\eta dx = \int_\Omega \operatorname{inc} T \cdot \eta dx + \int_{\partial\Omega} \mathcal{T}_1(T) \cdot \eta \, dS(x) + \int_{\partial\Omega} \mathcal{T}_0(T) \cdot \partial_N \eta \, dS(x) \quad (8.2)$$

with the trace operators defined as

$$\mathcal{T}_0(T) := (T \times N)^T \times N, \quad (8.3)$$

$$\mathcal{T}_1(T) := \left(\operatorname{Curl} (T \times N)^T \right)^S + ((\partial_N + k)T \times N)^t \times N + \left(\operatorname{Curl}^T T \times N \right)^S, \quad (8.4)$$

where k is twice the mean curvature of $\partial\Omega$ and $T^S = (T + T^T)/2$. In addition, it holds

$$\int_{\partial\Omega} \mathcal{T}_1(T) N dS(x) = 0. \quad (8.5)$$

Note that $(T \times N)^T \times N$ is made of permutations of the tangential components of T. Alternative expressions of $\mathcal{T}_1(T)$ are derived in [3]. Therefore, we can define the traces $\mathcal{T}_0(T) \in H^{-1/2}(\partial\Omega, \mathbb{S}^3)$ and $\mathcal{T}_1(T) \in H^{-3/2}(\partial\Omega, \mathbb{S}^3)/\mathcal{G}$ for every $T \in H^{\mathrm{inc}}(\Omega, \mathbb{S}^3)$ by

$$\langle \mathcal{T}_0(T), \varphi_0 \rangle = \int_\Omega T \cdot \operatorname{inc}\eta_0 dx - \int_\Omega \operatorname{inc} T \cdot \eta_0 dx, \qquad \forall \varphi_0 \in H^{1/2}(\partial\Omega, \mathbb{S}^3),$$

$$\langle \mathcal{T}_1(T), \varphi_1 \rangle = \int_\Omega T \cdot \operatorname{inc}\eta_1 dx - \int_\Omega \operatorname{inc} T \cdot \eta_1 dx, \qquad \forall \varphi_1 \in \tilde{H}^{3/2}(\partial\Omega, \mathbb{S}^3),$$

with $\eta_0 = \mathcal{L}_{\partial\Omega}(0, \varphi_0)$ and $\eta_1 = \mathcal{L}_{\partial\Omega}(\varphi_1, 0)$ (recall that $\mathcal{L}_{\partial\Omega}$ is the lifting operator defined in Theorem 8.2). In addition, by Lemma 8.1, $\mathcal{T}_1(T)$ admits a unique representative satisfying (8.5). By linearity of $\mathcal{L}_{\partial\Omega}$, this extends formula (8.2) to any functions $T \in H^{\mathrm{inc}}(\Omega, \mathbb{S}^3)$ and $\eta \in H^2(\Omega, \mathbb{S}^3)$.

From the two Green formulas recalled above one immediately infers that:

- $\operatorname{inc} \nabla^S v = 0$ in the sense of distributions for all $v \in H^1(\Omega, \mathbb{R}^3)$;
- $\operatorname{div} \operatorname{inc} E = 0$ in the sense of distributions for all $E \in H^{\mathrm{inc}}(\Omega, \mathbb{S}^3)$.

In particular, if $E \in H^{\mathrm{inc}}(\Omega, \mathbb{S}^3)$, then $\operatorname{inc} E N$ is defined in $H^{-1/2}(\partial\Omega, \mathbb{R}^3)$ by

$$\int_{\partial\Omega} \operatorname{inc} E N \cdot \varphi dx = \int_\Omega \operatorname{inc} E \cdot \nabla^S \varphi dx \qquad \forall \varphi \in H^1(\Omega, \mathbb{R}^3).$$

Let Γ be a smooth subset of $\partial\Omega$ and set

$$H_0^{\mathrm{inc}}(\Omega, \mathbb{S}^3) = \text{ the closure of } \mathcal{D}(\Omega, \mathbb{S}^3) \text{ in } H^{\mathrm{inc}}(\Omega, \mathbb{S}^3).$$

Further properties of the trace operators are given below.

Proposition 8.1 (Trace properties [2])

1. Let $v \in H^1(\Omega, \mathbb{R}^3)$ be such that $v = r$ on Γ in the sense of traces, with r a rigid displacement field. Then $\mathcal{T}_0(\nabla^S v) = \mathcal{T}_1(\nabla^S v) = 0$ on Γ.

2. We have the characterization

$$H_0^{\mathrm{inc}}(\Omega, \mathbb{S}^3) = \left\{ E \in H^{\mathrm{inc}}(\Omega, \mathbb{S}^3) : \mathcal{T}_0(E) = \mathcal{T}_1(E) = 0 \text{ on } \partial\Omega \right\}.$$

3. If $E \in H_0^{\mathrm{inc}}(\Omega, \mathbb{S}^3)$ *then* $\operatorname{inc} EN = 0$ *on* $\partial\Omega$.

8.2 Saint-Venant compatibility conditions and Beltrami decomposition

We state the following two results in the setting of L^p spaces for generality, although we will be concerned with $p = 2$ only.

Theorem 8.3 (Saint-Venant compatibility conditions [45]) *Assume that* Ω *is simply-connected. Let* $p \in (1, +\infty)$ *be a real number and let* $E \in L^p(\Omega, \mathbb{S}^3)$*. Then,*

$$\operatorname{inc} E = 0 \text{ in } W^{-2,p}(\Omega, \mathbb{S}^3) \Longleftrightarrow E = \nabla^S v$$

for some $v \in W^{1,p}(\Omega, \mathbb{R}^3)$*. Moreover, u is unique up to rigid displacements.*

Theorem 8.4 (Beltrami decomposition [45]) *Assume that* Ω *is simply-connected. Let* $p \in (1, +\infty)$ *be a real number and let* $E \in L^p(\Omega, \mathbb{S}^3)$*. Then, for any* $v_0 \in W^{1/p,p}(\partial\Omega)$*, there exists a unique* $v \in W^{1,p}(\Omega, \mathbb{R}^3)$ *with* $v = v_0$ *on* $\partial\Omega$ *and a unique* $F \in L^p(\Omega, \mathbb{S}^3)$ *with* $\operatorname{Curl} F \in L^p(\Omega, \mathbb{R}^{3\times 3})$*,* $\operatorname{inc} F \in L^p(\Omega, \mathbb{S}^3)$*,* $\operatorname{div} F = 0$ *and* $FN = 0$ *on* $\partial\Omega$ *such that*

$$E = \nabla^S v + \operatorname{inc} F. \tag{8.6}$$

A variant of Saint-Venant's compatibility conditions in the presence of boundary conditions is the following.

Proposition 8.2 (Saint-Venant with boundary condition [2]) *Assume that* Ω *is simply-connected. If* $E \in L^2(\Omega, \mathbb{S}^3)$ *satisfies*

$$\begin{cases} \operatorname{inc} E = 0 \text{ in } \Omega, \\ \mathcal{T}_0(E) = \mathcal{T}_1(E) = 0 \text{ on } \partial\Omega \end{cases} \tag{8.7}$$

then there exists $v \in H_0^1(\Omega, \mathbb{R}^3)$ *such that* $\nabla^S v = E$*. Moreover, the map* $E \in L^2(\Omega, \mathbb{S}^3) \mapsto v \in H_0^1(\Omega, \mathbb{R}^3)$ *is linear and continuous.*

We assume from now on that Ω is simply-connected.

8.3 Orthogonal decompositions

For Γ being a smooth subset of $\partial\Omega$, we define the sets

$$\begin{aligned}
\mathcal{V} &= \left\{E \in L^2(\Omega, \mathbb{S}^3) \ : \ \operatorname{inc} E = 0\right\}, \\
\mathcal{V}_\Gamma^0 &= \{E \in \mathcal{V} \ : \ \mathcal{T}_0(E) = \mathcal{T}_1(E) = 0 \text{ on } \Gamma\}, \\
\mathcal{V}_\Gamma^{00} &= \left\{\nabla^S v \ : \ v \in H^1(\Omega), v = 0 \text{ on } \Gamma\right\}, \\
\mathcal{W} &= \left\{E \in L^2(\Omega, \mathbb{S}^3) \ : \ \operatorname{div} E = 0\right\}, \\
\mathcal{W}_\Gamma^0 &= \{E \in \mathcal{W} \ : \ EN = 0 \text{ on } \Gamma\}.
\end{aligned}$$

From what precedes we infer the following relations [2]:

$$\mathcal{V} = \mathcal{V}^0_\emptyset = \mathcal{V}^{00}_\emptyset \qquad \mathcal{V}^{00}_\Gamma \subset \mathcal{V}^0_\Gamma, \qquad \mathcal{V}^{00}_{\partial\Omega} = \mathcal{V}^0_{\partial\Omega}.$$

A refinement of the Beltrami decomposition is obtained as follows.

Theorem 8.5 (Orthogonal decomposition of $L^2(\Omega, \mathbb{S}^3)$ [2]) *Assume that $\partial\Omega$ admits the partition $\partial\Omega = \Gamma_1 \cup \Gamma_2$ with $\Gamma_1 \cap \Gamma_2 = \emptyset$. We have the orthogonal decomposition*

$$L^2(\Omega, \mathbb{S}^3) = \mathcal{V}^{00}_{\Gamma_1} \oplus \mathcal{W}^0_{\Gamma_2}.$$

Related to this decomposition, the following lemma will be useful.

Lemma 8.3 (Boundary orthogonality relation [2]) *If $\mathbb{K} \in \mathcal{V}^{00}_{\Gamma_1}$ and $\operatorname{inc} \hat{F} \in \mathcal{W}^0_{\Gamma_2}$ then*

$$\int_{\Gamma_2} \left(\mathcal{T}_1(\mathbb{K}) \cdot \hat{F} + \mathcal{T}_0(\mathbb{K}) \cdot \partial_N \hat{F} \right) dS(x) = 0.$$

We now define the spaces with further differentiability properties

$$\begin{aligned} \mathbb{Z} &= \{E \in H^{\mathrm{inc}}(\Omega, \mathbb{S}^3) \ : \ \operatorname{div} E = 0 \text{ in } \Omega, EN = 0 \text{ on } \partial\Omega\}, \\ \mathbb{Z}_0 &= \{E \in \mathbb{Z} \ : \ \operatorname{inc} EN = 0 \text{ on } \partial\Omega\}, \\ \mathcal{F} &= \{E \in H^{\mathrm{inc}}(\Omega, \mathbb{S}^3) \ : \ \operatorname{inc} EN = 0 \text{ on } \partial\Omega\}. \end{aligned}$$

A straightforward consequence of Theorem 8.5 is the following.

Proposition 8.3 *We have the orthogonal decompositions*

$$H^{\mathrm{inc}}(\Omega, \mathbb{S}^3) = \mathbb{Z} \oplus \mathcal{V}, \qquad \mathcal{F} = \mathbb{Z}_0 \oplus \mathcal{V}.$$

8.4 Boundary value problems for the incompatibility

If $E \in H^{\mathrm{inc}}(\Omega, \mathbb{S}^3)$ is split into $E = E_i + E_c$, $E_i \in \mathbb{Z}$, $E_c \in \mathcal{V}$, then $\operatorname{inc} E = \operatorname{inc} E_i$. Thus, a Poincaré inequality for the incompatibility is naturally sought in $\mathbb{Z}$ or one of its subspaces. In fact, the following holds.

Proposition 8.4 (First Poincaré inequality [2]) *There exists $C > 0$ such that, for all $E \in \mathbb{Z}$,*

$$\|E\|_{H^1} \le C \|\operatorname{inc} E\|_{L^2}.$$

As a straightforward consequence, given $\mathbb{K} \in L^2(\Omega, \mathbb{S}^3)$ and $\mathbb{B}$ a symmetric uniformly positive definite fourth order tensor field, we infer by the Lax-Milgram theorem the existence of a unique $E \in \mathbb{Z}$ such that

$$\int_\Omega \mathbb{B} \operatorname{inc} E \cdot \operatorname{inc} \hat{E} dx = \int_\Omega \mathbb{K} \cdot \hat{E} dx \qquad \forall \hat{E} \in \mathbb{Z}. \tag{8.8}$$

The same result holds if $\mathbb{Z}_0$ is substituted for $\mathbb{Z}$. If $\operatorname{div}\mathbb{K} = 0$ in Ω, $\mathbb{K}N = 0$ on $\partial\Omega$ and $E \in \mathbb{Z}$ solves (8.8), then Proposition 8.3 shows that (8.8) holds actually for any $\hat{E}$ in $H^{\mathrm{inc}}(\Omega, \mathbb{S}^3)$. This yields the strong form

$$\begin{cases} \operatorname{inc}(\mathbb{B}\operatorname{inc}E) = \mathbb{K} \text{ in } \Omega \\ \operatorname{div}E = 0 \text{ in } \Omega \\ EN = 0 \text{ on } \partial\Omega \\ \mathcal{T}_0(\mathbb{B}\operatorname{inc}E) = \mathcal{T}_1(\mathbb{B}\operatorname{inc}E) = 0 \text{ on } \partial\Omega. \end{cases}$$

Dirichlet-type boundary conditions can be considered through the space

$$\mathcal{H}_0 = \{E \in H^2(\Omega, \mathbb{S}^3) \ : \ \operatorname{div}E = 0 \text{ in } \Omega, E = (\partial_N E \times N)^T \times N = 0 \text{ on } \partial\Omega\}.$$

Observe from the Green formula that the boundary conditions in $\mathcal{H}_0$ are dual to the trace operators $\mathcal{T}_1$ and $\mathcal{T}_0$. We have the following Poincaré inequality:

Proposition 8.5 (Second Poincaré inequality [3]) *There exists $C > 0$ such that, for all $E \in \mathcal{H}_0$,*

$$\|E\|_{H^2} \le C\|\operatorname{inc}E\|_{L^2}.$$

Obviously, given $\mathbb{K} \in L^2(\Omega, \mathbb{S}^3)$ and $\mathbb{B}$ a symmetric uniformly positive definite fourth order tensor field, there is a unique $E \in \mathcal{H}_0$ such that

$$\int_\Omega \mathbb{B}\operatorname{inc}E \cdot \operatorname{inc}\hat{E}dx = \int_\Omega \mathbb{K} \cdot \hat{E}dx \qquad \forall \hat{E} \in \mathcal{H}_0. \tag{8.9}$$

Nonhomogeneous boundary conditions can also be prescribed, using Theorem 8.2. In order to identify the strong form of (8.9), we first note that, if $E \in \mathcal{H}_0$ solves (8.9), then there exists a Lagrange multiplier (see e.g. [10]) $p \in L^2(\Omega, \mathbb{R}^3)$ such that

$$\int_\Omega \mathbb{B}\operatorname{inc}E \cdot \operatorname{inc}\hat{E}dx - \int_\Omega p \cdot \operatorname{div}\hat{E}dx = \int_\Omega \mathbb{K} \cdot \hat{E}dx$$

for all $\hat{E} \in H^2(\Omega, \mathbb{S}^3)$ with $\hat{E} = (\partial_N \hat{E} \times N)^T \times N = 0$ on $\partial\Omega$. Therefore the strong form reads

$$\begin{cases} \operatorname{inc}(\mathbb{B}\operatorname{inc}E) + \nabla^S p = \mathbb{K} \text{ in } \Omega \\ \operatorname{div}E = 0 \text{ in } \Omega \\ E = (\partial_N E \times N)^T \times N = 0 = 0 \text{ on } \partial\Omega. \end{cases}$$

9 Towards an intrinsic approach to linearized elasto-plasticity

9.1 Objectivity and principle of virtual powers

Consider a macroscopic solid represented by the domain Ω and subject to external loading. We place ourselves in a linearized setting, that is, our aim is to descibe the deformation of the solid between two close configurations (see Fig. 1), when the load admits a small increment. The evolution of this tangent modeling between increments will be discussed later, but an integrated nonlinear approach is currently out of our scope.

We will use the principle of virtual powers to describe the internal efforts acting within the solid and derive balance equations. In this framework, efforts are represented by powers, rather than forces. Two types of efforts are treated separately: the external efforts (the exterior medium acts on the body) and the internal efforts (the matter acts on itself). The corresponding powers are linear functionals that act on kinematical descriptors also called test or virtual fields. This is why we speak of virtual powers. The choice of these kinematical descriptors is of paramount importance.

A kinematical field is said objective if it is independent of the observer. The classical mathematical definition is the following: it is a scalar, vector, or tensor field that obeys the standard rules of transformation for such quantities through a roto-translation of the frame with arbitrary speed. In this setting, it is well-known that the velocity is not objective, whereas its symmetric gradient is. Nonetheless, we have seen that in the presence of defects the notion of velocity is not always well-defined. More general kinematical concepts are the geometric data of the solid seen as a Riemannian manifold. For us, only these fields (or their time rates) will be considered as objective. Hence, objective kinematical descriptors will be built upon the metric g. Typically, we will consider the strain $E := g - \mathbb{I}$ and its rate $\dot{E}$.

In our model we assume that the internal virtual power is a continuous linear functional on $L^2(\Omega, \mathbb{S}^3)$, the set of virtual strain rates. By Riesz representation, there exists a generalized force field $\Sigma \in L^2(\Omega, \mathbb{S}^3)$ such that

$$\mathcal{P}_{(i)}(\hat{E}) = \int_\Omega \Sigma \cdot \hat{E} dx \qquad \forall \hat{E} \in L^2(\Omega, \mathbb{S}^3). \tag{9.1}$$

Unlike its internal counterpart, the virtual external power is a linear functional against kinematical fields which may be non-objective (see, e.g., [46]). Typically, the velocity is considered in order to represent standard forces. However, in our framework, the velocity is only a byproduct of the strain rate, defined by orthogonal projection onto an appropriate function space. Therefore, it is natural to assume that the virtual external power is a linear functional on the set of virtual strain rates and we write

$$\mathcal{P}_{(e)}(\hat{E}) = \int_\Omega \mathbb{K} \cdot \hat{E} dx \qquad \forall \hat{E} \in L^2(\Omega, \mathbb{S}^3), \tag{9.2}$$

for some $\mathbb{K} \in L^2(\Omega, \mathbb{S}^3)$. The interpretation of $\mathbb{K}$ will be discussed in section 9.4. In the absence of inertial effects, the principle of virtual powers reads

$$\mathcal{P}_{(i)}(\hat{E}) = \mathcal{P}_{(e)}(\hat{E}), \tag{9.3}$$

for all $\hat{E}$ satisfying possible kinematical constraints.

9.2 Constitutive law

We assume that the generalized force Σ is a function of the local geometric data of the solid. This relation is called constitutive law. In our linearized framework, we assume that $\Sigma(x)$ is expressed as a linear function of the pair composed of $E(x)$, the strain at point x, and $\operatorname{inc} E(x)$, the linearized Riemannian curvature. Therefore we can write

$$\Sigma(x) = \mathbb{A}E(x) + \mathbb{B} \operatorname{inc} E(x),$$

for some fourth-order tensors $\mathbb{A}$ and $\mathbb{B}$.

In classical elasticity one has $\operatorname{inc} E = 0$, so that $\mathbb{A}$ is recognized as the Hooke tensor of the material. We place ourselves in the isotropic case where the standard expression

$$\mathbb{A} = \lambda \mathbb{I}_2 \otimes \mathbb{I}_2 + 2\mu \mathbb{I}_4$$

holds, with (λ, μ) the Lamé coefficients.

Tensor $\mathbb{B}$ is a new object. An assumption of consistency with linear elasticity will reduce its expression. First, let us emphasize that classical compatible elasticity formally corresponds to $\mathbb{B}$ 'large': indeed, the term $\mathbb{B} \operatorname{inc} E$ penalizes incompatible deformations. Second, the equations of linear elasticity are derived from the principle of virtual powers considering compatible test fields $\hat{E} = \nabla^S \hat{v}$. Thus, we want that following property be fulfilled: if $\mathbb{B}$ is homogeneous, then

$$\int_\Omega \mathbb{B} \operatorname{inc} E \cdot \hat{E} dx = 0 \qquad \forall \hat{E} = \nabla^S \hat{v}, \hat{v} \in \mathcal{D}(\Omega, \mathbb{R}^3).$$

From the Green formula, it turns out that $\mathbb{B} = \ell \mathbb{I}_4$, where ℓ is a scalar coefficient which we call incompatibility modulus, is a sufficient condition. Eventually, we arrive at the constitutive law

$$\Sigma = \mathbb{A}E + \ell \operatorname{inc} E. \tag{9.4}$$

Different derivations of (9.4), based on Mindlin's theory of gradient elasticity [50], are given in [4,43].

9.3 Equilibrium equations

Plugging (9.1), (9.2) and (9.4) into (9.3) entails that

$$\int_\Omega (\mathbb{A}E + \ell \operatorname{inc} E) \cdot \hat{E} dx = \int_\Omega \mathbb{K} \cdot \hat{E} dx, \tag{9.5}$$

for all admissible $\hat{E}$. In the following, we will not consider any kinematical constraint, whereby (9.5) reduces to

$$\mathbb{A}E + \ell \operatorname{inc} E = \mathbb{K}. \tag{9.6}$$

9.4 Interpretation of the external power and kinematical framework

Assume that $\partial\Omega = \Gamma_1 \cup \Gamma_2$, $\Gamma_1 \cap \Gamma_2 = \emptyset$, and take $\hat{E} \in L^2(\Omega, \mathbb{S}^3)$. In view of Theorem 8.5, we have the unique decomposition $\hat{E} = \nabla^S \hat{v} + \operatorname{inc} \hat{F}$ with $\nabla^S \hat{v} \in \mathcal{V}^{00}_{\Gamma_1}$ and $\operatorname{inc} \hat{F} \in \mathcal{W}^0_{\Gamma_2}$. The Green formula yields

$$\mathcal{P}_{(e)}(\hat{E}) = \int_\Omega \mathbb{K} \cdot \hat{E} dx = -\int_\Omega \operatorname{div} \mathbb{K} \cdot \hat{v} dx + \int_{\Gamma_2} \mathbb{K}N \cdot \hat{v} dS(x) + \int_\Omega \operatorname{inc} \mathbb{K} \cdot \hat{F} dx$$
$$+ \int_{\partial\Omega} \left(\mathcal{T}_1(\mathbb{K}) \cdot \hat{F} + \mathcal{T}_0(\mathbb{K}) \cdot \partial_N \hat{F} \right) dS(x). \tag{9.7}$$

Therefore, $f := -\operatorname{div} \mathbb{K}$ is identified with a body force, and $g := \mathbb{K}N$ is identified with a surface load on Γ_2. Now, if $\mathbb{K} \in \mathcal{V}^{00}_{\Gamma_1}$ the last two integrals of (9.7) vanish by virtue of Lemma 8.3. Then (9.7) rewrites as the classical expression of the external power in linear elasticity

$$\int_\Omega \mathbb{K} \cdot \hat{E} dx = \int_\Omega f \cdot \hat{v} dx + \int_{\Gamma_2} g \cdot \hat{v} dS(x). \tag{9.8}$$

To sum up, given $f \in L^2(\Omega, \mathbb{R}^3)$ and $g \in H^{-1/2}(\Gamma_2, \mathbb{R}^3)$, the standard external power is obtained after solving

$$\begin{cases} -\operatorname{div} \nabla^S w = f \text{ in } \Omega, \\ w = 0 \text{ on } \Gamma_1, \\ \nabla^S w N = g \text{ on } \Gamma_2, \end{cases} \tag{9.9}$$

and setting $\mathbb{K} = \nabla^S w \in L^2(\Omega, \mathbb{S}^3)$.

9.5 Existence results and elastic limit

The main result of this section is the following.

Theorem 9.6 (Well-posedness [2]) *Assume Ω is simply connected. Let $\mathbb{K} \in L^2(\Omega, \mathbb{S}^3)$. Let C be the Poincaré constant of Proposition 8.4. If $\mathbb{A}$ is uniformly positive definite and $|\ell| > C|\mathbb{A}|$ a.e., then there exists one and only one $E \in \mathcal{F}$ such that*

$$\mathbb{A}E + \ell \operatorname{inc} E = \mathbb{K}. \tag{9.10}$$

Moreover we have the a priori estimate

$$\| \operatorname{inc} E \|_{L^2} \leq \frac{\|\ell^{-1}\mathbb{A}\|_{L^\infty}}{1 - C\|\ell^{-1}\mathbb{A}\|_{L^\infty}} \|\mathbb{A}^{-1}\mathbb{K}\|_{L^2}. \tag{9.11}$$

The essential boundary condition $\operatorname{inc} EN = 0$ (no incompatibility flux) appears naturally in the proof, through integrations by parts. It models the fact that the outside of Ω has a purely elastic behavior, however future work should go further into this point, see the discussion in [2] where an alternative condition for the modeling of free boundaries is also analyzed. Inequality (9.11) shows that $\operatorname{inc} E$ tends to 0 as $|\ell| \to +\infty$. The following result is more precise.

Theorem 9.7 (Elastic limit [2]) *Assume that $\mathbb{A}$, $\mathbb{K}$ are fixed, ℓ is constant, $E^\ell \in \mathcal{F}$, $\mathbb{A}E^\ell + \ell \operatorname{inc} E^\ell = \mathbb{K}$ in Ω. There exists a unique $E^\infty \in \mathcal{V}$ such that*

$$\int_\Omega \mathbb{A}E^\infty \cdot \hat{E} dx = \int_\Omega \mathbb{K} \cdot \hat{E} dx \qquad \forall \hat{E} \in \mathcal{V}. \tag{9.12}$$

Moreover $\|E^\ell - E^\infty\|_{L^2} \to 0$ when $|\ell| \to +\infty$.

Theorem 9.7 shows that the standard linear elasticity problem with Neumann boundary condition is retrieved as a limit case when $|\ell| \to +\infty$. The modeling of Dirichlet type boundary conditions require further work.

From our derivations so far, the question of the sign of ℓ has not been fixed. The example of the bar in traction shown below suggests that ℓ be negative to obtain realistic solutions.

9.6 Example: bar in traction

We present a variant of an example treated in [2]. Consider the domain $\Omega = \mathbb{R}^2 \times (-h, h)$, for a given $h > 0$. Although the existence theory has been carried out for bounded domains, the semi-infinite case will allow analytical calculations through ordinary differential equations. We assume a uniform vertical traction on the planes $\{z = \pm h\}$. In view of (9.7)-(9.9) we obtain

$$\mathbb{K} = \begin{pmatrix} 0 & 0 & 0 \\ 0 & 0 & 0 \\ 0 & 0 & 1 \end{pmatrix}.$$

We search for a strain field of form

$$E = \begin{pmatrix} \varphi & 0 & 0 \\ 0 & \varphi & 0 \\ 0 & 0 & \psi \end{pmatrix},$$

where φ, ψ are functions of the z variable. One has

$$\mathbb{A}E = \begin{pmatrix} 2(\lambda+\mu)\varphi + \lambda\psi & 0 & 0 \\ 0 & 2(\lambda+\mu)\varphi + \lambda\psi & 0 \\ 0 & 0 & 2\lambda\varphi + (\lambda+2\mu)\psi \end{pmatrix}, \quad \operatorname{inc} E = \begin{pmatrix} \varphi'' & 0 & 0 \\ 0 & \varphi'' & 0 \\ 0 & 0 & 0 \end{pmatrix},$$

whereby $\mathbb{A}E + \ell \operatorname{inc} E = \mathbb{K}$ if and only if

$$\begin{cases} 2(\lambda+\mu)\varphi + \lambda\psi + \ell\varphi'' = 0 \\ 2\lambda\varphi + (\lambda+2\mu)\psi = 1. \end{cases} \tag{9.13}$$

Substitution leads to

$$\psi = \frac{1}{\lambda + 2\mu}(1 - 2\lambda\varphi), \tag{9.14}$$

$$2\mu(3\lambda + 2\mu)\varphi + \ell(\lambda + 2\mu)\varphi'' = -\lambda.$$

Due to the unboundedness of Ω the above equation has no unique solution. Therefore we prescribe $\varphi(\pm h) = 0$. This entails

$$\varphi(z) = \begin{cases} \dfrac{-\lambda}{2\mu(3\lambda + 2\mu)}\left(1 - \dfrac{\cos\frac{\omega z}{\sqrt{\ell}}}{\cos\frac{\omega h}{\sqrt{\ell}}}\right) & \text{if } \ell > 0, \\ \dfrac{-\lambda}{2\mu(3\lambda + 2\mu)}\left(1 - \dfrac{\cosh\frac{\omega z}{\sqrt{|\ell|}}}{\cosh\frac{\omega h}{\sqrt{|\ell|}}}\right) & \text{if } \ell < 0, \end{cases} \qquad \text{with } \omega = \sqrt{\frac{2\mu(3\lambda + 2\mu)}{\lambda + 2\mu}}.$$

The external work is obtained as

$$W = \int_{-h}^{h} \psi(z)dz,$$

with ψ given by (9.14), i.e.,

$$W = \begin{cases} \dfrac{2h}{\lambda + 2\mu}\left[1 + \dfrac{\lambda^2}{\mu(3\lambda + 2\mu)}\left(1 - \dfrac{\sqrt{\ell}}{\omega h}\tan\dfrac{\omega h}{\sqrt{\ell}}\right)\right] & \text{if } \ell > 0, \\ \dfrac{2h}{\lambda + 2\mu}\left[1 + \dfrac{\lambda^2}{\mu(3\lambda + 2\mu)}\left(1 - \dfrac{\sqrt{|\ell|}}{\omega h}\tanh\dfrac{\omega h}{\sqrt{|\ell|}}\right)\right] & \text{if } \ell < 0. \end{cases}$$

In the numerical outputs that follow we have used the data $Y = 10$, $\nu = 1/3$ for the Young modulus and the Poisson ratio of the material, respectively, and $h = 1$. Figure 2 displays the work W in function of ℓ. Clearly, choosing $\ell > 0$ is unphysical, since incompatible deformations produce less work that in the purely elastic case ($\ell \to \infty$). In contrast, choosing $\ell < 0$ is consistent with our expectation from the energetic point of view, the difference of work between the inelastic (irreversible) and elastic (reversible) transformations being the dissipation, see [2,4] for details.

Figure 3 shows the functions φ and ψ for some negative values of ℓ. For $\ell \to -\infty$, the classical elastic solution is retrieved.

Figure 4 shows the same functions calculated with ℓ variable in space, namely $\ell = -1$ in the interval $[-0.1, 0.1]$ and $\ell = -1000$ elsewhere. Although the domain is unbounded, the horizontal compression suggests that the model is able to predict necking phenomena. Let us stress once more that we have restricted ourselves to linearized equations. How to deal with finite deformations is briefly discussed thereafter.

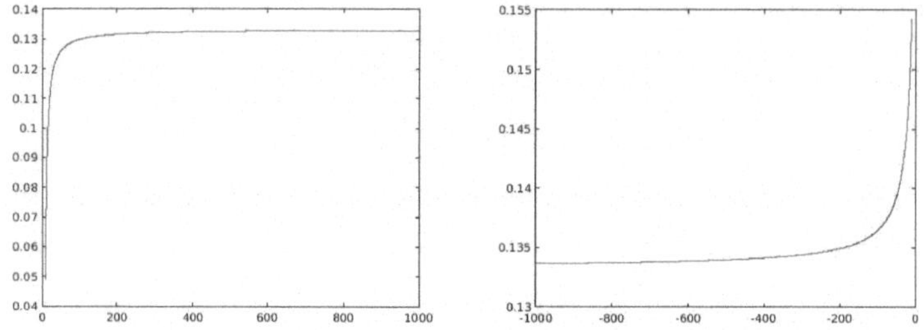

Fig. 2: External work as a function of ℓ for $\ell > 0$ (left) and $\ell < 0$ (right)

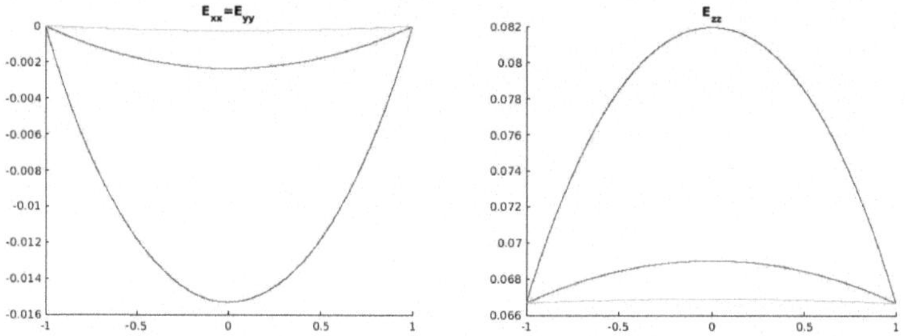

Fig. 3: Planar (left) and vertical (right) deformations for $\ell = -10$ (blue), $\ell = -100$ (red), $\ell = -1000$ (yellow).

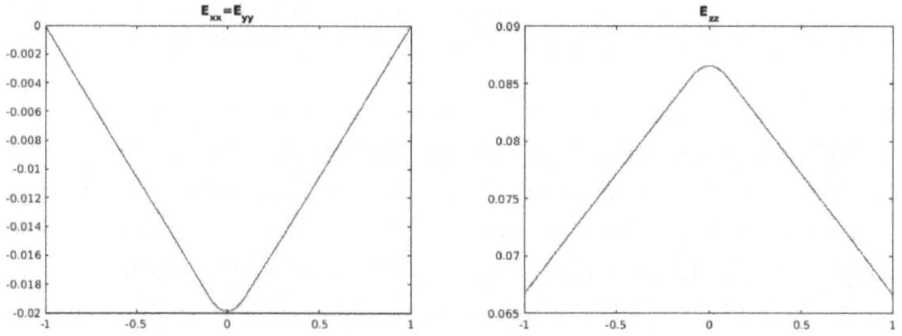

Fig. 4: Planar (left) and vertical (right) deformations for ℓ variable in space.

9.7 Incremental formulation of hardening problems

An incremental formulation consists in introducing a continuous family of loads, here denoted by $(\mathbb{K}_t)$, parameterized by a fictitious time t, starting from $\mathbb{K}_0 = 0$ and reaching the target value $\mathbb{K}_T = \mathbb{K}$ at final time T. The interval $[0, T]$ is split into subintervals $[t_k, t_{k+1}]$, and within each subinterval one solves the tangent problem (9.10). Here $\mathbb{A}$ and ℓ are tangent moduli, and E is the strain increment. Nonlinear phenomena occur when these moduli vary between two increments. In the first increment, for small load, the behavior is usually elastic: $|\ell|$ is taken very large, one can even solve the standard elasticity equations. At some point, according to some yield stress or energetic criterion, nonlinearity and irreversibility appear: $|\ell|$ should be locally decreased. As stipulated by the second principle of Thermodynamics, this modification of the incompatibility modulus must be associated with a dissipation of free energy. The procedure is repeated in the following increments. Of course, in case of elastic unloading, ℓ should be again taken everywhere "infinite".

The process of update of ℓ (and also $\mathbb{A}$) is obviously not completely determined. The point is to represent the complex hardening phenomenon. This will be investigated in future works. At least to comply with the Second Principle a sensitivity analysis of the free energy with respect to local perturbations of ℓ may be carried out. It has been done in [4] for the model (9.10) reduced to its principle part, i.e. (8.8). The full model is under scrutiny.

Acknowledgements

The second author was supported by national funding from FCT - Fundaçâo para a Ciência e a Tecnologia, under the project: UID/MAT/04561/2019 as well as by the FCT Starting Grant " Mathematical theory of dislocations: geometry, analysis, and modelling" (IF/00734/2013).

References

1. Al-Rub, R.A.: Continuum-based modeling of size effects in micro-and nanostructured materials. In: Handbook of Micromechanics and Nanomechanics. Pan Standford Publ. (2013)
2. Amstutz, S., Goethem, N.V.: Existence results for an intrinsic model of incompatible elasticity. hal report: hal-02045046 (2019)
3. Amstutz, S., Van Goethem, N.: Analysis of the incompatibility operator and application in intrinsic elasticity with dislocations. SIAM J. Math. Anal. **48**(1), 320–348 (2016)
4. Amstutz, S., Van Goethem, N.: Incompatibility-governed elasto-plasticity for continua with dislocations. Proc. R. Soc. A **473**(2199) (2017)
5. Anzellotti, G., Giaquinta, M.: On the existence of the fields of stresses and displacements for an elasto-perfectly plastic body in static equilibrium. J. Math. Pures Appl. **61**, 219–244 (1982)
6. Anzellotti, G., Luckhaus, S.: Dynamical evolution of elasto-perfectly plastic bodies. Appl. Math. Optim. **15**(6), 121–140 (1987)
7. Reddy, B.D., Ebobisse, F., McBride, A.T.: Well-posedness of a model of strain gradient plasticity for plastically irrotational materials. Int. J. of Plast. **24**(5) (2008)
8. Beltrami, E.: Sull'interpretazione meccanica delle formule di Maxwell. Mem. dell'Accad. di Bologna **7**, 1–38 (1886)
9. Bilby, B.A., Bullough, R., Smith, E.: Continuous distribution of dislocations: a new application of the methods of non-riemannian geometry. Proc. R. Soc. A **231**(1), 263–273 (1955)
10. Bonnans, J.F., Shapiro, A.: Perturbation analysis of optimization problems. Springer Series in Operations Research. Springer-Verlag, New York (2000)

11. Cartan, E.: Sur une generalisation de la notion de courbure de Riemann et les espaces a torsion. C. R. Acad. Sci. Paris **174**, 593–597 (1922)
12. Cesaro, E.: Sulle formole del Volterra, fondamentali nella teoria delle distorsioni elastiche. Rend. Accad. R. Napoli **12**, 311–321 (1906)
13. Ciarlet, P.G.: An introduction to differential geometry with applications to elasticity. J. Elasticity **78-79**(1-3), 3–201 (2005)
14. Ciarlet, P.G., Mardare, C.: Intrinsic formulation of the displacement-traction problem in linearized elasticity. Math. Models Methods Appl. Sci. **24**(6), 1197–1216 (2014)
15. Cosserat E. Cosserat, F.: Théorie des corps déformables. A. Hermann et Fils, Paris (1909)
16. Dal Maso, G., De Simone, A., Mora, M.G.: Quasistatic evolution problems for linearly elastic-perfectly plastic materials. Arch. Ration. Mech. Anal. **180**(2), 237–291 (2006)
17. Davoli, E., Francfort, G.A.: A critical revisiting of finite elasto-plasticity. SIAM J. Math. Anal. **47**(1), 526–565 (2015)
18. Dubrovin, B.A., Fomenko, A.T., Novikov, S.P.: Modern geometry - methods and applications, Part 1 (2nd edn). Cambridge studies in advanced mathematics. Springer-Verlag, New-York (1992)
19. Duvaut, G.: Mécanique des milieux continus. Collection Mathématiques appliquées pour la maîtrise. Masson (1990)
20. Ebobisse, F., Neff, P.: Existence and uniqueness for rate-independent infinitesimal gradient plasticity with isotropic hardening and plastic spin. Math. Mech. Solids **15**(6), 691–703 (2010)
21. Epstein, M.: The Geometrical Language of Continuum Mechanics. Cambridge University Press (2010)
22. Euler, L.: Recherces sur la courbure des surfaces. Mém. Acad. Sci. Berlin **16**, 119–143 (1767)
23. F. Ebobisse, P.N.: Existence and uniqueness for rate-independent infinitesimal gradient plasticity with isotropic hardening and plastic spin. Mathematics and Mechanic of Solids **14**(8) (2009)
24. Flaherty, F., do Carmo, M.: Riemannian Geometry. Mathematics: Theory & Applications. Birkhäuser Boston (2013)
25. Fleck, N.A., Hutchinson, J.W.: A reformulation of strain gradient plasticity. J. Mech. Phys. Solids **49**(10), 2245–2271 (2001)
26. Green, A.E., Zerna, W.: Theory of elasticity in general coordinates (finite strain). Phil. Mag. **41**, 313–336 (1950)
27. Greenberg, H.J.: Complementary minimum principles for an elastic -plastic material. Q. Appl. Math. **7**, 85–95 (1949)
28. Gurtin, M.E.: Configurational Forces as Basic Concepts of Continuum Physics. Applied Mathematical Sciences, 137. Springer (2000)
29. Gurtin, M.E., Anand, L.: A theory of strain-gradient plasticity for isotropic, plastically irrotational materials. i: small deformations. J. Mech. Phys. Solids **53**(7), 1624–1649 (2005)
30. Gurtin, M.E., Needleman, A.: Boundary conditions in small-deformation, single-crystal plasticity that account for the Burgers vector. J. Mech. Phys. Solids **53**(1), 1–31 (2005)
31. Han, W., Reddy, B.D.: Plasticity. Mathematical theory and numerical analysis. 2nd ed. Springer, New-York (2013)
32. Hardt, R., Kinderlehrer, D.: Elastic plastic deformation. Appl. Math. Optim. **10**, 213–246 (1983)
33. Hill, R.: Constitutive dual potentials in classical plasticity. Journal of the Mechanics and Physics of Solids **35**(1), 23 – 33 (1987)
34. Hull, D., Bacon, D.: Introduction to Dislocations. Materials science and technology. Elsevier Science (2011)
35. Johnson, C.: Existence theorems for plasticity problems. J. Math. Pures Appl. **55**, 431–444 (1976)
36. Kirchhoff, G.: Vorlesungen über Mechanik. Lecture notes in physics, 47. Birkhäuser, Basel, Leipzig (1876)
37. Kleinert, H.: Gauge fields in condensed matter, Vol.1. World Scientific Publishing, Singapore (1989)

38. Kondo, K.: Non-riemannian geometry of the imperfect crystal from a macroscopic viewpoint. in RAAG Memoirs of the unifying study of basic problems in engineering sciences by means of geometry, Vol.1, Division D, Gakuyusty Bunken Fukin-Day pp. 458–469 (1955)
39. Kozono, H., Yanagisawa, T.: L^r-variational inequality for vector fields and the Helmholtz-Weyl decomposition in bounded domains. Indiana Univ. Math. J. **58**(4), 1853–1920 (2009)
40. Kröner, E.: Continuum theory of defects. In: R. Balian (ed.) Physiques des défauts, Les Houches session XXXV (Course 3). North-Holland, Amsterdam (1980)
41. Kröner, J.E.: The internal mechanical state of solids with defects. Int. J. Solids and Structures **29**(14/15), 1849–1857 (1992)
42. Kuiper, N.H.: On C^1-isometric imbeddings. I, II. Nederl. Akad. Wet., Proc., Ser. A **58**, 545–556, 683–689 (1955)
43. Lazar, M., Maugin, G.A.: Nonsingular stress and strain fields of dislocations and disclinations in first strain gradient elasticity. Int. J. Eng. Sci. **43**, 1157–1184 (2005)
44. Love, A.E.H.: A treatise on the mathematical theory of elasticity. Cambridge University Press (1892)
45. Maggiani, G., Scala, R., Van Goethem, N.: A compatible-incompatible decomposition of symmetric tensors in L^p with application to elasticity. Math. Meth. Appl. Sci **38**(18), 5217–5230 (2015).
46. Maugin, G.: The method of virtual power in continuum mechanics: Application to coupled fields. Acta Mech. **35**, 1–70 (1980)
47. Michell, J.H.: On the direct determination of stress in an elastic solid, with application to the theory of plates. Proc. Lond. Math. Soc. **s1–31**, 100–124 (1906)
48. Mielke, A., Roubíček, T.: Rate-independent elastoplasticity at finite strains and its numerical approximation. Math. Models Methods Appl. Sci. **12**, 2203–2236 (2016)
49. Mielke, A., Stefanelli, U.: Linearized plasticity is the evolutionary γ-limit of finite plasticity. J. Eur. Math. Soc. (JEMS) **15**(3), 923–948 (2013)
50. Mindlin, R.D.: Micro-structure in linear elasticity. Arch. Ration. Mech. Anal. **16**, 51–78 (1964)
51. Moreau, J.J.: Application of convex analysis to the treatment of elastoplastic systems. In: Applications of methods of functional analysis to problems in mechanicss. Springer, Berlin (1976)
52. Müller, G., Friedrich, J.: Challenges in modeling of bulk crystal growth. J. Cryst. Growth **266**(1-3), 1–19 (2004)
53. Nash, J.: C^1 isometric imbeddings. Ann. Math. (2) **60**, 383–396 (1954)
54. Noll, W.: Materially uniform bodies with inhomogeneities. Arch. Rational Mech. Anal. **27**, 1–32 (1967)
55. Nunes, P.: Petri nonni salaciensis opera, basileae. Biblioteca Nacional de Portugal, Reediçãao e tradução, com comentários, pela Academia de Ciências de Lisboa – Fundação Calouste Gulbenkian, Lisboa, 2008 e 2011 (1566)
56. Rice, J.R.: On the structure of stress-strain relations for time-dependent plastic deformation in metals. Trans ASME. J. Appl. Mech p. 728 (1970)
57. Riemann, B.: Über die hypothesen, welche der geometrie zu grunde liegen. in Riemann's Gesamm. Math. Werke XIII Abhandl. Kgl. Ges. Wiss. Göttingen XIII, 272–287 (1868)
58. Barrè de Saint-Venant A. J. C., M.N.: Première section : De la résistance des solides par navier. - 3e é d. avec des notes et des appendices par m. barréde saint-venant. tome 1. In: Résumé des leçons donné es á l'Ecole des Ponts et Chaussées sur l'application de la mécanique á l'établissement des constructions et des machines. Dunod, Paris (1864)
59. Scala, R., Van Goethem, N.: Currents and dislocations at the continuum scale. Methods Appl. Anal. **23**(1), 1–34 (2016)
60. Scala, R., Van Goethem, N.: Geometric and analytic properties of dislocation singularities. Proc. Roy. Soc. Edinb Sect. A **149**(4) (2019)
61. Scala, R., Van Goethem, N.: A variational approach to single crystals with dislocations. SIAM J. Math. Anal. **51**(1), 489–531 (2019)
62. Schouten, J.A.: Ricci-Calculus (2nd edn). Springer Verlag, Berlin (1954)

63. Seugling, W.R.: Equations of compatibility for finite deformation of a continuous medium. The American Mathematical Monthly **57**(10), 679–681 (1950)
64. Simo, J., Hughes, T.: Computational inelasticity. Springer, Berlin (1998)
65. Sun, B.: Incompatible deformation field and riemann curvature tensor. Applied Mathematics and Mechanics **38**(3), 311–332 (2017)
66. Suquet, P.M.: Sur les équations de la plasticité: existence et régularité des solutions. J. Méc., Paris **20**, 3–39 (1981)
67. Tadmor, E., Miller, R., Elliott, R.: Continuum Mechanics and Thermodynamics: From Fundamental Concepts to Governing Equations. Cambridge University Press (2011)
68. Temam, R.: Problèmes mathématiques en plasticité. J. Phys. Chem. Solids **69**, 320–324 (2008)
69. Thom, R.: La vie et l'œuvre de Hassler Whitney. (Life and work of Hassler Whitney). C. R. Acad. Sci., Paris, Sér. Gén., Vie Sci. **7**(6), 473–476 (1990)
70. Van Goethem, N.: The non-Riemannian dislocated crystal: a tribute to Ekkehart Kröner's (1919-2000). J. Geom. Mech. **2**(3) (2010)
71. Van Goethem, N.: Strain incompatibility in single crystals: Kröner's formula revisited. J. Elast. **103**(1), 95–111 (2011)
72. Van Goethem, N.: The Frank tensor as a boundary condition in intrinsic linearized elasticity. J. Geom. Mech. **8**(4), 391–411 (2016)
73. Van Goethem, N.: Incompatibility-governed singularities in linear elasticity with dislocations. Math. Mech. Solids **22**(8), 1688–1695 (2017)
74. Van Goethem, N., Dupret, F.: A distributional approach to $2D$ Volterra dislocations at the continuum scale. Europ. Jnl. Appl. Math. **23**(3), 417–439 (2012)
75. Van Goethem, N., de Potter, A., den Bogaert, N.V., Dupret, F.: Dynamic prediction of point defects in Czochralski silicon growth. An attempt to reconcile experimental defect diffusion coefficients with the V/G criterion. J. Phys. Chem. Solids **69**, 320–324 (2008)
76. Volterra, V.: Sulle equazioni differenziali lineari: nota. Rend. Mat. Acc. Lincei **4**(3) (1887)
77. Volterra, V.: Sur l'équilibre des corps élastiques multiplement connexes. Ann. Sci. École Norm. Sup. **3**(24), 401–517 (1907)
78. Volterra, V.: Sulle equazioni integro-differenziali della teoria dell'elasticitá. Rend. Mat. Acc. Lincei **5**(18), 1295–301 (1909)
79. Wang, C.C.: On the geometric structure of simple bodies, a mathematical foundation for the theory of continuous distributions of dislocations. Arch. Rational Mech. Anal. **27**(1), 33–94 (1967)
80. Weyl, H.: Die Idee der Riemannschen Fläche. B. G. Teubner, Leipzig, 1 edition. 2 edn, B. G. Teubner, Leipzig, 1923; Reprint of 2 edn, Chelsea Co., New York, 1951; 3 edn, revised, B. G. Teubner, Leipzig, 1955. English translation of 3 edn, The Concept of a Riemann Surface, Addison-Wesley, 1964. Dover edition 2009. (1951)
81. Whitney, H.: Topological properties of differentiable manifolds. Bull. Am. Math. Soc. **43**, 785–805 (1937)
82. Yavari, A., Goriely, A.: Riemann–cartan geometry of nonlinear dislocation mechanics. Archive for Rational Mechanics and Analysis **205**(1), 59–118 (2012)
83. Yavari, A., Goriely, A.: Non-metricity and the nonlinear mechanics of distributed point defects. In: G.Q.G. Chen, M. Grinfeld, R.J. Knops (eds.) Differential Geometry and Continuum Mechanics, pp. 235–251. Springer International Publishing, Cham (2015).

Nonlocal Phase Field Models of Viscous Cahn–Hilliard Type

Pierluigi Colli, Gianni Gilardi, Jürgen Sprekels

Abstract

A nonlocal phase field model of viscous Cahn–Hilliard type is considered. This model constitutes a nonlocal version of a model for two-species phase segregation on an atomic lattice under the presence of diffusion that has been studied in a series of papers by P. Podio-Guidugli and the present authors. The resulting system of differential equations consists of a highly nonlinear parabolic equation coupled to a nonlocal ordinary differential equation, which has singular terms that render the analysis difficult. Some results are presented on the well-posedness and stability of the system as well as on the distributed optimal control problem.

1 Introduction

This note is concerned with a nonlocal variant of a model for phase segregation through atom rearrangement on a lattice proposed in [58] and fully discussed in [17] from the side of the thermodynamical derivation. The model turns out to offer an

Pierluigi Colli
Dipartimento di Matematica "F. Casorati", Università di Pavia, Via Ferrata 5, 27100 Pavia, Italy; e-mail: pierluigi.colli@unipv.it

Gianni Gilardi
Dipartimento di Matematica "F. Casorati", Università di Pavia, Via Ferrata 5, 27100 Pavia, Italy; e-mail: gianni.gilardi@unipv.it

Jürgen Sprekels
Department of Mathematics, Humboldt-Universität zu Berlin, Unter den Linden 6, 10099 Berlin, Germany and Weierstrass Institute for Applied Analysis and Stochastics, Mohrenstrasse 39, 10117 Berlin, Germany; e-mail: juergen.sprekels@wias-berlin.de

Key words: well-posedness, distributed optimal control, nonlinear phase field systems, nonlocal operators, first-order necessary optimality conditions.

AMS (MOS) Subject Classification: 35K55, 49K20, 74A15.

M. Hintermüller, J. F. Rodrigues (eds.), *Topics in Applied Analysis and Optimisation*, CIM Series in Mathematical Sciences, https://doi.org/10.1007/978-3-030-33116-0_3

alternative view with respect to the Fried–Gurtin approach to phase segregation processes (cf. [35], [48]). The nonlocal model has been proposed and discussed in [28–30], by showing results on well-posedness, regularity, and optimal control problems. The aim of this note is to recapitulate and review the contents of the contributions [28–30]. We start by presenting the model problem with origins and details.

1.1 About the model and related problems

The state variables are the *order parameter* ρ, interpreted as the (normalized) density of one of the two species, and the *chemical potential* μ. For physical reasons, μ is required to be nonnegative; on the other hand, the phase parameter ρ is expected to attain values in the interval $[0, 1]$.

Then, a local free energy density of the form

$$\psi = \widehat{\psi}(\rho, \nabla\rho, \mu) = -\mu\,\rho + F(\rho) + \frac{\sigma}{2}\,|\nabla\rho|^2, \tag{1.1}$$

is originally considered in the model. Note that the first term above combines $-\mu$ with the factor ρ, which is nonnegative when in the interval $[0, 1]$. Moreover, $\sigma > 0$ is a physical parameter, and F denotes a double-well potential. Typical examples for the double-well potential F are the regular potential

$$F_{reg}(r) := \frac{1}{4}r^2(r-1)^2, \quad r \in \mathbb{R}, \tag{1.2}$$

or the logarithmic potential

$$F_{log}(r) := r\ln r + (1-r)\ln(1-r) - cr(r-1), \quad r \in (0, 1), \tag{1.3}$$

where $c > 2$ in the latter case so that F_{log} is nonconvex. In view of (1.1), the derivation carried out in [17, 58] leads to the evolutionary system

$$2\rho\,\partial_t\mu + \mu\,\partial_t\rho - \Delta\mu = 0, \tag{1.4}$$

$$-\sigma\,\Delta\rho + F'(\rho) = \mu, \tag{1.5}$$

with the equations holding in $Q := \Omega\times(0, T)$. Here, Ω is a three-dimensional bounded and smooth domain, while $T > 0$ stands for a fixed final time. Equations (1.4)–(1.5) should be related to some proper boundary and initial conditions.

We note that the derivative F' of the potential intervenes in (1.5), where, in particular, the primitive F may behave as the classical regular potential F_{reg} or as the logarithmic potential F_{log}, respectively. Both potentials are rather smooth in their domains, but the derivative of the latter becomes singular at 0 and 1. However, also nondifferentiable potentials can be considered, and in this framework an important example is given by the so-called double-obstacle potential

$$F_{2obs}(r) := I_{[0,1]}(r) - c\,r\,(r-1), \quad r \in \mathbb{R}, \tag{1.6}$$

where $c > 0$ is a positive constant and $I : \mathbb{R} \longrightarrow [0, +\infty]$ denotes the indicator function of the interval $[0, 1]$, i.e., we have $I_{[0,1]}(r) = 0$ if $0 \le r \le 1$ and $I_{[0,1]}(r) = +\infty$ otherwise. In this case, the order parameter ρ is required to obey the constraint $0 \le \rho \le 1$ and (1.5) should be read as a differential inclusion with $F'(\rho)$ replaced by $\partial I_{[0,1]}(\rho) - c(2\rho + 1)$, that is, the subdifferential is used in place of the derivative for the convex and nonsmooth part of F.

The system (1.4)-(1.5) yields a variation of the Cahn–Hilliard system originally introduced in [8] and first studied mathematically in [34] (for a large list of references on the original Cahn–Hilliard system, see [50]). we note that (1.4)-(1.5) leads to ill-posed problems, in general. In fact, as it was noticed in [20], an associated initial-boundary value problem with zero Neumann boundary conditions for both ρ and μ may have infinitely many smooth and even nonsmooth solutions.

Hence, the next step was the insertion of two regularizing terms, depending on two paramenters $\varepsilon > 0$ and $\delta > 0$: this was done in [17] and the regularized model equations

$$(\varepsilon + 2\rho)\,\partial_t \mu + \mu\,\partial_t \rho - \Delta\mu = 0, \tag{1.7}$$

$$\delta\,\partial_t \rho - \sigma\,\Delta\rho + F'(\rho) = \mu \tag{1.8}$$

were considered. The positive coefficients ε and δ are intended to be small, of course. The presence of the ε–term is motivated by the desire to have a strictly positive coefficient as a factor of $\partial_t \mu$ in (1.4), in order to maintain the parabolic structure of equation (1.7). Actually, this new term involves an additional contribution in the free energy (1.1) in the form of a term just depending on μ. As to the δ–term, we point out that it transforms (1.5) into an Allen–Cahn equation with μ as source term; it is actually a regularization that has been already employed in various procedures involving the known *viscous Cahn–Hilliard system* [NC]. The system (1.7)–(1.8) was investigated in the series of papers [14, 17, 18, 21, 24] concerning well-posedness, regularity, optimal control and time discetization of the resulting Cauchy–Neumann problem.

By further reflections, the local free energy density (1.1) was generalized to the form

$$\psi = \widehat{\psi}(\rho, \nabla\rho, \mu) = -\mu\left(\frac{\varepsilon}{2} + g(\rho)\right) + F(\rho) + \frac{\sigma}{2}\,|\nabla\rho|^2, \tag{1.9}$$

with a function g having suitable properties (see the later assumptions (2.7)). Indeed, the behavior of the special choice $g_{spe}(\rho) = \rho$ in a right neighbourhood of 0 ($g_{spe}(\rho) \approx 0$) differs from that in a left neightbourhood of 1 ($g_{spe}(\rho) \approx 1$). Instead, by still assuming a nonnegativity property for the general function g, we may allow for many other situations like, e.g., a specular behavior around the extremal points of the domain of f.

Hence, if we take, without loss of generality, $\varepsilon = \delta = 1$ and insert a source term u in the equation corresponding to (1.7), in place of (1.7)–(1.8) the more general system

$$\big(1 + 2g(\rho)\big)\,\partial_t\mu + \mu\,g'(\rho)\,\partial_t\rho - \Delta\mu = u, \tag{1.10}$$
$$\partial_t\rho - \sigma\,\Delta\rho + F'(\rho) = \mu\,g'(\rho), \tag{1.11}$$

follows. The system (1.10)–(1.11) was investigated in the papers [15, 16, 19, 20, 22]. As you can expect, the right-hand side u in (1.10) is going to play the role of the control in the distributed optimal control problem.

The aim of the present note is the mathematical discussion of the system in which the local term $\frac{\sigma}{2}\,|\nabla\rho|^2$ in the local free energy density is substituted by a nonlocal expression. An interesting example results from considering a total free energy functional of the form

$$\mathcal{F}_{tot}[\rho] = \int_\Omega \Big[-\mu(x)\,g(\rho(x)) + F(\rho(x))\Big]\,dx \;+\; Q[\rho], \tag{1.12}$$

with the nonlocal contribution

$$Q[\rho] := \int_\Omega \rho(x) \int_\Omega k(|y-x|)(1-\rho(y))\,dy\,dx\,.$$

Then, following the derivation procedure detailed, e.g., in [20], we take the the variational derivative of the functional Q:

$$B[\rho](x) = \int_\Omega k(|y-x|)\,(1-2\,\rho(y))\,dy, \qquad x \in \Omega, \tag{1.13}$$

and obtain the following nonlocal variant of the system (1.10)–(1.11):

$$\big(1 + 2g(\rho)\big)\,\partial_t\mu + \mu\,g'(\rho)\,\partial_t\rho - \Delta\mu = 0, \tag{1.14}$$
$$\partial_t\rho + B[\rho] + F'(\rho) = \mu\,g'(\rho). \tag{1.15}$$

This is exactly the system which constitutes the main subject of study here. What is important is that we do not restrict ourselves to operators B of the exact form given in (1.13).

1.2 Nonlocal operators

We aim to deal with general operators B acting on functions defined in Q that enjoy suitable properties. Very simple examples that satisfy the conditions specified below (see (2.8)–(2.12) and (2.29)–(2.35)) are given by time convolution operators of the form

$$B[\rho](x,t) = \int_0^t k(t-s)\,\rho(x,s)\,ds\,, \tag{1.16}$$

and spatial convolutions of the form

$$B[\rho](x,t) = \int_\Omega k(x,y)\,\rho(y,t)\,dy\,, \tag{1.17}$$

provided that the respective integral kernels k are smooth enough. On the other hand, we are not able to include in our framework nonlocal-in-time nonlinearities of hysteresis type like the classical stop, play, Prandtl-Ishlinskii or Preisach operators (the reader may see [7] for the definition of these operators). The reason for this is that these operators carry a nonlocal memory with respect to time. For instance, the one-dimensional stop operator $\mathcal{S}$ (to take the simplest of these four operators) only enjoys (cf. [7]) the *nonlocal* Lipschitz property

$$|\mathcal{S}[\rho_1](t) - \mathcal{S}[\rho_2](t)| \le 2 \max_{0\le s\le t} |\rho_1(s) - \rho_2(s)|$$

for every $t \in [0,T]$ and every $\rho_1, \rho_2 \in C^0([0,T])$, and it is easily seen that the validity of the Lipschitz property, stated in the set of assumptions (2.8)–(2.12) given below, cannot be ensured.

As a further example for which the conditions (2.8)–(2.12) and (2.29)–(2.35) can be verified, we consider the integral operator

$$B[\rho](x) = \int_\Omega k(|y-x|)\,\rho(y)\,dy\,, \tag{1.18}$$

which acts on functions defined in Ω, and its extension to functions defined in Q, which turns out to be a special case of (1.17). Let $k \in C^0(0,+\infty)$ satisfy the condition

$$|k(r)| \le C_1\, r^{-\alpha} \quad \forall\, r > 0, \quad \text{for some } C_1 > 0 \text{ and } \alpha < 3. \tag{1.19}$$

Such kernels belong to the class of *weakly singular* kernels. Obviously, (2.9) holds, and since Ω is a bounded domain, it is well known that, for any $p \in (1,+\infty)$ such that $\alpha < \frac{3}{q}$, where $\frac{1}{p} + \frac{1}{q} = 1$, the linear operator B in (1.18) maps $L^p(\Omega)$ continuously (even compactly) into $C^0(\overline{\Omega})$ and thus into any $L^r(\Omega)$, $r \in [1,+\infty]$. Using Hölder's inequality, it is not difficult to show that for $\alpha < \frac{3}{2}$ the corresponding operator B satisfies all of the conditions (2.8), (2.10) and (2.11).

In order to fulfill also (2.12), we need additional assumptions, for instance, that k is continuously differentiable on $(0,+\infty)$ with

$$|k'(r)| \le C_2\, r^{-\beta} \quad \forall\, r > 0 \quad \text{with some } C_2 > 0 \text{ and } \beta < \tfrac{5}{2}. \tag{1.20}$$

Indeed, if (1.20) holds, for any $v \in L^2(0,T;V)$ we have that

$$\begin{aligned}\left|\int_{Q_t} \nabla v \cdot \nabla B[v]\right| &\le \int_{Q_t} |\nabla v|^2 + c \int_{Q_t} \left|\int_\Omega |y-x|^{-\beta}\, |v(y,s)|\, dy\right|^2 dx\, ds \\ &\le c \int_{Q_t} |\nabla v|^2 + c \int_{Q_t} \left[\int_\Omega \frac{dy}{|y-x|^{6\beta/5}}\right]^{5/3} \|v(s)\|_6^2\, dx\, ds \\ &\le c \int_{Q_t} (|v|^2 + |\nabla v|^2),\end{aligned}$$

due to the continuity of the embedding $V \subset L^6(\Omega)$ and the fact that $\frac{6\beta}{5} < 3$. Finally, we stress that in the important case of the (long-range) three-dimensional Newtonian potential $k(r) = \frac{C}{r}$, for which we have $\alpha = 1$ and $\beta = 2$, the kernel k satisfies both the conditions (1.19) and (1.20).

1.3 Overview of some related contribution

Free energies of the form (1.12) were proposed in [46, 47] and rigorously justified as macroscopic limits of microscopic phase segregation models with particle conserving dynamics (see also [9]). Indeed, in [46,47] Giacomin and Lebowitz, starting from a microscopic model, derived a macroscopic equation for phase segregation phenomena that turned out to be a nonlocal version of the well-known Cahn–Hilliard equation. From the mathematical viewpoint, this nonlocal Cahn–Hilliard equation is simpler than our system (1.14)–(1.15) and has received a good deal of attention over the past twenty years (see, e.g., [5, 6, 33, 40, 42, 49, 56]). Most of the theoretical results were devoted to well-posedness and some were concerned with the long-time behavior of solutions. Well-posedness and regularity issues were investigated for an equation with degenerate mobility and logarithmic potential in [42] (cf. also [33,40,41]). This required to show preliminarily that a solution stays eventually strictly away from the pure phases: the so-called separation property. For the case of a constant mobility and of regular potentials, some existence, uniqueness and regularity results were obtained in [5, 6, 49]. Nonsmooth potentials were considered in [33, 44]. The existence of a (connected) global attractor has been shown in [36] for a constant mobility and singular potentials, by exploiting the energy identity obtained in [13] as a by-product of results related to a phase separation model in binary fluids. Doubly nonlocal Cahn–Hilliard equations were approached in [43], where a class of nonlocal Cahn–Hilliard equations was considered and well-posedness, regularity and results on the long-time behavior were investigated in connection with the interaction between the two levels of nonlocality in the operators. The question, whether the global attractor has finite (fractal) dimension, was analyzed in [45], where the authors showed the existence of an exponential attractor. In [1], the authors addressed an equation that is connected to the gradient flow of a nonlocal total free energy functional.

1.4 About well-posedness and regularity results

At this point, we recall the system (1.14)–(1.15) and complement it with the Neumann homogeneous boundary condition for μ,

$$\partial_{\mathbf{n}}\mu = 0 \quad \text{on } \Sigma := \Gamma \times (0,T), \tag{1.21}$$

and with initial conditions for both ρ and μ,

$$\rho(\cdot,0) = \rho_0, \quad \mu(\cdot,0) = \mu_0, \quad \text{in } \Omega. \tag{1.22}$$

Here, Γ stands for the smooth boundary of the domain Ω, with the outward unit normal $\mathbf{n}$, and the outward normal derivative denoted by $\partial_{\mathbf{n}}$.

The state system (1.14)–(1.15), (1.21)–(1.22) is singular, with highly nonlinear and nonstandard coupling. In particular, unpleasant nonlinear terms involving time derivatives occur in (1.14). The nonlocal term $B[\rho]$ is a source for possible analytical difficulties, since the absence of the Laplacian in (1.15) may induce a low regularity for the order parameter ρ.

Despite the structure of the system, we can prove that the initial-boundary value problem (1.14)–(1.15), (1.21)–(1.22) is well posed under very general assumptions on the nonlinearity F, which allows for potentials like F_{log} and F_{2obs} (cf. (1.3) and (1.6)), and within a precise framework for the nonlocal operator B and other data. We follow a special technique for proving existence of solutions: indeed, the existence proof is based on an application of Tikhonov's fixed point theorem in a rather unusual separable and reflexive Banach space, namely the space $L^2(0,T;H^1(\Omega)) \cap L^{10/3}(Q)$.

1.5 The optimal control problem for a logarithmic potential

Next, we turn our interest to a distributed optimal control problem of the form

(CP) Minimize the cost functional

$$\begin{aligned} J(u,(\rho,\mu)) = &\frac{\beta_1}{2}\int_0^T\!\!\int_\Omega |\rho-\rho_Q|^2\,dx\,dt + \frac{\beta_2}{2}\int_0^T\!\!\int_\Omega |\mu-\mu_Q|^2\,dx\,dt \\ &+ \frac{\beta_3}{2}\int_0^T\!\!\int_\Omega |u|^2\,dx\,dt \end{aligned} \tag{1.23}$$

subject to the state system

$$(1+2g(\rho))\,\partial_t\mu + \mu\, g'(\rho)\,\partial_t\rho - \Delta\mu = u \quad \text{a.e. in } Q, \tag{1.24}$$

$$\partial_t\rho + B[\rho] + F'(\rho) = \mu\, g'(\rho) \quad \text{a.e. in } Q, \tag{1.25}$$

$$\partial_{\mathbf{n}}\mu = 0 \quad \text{a.e. on } \Sigma, \tag{1.26}$$

$$\rho(\cdot,0) = \rho_0, \quad \mu(\cdot,0) = \mu_0, \quad \text{a.e. in } \Omega, \tag{1.27}$$

and to the control constraints

$$u \in \mathcal{U}_{\mathrm{ad}} := \big\{ v \in H^1(0,T;L^2(\Omega)) \,:\, 0 \le v \le u_{\max} \ \text{ a.e. in } Q \\ \text{and } \|v\|_{H^1(0,T;L^2(\Omega))} \le R \big\}. \tag{1.28}$$

Here, we prescribe that the constants β_1, β_2, β_3 are nonnegative, R is positive and $\beta_1 + \beta_2 + \beta_3 > 0$. The threshold function $u_{\max} \in L^\infty(Q)$ is nonnegative. Moreover, ρ_Q, $\mu_Q \in L^2(Q)$ represent prescribed target functions of the tracking-type functional J. Although more general cost functionals could be admitted, we restrict ourselves to the above situation for the sake of simplicity.

The state system (1.24)–(1.27) comes exactly from (1.14)–(1.15), (1.21)–(1.22), and thus provides a nonlocal version of the original local model. Let us point out that the analogue of the control problem **(CP)** for the local case was studied in [18] for the special situation $g_{spe}(\rho) = \rho$, while the optimal boundary control problem was treated in [24].

The control function u on the right-hand side of (1.24) plays the role of a *microenergy source*. We remark at this place that the sign condition for the control u included in the definition (1.28) of $\mathcal{U}_{\mathrm{ad}}$ is crucial for the development of the theory: indeed, the property $u \ge 0$ a.e. in Q is needed to guarantee the nonnegativity of the chemical potential μ.

In our first approach to the optimal control problem, we assume that the nonlinearity F is a double-well potential defined at least in the open interval $(0,1)$, but with the derivative F' being singular at the endpoints $\rho = 0$ and $\rho = 1$: that is, we let $F = F_1 + F_2$, where F_2 is smooth and F_1 may behave for instance like the convex part of F_{log} in (1.3):

$$F_1(r) \approx r \log(r) + (1-r)\log(1-r), \qquad r \in (0,1). \tag{1.29}$$

The presence of the nonlocal term $B[\rho]$ in (1.25) constitutes the main difference to the local model. After stating the general assumptions and deriving new regularity and stability results for the state system, the directional differentiability of the control-to-state operator will be shown, and one can carry out the derivation of first-order necessary conditions of optimality.

1.6 Commenting on the optimal control problem

Optimal control problems of the above type often occur in industrial production processes. For instance, consider a metallic workpiece consisting of two different materials that tend to separate. Then a typical goal would be to monitor the production process in order that a desired distribution of the two materials (represented by the function ρ_Q) be realized during the time evolution; the deviation from the desired phase distribution is measured by the first summand in the cost J. The third summand of J represents the costs due to the control action u; the size of the factors $\beta_i \ge 0$

then reflects the relative importance that the two conflicting interests "realize the desired phase distribution as closely as possible" and "minimize the cost of the control action" have for the manufacturer.

The mathematical literature on control problems for phase field systems involving equations of viscous or nonviscous Cahn–Hilliard type is quite recent and still growing. We refer in this connection to the contributions [10, 11, 26, 27, 52, 61]. Control problems for convective Cahn–Hilliard systems were studied in [62,63], and a few analytical contributions were made to the coupled Cahn–Hilliard/Navier–Stokes system (cf. [37, 38, 51, 53, 54]). We mention the paper [59], in which a distributed optimal control problem is studied for a nonlocal convective Cahn–Hilliard equation with degenerate mobility and singular potential in three dimensions of space. The recent contribution [23] deals with the optimal control of a Cahn–Hilliard type system arising in the modeling of solid tumor growth. Control problems for Cahn–Hilliard type systems with dynamic boundary conditions also of Cahn–Hilliard type have been very recently investigated in [31, 32, 39].

1.7 The optimal control problem in the double-obstacle case

Now, we discuss the control problem in the case where the nonlinearity F takes the form (cf. (1.6)) $I_{[0,1]} + F_2$, where F_2 is smooth and $I_{[0,1]}$ denotes the indicator function of the interval $[0, 1]$. Hence, let us rewrite the control problem as follows.

(CP$_0$) Minimize the cost functional (1.23) subject to the control constraints (1.28) and to the state system (1.24)–(1.27) in which (1.25) is replaced by

$$\partial_t \rho + B[\rho] + \xi + F_2'(\rho) = \mu\, g'(\rho) \quad \text{a.e. in } Q, \tag{1.30}$$

$$\xi \in \partial I_{[0,1]}(\rho) \quad \text{a.e. in } Q. \tag{1.31}$$

The nonlinearity F_2 is assumed to be smooth (and concave), while $I_{[0,1]}$ is the indicator function of the interval $[0, 1]$, so that the subdifferential $\partial I_{[0,1]}$ appearing in (1.31) works as follows:

$$\xi \in \partial I_{[0,1]}(\rho) \quad \Longleftrightarrow \quad 0 \le \rho \le 1 \ \text{ and } \begin{cases} \xi \le 0 & \text{if } \rho = 0 \\ \xi = 0 & \text{if } 0 < \rho < 1 \\ \xi \ge 0 & \text{if } \rho = 1 \end{cases},$$

which is equivalent to the variational inequality

$$\rho \in [0, 1], \quad \xi(\rho - r) \ge 0 \quad \text{for all } r \in [0, 1].$$

The idea is to employ the results established for a logarithmic potential to deal with the nondifferentiable double-obstacle case when ξ satisfies the inclusions (1.31). In this case, it is well known that all of the classical constraint qualifications fail, so that the existence of suitable Lagrange multipliers cannot be ensured using standard methods of optimal control. Instead, our approach is guided by a strategy employed in [10] for viscous Cahn–Hilliard systems (see also [12] for for the simpler case of the Allen–Cahn equation) with dynamic boundary conditions: in [10], necessary optimality conditions for the double obstacle case are recovered by performing a so-called 'deep quench limit' in a family of optimal control problems with differentiable nonlinearities of a form like (1.3).

1.8 The deep quench limit procedure

Actually, we replace the inclusion (1.31) by

$$\xi = \varphi(\alpha)\, h'(\rho), \tag{1.32}$$

where h is defined as the logarithmic potential in (1.29), so that

$$h(\rho) = \rho \ln(\rho) + (1-\rho)\ln(1-\rho) \quad \text{if} \quad 0 < \rho < 1, \quad h(0) = h(1) = 0, \tag{1.33}$$

and $\varphi \in C^0((0,1])$ is a positive function satisfying

$$\lim_{\alpha \searrow 0} \varphi(\alpha) = 0. \tag{1.34}$$

We notice that the simple choice $\varphi(\alpha) = \alpha^p$, for some $p > 0$, can be made; however, there might be situations (e. g.., in the numerical approximation) in which it is advantageous to let φ have a different behavior as $\alpha \searrow 0$. We observe that

$$h'(\rho) = \ln\left(\frac{\rho}{1-\rho}\right) \quad \text{and} \quad h''(\rho) = \frac{1}{\rho(1-\rho)} > 0 \quad \text{for } \rho \in (0,1).$$

Hence, in particular, we have that

$$\lim_{\alpha \searrow 0} \varphi(\alpha)\, h'(\rho) = 0 \quad \text{for } 0 < \rho < 1,$$

$$\lim_{\alpha \searrow 0}\Big(\varphi(\alpha) \lim_{\rho \searrow 0} h'(\rho)\Big) = -\infty, \quad \lim_{\alpha \searrow 0}\Big(\varphi(\alpha) \lim_{\rho \nearrow 1} h'(\rho)\Big) = +\infty\,.$$

We thus may regard the graphs of the functions $\varphi(\alpha)\, h'$ as approximations to the graph of the subdifferential $\partial I_{[0,1]}$.

The next step is to consider for any $\alpha \in (0,1]$ the optimal control problem (later to be denoted by $(\mathbf{CP}_\alpha)$), which results if in $(\mathbf{CP}_0)$ the relation (1.31) is replaced by (1.32). For this type of problem, the existence of optimal controls $u^\alpha \in \mathcal{U}_{\text{ad}}$, as well as first-order necessary optimality conditions, are known. Proving a priori estimates (uniform in $\alpha \in (0,1]$), and employing compactness and monotonicity arguments, we will be able to show the following existence and approximation result: whenever $\{u^{\alpha_n}\} \subset \mathcal{U}_{\text{ad}}$ is a sequence of optimal controls for $(\mathbf{CP}_{\alpha_n})$, where $\alpha_n \searrow 0$ as

$n \to \infty$, then there are a subsequence of $\{\alpha_n\}$, still indexed by n, and an optimal control $\overline{u} \in \mathcal{U}_{\rm ad}$ of $(\mathbf{CP}_0)$ such that

$$u^{\alpha_n} \to \overline{u} \quad \text{weakly-star in } \mathcal{X} \text{ as } n \to \infty, \tag{1.35}$$

where

$$\mathcal{X} := H^1(0,T;L^2(\Omega)) \cap L^\infty(Q) \tag{1.36}$$

denotes the control space. In other words, optimal controls for $(\mathbf{CP}_\alpha)$ are for small $\alpha > 0$ likely to be 'close' to optimal controls for $(\mathbf{CP}_0)$.

It is natural to ask whether also the reverse holds, i.e., whether every optimal control for $(\mathbf{CP}_0)$ can be approximated by a sequence $\{u^{\alpha_n}\}$ of optimal controls for $(\mathbf{CP}_{\alpha_n})$, for some sequence $\alpha_n \searrow 0$. Unfortunately, we are not able to prove such a global result that applies to all optimal controls of $(\mathbf{CP}_0)$. However, a local type of result can be established. To this end, let $\overline{u} \in \mathcal{U}_{\rm ad}$ be any optimal control for $(\mathbf{CP}_0)$. We introduce the 'adapted' cost functional

$$\widetilde{J}(u,(\rho,\mu)) := J(u,(\rho,\mu)) + \frac{1}{2}\|u - \overline{u}\|^2_{L^2(Q)} \tag{1.37}$$

and consider for every $\alpha \in (0,1]$ the *adapted control problem* $(\widetilde{\mathbf{CP}}_\alpha)$ of minimizing $\widetilde{J}$ subject to $u \in \mathcal{U}_{\rm ad}$ and to the constraint that (ρ,μ) solves the approximating system (1.24), (1.30), (1.32), (1.26), (1.27). It will then turn out that the following is true:

(i) There are some sequence $\alpha_n \searrow 0$ and minimizers $\overline{u}^{\alpha_n} \in \mathcal{U}_{\rm ad}$ of the adapted control problem associated with α_n, $n \in \mathbb{N}$, such that

$$\overline{u}^{\alpha_n} \to \overline{u} \quad \text{strongly in } L^2(Q) \text{ as } n \to \infty. \tag{1.38}$$

(ii) It is possible to pass to the limit as $\alpha \searrow 0$ in the first-order necessary optimality conditions corresponding to the adapted control problems associated with $\alpha \in (0,1]$ in order to derive first-order necessary optimality conditions for problem $(\mathbf{CP}_0)$.

Of course, it will be interesting to see which kind of first-order necessary optimality conditions can be derived by adopting the strategy outlined in (ii).

2 Results

In this section, we will describe our results, focusing in particular on the optimal control problem for logarithmic type potentials. Let us start with some notation and tools.

2.1 The mathematical framework

We recall that Ω is the body where the evolution takes place. We assume $\Omega \subset \mathbb{R}^3$ to be open, bounded, connected, and smooth, and we write $|\Omega|$ for its Lebesgue measure.

Next, in order to list our assumptions on the nonlocal operator B and even for future convenience, we set

$$V := H^1(\Omega), \quad H := L^2(\Omega), \quad \text{and} \quad W := \{v \in H^2(\Omega) : \partial_{\mathbf{n}} v = 0\}, \tag{2.1}$$

and

$$Q_t := \Omega \times (0,t), \quad Q^t := \Omega \times (t,T) \quad \text{for } t \in [0,T], \quad Q = Q_T = Q^0. \tag{2.2}$$

We point out, for three dimensions of space and smooth domains, the embeddings $V \subset L^p(\Omega)$, $1 \le p \le 6$, and $H^2(\Omega) \subset C^0(\overline{\Omega})$, which are continuous and (in the first case only for $1 \le p < 6$) compact. In particular, we have that

$$\|v\|_p \le C_\Omega \|v\|_{H^1(\Omega)} \quad \text{for every } v \in H^1(\Omega) \text{ and } p \in [1,6], \tag{2.3}$$

where C_Ω depends only on Ω. We also recall the continuous embedding

$$\begin{aligned} &\Big(L^\infty(0,T;L^2(\Omega)) \cap L^2(0,T;H^1(\Omega))\Big) \\ &\qquad \subset \Big(L^{10/3}(Q) \cap L^{7/3}(0,T;L^{14/3}(\Omega))\Big), \end{aligned} \tag{2.4}$$

which is a consequence of the Young, Sobolev and interpolation inequalities. Finally, in order to avoid a boring notation, we follow a general rule to denote constants. The small-case symbol c stands for different constants which depend only on Ω, on the final time T, the shape of the nonlinearities and on the constants and the norms of the functions involved in the assumptions of our statements. About time derivatives of a time-dependent function v, we use both the notations $\partial_t v$, $\partial_t^2 v$ and the shorter ones v_t, v_{tt}, just according to our convenience.

About the structure of our system, we assume that

$$\beta := \partial F_1 \; : \mathbb{R} \to 2^{\mathbb{R}} \quad \text{is maximal monotone with} \quad 0 \in \beta(0). \tag{2.5}$$

$$\pi := F_2' \; : \mathbb{R} \to \mathbb{R} \quad \text{is Lipschitz continuous.} \tag{2.6}$$

$$\begin{aligned} &g : \overline{D(\beta)} \to [0,+\infty) \quad \text{is } C^2\text{, bounded and concave, and} \\ &\quad g' \text{ is bounded and Lipschitz continuous.} \end{aligned} \tag{2.7}$$

In (2.7), $D(\beta)$ is the effective domain of β. For $r \in D(\beta)$, we also use the symbol $\beta^\circ(r)$ for the element of $\beta(r)$ having minimum modulus. Notice that, with respect to the notation used before, $F = F_1 + F_2$, and F' is replaced by the sum of the subdifferential β of F_1 and the derivative π of F_2.

As for the nonlocal operator B, we assume that it maps $L^2(0,T;H) = L^2(Q)$ into itself, is causal, and enjoys the following properties:

$$B : L^2(0,T;H) \to L^2(0,T;H); \tag{2.8}$$

$$B[u]|_{Q_t} = B[v]|_{Q_t} \quad \text{whenever } u|_{Q_t} = v|_{Q_t}, \quad \text{for every } t \in (0,T]; \tag{2.9}$$

$$B(L^p(Q_t)) \subset L^p(Q_t) \quad \text{and} \quad \|B[v]\|_{L^p(Q_t)} \le C_{B,p}\big(1 + \|v\|_{L^p(Q_t)}\big)$$
$$\text{for every } v \in L^p(Q),\ t \in (0,T], \text{ and } p \in [2,6]; \tag{2.10}$$

$$\|B[u] - B[v]\|_{L^2(Q_t)} \le C_B\|u - v\|_{L^2(Q_t)}$$
$$\text{for every } u, v \in L^2(Q) \text{ and } t \in (0,T]; \tag{2.11}$$

$$B(L^2(0,T;V)) \subset L^2(0,T;V) \text{ and, for every } v \in L^2(0,T;V) \text{ and } t \in (0,T],$$
$$\Big|\int_0^t\!\!\int_\Omega \nabla B[v] \cdot \nabla v \,dx\,ds\Big| \le C_B\Big(1 + \int_0^t\!\!\int_\Omega (|v|^2 + |\nabla v|^2)\,dx\,ds\Big)\,. \tag{2.12}$$

In the above formulas, $C_{B,p}$ and C_B are given structural constants, and, for any Banach space X, the symbol $\|\cdot\|_X$ denotes its norm. The same notation is then used also for powers of X. However, in the following we simply write $\|\cdot\|_p$ for the standard norm in $L^p(\Omega)$, for $1 \le p \le +\infty$.

2.2 Mathematical problem and general results

At this point, we make precise the problem under investigation. Let

$$\mu_0 \in V \quad \text{and} \quad \mu_0 \ge 0 \quad \text{a.e. in } \Omega, \tag{2.13}$$

$$\rho_0 \in V, \quad \rho_0 \in D(\beta) \quad \text{a.e. in } \Omega, \quad \text{and} \quad \rho_0|\beta^\circ(\rho_0)|^{7/3} \in L^1(\Omega), \tag{2.14}$$

and look for a triplet (μ, ρ, ξ) satisfying

$$\mu \in H^1(0,T;H) \cap L^\infty(0,T;V) \cap L^2(0,T;W^{2,3/2}(\Omega)), \tag{2.15}$$

$$\mu \ge 0 \quad \text{a.e. in } Q, \tag{2.16}$$

$$\rho \in L^\infty(0,T;V) \quad \text{and} \quad \partial_t\rho \in L^{10/3}(Q), \tag{2.17}$$

$$\xi \in L^2(0,T;H), \tag{2.18}$$

and solving the initial-boundary value problem

$$\big(1 + 2g(\rho)\big)\,\partial_t\mu + \mu\,g'(\rho)\,\partial_t\rho - \Delta\mu = u \quad \text{a.e. in } Q, \tag{2.19}$$

$$\partial_t\rho + B[\rho] + \xi + \pi(\rho) = \mu\,g'(\rho) \quad \text{and} \quad \xi \in \beta(\rho) \quad \text{a.e. in } Q, \tag{2.20}$$

$$\partial_{\mathbf{n}}\mu = 0 \quad \text{a.e. on } \Sigma, \tag{2.21}$$

$$\mu(0) = \mu_0 \quad \text{and} \quad \rho(0) = \rho_0\,, \tag{2.22}$$

where $\Sigma := \Gamma \times (0,T)$. Here are our results.

Theorem 2.1 *With the assumptions and notations* (2.5)–(2.12) *on the structure and* (2.13)–(2.14) *on the initial data, assume $u \in L^\infty(Q)$ and $u \ge 0$ a.e. in Q. Then, problem* (2.19)–(2.22) *has at least one solution satisfying* (2.15)–(2.18).

Theorem 2.2 *Under the assumptions of Theorem 2.1, suppose in addition that*

$$\mu_0 \in L^\infty(\Omega) \quad \textit{and} \quad \rho_0\,\big(\beta^\circ(\rho_0)\big)^5 \in L^1(\Omega). \tag{2.23}$$

Then the solution to problem (2.19)–(2.22) *is unique and also satisfies*

$$\mu \in L^\infty(Q), \quad \partial_t\rho \in L^6(Q) \quad \textit{and} \quad \xi \in L^6(Q). \tag{2.24}$$

Remark 1 One can also prove the existence of a solution to the more general problem obtained by replacing equation (2.19) by

$$\big(1 + 2g(\rho)\big)\,\partial_t\mu + \mu\,g'(\rho)\,\partial_t\rho - \operatorname{div}\big(\kappa(\mu)\nabla\mu\big) = u, \tag{2.25}$$

where $\kappa : [0,+\infty) \to (0,+\infty)$ is a bounded continuous function such that $1/\kappa$ is also bounded (like the uniformly parabolic case discussed in [16, 20, 22], while the degenerate case treated in [22] is more delicate).

The requirement $u \ge 0$ in the assumptions is needed to ensure that $\mu \ge 0$, as one can see by testing the equation by the negative part of μ (like in the proof of [22, Lemma 4.1]). In fact, we emphasize that (2.16) is a straightforward consequence of (2.19) and the non-negativity of u. Now, let us spend some words about the proof of the two theorems, by considering the simpler case $u \equiv 0$. Observe that

$$(1 + 2g(\rho))\,\mu\,\mu_t + g'(\rho)\,\rho_t\,\mu^2 \;=\; \partial_t\left(\left(\tfrac{1}{2} + g(\rho)\right)\,\mu^2\right).$$

whence, testing (2.19) formally by 2μ, we obtain that, for all ρ,

$$\int_\Omega (1 + 2g(\rho(t)))|\mu(t)|^2\,dx \;+\; 2\int_0^t\!\!\int_\Omega |\nabla\mu|^2\,dx\,ds \;=\; \int_\Omega (1 + 2g(\rho_0))|\mu_0|^2\,dx\,.$$

From this kind of energy equality, we immediately deduce that

$$\max\left\{\|\mu\|^2_{L^\infty(0,T;H)}\,,\; \|\nabla\mu\|^2_{L^2(Q)}\right\} \;\le\; (1 + 2\,\sup\,g)\,\|\mu_0\|^2_H\,.$$

Then the embedding (2.4) yields that

$$\|\mu\|_{L^{10/3}(Q)\cap L^2(0,T;V)} \le \hat{M} := C_0\,(1+2\,\sup g)^{1/2}\,\|\mu_0\|_H\,.$$

Next, in the separable and reflexive Banach space

$$\mathcal{M} := \left(L^{10/3}(Q)\cap L^2(0,T;V)\right)$$

consider the nonempty, closed, convex, bounded (and thus weakly compact) set

$$\mathcal{M}_0 := \left\{v\in\mathcal{M} : \|v\|_{\mathcal{M}} \le \hat{M} \ \text{ and } \ v\ge 0 \ \text{ a.e. in } Q\right\},$$

and let

$$\mathcal{R} := W^{1,10/3}(0,T;L^{10/3}(\Omega))\cap L^\infty(0,T;V)\,.$$

Define the mappings

$$\mathcal{T}_1 : \mathcal{M}_0 \to \mathcal{R}, \quad \mu \mapsto \ \text{unique solution } \rho \text{ to (2.20)} + \boxed{\rho(0)=\rho_0}\,,$$

$$\mathcal{T}_2 : \mathcal{R} \to \mathcal{M}_0, \quad \rho \mapsto \ \text{unique solution } \mu \text{ to (2.19)} + \text{(2.21)} + \boxed{\mu(0)=\mu_0}\,,$$

$$\mathcal{T} : \mathcal{M}_0 \to \mathcal{M}_0, \ \mu \mapsto \ \mathcal{T}[\mu] := (\mathcal{T}_2\circ\mathcal{T}_1)[\mu]\,.$$

In the proof, we show that the mapping $\mathcal{T} : \mathcal{M}_0 \to \mathcal{M}_0$ is weakly sequentially continuous. Then, we are in a condition to apply the Tikhonov's Fixed Point Theorem, which implies that $\mathcal{T}$ has a fixed point $\overline{\mu}\in\mathcal{M}_0$ and $(\overline{\rho},\overline{\mu},\overline{\xi})$, where $\overline{\rho}=\mathcal{T}_1(\overline{\mu})$ and $\overline{\xi}\in\beta(\overline{\rho})$ a.e. in Q, is the sought solution. The inherent difficulties are:

- the variable ρ does not enjoy much spatial regularity;
- it is not obvious how to solve (2.20) + initial condition for given $\mu\in\mathcal{M}_0$, since μ needs not be bounded;
- it is not obvious how to solve (2.19) + (2.21) + initial condition for given $\rho\in\mathcal{R}$, since the coefficient functions are nonsmooth.

To overcome these difficulties, the remedies are

- replace (2.20) by the solvable approximating equation

  $$\rho^\varepsilon_t + B[\rho^\varepsilon] + \beta_\varepsilon(\rho^\varepsilon) + \pi(\rho^\varepsilon) = Q_\varepsilon(\mu)\,g'(\rho^\varepsilon)\,,$$

  with the cutoff function

  $$Q_\varepsilon(r) = \max\{0, \min\{r, 1/\varepsilon\}\}$$

  and the Yosida approximation β_ε of β. Show uniform a priori estimates, and let $\varepsilon\searrow 0$;
- approximate ρ in (2.19) by the smooth functions ρ^ε, solve (2.19) + (2.21) + initial condition with these, show uniform a priori estimates, and let $\varepsilon\searrow 0$.

About the proof of Theorem 2.2, we say that first one shows, using again an approximation argument, that

$$\partial_t \rho \in L^6(Q), \quad \xi \in L^6(Q).$$

Then we infer, in particular, that

$$\partial_t \rho \in L^{7/3}(0,T; L^{14/3}(\Omega)).$$

Hence, by a Moser iteration type argument from the paper [17], one concludes that

$$\mu \in L^\infty(Q),$$

and, with this, the uniqueness proof follows from an L^1 – type estimation technique, which has become standard, for instance, in the analysis of phase field systems containing hysteresis operators (see, e.g., [15, 16]).

2.3 Special case for the existence result

Consider the optimal control problem (1.23)–(1.28). We make the following assumptions on the data:

$$F = F_1 + F_2, \text{ where } F_1 \in C^3(0,1) \text{ is convex, } F_2 \in C^3[0,1], \text{ and}$$
$$\lim_{r \searrow 0} F_1'(r) = -\infty, \quad \lim_{r \nearrow 1} F_1'(r) = +\infty; \tag{2.26}$$

$$\rho_0 \in V,\ F'(\rho_0) \in H,\ \mu_0 \in W, \text{ where } \mu_0 \geq 0 \text{ a.e. in } \Omega,$$
$$\inf \{\rho_0(x) : x \in \Omega\} > 0, \quad \sup \{\rho_0(x) : x \in \Omega\} < 1; \tag{2.27}$$

$$g \in C^3[0,1] \text{ satisfies } g(\rho) \geq 0 \text{ and } g''(\rho) \leq 0 \text{ for all } \rho \in [0,1]. \tag{2.28}$$

Note that (2.27) implies $\mu_0 \in C(\overline{\Omega})$, and (2.26) and (2.27) ensure that both $F(\rho_0)$ and $F'(\rho_0)$ are in $L^\infty(\Omega)$, whence in H. In addition, let us point out that the logarithmic potential (1.29) fulfills the conditions in (2.26).

Here, we ask the nonlocal operator $B\colon L^1(Q) \to L^1(Q)$ to satisfy the following conditions:

$$\text{For every } t \in (0,T], \text{ we have } \quad B[v]|_{Q_t} = B[w]|_{Q_t} \quad \text{whenever } v|_{Q_t} = w|_{Q_t}. \tag{2.29}$$

$$\text{For all } p \in [2,+\infty], \text{ we have } B(L^p(Q_t)) \subset L^p(Q_t) \text{ and}$$
$$\|B[v]\|_{L^p(Q_t)} \leq C_{B,p} \left(1 + \|v\|_{L^p(Q_t)}\right) \quad \text{for } v \in L^p(Q),\ t \in (0,T]. \tag{2.30}$$

For $v, w \in L^1(0,T;H)$ and $t \in (0,T]$, it holds that

$$\int_0^t \|B[v](s) - B[w](s)\|_6 \, ds \leq C_B \int_0^t \|v(s) - w(s)\|_H \, ds. \tag{2.31}$$

It holds, for every $v \in L^2(0,T;V)$ and $t \in (0,T]$, that

$$\|\nabla B[v]\|_{L^2(0,t;H)} \le C_B\big(1 + \|v\|_{L^2(0,t;V)}\big). \tag{2.32}$$

For every $v \in H^1(0,T;H)$, we have $\partial_t B[v] \in L^2(Q)$ and

$$\|\partial_t B[v]\|_{L^2(Q)} \le C_B\big(1 + \|\partial_t v\|_{L^2(Q)}\big). \tag{2.33}$$

Moreover, B is continuously Fréchet differentiable from $L^2(Q)$ into $L^2(Q)$, and the Fréchet derivative $DB[\overline{v}] \in \mathcal{L}(L^2(Q), L^2(Q))$ of B at $\overline{v}$ fulfills

$$\|DB[\overline{v}](w)\|_{L^p(Q_t)} \le C_B\,\|w\|_{L^p(Q_t)} \quad \forall\, w \in L^p(Q), \quad \forall\, p \in [2,6], \tag{2.34}$$

$$\|\nabla(DB[\overline{v}](w))\|_{L^2(Q_t)} \le C_B\,\|w\|_{L^2(0,t;V)} \quad \forall\, w \in L^2(0,T;V), \tag{2.35}$$

for every $\overline{v} \in L^2(Q)$ and $t \in (0,T]$.

In the above formulas, $C_{B,p}$ and C_B again denote given positive structural constants. We also underline that (2.34) implicitly requires that $DB[\overline{v}](w)|_{Q_t}$ depends only on $w|_{Q_t}$, but this is a consequence of (2.29).

We recall that J and $\mathcal{U}_{\rm ad}$ are defined by (1.23) and (1.28), respectively, where

$$\beta_1\,,\beta_2\,,\beta_3 \ge 0, \quad \beta_1 + \beta_2 + \beta_3 > 0, \quad \text{and} \quad R > 0, \tag{2.36}$$

$$\rho_Q, \mu_Q \in L^2(Q), \quad u_{\max} \in L^\infty(Q) \quad \text{and} \quad u_{\max} \ge 0 \quad \text{a.e. in } Q. \tag{2.37}$$

Remark 2 In view of (2.34), for every $t \in [0,T]$ it holds that

$$\|B[v] - B[w]\|_{L^2(Q_t)} \le C_B\,\|v - w\|_{L^2(Q_t)} \quad \forall\, v, w \in L^2(Q), \tag{2.38}$$

that is, the condition (2.11) is fulfilled. Moreover, (2.30) and (2.32) imply that B maps $L^2(0,T;V)$ into itself and that (2.12) is fulfilled. Moreover, thanks to (2.34) and (2.35), there is some constant $\widetilde{C}_B > 0$ such that

$$\|DB[\overline{v}](w)\|_{L^2(0,t;V)} \le \widetilde{C}_B\,\|w\|_{L^2(0,t;V)} \quad \forall\, \overline{v} \in L^2(Q), \ \forall\, w \in L^2(0,T;V). \tag{2.39}$$

Remark 3 The integral operator (1.18) satisfies the conditions (2.29) and (2.30), provided that the integral kernel k belongs to $C^1(0,+\infty)$ and satisfies, with suitable constants $C_1 > 0$, $C_2 > 0$, $0 < \alpha < \frac{3}{2}$, $0 < \beta < \frac{5}{2}$, the growth conditions

$$|k(r)| \le C_1\, r^{-\alpha}, \quad |k'(r)| \le C_2\, r^{-\beta}, \quad \forall\, r > 0.$$

In fact, it holds $2\alpha < 3$ and thus, for all $v, w \in L^1(0,T;H)$ and $t \in (0,T]$,

$$
\begin{aligned}
&\int_0^t \|B[v](s) - B[w](s)\|_6 \, ds \\
&\le C_1 \int_0^t \Big(\int_\Omega \Big| \int_\Omega |y-x|^{-\alpha} |v(y,s) - w(y,s)| \, dy \Big|^6 dx \Big)^{1/6} ds \\
&\le C_3 \int_0^t \Big(\int_\Omega \Big| \Big(\int_\Omega |y-x|^{-2\alpha} \, dy \Big)^{1/2} \|v(s) - w(s)\|_H \Big|^6 dx \Big)^{1/6} ds \\
&\le C_4 \int_0^t \|v(s) - w(s)\|_H \, ds,
\end{aligned}
$$

with global constants C_i, $3 \le i \le 4$; the condition (2.31) is thus verified. Condition (2.32) holds true as well: indeed, as $\frac{6\beta}{5} < 3$, for every $t \in (0,T]$ and $v \in L^2(0,T;V)$ we have

$$
\begin{aligned}
\|\nabla B[v]\|^2_{L^2(0,t;H)} &\le C_2 \int_0^t \int_\Omega \Big| \int_\Omega |y-x|^{-\beta} |v(y,s)| \, dy \Big|^2 dx \, ds \\
&\le C_5 \int_0^t \int_\Omega \Big(\int_\Omega |y-x|^{-6\beta/5} dy \Big)^{5/3} \|v(s)\|_6^2 \, dx \, ds \\
&\le C_6 \int_0^t \|v(s)\|_V^2 \, ds \, .
\end{aligned}
$$

Finally, since the operator B is linear in this case, we have $DB[\overline{v}] = B$ for every $\overline{v} \in L^2(Q)$ and, consequently, (2.34), (2.35), (2.39) are fulfilled.

We have the following existence and regularity result for the state system.

Theorem 2.3 *Suppose that* (2.26)–(2.37) *are satisfied. Then the state system* (1.24)–(1.27) *has for every* $u \in \mathcal{U}_{ad}$ *a unique solution* (ρ, μ) *such that*

$$
\rho \in H^2(0,T;H) \cap W^{1,\infty}(0,T;L^\infty(\Omega)) \cap H^1(0,T;V), \tag{2.40}
$$

$$
\mu \in W^{1,\infty}(0,T;H) \cap H^1(0,T;V) \cap L^\infty(0,T;W) \subset L^\infty(Q). \tag{2.41}
$$

Moreover, there are constants $0 < \rho_* < \rho^* < 1$, $\mu^* > 0$, *and* $K_1^* > 0$, *which depend only on the given data, such that for every* $u \in \mathcal{U}_{ad}$ *the corresponding solution* (ρ, μ) *satisfies*

$$
0 < \rho_* \le \rho \le \rho^* < 1, \quad 0 \le \mu \le \mu^*, \quad \text{a.e. in } Q, \tag{2.42}
$$

$$
\begin{aligned}
&\|\mu\|_{W^{1,\infty}(0,T;H) \cap H^1(0,T;V) \cap L^\infty(0,T;W) \cap L^\infty(Q)} \\
&+ \|\rho\|_{H^2(0,T;H) \cap W^{1,\infty}(0,T;L^\infty(\Omega)) \cap H^1(0,T;V)} \le K_1^* .
\end{aligned} \tag{2.43}
$$

Proof Theorems 2.1 and 2.2 ensure that for every $u \in \mathcal{U}_{ad}$ there exists a unique solution (ρ, μ) with the prescribed regularities. Moreover, we have that

$$\|\mu\|_{H^1(0,T;H)\cap L^\infty(0,T;V)\cap L^\infty(Q)\cap L^2(0,T;W^{2,3/2}(\Omega))} + \|\rho\|_{L^\infty(0,T;V)} + \|\partial_t\rho\|_{L^6(Q)} \le c \quad \forall u \in \mathcal{U}_{\mathrm{ad}}. \tag{2.44}$$

Next, as $0 < \rho < 1$ a.e. in Q, the condition (2.30) for $p = +\infty$ implies that $\|B[\rho]\|_{L^\infty(Q)} \le C_2$ for all $u \in \mathcal{U}_{\mathrm{ad}}$, and it follows from (2.44) and the general assumptions on ρ_0, g, and F, that there are constants ρ_*, ρ^* such that, for every $u \in \mathcal{U}_{\mathrm{ad}}$,

$$\begin{aligned}
&0 < \rho_* \le \inf\{\rho_0(x) : x \in \Omega\} \le \sup\{\rho_0(x) : x \in \Omega\} \le \rho^* < 1,\\
&F'(\rho) + B[\rho] - \mu\, g'(\rho) \le 0 \ \text{ if } 0 < \rho \le \rho_*,\\
&F'(\rho) + B[\rho] - \mu\, g'(\rho) \ge 0 \ \text{ if } \rho^* \le \rho < 1.
\end{aligned}$$

Hence, multiplying (1.25) by the positive part $(\rho - \rho^*)^+$ of $\rho - \rho^*$, and integrating over Q_t, we find that

$$\begin{aligned}
0 &= \int_0^t\!\!\int_\Omega \partial_t\rho\,(\rho - \rho^*)^+\,dx\,ds + \int_0^t\!\!\int_\Omega (F'(\rho) + B[\rho] - \mu\, g'(\rho))(\rho - \rho^*)^+\,dx\,ds\\
&\ge \frac{1}{2}\int_\Omega \left|(\rho(t) - \rho^*)^+\right|^2 dx,
\end{aligned}$$

whence we conclude that, a.e. in Ω, $(\rho - \rho^*)^+(t) = 0$, and thus $\rho(t) \le \rho^*$ for almost every $t \in (0,T)$. The other bound for ρ in (2.42) is proved similarly. One now has to recover the missing bounds in (2.43), which then also imply the missing regularity in (2.40)–(2.41): the related procedure is precisely carried out in [29] by using a bootstrapping argument. □

Note that the estimates (2.42) and (2.43), as well as the assumptions on the data and the continuity of the embedding $V \subset L^6(\Omega)$, entail that also

$$\begin{aligned}
&\max_{0\le i\le 3} \|F^{(i)}(\rho)\|_{L^\infty(Q)} + \max_{0\le i\le 3} \|g^{(i)}(\rho)\|_{L^\infty(Q)}\\
&+ \|\nabla\mu\|_{L^\infty(0,T;L^6(\Omega)^3)} + \|\partial_t\mu\|_{L^2(0,T;V)}\\
&+ \|B[\rho]\|_{H^1(0,T;L^2(\Omega))\cap L^\infty(Q)\cap L^2(0,T;V)} \le K_1^* \quad \forall u \in \mathcal{U}_{\mathrm{ad}},
\end{aligned} \tag{2.45}$$

by possibly choosing a larger constant K_1^*.

According to Theorem 2.3, the control-to-state mapping $\mathcal{S} : \mathcal{U}_{\mathrm{ad}} \ni u \mapsto (\rho, \mu)$ is well defined. We now study its stability properties. We have the following result.

Theorem 2.4 *Suppose that* (2.26)–(2.37) *are fulfilled, and let* $u_i \in \mathcal{U}_{\mathrm{ad}}$, $i = 1,2$, *be given and* $(\rho_i, \mu_i) = \mathcal{S}(u_i)$, $i = 1,2$, *be the associated solutions to the state system* (1.24)–(1.27). *Then there exists a contant* $K_2^* > 0$, *which depends only on the data of the problem, such that, for every* $t \in (0,T]$,

$$\|\rho_1 - \rho_2\|_{H^1(0,t;H)\cap L^\infty(0,t;L^6(\Omega))} + \|\mu_1 - \mu_2\|_{H^1(0,t;H)\cap L^\infty(0,t;V)\cap L^2(0,t;W)}$$
$$\le K_2^* \|u_1 - u_2\|_{L^2(0,t;H)} . \qquad (2.46)$$

It turns out that (2.46) is exactly the right kind of stability estimate useful for showing the directional differentiability of $\mathcal{S}$. The proof of Theorem 2.4 is very technical (see [29, Section 2]) and requires a bootstrapping procedure with several steps. Crucial is, in view of the low spatial regularity of ρ_i, the estimate in $L^\infty(0,t;L^6(\Omega))$, which cannot be established by deriving an $L^\infty(0,t;V)$ – estimate.

2.4 Directional differentiability of the control-to-state mapping

In this subsection, we discuss the differentiability properties of the solution operator $\mathcal{S}$. To this end, we introduce the spaces

$$\mathcal{X} := H^1(0,T;H) \cap L^\infty(Q),$$
$$\mathcal{Y} := H^1(0,T;H) \times \left(L^\infty(0,T;H) \cap L^2(0,T;V)\right),$$

endowed with their natural norms, and consider the control-to-state operator $\mathcal{S}$ as a mapping between $\mathcal{U}_{\text{ad}} \subset \mathcal{X}$ and $\mathcal{Y}$. Now let $\overline{u} \in \mathcal{U}_{\text{ad}}$ be fixed and put $(\overline{\rho}, \overline{\mu}) := \mathcal{S}(\overline{u})$. We then study the linearization of the state system (1.24)–(1.27) at $\overline{u}$, which is given by:

$$(1 + 2g(\overline{\rho}))\,\eta_t + 2g'(\overline{\rho})\,\overline{\mu}_t\,\xi + g'(\overline{\rho})\,\overline{\rho}_t\,\eta$$
$$+ \overline{\mu}\,g''(\overline{\rho})\,\overline{\rho}_t\,\xi + \overline{\mu}g'(\overline{\rho})\,\xi_t - \Delta\eta = h \quad \text{a.e. in } Q, \qquad (2.47)$$

$$\xi_t + F''(\overline{\rho})\,\xi + DB[\overline{\rho}](\xi) = \overline{\mu}\,g''(\overline{\rho})\,\xi + g'(\overline{\rho})\,\eta \quad \text{a.e. in } Q, \qquad (2.48)$$

$$\partial_{\mathbf{n}}\eta = 0 \quad \text{a.e. on } \Sigma, \qquad (2.49)$$

$$\eta(0) = \xi(0) = 0 \quad \text{a.e. in } \Omega. \qquad (2.50)$$

Here, $h \in \mathcal{X}$ must satisfy $\overline{u} + \overline{\lambda} h \in \mathcal{U}_{\text{ad}}$ for some $\overline{\lambda} > 0$. If the system (2.47)–(2.50) has for any such h a unique solution pair (ξ, η), we expect that the directional derivative $\delta\mathcal{S}(\overline{u}; h)$ of $\mathcal{S}$ at $\overline{u}$ in the direction h coincides with (ξ, η). Actually, the problem (2.47)–(2.50) makes sense for every $h \in L^2(Q)$, and it is uniquely solvable under this weaker assumption.

Theorem 2.5 *Suppose that the general hypotheses* (2.26)–(2.37) *are satisfied and let* $h \in L^2(Q)$. *Then the linearized problem* (2.47)–(2.50) *has a unique solution* (ξ, η) *satisfying*

$$\xi \in H^1(0,T;H) \cap L^\infty(0,T;L^6(\Omega)), \qquad (2.51)$$
$$\eta \in H^1(0,T;H) \cap L^\infty(0,T;V) \cap L^2(0,T;W). \qquad (2.52)$$

Proof Uniqueness can easily be shown. About existence, the proof is split in several steps (see [29, Section 3]). An approximating problem depending on a parameter $\varepsilon \in (0,1)$ is introduced and well-posedness for this problem is proved by performing suitable a priori estimates. Finally, a solution to problem (2.47)–(2.50) is constructed by letting ε tend to zero. □

Now, we can show that $\mathcal{S}$ is directionally differentiable. The following result holds true.

Theorem 2.6 *Assume that* (2.26)–(2.37) *are satisfied, let* $\overline{u} \in \mathcal{U}_{\rm ad}$ *be given and* $(\overline{\rho}, \overline{\mu}) = \mathcal{S}(\overline{u})$. *Moreover, if* $h \in \mathcal{X}$ *is a function such that* $\overline{u} + \overline{\lambda} h \in \mathcal{U}_{\rm ad}$ *for some* $\overline{\lambda} > 0$, *then the directional derivative* $\delta\mathcal{S}(\overline{u}; h)$ *of* $\mathcal{S}$ *at* $\overline{u}$ *in the direction* h *exists in the space* $(\mathcal{Y}, \|\cdot\|_{\mathcal{Y}})$, *and we have* $\delta\mathcal{S}(\overline{u}; h) = (\xi, \eta)$, *where* (ξ, η) *is the unique solution to the linearized system* (2.47)–(2.50).

Proof The argument is very technical and fully uses the stability result of Theorem 2.5. The interested reader can see [29, Section 3] for details. Let us just give an idea how it works. For $0 < \lambda \le \overline{\lambda}$ we have $\overline{u} + \lambda h \in \mathcal{U}_{\rm ad}$ and put

$$u^\lambda := \overline{u} + \lambda h, \quad (\rho^\lambda, \mu^\lambda) := \mathcal{S}(u^\lambda),$$
$$y^\lambda := \rho^\lambda - \overline{\rho} - \lambda\xi, \quad z^\lambda := \mu^\lambda - \overline{\mu} - \lambda\eta\,.$$

We need to show that

$$0 = \lim_{\lambda \searrow 0} \frac{\|\mathcal{S}(\overline{u} + \lambda h) - \mathcal{S}(\overline{u}) - \lambda\, \delta\mathcal{S}(\overline{u}; h)\|_{\mathcal{Y}}}{\lambda}$$
$$= \lim_{\lambda \searrow 0} \frac{\|y^\lambda\|_{H^1(0,T;H)} + \|z^\lambda\|_{L^\infty(0,T;H)} + \|z^\lambda\|_{L^2(0,T;V)}}{\lambda}\,.$$

This is proved by using the fact that (z^λ, y^λ) solves the system

$$\begin{aligned}
&(1 + 2g(\overline{\rho}))\, z^\lambda_t + g'(\overline{\rho})\,\overline{\rho}_t\, z^\lambda + \overline{\mu}\, g'(\overline{\rho})\, y^\lambda_t - \Delta z^\lambda \\
&= -2\left(g(\rho^\lambda) - g(\overline{\rho})\right)\left(\mu^\lambda_t - \overline{\mu}_t\right) - 2\,\overline{\mu}_t \left(g(\rho^\lambda) - g(\overline{\rho}) - \lambda g'(\overline{\rho})\xi\right) \\
&\quad - \overline{\mu}\,\overline{\rho}_t \left(g'(\rho^\lambda) - g'(\overline{\rho}) - \lambda g''(\overline{\rho})\xi\right) - \overline{\mu}\left(g'(\rho^\lambda) - g'(\overline{\rho})\right)\left(\rho^\lambda_t - \overline{\rho}_t\right) \\
&\quad - \left(\mu^\lambda - \overline{\mu}\right)\left[\left(g'(\rho^\lambda) - g'(\overline{\rho})\right)\overline{\rho}_t + g'(\rho^\lambda)\left(\rho^\lambda_t - \overline{\rho}_t\right)\right] \quad \text{a.e. in } Q,
\end{aligned} \tag{2.53}$$

$$\begin{aligned}
y^\lambda_t &= -\left(F'(\rho^\lambda) - F'(\overline{\rho}) - \lambda F''(\overline{\rho})\xi\right) - \left(B[\rho^\lambda] - B[\overline{\rho}] - \lambda DB[\overline{\rho}](\xi)\right) \\
&\quad + g'(\overline{\rho}) z^\lambda + \overline{\mu}\left(g'(\rho^\lambda) - g'(\overline{\rho}) - \lambda g''(\overline{\rho})\xi\right) \\
&\quad + \left(\mu^\lambda - \overline{\mu}\right)\left(g'(\rho^\lambda) - g'(\overline{\rho})\right) \quad \text{a.e. in } Q,
\end{aligned} \tag{2.54}$$

$$\partial_{\mathbf{n}} z^\lambda = 0 \quad \text{a.e. on } \Sigma, \tag{2.55}$$

$$z^\lambda(0) = y^\lambda(0) = 0 \quad \text{a.e. in } \Omega. \tag{2.56}$$

We test (2.53) by z^λ and (2.54) by y_t^λ. Let us just show a pair of critical estimates. This is the first:

$$\begin{aligned}
&\int_0^t \int_\Omega |g(\rho^\lambda) - g(\overline{\rho})| \, |\mu_t^\lambda - \overline{\mu}_t| \, |z^\lambda| \, dx \, ds \\
&\leq C \int_0^t \|(\rho^\lambda - \overline{\rho})(s)\|_3 \, \|(\mu_t^\lambda - \overline{\mu}_t)(s)\|_2 \, \|z^\lambda(s)\|_6 \, ds \\
&\leq C \, \|\rho^\lambda - \overline{\rho}\|_{L^\infty(0,t;L^3(\Omega))} \, \|\mu^\lambda - \overline{\mu}\|_{H^1(0,t;H)} \, \|z^\lambda\|_{L^2(0,t;V)} \\
&\leq \gamma \, \|z^\lambda\|^2_{L^2(0,t;V)} + \tfrac{C}{\gamma} \, \lambda^4,
\end{aligned}$$

and here one can see why the stability estimate for $\|\mu_t\|_{L^2(Q)}$ is needed. The second critical estimate is given by:

$$\begin{aligned}
&\int_0^t \int_\Omega |\overline{\mu}_t| \, |g(\rho^\lambda) - g(\overline{\rho}) - \lambda \, g'(\overline{\rho}) \, \xi| \, |z^\lambda| \, dx \, ds \\
&\leq C \int_0^t \int_\Omega |\overline{\mu}_t| \left(|y^\lambda| + |\rho^\lambda - \overline{\rho}|^2 \right) |z^\lambda| \, dx \, ds \\
&\leq C \int_0^t \|\overline{\mu}_t(s)\|_6 \left(\|y^\lambda(s)\|_2 \, \|z^\lambda(s)\|_3 + \|(\rho^\lambda - \overline{\rho})(s)\|_6^2 \, \|z^\lambda(s)\|_2 \right) ds \\
&\leq \gamma \int_0^t \|z^\lambda(s)\|_V^2 + \left(1 + \tfrac{C}{\gamma}\right) \int_0^t \|\overline{\mu}_t(s)\|_V^2 \left(\|y^\lambda(s)\|_H^2 + \|z^\lambda(s)\|_H^2 \right) ds \\
&\quad + C \, \|\rho^\lambda(s) - \overline{\rho}(s)\|^4_{L^\infty(0,t;L^6(\Omega))} \, .
\end{aligned}$$

We can realize here why one needs the stability estimate for ρ in $L^\infty(0,T;L^6(\Omega))$.□

We are now in the position to establish the following result.

Corollary 2.1 *Let the general hypotheses* (2.26)–(2.37) *be fulfilled and assume that* $\overline{u} \in \mathcal{U}_{\text{ad}}$ *is a solution to the control problem* **(CP)** *with associated state* $(\overline{\rho}, \overline{\mu}) = \mathcal{S}(\overline{u})$. *Then we have, for every* $v \in \mathcal{U}_{\text{ad}}$,

$$\begin{aligned}
&\beta_1 \int_0^T \int_\Omega (\overline{\rho} - \rho_Q) \, \xi \, dx \, dt + \beta_2 \int_0^T \int_\Omega (\overline{\mu} - \mu_Q) \, \eta \, dx \, dt \\
&+ \beta_3 \int_0^T \int_\Omega \overline{u} \, (v - \overline{u}) \, dx \, dt \geq 0,
\end{aligned} \tag{2.57}$$

where (ξ,η) *denotes the (unique) solution to the linearized system* (2.47)–(2.50) *for* $h = v - \overline{u}$.

Proof Let $v \in \mathcal{U}_{\rm ad}$ be arbitrary. Then $h = v - \overline{u}$ is an admissible direction, since $\overline{u} + \lambda h \in \mathcal{U}_{\rm ad}$ for $0 < \lambda \le 1$. For any such λ, we have

$$
\begin{aligned}
0 &\le \frac{J(\overline{u} + \lambda h, \mathcal{S}(\overline{u} + \lambda h)) - J(\overline{u}, \mathcal{S}(\overline{u}))}{\lambda} \\
&\le \frac{J(\overline{u} + \lambda h, \mathcal{S}(\overline{u} + \lambda h)) - J(\overline{u}, \mathcal{S}(\overline{u} + \lambda h))}{\lambda} \\
&\quad + \frac{J(\overline{u}, \mathcal{S}(\overline{u} + \lambda h)) - J(\overline{u}, \mathcal{S}(\overline{u}))}{\lambda} .
\end{aligned}
$$

It follows immediately from the definition of the cost functional J that the first summand on the right-hand side of this inequality converges to $\int_0^T\int_\Omega \beta_3\, \overline{u}\, h\, dx\, dt$ as $\lambda \searrow 0$. For the second summand, we obtain from Theorem 2.6 that

$$
\begin{aligned}
&\lim_{\lambda \searrow 0} \frac{J(\overline{u}, \mathcal{S}(\overline{u} + \lambda h)) - J(\overline{u}, \mathcal{S}(\overline{u}))}{\lambda} \\
&= \beta_1 \int_0^T\int_\Omega (\overline{\rho} - \rho_Q)\xi\, dx\, dt + \beta_2 \int_0^T\int_\Omega (\overline{\mu} - \mu_Q)\eta\, dx\, dt,
\end{aligned}
$$

whence the assertion follows. □

2.5 Existence and first-order necessary conditions of optimality

This subsection is devoted to the derivation of first-order optimality conditions for problem **(CP)**. At first,we state the existence of optimal controls.

Theorem 2.7 *Suppose that the conditions* (2.26)–(2.37) *are satisfied. Then the problem* **(CP)** *has a solution* $\overline{u} \in \mathcal{U}_{\rm ad}$.

Proof It uses the direct method of the calculus of variations. Let $\{u_n\}_{n\in\mathbb{N}} \subset \mathcal{U}_{\rm ad}$ be a minimizing sequence for **(CP)**, and consider the sequence $\{(\rho_n, \mu_n)\}_{n\in\mathbb{N}}$ of the associated solutions to (1.24)–(1.27). We then can infer from the global estimate (2.43) the existence of a triple $(\overline{u}, \overline{\rho}, \overline{\mu})$ such that, possibly taking a subsequence again indexed by n,

$$
\begin{aligned}
u_n &\to \bar{u} \quad \text{weakly star in } H^1(0,T;H) \cap L^\infty(Q), \\
\rho_n &\to \bar{\rho} \quad \text{weakly star in } H^2(0,T;H) \cap W^{1,\infty}(0,T;L^\infty(\Omega)) \cap H^1(0,T;V), \\
\mu_n &\to \bar{\mu} \quad \text{weakly star in } W^{1,\infty}(0,T;H) \cap H^1(0,T;V) \cap L^\infty(0,T;W).
\end{aligned}
$$

Clearly, we have that $\overline{u} \in \mathcal{U}_{\rm ad}$ and, by virtue of the Aubin–Lions lemma (cf. [55, Thm. 5.1, p. 58]) and similar compactness results (cf. [60, Sect. 8, Cor. 4]), we derive a strong convergence of ρ_n to $\overline{\rho}$ in $L^2(Q)$. Now, it suffices to show that $(\overline{\rho},\overline{\mu})$ solves (1.24)–(1.27) for $\overline{u}$, which is a straightforward check. Indeed, from the weak sequential lower semicontinuity of the cost functional J it follows that $\bar{u}$, along with $(\overline{\rho},\overline{\mu}) = \mathcal{S}(\overline{u})$, is a solution to **(CP)**. □

We now turn our interest to the derivation of first-order necessary optimality conditions for problem **(CP)** and generally assume in the following that $\overline{u} \in \mathcal{U}_{\rm ad}$ is an optimal control with associated state $(\overline{\rho},\overline{\mu})$, which has the properties (2.42)–(2.45). Our aim is to eliminate ξ and η from the variational inequality (2.57). To this end, we employ the adjoint state system associated with (1.24)–(1.27) for $\overline{u}$, which is formally given by

$$\begin{aligned} &-(1+2g(\overline{\rho}))\, p_t - g'(\overline{\rho})\,\overline{\rho}_t\, p - \Delta p \\ &\quad - g'(\overline{\rho})\, q = \beta_2(\overline{\mu} - \mu_Q) \quad \text{in } Q, \end{aligned} \tag{2.58}$$

$$\begin{aligned} &- q_t + F''(\overline{\rho})\, q - \overline{\mu}\, g''(\overline{\rho})\, q + g'(\overline{\rho})\left(\overline{\mu}_t\, p - \overline{\mu}\, p_t\right) \\ &\quad + DB[\overline{\rho}]^*(q) = \beta_1(\overline{\rho} - \rho_Q) \quad \text{in } Q, \end{aligned} \tag{2.59}$$

$$\partial_{\mathbf{n}} p = 0 \quad \text{on } \Sigma, \tag{2.60}$$

$$p(T) = q(T) = 0 \quad \text{in } \Omega\,. \tag{2.61}$$

In (2.59), $DB[\overline{\rho}]^* \in \mathcal{L}(L^2(Q), L^2(Q))$ denotes the adjoint operator associated with the operator $DB[\overline{\rho}] \in \mathcal{L}(L^2(Q), L^2(Q))$, thus defined by the identity

$$\int_0^T\!\!\int_\Omega DB[\overline{\rho}]^*(v)\, w\, dx\, dt \;=\; \int_0^T\!\!\int_\Omega v\, DB[\overline{\rho}](w)\, dx\, dt \quad \forall\, v, w \in L^2(Q)\,. \tag{2.62}$$

As, for every $v \in L^2(Q)$, the restriction of $DB[\overline{\rho}](v)$ to Q_t depends only on $v|_{Q_t}$, it follows that, for every $w \in L^2(Q)$, the restriction of $DB[\overline{\rho}]^*(w)$ to $Q^t = \Omega \times (t,T)$ (see (2.2)) depends only on $w|_{Q^t}$. Moreover, (2.34) implies that

$$\|DB[\overline{\rho}]^*(w)\|_{L^2(Q^t)} \le C_B \|w\|_{L^2(Q^t)} \quad \forall\, w \in L^2(Q). \tag{2.63}$$

We also note that in the case of the integral operator (1.18) it follows from Fubini's theorem that $DB[\overline{\rho}]^* = DB[\overline{\rho}] = B$.

The following existence and uniqueness result holds true for the adjoint system.

Theorem 2.8 *Assume* (2.26)–(2.37), *and let* $\overline{u} \in \mathcal{U}_{\rm ad}$ *be a solution to the control problem* **(CP)** *with associated state* $(\overline{\rho},\overline{\mu}) = \mathcal{S}(\overline{u})$. *Then the adjoint system* (2.58)–(2.61) *has a unique solution* (p,q) *satisfying*

$$p \in H^1(0,T;H) \cap L^\infty(0,T;V) \cap L^2(0,T;W) \quad \textit{and} \quad q \in H^1(0,T;H)\,. \tag{2.64}$$

Moreover, we have the variational inequality

$$\int_0^t \int_\Omega (p + \beta_3 \overline{u})(v - \overline{u})\, dx\, dt \ge 0 \quad \forall\, v \in \mathcal{U}_{\rm ad}\,. \tag{2.65}$$

Proof While the derivation of (2.65) and of the uniqueness result is standard, the proof of existence is again not. An idea for the existence proof is to approximate $\overline{\rho}, \overline{\mu}$ by smooth functions $\rho^\varepsilon, \mu^\varepsilon \in C^\infty(\overline{Q})$, show existence for the corresponding system using an abstract result for general evolution equations in [2], deduce uniform a priori estimates, and perform the limit process as $\varepsilon \searrow 0$. Details can be found in [29, Section 4].

About the variational inequality, we fix $v \in \mathcal{U}_{\rm ad}$ and choose $h = v - \overline{u}$. Then, we write the linearized system (2.47)–(2.50) and multiply the equations (2.47) and (2.48) by p and q, respectively. At the same time, we consider the adjoint system and multiply the equations (2.58) and (2.59) by $-\eta$ and $-\xi$, respectively. Then, we add all the equalities obtained in this way and integrate over Q. Many terms cancel out, in particular, the contributions given by the Laplace operators, due to the boundary conditions (2.49) and (2.60), as well as the terms involving $DB[\overline{\rho}]$ and $DB[\overline{\rho}]^*$, by the definition of adjoint operator (see (2.62)).. What remains is:

$$\begin{aligned}
&\int_0^T \int_\Omega \big(2g'(\overline{\rho})\overline{\rho}_t\, \eta\, p + (1 + 2g(\overline{\rho}))\, \eta_t\, p + (1 + 2g(\overline{\rho}))\, \eta\, p_t\big)\, dx\, dt \\
&\quad + \int_0^T \int_\Omega \big(\overline{\mu}_t\, g'(\overline{\rho})\, \xi\, p + \overline{\mu}\, g''(\overline{\rho})\overline{\rho}_t\, \xi\, p + \overline{\mu}\, g'(\overline{\rho})\, \xi_t\, p + \overline{\mu}\, g'(\overline{\rho})\, \xi\, p_t\big)\, dx\, dt \\
&\quad + \int_0^T \int_\Omega (\xi_t\, q + \xi\, q_t)\, dx\, dt \\
&= \int_0^T \int_\Omega \big((v - \overline{u})\, p - \beta_2(\overline{\mu} - \mu_Q)\eta - \beta_1(\overline{\rho} - \rho_Q)\xi\big)\, dx\, dt\,.
\end{aligned}$$

Note that the expression on the left-hand side reduces to

$$\int_0^T \int_\Omega \partial_t \big\{(1 + 2g(\overline{\rho}))\, \eta\, p + \overline{\mu}\, g'(\overline{\rho})\, \xi\, p + \xi q\big\}\, dx\, dt\,,$$

and, consequently, it vanishes, due to the initial conditions (2.50) and final conditions (2.61). This entails that

$$\int_0^T \int_\Omega \big(\beta_1(\overline{\rho} - \rho_Q)\eta + \beta_2(\overline{\mu} - \mu_Q)\xi\big)\, dx\, dt = \int_0^T \int_\Omega (v - \overline{u})\, p\, dx\, dt\,,$$

and (2.65) becomes an easy consequence of (2.57). □

Remark 4 The variational inequality (2.65), along with the state system (1.24)–(1.27) and the adjoint system (2.58)–(2.61), provides the first-order necessary optimality conditions for the control problem **(CP)**. Notice that in the case $\beta_3 > 0$ the function $\overline{u}$ is nothing but the $L^2(Q)$-orthogonal projection of $-\beta_3^{-1} p$ onto $\mathcal{U}_{\rm ad}$.

2.6 The double-obstacle case

We refer to Subsections 1.7 and 1.8 for a general discussion about this case and for the notation. In addition, we recall that here the system under investigation can be stated as (2.19)–(2.22), with $\beta = \partial I_{[0,1]}$ and $\pi = F_2'$ with the regularity $F_2 \in C^3([0,1])$ as in (2.26).

By our approach (see [30] for the full detail), the result stated below is completely proved; we apologize for the long statement, but it collects all things and allows the reader to understand the full construction in a synthetic way.

Theorem 2.9 *Assume* $\beta = \partial I_{[0,1]}$, $\pi = F_2' \in C^2([0,1])$, (1.33), (1.34), *and* (2.27)–(2.37). *Let* $\bar{u} \in \mathcal{U}_{\text{ad}}$ *be an optimal control for* $(\mathbf{CP}_0)$ *with the associated solution* $(\bar{\mu}, \bar{\rho}, \bar{\xi})$ *to the state system* (2.19)–(2.22).

Moreover, let $\{\alpha_n\} \subset (0,1]$, *with* $\alpha_n \searrow 0$ *as* $n \to \infty$, *be such that there are optimal pairs* $((\bar{\mu}^{\alpha_n}, \bar{\rho}^{\alpha_n}), \bar{u}^{\alpha_n})$ *for the adapted problem* $(\widetilde{\mathbf{CP}}_{\alpha_n})$ *satisfying*

$$u^{\alpha_n} \to \bar{u} \quad \text{strongly in } L^2(Q), \tag{2.66}$$

$$\mu^{\alpha_n} \to \bar{\mu} \quad \text{weakly-star in } L^\infty(Q) \cap H^1(0,T;H) \cap L^2(0,T;W), \tag{2.67}$$

$$\rho^{\alpha_n} \to \bar{\rho} \quad \text{weakly-star in } L^\infty(Q) \cap H^1(0,T;H) \cap L^\infty(0,T;V), \tag{2.68}$$

$$\rho_t^{\alpha_n} \to \bar{\rho}_t \quad \text{weakly in } L^6(Q), \tag{2.69}$$

$$\varphi(\alpha_n) h'(\rho^{\alpha_n}) \to \bar{\xi} \quad \text{weakly in } L^6(Q), \tag{2.70}$$

$$\widetilde{J}(u^{\alpha_n}, (\mu^{\alpha_n}, \rho^{\alpha_n})) \to J(\bar{u}, (\bar{\mu}, \bar{\rho})), \tag{2.71}$$

and having the associated adjoint variables $\{(p^{\alpha_n}, q^{\alpha_n})\}$. *Then, for any subsequence* $\{n_k\}_{k \in \mathbb{N}}$ *of* $\mathbb{N}$, *there are a subsequence* $\{n_{k_\ell}\}_{\ell \in \mathbb{N}}$ *and some triple* (p, q, λ) *such that*

- $p \in H^1(0,T;H) \cap L^\infty(0,T;V) \cap L^2(0,T;W)$, $q \in L^\infty(0,T;H)$, *and* $\lambda \in Y'$, *where*

$$Y := \{v \in H^1(0,T;H) : v(0) = 0\};$$

- *the relations*

$$p^{\alpha_n} \to p \quad \text{weakly-star in } H^1(0,T;H) \cap L^\infty(0,T;V) \cap L^2(0,T;W), \tag{2.72}$$

$$q^{\alpha_n} \to q \quad \text{weakly-star in } L^\infty(0,T;H), \tag{2.73}$$

$$\lambda^{\alpha_n} := \varphi(\alpha_n)\, h''(\bar{\rho}^{\alpha_n})\, q^{\alpha_n} \to \lambda \quad \text{weakly in } Y', \tag{2.74}$$

$$\liminf_{n\to\infty} \int_0^T \int_\Omega \lambda^{\alpha_n} q^{\alpha_n}\, dx\, dt = \liminf_{n\to\infty} \int_0^T \int_\Omega \varphi(\alpha_n)\, h''(\bar{\rho}^{\alpha_n})\, |q^{\alpha_n}|^2\, dx\, dt \ge 0, \tag{2.75}$$

$$\lim_{n\to\infty} \int_0^T \int_\Omega \lambda^{\alpha_n}\, \bar{\rho}^{\alpha_n}(1 - \bar{\rho}^{\alpha_n})\, \phi\, dx\, dt = \lim_{n\to\infty} \int_0^T \int_\Omega \varphi(\alpha_n)\, q^{\alpha_n}\, \phi\, dx\, dt = 0 \tag{2.76}$$

are valid with the sequences indexed by n_{k_ℓ} *and the limits taken as* $\ell \to \infty$*;*

- *the variational inequality* (2.65) *and the adjoint system equations*

$$
\begin{aligned}
&-(1+2g(\bar\rho))\,p_t - g'(\bar\rho)\,\bar\rho_t\,p - \Delta p - g'(\bar\rho)\,q \\
&= \beta_2(\bar\mu - \mu_Q) \quad \text{a.e. in } Q,
\end{aligned}
\tag{2.77}
$$

$$
\partial_{\mathbf n} p = 0 \quad \text{a.e. on } \Sigma, \quad p(T) = 0 \quad \text{a.e. in } \Omega,
\tag{2.78}
$$

$$
\begin{aligned}
&\langle \lambda, v\rangle_Y + \int_0^T\!\!\int_\Omega q\,v_t\,dx\,dt + \int_0^T\!\!\int_\Omega F''(\bar\rho)\,q\,v\,dx\,dt \\
&- \int_0^T\!\!\int_\Omega \bar\mu\,g''(\bar\rho)\,q\,v\,dx\,dt + \int_0^T\!\!\int_\Omega g'(\bar\rho)\,\bar\mu_t\,p\,v\,dx\,dt \\
&= \int_0^T\!\!\int_\Omega g'(\bar\rho)\,\bar\mu\,p_t\,v\,dx\,dt - \int_0^T\!\!\int_\Omega DB[\bar\rho]^*(q)\,v\,dx\,dt \\
&\quad + \int_0^T\!\!\int_\Omega \beta_1(\bar\rho - \rho_Q)\,v\,dx\,dt \qquad \forall\, v \in Y
\end{aligned}
\tag{2.79}
$$

are satisfied.

Remark 5 We are unable to show that the limit triple (p, q, λ) solving (2.77)–(2.79) is uniquely determined for a fixed optimal pair $((\bar\mu, \bar\rho), \bar u)$. Therefore, it may well happen that the limits differ for different subsequences. However, it follows from the variational inequality (2.65) that, for any such limit, it holds with the orthogonal projection $\mathbb{P}_{\mathcal{U}_{\rm ad}}$ onto $\mathcal{U}_{\rm ad}$ with respect to the standard inner product in $L^2(Q)$ that in the case $\beta_3 > 0$ we have $\bar u = \mathbb{P}_{\mathcal{U}_{\rm ad}}\left(-\beta_3^{-1} p\right)$.

Acknowledgments

This research was supported by the Italian Ministry of Education, University and Research (MIUR): Dipartimenti di Eccellenza Program (2018–2022) – Dept. of Mathematics “F. Casorati”, University of Pavia. In addition, PC and GG gratefully acknowledge some financial support from the MIUR-PRIN Grant 2015PA5MP7 “Calculus of Variations”, the GNAMPA (Gruppo Nazionale per l’Analisi Matematica, la Probabilità e le loro Applicazioni) of INdAM (Istituto Nazionale di Alta Matematica) and the IMATI – C.N.R. Pavia.

References

1. H. Abels, S. Bosia, M. Grasselli, *Cahn–Hilliard equation with nonlocal singular free energies*, Ann. Mat. Pura Appl. (4) **194** (2015), 1071–1106.
2. C. Baiocchi, *Sulle equazioni differenziali astratte lineari del primo e del secondo ordine negli spazi di Hilbert*, Ann. Mat Pura Appl. (4) **76** (1967), 233–304.

3. V. Barbu, *Necessary conditions for nonconvex distributed control problems governed by elliptic variational inequalities*, J. Math. Anal. Appl. **80** (1981), 566–597.
4. V. Barbu, M.L. Bernardi, P. Colli, G. Gilardi, *Optimal control problems of phase relaxation models*, J. Optim. Theory Appl. **109** (2001), 557–585.
5. P.W. Bates, J. Han, *The Neumann boundary problem for a nonlocal Cahn–Hilliard equation*, J. Differential Equations **212** (2005), 235–277.
6. P.W. Bates, J. Han, *The Dirichlet boundary problem for a nonlocal Cahn–Hilliard equation*, J. Math. Anal. Appl. **311** (2005), 289–312.
7. M. Brokate, J. Sprekels, *"Hysteresis and phase transitions"*, Applied Mathematical Sciences **121**, Springer, New York, 1996.
8. J. W. Cahn, J. E. Hilliard, *Free energy of a nonuniform system I. Interfacial free energy*, J. Chem. Phys. **2** (1958), 258–267.
9. C.K. Chen, P.C. Fife, *Nonlocal models of phase transitions in solids*, Adv. Math. Sci. Appl. **10** (2000), 821–849.
10. P. Colli, M.H. Farshbaf-Shaker, G. Gilardi, J. Sprekels, *Optimal boundary control of a viscous Cahn–Hilliard system with dynamic boundary condition and double obstacle potentials*, SIAM J. Control Optim. **53** (2015), 2696–2721.
11. P. Colli, M.H. Farshbaf-Shaker, G. Gilardi, J. Sprekels, *Second-order analysis of a boundary control problem for the viscous Cahn–Hilliard equation with dynamic boundary conditions*, Ann. Acad. Rom. Sci. Math. Appl. **7** (2015), 41–66.
12. P. Colli, M.H. Farshbaf-Shaker, J. Sprekels, *A deep quench approach to the optimal control of an Allen–Cahn equation with dynamic boundary conditions and double obstacles*, Appl. Math. Optim. **71** (2015), 1–24.
13. P. Colli, S. Frigeri, M. Grasselli, Global existence of weak solutions to a nonlocal Cahn–Hilliard–Navier–Stokes system, *J. Math. Anal. Appl.* 386 (2012) 428-444.
14. P. Colli, G. Gilardi, P. Krejčí, P. Podio-Guidugli, J. Sprekels, *Analysis of a time discretization scheme for a nonstandard viscous Cahn–Hilliard system*, ESAIM Math. Model. Numer. Anal. **48** (2014), 1061–1087.
15. P. Colli, G. Gilardi, P. Krejčí, J. Sprekels, *A vanishing diffusion limit in a nonstandard system of phase field equations*, Evol. Equ. Control Theory **3** (2014), 257–275.
16. P. Colli, G. Gilardi, P. Krejčí, J. Sprekels, *A continuous dependence result for a nonstandard system of phase field equations*, Math. Methods Appl. Sci. **37** (2014), 1318–1324.
17. P. Colli, G. Gilardi, P. Podio-Guidugli, J. Sprekels, *Well-posedness and long-time behavior for a nonstandard viscous Cahn–Hilliard system*, SIAM J. Appl. Math. **71** (2011), 1849–1870.
18. P. Colli, G. Gilardi, P. Podio-Guidugli, J. Sprekels, *Distributed optimal control of a nonstandard system of phase field equations*, Contin. Mech. Thermodyn. **24** (2012), 437–459.
19. P. Colli, G. Gilardi, P. Podio-Guidugli, J. Sprekels, *Global existence for a strongly coupled Cahn–Hilliard system with viscosity*, Boll. Unione Mat. Ital. (9) **5** (2012), 495–513.
20. P. Colli, G. Gilardi, P. Podio-Guidugli, J. Sprekels, *Continuous dependence for a nonstandard Cahn–Hilliard system with nonlinear atom mobility*, Rend. Sem. Mat. Univ. Politec. Torino **70** (2012), 27–52.
21. P. Colli, G. Gilardi, P. Podio-Guidugli, J. Sprekels, *An asymptotic analysis for a nonstandard Cahn–Hilliard system with viscosity*, Discrete Contin. Dyn. Syst. Ser. S **6** (2013), 353–368.
22. P. Colli, G. Gilardi, P. Podio-Guidugli, J. Sprekels, *Global existence and uniqueness for a singular/degenerate Cahn–Hilliard system with viscosity*, J. Differential Equations **254** (2013), 4217–4244.
23. P. Colli, G. Gilardi, E. Rocca, J. Sprekels, *Optimal distributed control of a diffuse interface model of tumor growth*, Nonlinearity **30** (2017), 2518–2546.
24. P. Colli, G. Gilardi, J. Sprekels, *Analysis and optimal boundary control of a nonstandard system of phase field equations*, Milan J. Math. **80** (2012), 119–149.
25. P. Colli, G. Gilardi, J. Sprekels, *Regularity of the solution to a nonstandard system of phase field equations*, Rend. Cl. Sci. Mat. Nat. **147** (2013), 3–19.
26. P. Colli, G. Gilardi, J. Sprekels, *A boundary control problem for the pure Cahn–Hilliard equation with dynamic boundary conditions*, Adv. Nonlinear Anal. **4** (2015), 311–325.

27. P. Colli, G. Gilardi, J. Sprekels, *A boundary control problem for the viscous Cahn–Hilliard equation with dynamic boundary conditions*, Appl. Math. Optim. **73** (2016), 195–225.
28. P. Colli, G. Gilardi, J. Sprekels, *On an application of Tikhonov's fixed point theorem to a nonlocal Cahn–Hilliard type system modeling phase separation*, J. Differential Equations **260** (2016), 7940–7964.
29. P. Colli, G. Gilardi, J. Sprekels, *Distributed optimal control of a nonstandard nonlocal phase field system*, AIMS Mathematics **1** (2016), 225–260.
30. P. Colli, G. Gilardi, J. Sprekels, *Distributed optimal control of a nonstandard nonlocal phase field system with double obstacle potential*, Evol. Equ. Control Theory **6** (2017), 35–58..
31. P. Colli, G. Gilardi, J. Sprekels, *Optimal velocity control of a viscous Cahn–Hilliard system with convection and dynamic boundary conditions*, SIAM J. Control Optim. **56** (2018), 1665–1691.
32. P. Colli, G. Gilardi, J. Sprekels, *Optimal velocity control of a convective Cahn–Hilliard system with double obstacles and dynamic boundary conditions: a 'deep quench' approach*, J. Convex Anal., to appear (2019) (see also preprint arXiv:1709.03892 [math.AP] (2017), pp. 1–30).
33. P. Colli, P. Krejčí, E. Rocca, J. Sprekels, *Nonlinear evolution inclusions arising from phase change models*, Czechoslovak Math. J. **57** (2007), 1067–1098.
34. C.M. Elliott, S. Zheng, *On the Cahn–Hilliard equation*, Arch.. Rational Mech. Anal. **96** (1986), 339–357.
35. E. Fried, M.E. Gurtin, *Continuum theory of thermally induced phase transitions based on an order parameter*, Phys. D **68** (1993), 326–343.
36. S. Frigeri, M. Grasselli, *Nonlocal Cahn–Hilliard–Navier–Stokes systems with singular potentials*, Dyn. Partial Differ. Equ. **9** (2012), 273–304.
37. S. Frigeri, M. Grasselli, J. Sprekels, *Optimal distributed control of two-dimensional nonlocal Cahn–Hilliard–Navier–Stokes systems with degenerate mobility and singular potential*, Appl. Math. Optim., https://doi.org/10.1007/s00245-018-9524-7 (see also preprint arXiv:1801.02502 [math.AP] (2018), pp. 1–32).
38. S. Frigeri, E. Rocca, J. Sprekels, *Optimal distributed control of a nonlocal Cahn–Hilliard/Navier–Stokes system in two dimensions*, SIAM J. Control Optim. **54** (2016), 221–250.
39. T. Fukao and N. Yamazaki, *A boundary control problem for the equation and dynamic boundary condition of Cahn–Hilliard type, in "Solvability, Regularity, Optimal Control of Boundary Value Problems for PDEs"*, P. Colli, A. Favini, E. Rocca, G. Schimperna, J. Sprekels (ed.), Springer INdAM Series **22**, Springer, Milan, 2017, pp. 255–280.
40. H. Gajewski, *On a nonlocal model of non-isothermal phase separation*, Adv. Math. Sci. Appl. **12** (2002), 569–586.
41. H. Gajewski, J.A. Griepentrog, *A descent method for the free energy of multicomponent systems*, Discrete Contin. Dyn. Syst. 15 (2006), 505–528.
42. H. Gajewski, K. Zacharias, *On a nonlocal phase separation model*, J. Math. Anal. Appl. **286** (2003), 11–31.
43. C.G. Gal, *Doubly nonlocal Cahn–Hilliard equations*, Ann. Inst. H. Poincaré Anal. Non Linéaire **35** (2018), 357–392.
44. C.G. Gal, A. Giorgini, M. Grasselli, *The nonlocal Cahn–Hilliard equation with singular potential: well-posedness, regularity and strict separation property*, J. Differential Equations **263** (2017), 5253–5297.
45. C.G. Gal, M. Grasselli, *Longtime behavior of nonlocal Cahn–Hilliard equations*, Discrete Contin. Dyn. Syst. **34** (2014), 145–179.
46. G. Giacomin, J.L. Lebowitz, *Phase segregation dynamics in particle systems with long range interactions. I. Macroscopic limits*, J. Statist. Phys. **87** (1997), 37–61.
47. G. Giacomin, J.L. Lebowitz, *Phase segregation dynamics in particle systems with long range interactions. II. Phase motion*, SIAM J. Appl. Math. **58** (1998), 1707–1729.
48. M.E. Gurtin, *Generalized Ginzburg–Landau and Cahn–Hilliard equations based on a microforce balance*, Phys. D **92** (1996), 178–192.
49. J. Han, *The Cauchy problem and steady state solutions for a nonlocal Cahn–Hilliard equation*, Electron. J. Differential Equations **113** (2004), 9 pp.

50. M. Heida, *Existence of solutions for two types of generalized versions of the Cahn–Hilliard equation*, Appl. Math. **60** (2015), 51–90.
51. M. Hintermüller, T. Keil, D. Wegner, *Optimal control of a semidiscrete Cahn–Hilliard–Navier–Stokes system with non-matched fluid densities*, SIAM J. Control Optim. **55** (2017), 1954–1989.
52. M. Hintermüller, D. Wegner, *Distributed optimal control of the Cahn–Hilliard system including the case of a double-obstacle homogeneous free energy density*, SIAM J. Control Optim. **50** (2012), 388–418.
53. M. Hintermüller, D. Wegner, *Optimal control of a semidiscrete Cahn–Hilliard–Navier–Stokes system*, SIAM J. Control Optim. **52** (2014), 747–772..
54. M. Hintermüller, D. Wegner, *Distributed and boundary control problems for the semidiscrete Cahn–Hilliard/Navier–Stokes system with nonsmooth Ginzburg–Landau energies*, Isaac Newton Institute Preprint Series No. NI14042-FRB (2014), 1–29.
55. J.L. Lions, *"Quelques méthodes de résolution des problèmes aux limites non linéaires"*, Dunod Gauthier-Villars, Paris, 1969.
56. S..-O. Londen, H. Petzeltová, *Convergence of solutions of a non-local phase-field system*, Discrete Contin. Dyn. Syst. Ser. S **4** (2011), 653–670.
57. A. Novick-Cohen, *On the viscous Cahn–Hilliard equation, in "Material instabilities in continuum mechanics" (Edinburgh, 1985–1986)*, Oxford Sci. Publ., Oxford Univ. Press, New York, 1988, pp. 329–342.
58. P. Podio-Guidugli, *Models of phase segregation and diffusion of atomic species on a lattice*, Ric. Mat. **55** (2006), 105–118.
59. E. Rocca, J. Sprekels, *Optimal distributed control of a nonlocal convective Cahn–Hilliard equation by the velocity in three dimensions*, SIAM J. Control Optim. **53** (2015), 1654–1680.
60. J. Simon, *Compact sets in the space $L^p(0,T;B)$*, Ann. Mat. Pura Appl. (4) **146** (1987), 65–96.
61. Q.-F. Wang, S.-i. Nakagiri, *Weak solutions of Cahn–Hilliard equations having forcing terms and optimal control problems, Mathematical models in functional equations (Japanese) (Kyoto, 1999)*, Sūrikaisekikenkyūsho Kōkyūroku No. 1128 (2000), pp. 172–180.
62. X. Zhao, C. Liu, *Optimal control of the convective Cahn–Hilliard equation*, Appl. Anal. **92** (2013), 1028–1045.
63. X. Zhao, C. Liu, *Optimal control for the convective Cahn–Hilliard equation in 2D case*, Appl. Math. Optim. **70** (2014), 61–82.

Invariant and Quasi-invariant Measures for Equations in Hydrodynamics

Ana Bela Cruzeiro and Alexandra Symeonides

1 Introduction

Statistical solutions for differential equations, as opposed to pointwise classical ones, are solutions where the initial data is taken in a measure one set in some probability space. The statistical approach to differential equations was partly motivated by A.N. Kolmogorov's ideas on turbulence ([28]), a subject where the interest relies on computation of mean quantities, and is suitable for situations where initial conditions are not precisely known. The use of probability theory to describe the concept of ensemble average, as far as Hydrodynamical equations are concerned, can be traced back to the work of E. Hopf [23]. Later on statistical solutions were introduced by C. Foias [20, 21] and studied also by M.I. Vishik and A.V. Fursikov [33].

Statistical solutions involve probability measures on (infinite-dimensional) spaces of functions or of distributions. They can be time dependent or stationary in time; this last case applies to statistical equilibrium regimes and correspond to invariant probability measures. Invariant measures of Gibbsian type for hydrodynamical systems, with Gibbs density determined by the invariants of the underlying motion, have been discussed in [5], [3] or [9] for example.

The search for invariant measures is not only important for physical reasons. Mathematically, they provide a particular probability space where analysis can be performed. They allow, for example, to extend local to global solutions of the corresponding equations of motion (as it is done in [10], for example). In the same spirit Gibbs measures have been very successful in the construction of global solutions to dispersive equations of low regularity. There is an intensive activity in this area; we refer to the works of N. Tzvetkov and collaborators, for example (c.f. [11]).

Ana Bela Cruzeiro
GFMUL and Dep. Mat. Instituto Superior Técnico Univ. Lisboa, Av. Rovisco Pais 1049-001 Lisboa, Portugal e-mail: abcruz@math.ist.utl.pt
Alexandra Symeonides
GFMUL and Dep. Mat. Faculdade de Ciências Univ. Lisboa, Campo Grande, Ed. C6, 1749-016 Lisboa, Portugal e-mail: asymeonides@fc.ul.pt

M. Hintermüller, J. F. Rodrigues (eds.), *Topics in Applied Analysis and Optimisation*, CIM Series in Mathematical Sciences, https://doi.org/10.1007/978-3-030-33116-0_4

Our main purpose is to show how invariant Gaussian measures are helpful to prove existence of solutions for some equations in Hydrodynamics starting (almost everywhere) from their support. The corresponding functional spaces, endowed with probability measures that we know a priori to be invariant or quasi-invariant for the equations or, more precisely, that we know to be infinitesimally invariant, are, in some sense, similar to the Wiener space (the space of the Brownian motion). They are particular examples of what was defined and named in [22] abstract Wiener spaces.

One can consider on an abstract Wiener space a differential calculus, the Malliavin calculus, c.f. [30], which is adapted to deal with non regular functionals. Once formulated in functional spaces, some partial differential equations such as the ones we consider here become ordinary differential equations. Flows for ordinary differential equations in context of Wiener spaces were first studied in [14], using Malliavin calculus techniques. We describe the general methods and show how they can be applied to construct flows for some equations in Hydrodynamics.

In this work we shall consider the standard incompressible Euler equation, a modified Euler equation and also the averaged-Euler ones, all in the two dimensional space.

In Malliavin calculus it is possible, under sufficient regularity assumptions, to consider surfaces (of finite corank) of the corresponding Wiener spaces. Then surface measures can be defined. In our work, where the Gaussian measures are associated to invariant quantities of the motion, it is of interest to study surfaces which correspond to level sets of some other invariant quantities. We shall describe this construction in the more regular case of the averaged-Euler equations.

2 The two-dimensional Euler equation (periodic case)

2.1 Formulation of the equation

Let us consider the two-dimensional incompressible Euler equation with periodic boundary conditions, namely

$$\partial_t u + (u.\nabla)u = \nabla p, \qquad \text{div}\, u(t,\cdot) = 0, \tag{2.1}$$

where $u = u(t,x), x \in \mathbb{T}^2 = [0,2\pi] \times [0,2\pi]$, the two-dimensional flat torus, $(u.\nabla) = u^1\partial_1 + u^2\partial_2$. This equation models the velocity $u : \mathbb{R}^+ \times \mathbb{T}^2 \to \mathbb{T}^2$ of incompressible non viscous fluids. The function p represents the pressure and is also an unknown of the system. The first equation above corresponds to Newton's second law and the second equation is the incompressibility condition.

It is known that solutions do not blow up starting from smooth data with finite kinetic energy (T. Kato ([27]), C. Bardos ([7]) among others). Local existence of smooth solutions dates back from L. Lichtenstein ([29]). In two dimensional bounded domains existence, uniqueness and global regularity of solutions with bounded initial vorticity was shown (V.I. Judovič [25]); these results were extended, in the framework of weak solutions, to the case where the initial vorticity is in L^p, with $p > 1$ and also for $p = 1$, when the vorticity is some finite measure. There is an extensive literature about this equation.

Incompressibility implies the existence of a function φ such that $u = (-\partial_2\varphi, \partial_1\varphi)$. We use the notation $\nabla^\perp\varphi = (-\partial_2\varphi, \partial_1\varphi)$. Replacing this expression in Euler equation and applying the operator rot $v = -\partial_2 v^1 + \partial_1 v^2$ to both members, we derive the following formulation of Euler equation in terms of φ

$$\partial_t(\Delta\varphi) = -\nabla^\perp\varphi.\nabla\Delta\varphi \tag{2.2}$$

where Δ is the Laplacian; this equation is to be considered under the periodic boundary conditions

$$\varphi(t, 0, y) = \varphi(t, 2\pi, y), \qquad \varphi(t, x, 0) = \varphi(t, x, 2\pi).$$

Let us consider the functions $e_k(x) = \frac{1}{2\pi}e^{ik.x}$ for $k \in \mathbb{Z}^2$, $k.x = k_1x_1 + k_2x_2$. They form a complete set of orthonormal functions in $L^2(\mathbb{T})$ which are eigenfunctions of the Laplacian operator, having $-k^2 = -(k_1^2 + k_2^2)$ for eigenvalues. We shall write $|k| = \sqrt{k_1^2 + k_2^2}$, but, with a slight abuse of notation, k^β instead of $|k|^\beta$ when β is even.

We write the decomposition $\varphi(t, x) = \sum_{k>0}\varphi_k(t)e_k(x)$, where $k > 0$ means $k_1 > 0$ or $k_1 = 0$ and $k_2 > 0$, a condition which may be assumed since we deal with real-valued functions and that we can consider, without losing generality, to have zero mean. Then equation (2.2) can be written in the following form,

$$\frac{d}{dt}\varphi_k = B_k(\varphi) \tag{2.3}$$

for

$$B_k(\varphi) = \frac{1}{2\pi}\sum_l \left[\frac{1}{k^2}(l^\perp.k)(l.k) - \frac{1}{2}(l^\perp.k)\right]\varphi_l\varphi_{k-l}, \tag{2.4}$$

and where $k^\perp = (-k_2, k_1)$.

In a more condensed way, if $B = \sum_k B_k$, we can write

$$\frac{d}{dt}\varphi(t, \cdot) = B(\varphi(t, \cdot)). \tag{2.5}$$

Notice the quadratic nature of this (infinite-dimensional) equation.

The explicit Fourier expression (2.4) can be shown directly. In a more geometric approach, it can also be derived using the remarkable fact that Euler equation corresponds to a geodesic one on the space of diffeomorphisms (of the torus, in this case) with respect the L^2 metric. The geodesic nature of the Euler flow was shown by V. Arnold in [6]. Euler equation can be therefore written as

$$\frac{d}{dt}u_k = -\sum_{i,j}\Gamma^k_{i,j}u_iu_j,$$

where $\Gamma^k_{i,j}$ are the Christoffel symbols for the Levi-Civita connection associated with the metric. The explicit expression of these Christoffel symbols can be found in [15], for example, and from them one can derive Euler equation written in Fourier modes.

2.2 Invariant quantities and Gibbs measures

Among the well known invariant quantities for the Euler equation we have the energy, namely

$$E = \frac{1}{2}\int_{\mathbb{T}^2} u^2 dx = -\int_{\mathbb{T}^2} \varphi\Delta\varphi dx$$

and the enstrophy, that will play here a special rôle:

$$S = \frac{1}{2}\int_{\mathbb{T}^2} (\text{rot } u)^2 dx = \frac{1}{2}\int_{\mathbb{T}^2} (\Delta\varphi)^2 dx.$$

For $\gamma > 0$ let μ^k_γ be the Gaussian probability measures on $\mathbb{C}$

$$d\mu^k_\gamma(z) = \frac{\gamma k^4}{2\pi} \exp\Big(-\frac{1}{2}\gamma k^4 |z|^2\Big) dxdy$$

where $z = x + iy$, and consider

$$d\mu_\gamma(\varphi) = \Pi_{k>0}\, d\mu^k_\gamma(\varphi_k). \tag{2.6}$$

These a priori formal measures can be realised as true probability measures on the Sobolev function spaces

$$H^\beta(\mathbb{T}^2) = \{\varphi = \sum_{k>0} \varphi_k e_k : \sum_k k^{2\beta} |\varphi_k|^2 < +\infty\}$$

for $\beta < 1$. Indeed, we have

$$\int \sum_{k>0} k^{2\beta} |\varphi_k|^2 d\mu_\gamma(\varphi) = \sum_{k>0} \int k^{2\beta} |\varphi_k|^2 d\mu^k_\gamma(\varphi_k) = \frac{2}{\gamma}\sum_{k>0} \frac{1}{k^{2(2-\beta)}},$$

which is finite for $\beta < 1$. The space $H^\beta(\mathbb{T}^2)$ is a Hilbert space with inner product given by $< \varphi, \psi >_\beta = \sum_{k>0} k^{2\beta} \varphi_k \psi_k$. Moreover it coincides with the Sobolev space

$$\{\varphi : \mathbb{T}^2 \to \mathbb{R} : (I - \Delta)^{\frac{\beta}{2}} \varphi \in L^2(\mathbb{T}^2)\}.$$

For negative values of the exponent β, $(I - \Delta)^{\frac{\beta}{2}}$ is considered as a pseudo-differential operator.

The measures μ_γ are in fact Gibbs-type measures associated with the enstrophy. We can regard them as

$$d\mu_\gamma = \frac{1}{Z} e^{-\gamma S(\varphi)} \mathrm{d}(\varphi),$$

where Z is a normalizing constant and $\mathrm{d}(\varphi) = \Pi_{k>0} d\varphi_k$. This is of course a formal expression, although very intuitive, since d is not a well-defined measure (nor Z could be a finite constant).

We remark that the Gaussian measures μ_γ can also be realised as laws of some Wiener stochastic processes (c.f. [17]).

2.3 The vorticity flow

For $\beta < 1$ the triple $(H^\beta(\mathbb{T}^2), H^2(\mathbb{T}^2), \mu_\gamma)$ constitutes an abstract Wiener space in the sense of Gross [22], with measurable norm $\|\varphi\|_\beta^2 =< \varphi, \varphi >_\beta = \sum_{k>0} k^{2\beta} \varphi_k^2$. This means that H^2 is densely embedded in H^β and the inclusion radonifies the canonical Gaussian cylinder measure on H^2 with covariance given by the $\|\cdot\|_2$ norm.

In such spaces one can consider the (infinite dimensional) Malliavin calculus ([30]), in which H^2 plays the rôle of the tangent space to H^β and is called the Cameron-Martin space. Notice that H^2 is dense in H^β and that $\mu_\gamma(H^2) = 0$. In particular, for a functional F defined in H^β and $h \in H^2$, the Malliavin derivative in the direction of h is defined by the μ_γ almost-everywhere limit

$$D_h F(\varphi) := \lim_{\epsilon \to 0} \frac{1}{\epsilon} [F(\varphi + \epsilon h) - F(\varphi)]$$

when such limit exists. The gradient operator is defined as

$$\nabla F(\varphi)(h) := D_h F(\varphi). \tag{2.7}$$

By Riesz representation theorem ∇F can be identified with an element of the Hilbert space H^2 and higher order derivatives are defined recurrently, in a similar way. In this framework, if $B : H^\beta \to H^2$, the divergence of B is the dual of the gradient, that we denote by $\delta_{\mu_\gamma} B$, and satisfies

$$\int_{H^\beta} G \, \delta_{\mu_\gamma} B d\mu_\gamma = \int_{H^\beta} < B, \nabla G >_{H^2} d\mu_\gamma \tag{2.8}$$

for every "test functional" G. Usually test functionals are chosen to be the cylindrical ones, namely those which only depend on a finite number of Fourier modes and constitute a dense subset of $L^2_{\mu_\gamma}(H^\beta)$. In this work we shall not be too specific on the chosen set of test functionals.

It is important to stress, since this will be used in the sequel, that in some cases we can define a divergence even if the vector field B does not take values in the Cameron-Martin space. The formula above has to be well-defined for every test functional G and the corresponding divergence operator has to be closable (say, in some L^p_μ). This is obviously the case when $\int_{H^\beta} < B, \nabla G >_{H^2} d\mu_\gamma = 0$ for every G.

Coming back to the vorticity equation (2.5) (or (2.3)), and with the definitions above, one can prove the following statements:

Theorem 2.1 *(regularity of B)* If $\beta < -1$ the functional B belongs to $L^p_{\mu_\gamma}(H^\beta; H^\beta)$ for all $p \geq 1$.

Theorem 2.2 The divergence of B with respect to μ_γ, $\delta_{\mu_\gamma} B$, is equal to zero.

The second result means that μ_γ is infinitesimally invariant for B, namely that

$$\int_{H^\beta} D_h B \, d\mu_\gamma = 0$$

for all $h \in H^2$.

If we can solve the equation for the vorticity, namely if there exists a flow φ_t solving (2.5) with $\varphi_0 = \varphi$, this also means that $\frac{d}{dt}\big|_{t=0} \int F(\varphi_t) d\mu_\gamma(\varphi) = 0$ for every test functional F. Actually, the weaker property given by Theorem 2.2, combined with the regularity results for B, allows to prove existence of the integral flow.

We refer to [2] for the proofs. They rely upon the explicit expressions of the probability measures μ_γ and of the vorticity vector field B. Remark that, since the measures are Gaussian with covariance given by the $H^2(\mathbb{T}^2)$ norm, we formally expect the divergence to be

$$\delta_{\mu_\gamma} B(\varphi) = -\operatorname{div} B(\varphi) + < B(\varphi), \varphi >_2$$

where div would denote a standard divergence with respect to the flat "measure" d. It turns out that, indeed, both of those terms are meaningful and equal zero (the second one as a result of the invariance of the enstrophy, the first one because B_k does not depend on φ_k.

In order to prove the existence of a solution to (2.5) we proceed by defining Galerkin approximations. More precisely we consider the finite dimensional approximation of $B = \sum_{k>0} B_k e_k$, obtained by truncating (2.4) after a finite number of Fourier modes,

$$B^n_k(\varphi^n) = \frac{1}{2\pi} \sum_{0<l^2\leq n} \left[\frac{(l^\perp.k)(k.l)}{k^2} - \frac{h^\perp \cdot k}{2} \right] \varphi^n_l \varphi^n_{k-l}.$$

for $\varphi^n = \sum_{0<k^2\leq n} (\varphi.e_k) e_k$. Then, writing $B^n = \sum_{k>0} B^n_k e_k$, the corresponding finite-dimensional quadratic ordinary differential equations

$$\frac{d}{dt} \varphi^n_t = B^n(\varphi^n_t)$$

are globally (in time) well-posed. This is a consequence of their regularity and the conservation of energy.

For $\varphi \in H^\beta$ we write the decomposition $\varphi = \Pi_n \varphi + \Pi_n^\perp \varphi$, where Π_n is the orthogonal projection on the subspace spanned by the finite number of functions e_k on which B^n depends upon. Then extend the flows φ^n_t as

$$U_t^n(\varphi) = \varphi_t^n + \Pi_n^\perp \varphi$$

with $\varphi_0^n = \Pi_n \varphi_0$.

We can construct solutions of the periodic Euler equation starting μ_γ almost everywhere from functions of H^β, $\beta < -1$ and the corresponding flows keep the measure invariant. More precisely we have the following,

Theorem 2.3 Let $\beta < -1$. There exists a global flow U_t defined for $\varphi - \mu_\gamma$-almost everywhere on a probability space $(\Omega, \mathcal{F}, \mathbb{P})$ and with values in the space $H^\beta(\mathbb{T}^2)$ such that

(i) $U_t(\omega) = \varphi + \int_0^t B(U_s(\omega))ds \quad \mathbb{P}$ a.e. $\forall t \in \mathbb{R}$;

(ii) the measure μ_γ is invariant for the flow in the following sense:

$$\int_{H^\beta} f(U_t(\omega))d\mathbb{P}(\omega) = \int_{H^\beta} f(\varphi)d\mu_\gamma(\varphi) \quad \forall f \in C_b(H^\beta),\ \forall t \in \mathbb{R};$$

(iii) The flow is continuous from $H^\beta(\mathbb{T}^2)$ to $H^\beta(\mathbb{T}^2)$.

For the proof we first consider the sequence $U_t^n(\varphi)$ as a sequence of stochastic processes with laws defined on the space $C(\mathbb{R}; H^\beta(\mathbb{T}^2))$. Using the invariance of the measures μ_γ^n for these flows together with the uniform bound of B^n in $L^2_{\mu_\gamma}$, we show that their laws are tight in the weak topology of the space of probability measures over $C(\mathbb{R}; H^\beta(\mathbb{T}^2))$. This allows to extract a weakly convergence subsequence and a corresponding limit stochastic process by Skorohod's theorem. The limiting process will be defined in some probability space, not necessarily $H^\beta(\mathbb{T}^2)$.

The remaining properties follow from convergence arguments. We refer to [2] and [17] for the complete proofs of this Theorem.

Remark 1. Concerning uniqueness, we refer to the work [4]. There a certain operator is naturally associated with the vector field B and uniqueness is proved in the sense that such operator is essentially self-adjoint in $L^2_{\mu_\gamma}$.

Remark 2. The reader may wonder why one does not consider Gibbs measures associated with the energy instead of the enstrophy, which would be a more natural invariant quantity and would allow to have Euler solutions with finite energy. The reason is that the vector field B does not have enough regularity to prove existence of the corresponding solutions, at least with the methods we use. We point out that in [13] a sort of "projection" of this construction onto the (infinite dimensional) submanifold of renormalized-energy functions was studied. In particular, [13] defines other invariant measures which are not Gaussian. They are surface type measures (c.f. section 5 for a construction of invariant surface measures in the case of the averaged-Euler equations).

2.4 Lagrangian flows

The Lagrangian approach to Hydrodynamics studies positions of the fluid particles instead of their velocities. It concerns the integral flows g_t verifying

$$\partial_t g_t(x) = u(u, g_t(x)), g_0(x) = x.$$

Lagrangian flows, as mentioned before, can be interpreted as geodesics on the space of diffeomorphisms of the underlying configuration space, following [6].

An interesting question is whether there exist invariant measures for these flows. The answer is negative: this was proved in the case of the torus (in any dimension) in [16], revealing a striking difference between the Lagrangian and the Eulerian picture as far as ergodicity is concerned. Lack of compactness is due to energy dissipation from low modes to high modes in the Fourier decomposition and the non existence of an invariant probability measure for the Lagrangian flow can be heuristically explained by the passage of the energy to smaller and smaller scales during the time evolution. We refer to [16] for these results.

3 The two-dimensional Euler equation (non periodic case)

The construction of last section can be extended to the non periodic setting. For $\beta \in \mathbb{R}$ we consider the local Sobolev spaces $H^\beta_{loc}(\mathbb{R}^2)$,

$$H^\beta_{loc}(\mathbb{R}^2) := \{u : \forall K \subset \mathbb{R}^2 \text{ compact, } D^\beta u \in L^2(K)\}.$$

We may assume that the compact sets K are of type $K = [0, L] \times [0, L]$ for $L \in \mathbb{N}$. The spaces $H^\beta_{loc}(\mathbb{R}^2)$ are not normed spaces, however it is possible to equip them with the topology induced by the distances $d_{\beta,2}$ defined by

$$d_{\beta,2}(u, v) := \sum_L \frac{1}{2^L} \frac{\|D^\beta(u - v)\|_{L^2([0,L]^2)}}{1 + \|D^\beta(u - v)\|_{L^2([0,L]^2)}}. \tag{3.1}$$

These are complete spaces.

Naturally, the construction of last section can be performed on $[0, L] \times [0, L]$, suitably adapting the constants involved. In particular we use the orthonormal functions $e^L_k(x) = \frac{1}{L} e^{i\frac{2\pi}{L}k.x}$ and the enstrophy Gaussian measures $\mu_{\gamma,L}$ whose formal expression are

$$d\mu_{L,\gamma}(\varphi^L) := e^{-\frac{\gamma}{2}\int_{\mathbb{T}^2} |\Delta\varphi^L|^2 dx} \mathcal{D}\varphi^L,$$

$$\mathcal{D}\varphi^L = \Pi_{k>0}\gamma \left(\frac{2\pi k}{L}\right)^4 \frac{d\varphi^L_k}{2\pi}.$$

Then one can prove ([17]) that

$$\mu_{\gamma,L} \longrightarrow \sigma_\gamma$$

weakly in $H^\beta_{loc}(\mathbb{R}^2)$ when $L \longrightarrow \infty$ and, moreover, that

$$\text{supp}\,(\sigma_\gamma) = H^\beta_{loc}(\mathbb{R}^2), \quad \text{for } \beta < 1.$$

Notice that, for fixed L, the measure σ_γ coincides with $\mu_{\gamma,L}$ on functions restricted to $[0, L] \times [0, L]$. Accordingly, the vector fields B_L defined in the last section (with respect to $[0, L] \times [0, L]$), namely

$$B^L_k(\varphi^L) = \frac{1}{L}\left(\frac{2\pi}{L}\right)^2 \sum_{h>0, h\neq k} \left[\frac{(h^\perp \cdot k)(k \cdot h)}{k^2} - \frac{h^\perp \cdot k}{2}\right] \varphi^L_h \varphi^L_{k-h},$$

converge to $\tilde{B} \in L^p_{\mu_\gamma}(H^\beta_{loc}(\mathbb{R}^2); H^\beta_{loc}(\mathbb{R}^2))$ for all $p \geq 1$, when $\beta < -1$. Using these approximations we can construct a vorticity flow for the Euler equation in $\mathbb{R}^2$, i.e., we have the following,

Theorem 3.4 Let $\beta < -1$. There exists a global flow U_t defined for $\varphi - \sigma_\gamma$-almost everywhere on a probability space $(\Omega, \mathcal{F}, \mathcal{P})$ and with values in the space $H^\beta_{loc}(\mathbb{R}^2)$ such that

(i) $U_t(\omega) = \varphi + \int_0^t \tilde{B}(U_s(\omega))ds \quad \mathcal{P}$ a.e. $\forall t \in \mathbb{R}$;
(ii) the measure σ_γ is invariant for the flow in the following sense:

$$\int_{H^\beta_{loc}} f(U_t(\omega))d\mathcal{P}(\omega) = \int_{H^\beta_{loc}} f(\varphi)d\mu_\gamma(\varphi) \quad \forall f \in C_b(H^\beta_{loc}),\ \forall t \in \mathbb{R};$$

(iii) The flow is continuous from $H^\beta_{loc}(\mathbb{R}^2)$ to $H^\beta_{loc}(\mathbb{R}^2)$.

4 A modified Euler equation

In this paragraph we consider a modification of the Euler equation. It corresponds to replacing the Laplacian by the Ornstein-Ulhenbeck operator in the vorticity equation, which induces a modification of the pressure term in the Euler one. This operator, associated to the space $L^2(\mathbb{R}^2)$ with respect to the Gaussian measure, allows to construct statistical solutions which are much more regular then those constructed for the Euler equation: not only distributions but functions. In this case their laws are quasi-invariant (not invariant) probability measures, i.e., their laws at each time are absolutely continuous with respect to the (fixed) initial one.

The modified Euler equation is the following

$$\partial_t u + (u.\nabla)u = -\nabla p + \epsilon p x, \qquad \text{div}\, u(t,\cdot) = 0, \tag{4.1}$$

with $u : \mathbb{R}\times\mathbb{R}^2 \longrightarrow \mathbb{R}^2$ and $p : \mathbb{R}\times\mathbb{R}^2 \longrightarrow \mathbb{R}$. The pressure p (as well as the velocity) may depend on the constant ϵ. After the change of variables $\tilde{u}(t,x) = \rho(x)u(t,x)$, $\rho(x) = \frac{1}{2\pi}e^{-\frac{\epsilon}{2}|x|^2}$, this equation is seen to be equivalent to the following one

$$\partial_t\tilde{u} + (\tilde{u}.\nabla)(\rho^{-1}\tilde{u}) = \nabla(\rho\, p), \qquad \delta_{\rho^{-1}dx}\tilde{u}(t,\cdot) = 0. \tag{4.2}$$

Here δ denotes the dual of the gradient operator in $L^2(\rho^{-1}dx)$.

Consider, for $\beta \in \mathbb{R}$, the Sobolev spaces

$$\mathcal{H}^\beta_\rho(\mathbb{R}^2) = \left\{v : \mathbb{R}^2 \longrightarrow \mathbb{R}, v \in L^2_{\rho dx}(\mathbb{R}^2) \ : \ (I - \mathcal{L}^\epsilon)^{\beta/2}v \in L^2_{\rho dx}(\mathbb{R}^2)\right\}. \tag{4.3}$$

where $\mathcal{L}^\epsilon$ is the Ornstein-Ulhenbeck operator

$$\mathcal{L}^\epsilon\varphi = \Delta\varphi(x) - \epsilon x.\nabla\varphi(x).$$

Since ϵ will be fixed, we will not explicitly write it unless necessary; for example we simply write ρ instead of ρ^ϵ.

4.1 The vorticity flow for the modified Euler equation

Proceeding as in the Euler case, we can deduce the vorticity equation associated with the modified Euler model. It is the following:

$$\partial_t(\mathcal{L}^\epsilon\varphi) = -\nabla^\perp\varphi.\nabla\mathcal{L}^\epsilon\varphi. \tag{4.4}$$

Concerning classical (as opposed to statistical) solutions, we have shown in [18] that there exist bounded weak solutions for this equation. More precisely, denoting by ω the vorticity of $\tilde{u}$, we have shown the following

Theorem 4.5 Given ω_0 such that $\rho^{-1}\omega_0 \in L^1 \cap L^\infty$, there exists $T > 0$ and a weak solution of equation (4.4) ω_t such that $\rho^{-1}\omega \in L^\infty([-T,T]; L^1 \cap L^\infty)$.

For the proof we observe that the quantity $\mathcal{L}^\epsilon\varphi$ is conserved along the particle's trajectories with velocity u. This fact implies that a solution of the ordinary differential equation which defines the particle's trajectories induces a weak solution for (4.4). By Osgood theorem in Banach spaces such equation is solvable, since its field has quasi-Lipschitz regularity.

In order to study statistical solutions of the vorticity equation, we recall that Hermite polynomials form an orthonormal basis of the space L^2 endowed with a Gaussian measure on $\mathbb{R}^n$. When $n = 2$ the explicit expressions of the Hermite polynomials are

$$\mathcal{H}_k(x) = \Pi_{i=1}^2\mathcal{H}_{k_i}(x_i), \quad k \in \mathbb{Z}^2, k \geq 0 \tag{4.5}$$

$$\mathcal{H}_{k_i}(x_i) = \frac{1}{\sqrt{\epsilon^{k_i} k_i!}} e^{\frac{\epsilon x_i^2}{2}} \frac{\partial^{k_i}}{\partial x_i^{k_i}} (e^{-\frac{\epsilon x_i^2}{2}}) \tag{4.6}$$

Among other properties (c.f [19]) Hermite polynomials are eigenfunctions of the Ornstein-Ulhenbeck operator $\mathcal{L}^\epsilon$ with eigenvalues equal to $-\epsilon|k|$, $\forall k \geq 0$. Also the Sobolev spaces defined in (4.3) can be characterised in the following way:

$$\mathcal{H}_\rho^\beta(\mathbb{R}^2) = \left\{ v = \sum_{k\geq 0} v_k \mathcal{H}_k \ : \ \sum_{k\geq 0} (1+\epsilon|k|)^\beta |u_k|^2 < +\infty \right\}.$$

These are Hilbert spaces with inner product given by

$$< u, v >_{\beta,\rho} = \sum_{k\geq 0} (1+\epsilon|k|)^\beta u_k \bar{v}_k.$$

4.2 Quasi-invariant measures

We consider the Gaussian measures

$$d\mu^n_{\rho,\gamma}(\varphi) \simeq \prod_{k\in\{\alpha_1,\dots,\alpha_{d(n)}\}} \frac{\gamma(1+\epsilon|k|)^2}{2\pi} e^{-\frac{\gamma}{2}(1+\epsilon|k|)^2|\varphi_k|^2} d\varphi_k,$$

for $\varphi(t,x) = \sum_{k\geq 0} \varphi_k(t)\mathcal{H}^\epsilon_k(x) \in L^2_\rho$. The $\alpha_1, \dots, \alpha_{d(n)}$ are non-negative pairs of $\mathbb{Z}^2$. These measures converge when n goes to infinity to a probability measure

$$d\mu_{\rho,\gamma}(\varphi) \simeq \prod_{k\geq 0} \frac{\gamma(1+\epsilon|k|)^2}{2\pi} e^{-\frac{\gamma}{2}(1+\epsilon|k|)^2|\varphi_k|^2} d\varphi_k \tag{4.7}$$

or, in a formal way,

$$d\mu_{\rho,\gamma}(\varphi) \simeq \frac{1}{Z_\gamma} e^{-\frac{\gamma}{2}\|(I-L^\epsilon)\varphi\|^2_{L^2_\rho}} \mathcal{D}\varphi.$$

Thus $\mu_{\rho,\gamma}$ is the Gibbs-type measure associated to the quantity $\frac{1}{2}\|(I-L^\epsilon)\varphi\|^2_{L^2_\rho}$. We refer to [18] for details.

Using some dispersive bounds for Hermite functions (c.f. [18] and references within) we can characterise the support of these measures, which are not only distributions, but actually regular functions. More precisely,

Theorem 4.6 If $\iota > 0$ and $p \in]2, \frac{10}{3}[$, then

$$\text{supp}\ \mu_{\rho,\gamma} = \mathcal{H}_\rho^{-\iota}(\mathbb{R}^2) \cap L^p_{loc}(\mathbb{R}^2.)$$

Similarly to the construction we have described for the Euler case, we can write the vorticity equation for the modified Euler one as an infinite system of ordinary differential equations, using the orthonormal basis of $L^2_{\rho dx}$ that consists of the Hermite polynomials defined above. It can be written as,

$$\frac{d}{dt}\varphi_k(t) = B^\epsilon_k(\varphi), \qquad \forall k > 0, \tag{4.8}$$

where B^ϵ_k denotes the k-th component in the Hermite basis of B^ϵ, namely

$$B^\epsilon(\varphi) = \sum_{k>0} B^\epsilon_k(\varphi)\mathcal{H}^\epsilon_k(x), \tag{4.9}$$

where

$$B^\epsilon_k(\varphi) = -\frac{1}{\epsilon^2|k|}\sum_{p\geq|k|/2}\ \sum_{\substack{q\geq 0\\ |q|<|p|}} (|p|-|q|)A(p,q,k)\varphi_p\varphi_q. \tag{4.10}$$

The coefficients are

$$\begin{aligned} A(p,q,k) := \Big[&-\sqrt{p_2q_1}\,\Theta\Big(p_1, q_1-1, \tfrac{p_1+q_1-1-k_1}{2}\Big)\,\Theta\Big(p_2-1, q_2, \tfrac{p_2+q_2-1-k_2)}{2}\Big) \\ &+\sqrt{p_1q_2}\,\Theta\Big(p_1-1, q_1, \tfrac{p_1+q_1-1-k_1}{2}\Big)\,\Theta\Big(p_2, q_2-1, \tfrac{p_2+q_2-1-k_2}{2}\Big)\Big], \end{aligned} \tag{4.11}$$

$$\Theta(n,m,r) = \left[\binom{n}{r}\binom{m}{r}\binom{n+m-2r}{n-r}\right]^{1/2}. \tag{4.12}$$

Although the expressions seem more complicated, this vector field is much more regular then the Euler one of section 2. We have (c.f. [18] for the proofs):

Theorem 4.7 For every $\beta \in \mathbb{R}$ and $\iota > 0$,

(i) B^ϵ belongs to the space $L^r_{\mu_{\rho,\gamma}}(\mathcal{H}^{-\iota}_\rho(\mathbb{R}^2); \mathcal{H}^\beta_\rho(\mathbb{R}^2))$ for all $r \geq 1$;

(ii) $\nabla B^\epsilon \in L^r_{\mu_{\rho,\gamma}}(\mathcal{H}^{-\iota}_\rho(\mathbb{R}^2); H.S.(\mathcal{H}^2_\rho(\mathbb{R}^2); \mathcal{H}^\beta_\rho(\mathbb{R}^2)))$;

(iii) $\nabla^2 B^\epsilon \in L^r_{\mu_{\rho,\gamma}}(\mathcal{H}^{-\iota}_\rho(\mathbb{R}^2); H.S.(\mathcal{H}^2_\rho(\mathbb{R}^2)\otimes\mathcal{H}^2_\rho(\mathbb{R}^2); \mathcal{H}^\beta_\rho(\mathbb{R}^2)))$ for all $r \geq 1$.

Here $H.S.$ referes to the Hilbert-Schmidt operator norm.

4.3 Statistical solutions

We consider the Malliavin calculus as summarised at the beginning of paragraph 2.3 applied now to the abstract Wiener space $(\mathcal{H}^{-\epsilon}_\rho(\mathbb{R}^2), \mathcal{H}^2_\rho(\mathbb{R}^2), \mu_{\rho,\gamma})$. From the regularity results stated in Theorem 4.7 we see in particular that, contrary to the Euler case, the vector field B^ϵ takes values in the Cameron-Martin space $(\mathcal{H}^2_\rho)$.

In this case the divergence operator does not vanish. It is given by the following expression:

$$\delta_{\mu_{\gamma,\rho}} B^\epsilon(\varphi) = \sum_k \frac{1}{\epsilon^2|k|} \sum_{p\geq 0} (|p|-|k|)A(p,k,k)\varphi_p \tag{4.13}$$

$$+\gamma \sum_k \frac{(1+\epsilon|k|)^2}{\epsilon^2|k|} \sum_{|p|\geq|k|/2} \sum_{\substack{q\geq 0\\ |q|<|p|}} (|p|-|q|)A(p,q,k)\varphi_p\varphi_q\bar{\varphi}_k. \tag{4.14}$$

Concerning the regularity of the divergence, we have

Theorem 4.8 If $\beta \in \mathbb{R}$ and $\iota > 0$, for all $r \geq 1$ the divergence $\delta_{\mu_{\rho,\gamma}} B^\epsilon$ belongs to the space $L^r_{\mu_{\rho,\gamma}}(\mathcal{H}_\rho^{-\iota};\mathbb{R})$.

Theorems 4.7 and 4.8 provide enough regularity for the vector field B^ϵ to define an associated flow using Malliavin calculus techniques. The divergence estimates are essencial, since the divergence allows to control the action of the flow on the measure. This can be seen by looking at formula (4.18). The corresponding expression for the finite dimensional flows is used to control, uniformly, the densities of their laws. We have, in our case, the existence of a positive λ such that

$$\int \left[\exp(\lambda|\delta_{\mu_{\rho,\gamma}} B^\epsilon(\varphi)|) + \exp(\lambda\|\nabla B^\epsilon(\varphi)\|)\right] d\mu_{\rho,\gamma}(\varphi) < +\infty. \tag{4.15}$$

Another central formula in the proofs, allowing to show that approximations of flows form Cauchy sequences, can be found in [14] and is the following (here we do not precise the assumptions on vector fileds A and B):

$$U_t^A - U_t^B = -\int_0^t ((U_{t+s})'.[(A-B)(U^B_{-t-s})]) \circ \Phi_{s,t}\, ds$$

where

$$\Phi_{s,t} = U^A_{-s} \circ U^B_{t+s}$$

and $(U_t)'$ stands for the Jacobian matrix of the flow.

The construction of flows on abstract Wiener spaces was initiated in [14] and extended by [34] with less restrictive hypothesis. The results can be applied in our context and the following holds,

Theorem 4.9 Let $\iota > 0$; then to $B^\epsilon : \mathcal{H}_\rho^{-\iota} \to \mathcal{H}_\rho^2$ we can associate an almost-surely unique flow $U_t^\epsilon(\varphi)$ such that

$$U_t^\epsilon(\varphi) = \varphi + \int_0^t B^\epsilon(U_s^\epsilon(\varphi))ds \qquad \mu_{\rho,\gamma} - a.e.\ \varphi \in \mathcal{H}_\rho^{-\iota} \cap L^p_{loc}, \quad \forall t \in \mathbb{R}. \tag{4.16}$$

Moreover the measure $\mu_{\rho,\gamma}$ is quasi-invariant under U_t^ϵ and

$$k_t(\varphi) = \exp\left(\int_0^t \operatorname{div}_{\mu_{\rho,\gamma}} B^\epsilon(U^\epsilon_{-s}(\varphi))ds\right) \tag{4.17}$$

is the corresponding Radon-Nikodym density, namely $k_t := \frac{d(U_t^\epsilon)*\mu_{\rho,\gamma}}{d\mu_{\rho,\gamma}}$.

We have $k_t \in L^r_{\mu_{\rho,\gamma}}$, for all $r \geq 1$.

We observe that, when ϵ tends to zero, equations (4.1) converge to the standard Euler equation. However we cannot rigorously consider such limit for statistical solutions since the measures $\mu_{\rho,\gamma}$ for each ϵ are mutually singular.

5 The two-dimensional averaged-Euler equations

The averaged-Euler equations were introduced by D.D. Holm, J.E. Marsden and T.S. Ratiu in [24]. They are modifications of the Euler equations such that non linear effects at small scales of the motion are negligible; the dynamics remains turbulent, but non dissipative. In the incompressible case, they are the following

$$\partial_t Au + (u \cdot \nabla)Au + (\nabla u)^T \cdot Au = -\nabla p, \qquad \operatorname{div} u(t,\cdot) = 0 \tag{5.1}$$

where $A = (1 - a^2\Delta)^s$, with a a real parameter and s a positive number.

Global existence and uniqueness for solutions of the two-dimensional averaged-Euler equation have been studied in $\mathbb{R}^2$ and in bounded domains for H^3 initial velocities, see [12] and [31], respectively.

O. Bell, A. Chorin and W. Crutchfield pointed out in [8] that invariant Gibbs measures can be defined for this system, as the equations conserve the energy and the enstrophy. In their perspective the invariant measures were essentially used to perform numerical predictions of the dynamics. In this section we report on the work [32], where invariant measures and corresponding statistical solutions for the averaged-Euler equations were constructed.

5.1 Formulation of the equations and invariant quantities

Applying the operator rot to equation (5.1) and writing, as we did for the Euler equation, $u = \nabla^\perp \varphi$, we obtain its vorticity formulation

$$\partial_t A\Delta\varphi + (\nabla^\perp \varphi . \nabla)A\Delta\varphi = 0. \tag{5.2}$$

This equation will be considered in the two-dimensional torus $\mathbb{T}^2$, or equivalently, under the periodic boundary conditions

$$\varphi(0, y, t) = \varphi(2\pi, y, t) \text{ and } \varphi(x, 0, t) = \varphi(x, 2\pi, t), \qquad \forall (x, y) \in \mathbb{T}^2.$$

The following quantities, the energy and enstrophy for the system, are invariants of the averaged-Euler motion

$$E = -\frac{1}{2}\int_{\mathbb{T}^2} \varphi A\Delta\varphi dx$$

and

$$S = \frac{1}{2}\int_{\mathbb{T}^2} (A\Delta\varphi)^2 dx.$$

We consider the Fourier decomposition of equation (5.2). If $e_k, k \in \mathbb{Z}^2$ are the functions considered in paragraph 2.1., that form an orthonormal basis of $L^2(\mathbb{T}^2)$, and if we write $\varphi(t,x) = \sum_{k>0} \varphi_k(t) e_k(x)$, we have,

$$\begin{aligned} 2\pi k^2(1+a^2k^2)^s \frac{d\varphi_k}{dt} &= -\sum_{\substack{h+j=k,\\ h>0}} (h\cdot j^\perp) j^2 (1+a^2j^2)^s \varphi_h\varphi_j \\ &= \frac{1}{2}\sum_{\substack{h+j=k,\\ h>0}} (h.j^\perp)[h^2-j^2](1+a^2j^2)^s\varphi_h\varphi_j, \quad \forall k>0. \quad (5.3) \end{aligned}$$

where, we recall, $h^\perp = (-h_2, h_1)$ for $h \in \mathbb{Z}^2$. We can therefore write the averaged-Euler equations as infinite systems of ordinary differential equations

$$\frac{d}{dt}\varphi_k = \mathcal{B}_k(\varphi)$$

$k \in \mathbb{Z}^2$, $k > 0$, for

$$\mathcal{B}_k(\varphi) = \frac{1}{2\pi}\sum_{l>0}\left[\frac{1}{k^2}(l^\perp.k)(l-k) - \frac{1}{2}(l^\perp.k)\right]\frac{(1+a^2(k-l)^2)^s}{(1+a^2k^2)^s}\varphi_l\varphi_{k-l}$$

or, in a more concise form, $\frac{d}{dt}\varphi = \mathcal{B}(\varphi)$ with $\mathcal{B}(\varphi) = \sum_k B_k(\varphi)e_k$.

5.2 Gaussian invariant measures and statistical solutions

Let us consider the Sobolev spaces

$$H^{p,s} = \left\{ v = \sum_{k\in\mathbb{Z}^2} v_k e_k : \sum_{k>0} k^{2p}(1+a^2k^2)^{ps}|v_k|^2 < +\infty \right\}$$

with inner product $< u, v >_{2,s} = \sum_{k>0} k^4(1+a^2k^2)^{2s} u_k \bar{v}_k$. The energy and the enstrophy can be written, respectively, as

$$E = \frac{1}{2}||\varphi||^2_{1,s}$$

and

$$S = \frac{1}{2}||\varphi||^2_{2,s}.$$

Consider the probability measures defined for $\gamma \in \mathbb{R}^+$ by

$$d\mu^k_{\gamma,s}(z) = \frac{\gamma k^4(1+a^2k^2)^{2s}}{2\pi} \exp\left\{-\frac{1}{2}\gamma k^4(1+a^2k^2)^{2s}|z|^2\right\} dxdy \tag{5.4}$$

where $z = x + iy$. Then

$$d\mu_{\gamma,s}(\varphi) = \Pi_{k>0} d\mu^k_{\gamma,s}(\varphi_k) \tag{5.5}$$

is a well defined probability measure with support in $H^{1-\alpha,s}$ for any $\alpha > -\frac{s}{s+1}$; indeed, we have

$$\int \|\varphi\|^2_{1-\alpha,s} d\mu_{\gamma,s}(\varphi) = \frac{2}{\gamma} \sum_{k>0} \frac{1}{k^{2(1+\alpha)}(1+a^2k^2)^{s(1+\alpha)}} < +\infty, \qquad \forall \alpha > -\frac{s}{s+1}.$$

Formally, we have

$$d\mu_{\gamma,s} = \frac{1}{Z} e^{-\gamma S(\varphi)} \mathrm{d}(\varphi).$$

The triple $(H^{1-\alpha,s}, H^{2,s}, \mu_\gamma)$ is an abstract Wiener space with measurable norm $\|\cdot\|_{1-\alpha,s}$ for any $\alpha > -\frac{s}{s+1}$ and the averaged-Euler vector field $\mathcal{B}$ satisfies the following properties (c.f. [32]),

Theorem 5.10 For every $\alpha > 2$ and $p \geq 1$, $\mathcal{B} \in L^p_{\mu_{\gamma,s}}(H^{1-\alpha,s}; H^{1-\alpha,s})$.
Moreover, for every $\alpha > 2$ and $p \geq 1$,

$$\nabla \mathcal{B} \in L^p_{\mu_{\gamma,s}}(H^{1-\alpha,s}; H.S.(H^{2,s}, H^{1-\alpha,s}))$$

and

$$\nabla^2 \mathcal{B} \in L^p_{\mu_{\gamma,s}}(H^{1-\alpha,s}; H.S.(H^{2,s} \otimes H^{2,s}, H^{1-\alpha,s})).$$

Defining the divergence operator with respect to the measure μ_γ analogously to definition 2.8, we have, as expected since the measures are associated to the enstrophy,

Theorem 5.11 The vector field $\mathcal{B}$ is infinitesimally invariant with respect to the measures $\mu_{\gamma,s}$, namely we have

$$\delta_{\mu_{\gamma,s}} \mathcal{B} = 0$$

Using the regularity properties of the vector field $\mathcal{B}$ stated in the last theorem, and using analogous methods to those described in section 2, one proves existence of a $\mu_{\gamma,s}$-a.e. solution of the vorticity equation for the averaged-Euler system. Moreover the measures are invariant for the associated flows. More precisely, we have the following,

Theorem 5.12 For $\alpha > 2$ there exists a global flow $U_t(\omega)$ defined for φ $\mu_{\gamma,s}$- almost everywhere on a probability space $(\Omega, \mathcal{F}, P_\gamma)$ and with values in the space $H^{1-\alpha,s}(\mathbb{T}^2)$ such that

(i) $U_t(\omega) = \varphi + \int_0^t \mathcal{B}(U_s(\omega))ds \quad P_\gamma$ a.e. $\forall t \in \mathbb{R}$;

(ii) the measure $\mu_{\gamma,s}$ is invariant for the flow in the following sense:

$$\int_{H^{1-\alpha,s}} f(U_t(\omega))dP_\gamma(\omega) = \int_{H^{1-\alpha,s}} f(\varphi)d\mu_{\gamma,s}(\varphi) \quad \forall f \in C_b(H^{1-\alpha,s}),\ \forall t \in \mathbb{R}.$$

5.3 Surface measures

It is not possible to define the averaged-Euler flow constructed in the last section if we start from finite energy functions, since the measures $\mu_{\gamma,s}$ are supported by infinite energy ones. Nevertheless, and contrary to the Euler case studied in [13], the energy E for the averaged-Euler equations belongs to the space $L^2_{\mu_{\gamma,s}}$. Actually we have much more: the Malliavin derivatives of all orders $\nabla^r E$ belong to all $L^p_{\mu_{\gamma,s}}$ (as proved in [32]). This property is written as

$$E \in \mathcal{W}_\infty(H^{1-\alpha,s}). \tag{5.6}$$

Note that such regularity does not imply smoothness of the energy functional, since we do not have Sobolev immersion theorems in infinite dimensions. Still, (5.6) together with the property

$$\|\nabla E\|_{2,s}^{-1} \in \mathcal{W}_\infty(H^{1-\alpha,s}) \tag{5.7}$$

are enough to consider surface-type measures on the level sets of E by a desintegration procedure. Such measures are also invariant for the flow and allow to define the flow in these submanifolds, almost-everywhere with respect to the new measures.

The construction of surface measures in abstract Wiener spaces was made in [1] (c.f. also [30]). We present it here, in a summarised way and in our context. For the rest of the section, and for sake of simplicity, we shall simply write μ to denote our invariant measure $\mu_{\gamma,s}$.

Recall the notion of capacity on a Wiener space: if O is an open subset of $H^{1-\alpha,s}$,

$$\text{cap}_{p,r}(O) = \inf\{\|\nabla^r F\|_{L^p(\mu)} : F > 0, F \geq 1\ \mu - \text{ a.e. in } O\}$$

and, for an arbitrary set $A \subset H^{1-\alpha,s}$, $\text{cap}_{p,r}(A) = \inf\{\text{cap}_{p,r}(O) : O \text{ open},\ A \subset O\}$.

Let $g \in \mathcal{W}_\infty$. Then one can prove that there exists a functional g^* (called a redefinition of g) and an associated sequence of open sets O_n such that $\text{cap}_{p,r} \cap_n O_n = 0$ for every p, r, g^* is continuous in $H^{1-\alpha,s} \setminus (\cap_n O_n)$ and, for each n, g^* and $\nabla^r g^*$ are also continuous in $H^{1-\alpha,s} \setminus O_n$.

Following [1,30], when the conditions (5.6) and (5.7) hold, the measures $(E)_*\mu$ and $(E)_*(g\mu)$ are absolutely continuous with respect to Lebesgue measure and have C^∞ Radon-Nikodym densities. Denote these densities by $\eta(y) = \frac{d(E)_*\mu}{dy}$ and $\eta_g(y) =$

$\frac{d(E)_*(g\mu)}{dy}$, respectively. Moreover the function $E_g = \frac{\eta_g}{\eta}$ is smooth in the open set $O = \{y \in \mathbb{R} : \eta(y) > 0\}$.

Now consider the linear functional

$$T_E : \mathcal{W}_\infty \to C^\infty(O), \quad T_E(g)(\cdot) = E_g(\cdot)$$

and denote by T_E^* its dual. Computed on the Dirac delta function δ_y, we have $< T_E^*(\delta_y), g >= E_g(y)$.

For each y it is possible to prove the existence of a measure Ξ_y with support in the energy level set $S_y = \{\varphi \in H^{1-\alpha,s} : E(\varphi) = y\}$ such that g^* is Ξ_y-integrable and

$$\int g^* d\Xi_y = \frac{\eta_g(y)}{\eta(y)}. \tag{5.8}$$

In particular the following desintegration formula holds,

$$\int_{H^{1-\alpha,s}} < v(E(\varphi))\mathcal{B}, \nabla f >_{H^{2,s}} d\mu = \int_{\mathbb{R}} v(y)\eta(y) \int_{S_y} < \mathcal{B}, \nabla f >^*_{H^{2,s}} d\Xi_y dy \tag{5.9}$$

for every test functional f and every smooth function v.

We know that the divergence of $\mathcal{B}$ with respect to μ is zero (Theorem 5.11) or, equivalently, that $\int < \mathcal{B}, \nabla f >_{H^{2,s}} (\varphi)d\mu(\varphi) = 0$ for every test functional f on $H^{1-\alpha,s}$. Using the desintegration formula (5.9) and the conservation of energy, we deduce that the surface measures Ξ_y are also infinitesimally invariant for $\mathcal{B}$, or, more precisely, that

$$\int < \mathcal{B}, \nabla f >^*_{H^{2,s}} (\varphi)d\Xi_y(\varphi) = 0$$

Considering the level surfaces for the energy and the corresponding surface measures whose construction we have described, we can prove the following (c.f. [32]),

Theorem 5.13 Let $\alpha > 2$ and let $y > 0$ be such that $\eta(y) > 0$. Then there exists a flow V defined on a probability space $(\Omega, \mathcal{F}, P^y)$ with $V(\cdot, \omega) \in C(\mathbb{R}; S_y)$, $\omega \in \Omega$ such that, for every redefinition $\mathcal{B}^*$ of $\mathcal{B}$,

$$(i)\ \ V(t,\omega) = \varphi + \int_0^t \mathcal{B}^*(V(s,\omega))ds \quad P^y \text{ a.e.-}\omega, \ \ \forall t \in \mathbb{R}$$

$$(ii)\ \ \int_{S_y} f(V(t,\omega))dP^y(\omega) = \int_{S_y} f d\Xi_y \forall t \in \mathbb{R}$$

for every test functional f, i.e., the flow is invariant under the action of the surface measure Ξ_y.

Finally we remark that Poincaré recurrence theorem (c.f. for example [26]) holds for the flows above. We can indeed prove that the global flow returns infinitely many times to its initial (a.e.) state. More precisely, let $\alpha > 2$; then, for $\epsilon > 0$ sufficiently small, and for (Ξ_y-a.e.) φ, φ_0 with $\|\varphi - \varphi_0\|_{1-\alpha,s} < \epsilon$, there exists a sequence $t_n \to +\infty$ such that $\mathbb{E}_{P^y} \|V(t_n, \omega) - \varphi_0\|_{1-\alpha,s} < \epsilon$.

Acknowledgements The authors acknowledge the support of the FCT Portuguese grant PTDC/MAT-STA/0975/2014. The second author was also supported by the LisMath fellowship PD/BD/52641/2014, FCT, Portugal.

References

1. H. Airault, P. Malliavin, Intégration géométrique sur l'espace de Wiener, *Bull. Sci. Math.*, 112 (1988) 3–52.
2. S. Albeverio, A.B. Cruzeiro, Global flows with invariant (Gibbs) measures for the Euler and Navier-Stokes two dimensional fluids, *Comm. Math. Phys.* 129 (1990), 431–444.
3. S. Albeverio, M. R. de Faria, R. H¯egh-Krohn, Stationary measures for the periodic Euler flow in two dimensions, *J. Stat. Phys.*, 20 (1979)
4. S. Albeverio, B. Ferrario, Uniqueness results for the generators of the two-dimensional Euler and Navier-Stokes flows. The case of Gaussian invariant measures, *J. Funct. Anal.*, 193 (2002), 77–93.
5. S. Albeverio, R. Hoegh Krohn, Stochastic flows with stationary distribution for two dimensional inviscid fluids, *Stoch. Proc. Appl.* 31, (1989) 1–31.
6. V. I. Arnold, Sur la géométrie différentielle des groupes de Lie de dimension infinie et ses applications à l' hydrodynamique des fluides parfaits, *Ann. Inst. Fourier*, 16 (1966), 316–361.
7. C. Bardos, Existence et unicité de la solution de l'équation d'Euler en dimension deux, *J. Math. Anal. Appl.*, 40 (1972), 769–790.
8. J. Bell, A. Chorin, W. Crutchfield, Stochastic optimal prediction with application to averaged Euler equations, *Proc. 7th Nat. Conf. CDF, Taiwan*, (2000) 1–13.
9. C. Boldrighini, S. Frigio, Equilibrium states for the two dimensional incompressible Euler fluid, *Atti Sem. Mat. Fis. Univ. Modena XXVII*, (1978) 106–125.
10. J. Bourgain, Invariant Measures for NLS in Infinite Volume, *Comm. Math. Phys. 210*, (2000) 605–620.
11. N. Burq, L. Thomann, N. Tzvetkov, Remarks on the Gibbs measures for nonlinear dispersive equations, *arXiv:1412.7499*.
12. V. Busuioc, On second grade fluids with vanishing viscosity, *C. R. Acad. Sci. Paris Ser. I*, 328 (1999)1241–1246.
13. F. Cipriano, The two dimensional Euler equations: a statistical study, *Comm. Math. Phys.*, 201 (1999) 139–154.
14. A.B. Cruzeiro, Équations différentielles sur l'espace de Wiener et formules de Cameron-Martin non-linéaires. *J. Funct. Anal.*, 54, 2 (1983) 206–227.
15. A.B. Cruzeiro, F. Flandoli, P. Malliavin, Brownian motion on volume preserving diffeomorphisms group and existence of global solutions of 2D stochastic Euler equation. *J. Funct. Anal.* 242, (2007) 304–326.
16. A.B. Cruzeiro, P. Malliavin, Nonergodicity of Euler fluid dynamics on tori versus positivity of the Arnold-Ricci tensor, *J. Funct. Anal.*, 254 (2008),7, 1903–1925.
17. A.B. Cruzeiro, A. Symeonides, Invariant measures for the non-periodic two-dimensional Euler equation, *Bull. Sci. Math. 148 (2018), 33–52.*
18. A.B. Cruzeiro, A. Symeonides, On a non-periodic modified Euler equation: existence and quasi-invariant measures, *arXiv:1711.09045*.

19. A. Erdérlyi, W. Magnus, F. Oberhettinger, F. G. Tricomi, Higher trascendental functions, *McGraw-Hill*, vol. II (1955).
20. C. Foias, Statistical study of Navier-Stokes equations I, *Rend. Sem. Mat. Univ. Padova* 48, (1972) 219–348.
21. C. Foias, Statistical study of Navier-Stokes equations II, *Rend. Sem. Mat. Univ. Padova* 49, (1973) 9–123.
22. L. Gross, Abstract Wiener spaces, *Proc. Fifth Berkeley Sympos. Math. Statist. and Probability (Berkeley, Calif., 1965/66), Vol. II: Contributions to Probability Theory, Part 1, Univ. California Press, Berkeley, Calif.* (1967) 31–42.
23. E. Hopf, Statistical hydromechanics and functional calculus, *J. Rational Mech. Anal.* 1, (1952) 87–123.
24. D. D. Holm, J. E. Marsden, T. S. Ratiu, Euler-Poincaré models of ideal fluids with non-linear dispersion, *Phys. Rev. Lett.*, 80 (1998), 4173–4176.
25. V. I. Judovič, Non-stationary flows of an ideal incompressible fluid, *Vyčisl. Mat. i Mat. Fiz.*, 3 (1963), 1032–1066.
26. M. Kac, Probability and related topics in physical sciences, *Wiley-Interscience New York*, (1959).
27. T. Kato, On classical solutions of the two-dimensional nonstationary Euler equation, *Arch. Rational Mech. Anal. 25*, (1967) 188–200.
28. A.N. Kolmogorov, The local structure of turbulence in incompressible viscous fluid for very large Reynold's numbers, *C. R. (Doklady) Acad. Sci. URSS (N.S.)* 30, (1941) 301–305.
29. L. Lichtenstein, Uber einige existenz problem der hydrodynamik homo gener unzusammendrückbarer, reibunglosser Flšsikeiten und die Helmholtzschen wirbelsalitze, *Mat. Zeit. Phys.*, 23 (1925), 89–154.
30. P. Malliavin, Stochastic Analysis, *Grund. der mathem. Wissen.*, 313, Springer-Verlag, New York (1997).
31. S. Shkoller, Analysis on groups of diffeomorphisms of manifolds with boundary and the averaged motion of a fluid, *J. Diff. Geom.*, 55 (2000) 145–191.
32. A. Symeonides, Invariant measures for the two-dimensional averaged-Euler equations, *arXiv:1605.06974*.
33. M. I. Vishik, A.V. Fursikov, Mathematical Problems of Statistical Hydrodynamics, *Kluwer, Dordrecht*, (1988).
34. A.S. Üstünel, M. Zakai, Trasformation of measure on Wiener space, *Springer - Verlag Berlin Heidelberg* (2000).

Long-range Phase Coexistence Models: Recent Progress on the Fractional Allen-Cahn Equation

Serena Dipierro and Enrico Valdinoci

Abstract In this set of notes, we present some recent developments on the fractional Allen-Cahn equation

$$(-\Delta)^s u = u - u^3,$$

with special attention to Γ-convergence results, energy and density estimates, convergence of level sets, Hamiltonian estimates, rigidity and symmetry results.

1 Prelude

The study of nonlocal models for phase coexistence equations is an interesting, and remarkably challenging, research topic which has experienced a rapid growth in the recent mathematical literature. The goal of this paper is to collect several results and present them in a unified and easily accessible way, with a style which tries to combine, as much as possible, rigorous presentations and intuitive descriptions of the problems under consideration and of the methods used in some of the proofs.

The model will be also somehow described "from scratch" and, in spite of the necessary simplifications which make the topic mathematically treatable, we hope that we managed to preserve some important treats from the physical model in view of the applications in material sciences, using also these real-world motivations as a hint towards the development of rigorous and quantitative theories.

In what follows, we will discuss specifically Γ-convergence results, energy and density estimates, convergence of level sets, Hamiltonian estimates, rigidity and symmetry results. To this end, we will first recall the classical Allen-Cahn phase coexistence model in Section 2, where we will also present some of the classical

Serena Dipierro and Enrico Valdinoci
School of Mathematics and Statistics, University of Western Australia, 35 Stirling Hwy, Crawley WA 6009, Australia. e-mail: serena.dipierro@uwa.edu.au, enrico.valdinoci@uwa.edu.au

M. Hintermüller, J. F. Rodrigues (eds.), *Topics in Applied Analysis and Optimisation*,
CIM Series in Mathematical Sciences, https://doi.org/10.1007/978-3-030-33116-0_5

results and conjectures about it. Then, we will consider the nonlocal counterpart of these problems in Section 3.

2 The classical Allen-Cahn equation

The so-called Allen-Cahn equation is a semilinear, scalar equation, originally introduced by John W. Cahn and Sam Allen in the 1970s. In the stationary case, this equation is of elliptic type and can be written in the form

$$-\Delta u = u - u^3 \quad \text{in } \Omega. \tag{2.1}$$

Equation (2.1) is typically endowed with either Dirichlet or Neumann conditions. In our setting, we mainly focus on critical points of the associated energy functional and in particular on local minima: in this case, one can think that the energy candidates are chosen among the smooth and compactly supported perturbations of u and thus (2.1) is naturally complemented by the Dirichlet datum of the solution itself.

In (2.1), the function $u = u(x)$ represents an order parameter that determines the phase of the medium at a given point $x \in \Omega$. In this setting, the "pure phases" are denoted by the state parameters -1 and 1, and the Laplacian term can be considered as a surface tension or interfacial energies, which ends up preventing abrupt phase changes from point to point and "wild" phase oscillations. The set $\Omega \subseteq \mathbb{R}^n$ can be viewed as the "container" and then equation (2.1) aims at giving a simple, but effective, description of phase coexistence.

One sees that equation (2.1) possesses a variational structure and solutions of (2.1) correspond to the critical points of the energy functional

$$\begin{gathered} \mathcal{J}_\Omega(u) := \int_\Omega \left(\frac{1}{2} |\nabla u(x)|^2 + W(u(x)) \right) dx, \\ \text{where } W(r) := \frac{(1-r^2)^2}{4}. \end{gathered} \tag{2.2}$$

It is of course tempting to look at the "big picture" offered by this scenario: namely, given a candidate function $u_\star$, one can consider the blow-down sequence

$$u_\varepsilon(x) := u_\star\left(\frac{x}{\varepsilon}\right) \qquad \text{as } \varepsilon \searrow 0, \tag{2.3}$$

and one remarks that

$$\begin{aligned}\mathcal{J}_{\Omega/\varepsilon}(u_\star) &= \int_{\Omega/\varepsilon}\left(\frac{\varepsilon^2}{2}\,|\nabla u_\varepsilon(\varepsilon x)|^2 + W\big(u_\varepsilon(\varepsilon x)\big)\right)dx\\ &= \frac{1}{\varepsilon^n}\int_{\Omega}\left(\frac{\varepsilon^2}{2}\,|\nabla u_\varepsilon(x)|^2 + W\big(u_\varepsilon(x)\big)\right)dx\\ &= \frac{1}{\varepsilon^{n-1}}\,\mathcal{J}_{\Omega,\varepsilon}(u_\varepsilon),\end{aligned}$$

where

$$\mathcal{J}_{\Omega,\varepsilon}(v) := \int_{\Omega}\left(\frac{\varepsilon}{2}\,|\nabla v(x)|^2 + \frac{1}{\varepsilon}\,W\big(v(x)\big)\right)dx.$$

In particular, $u_\star$ is a local minimizer for $\mathcal{J}_{\Omega/\varepsilon}$ (that is, $\mathcal{J}_{\Omega/\varepsilon}(u_\star) \le \mathcal{J}_{\Omega/\varepsilon}(u_\star + \varphi)$ for all $\varphi \in C_0^\infty(\Omega/\varepsilon)$) if and only if u_ε is a local minimizer for $\mathcal{J}_{\Omega,\varepsilon}$ (that is, $\mathcal{J}_{\Omega,\varepsilon}(u_\varepsilon) \le \mathcal{J}_{\Omega,\varepsilon}(u_\varepsilon + \varphi)$ for all $\varphi \in C_0^\infty(\Omega)$). In this setting, the following results are classical:

- Γ-*convergence.* If Ω is a smooth domain and $u_\varepsilon : \Omega \to [-1,1]$ is a sequence of minimizers for $\mathcal{J}_{\Omega,\varepsilon}$ such that

  $$\sup_{\varepsilon\in(0,1)} \mathcal{J}_{\Omega,\varepsilon}(u_\varepsilon) < +\infty,$$

  then, up to a subsequence,

  $$\lim_{\varepsilon\searrow 0} u_\varepsilon = \chi_E - \chi_{\mathbb{R}^n\setminus E} \quad \text{in } L^1(\Omega), \tag{2.4}$$

  for some set $E \subseteq \mathbb{R}^n$ which minimizes the perimeter in $\overline{\Omega}$ with respect to its boundary datum, see [43].
- *Energy and density estimates.* If $R \ge 1$ and $u : B_{R+1} \to [-1,1]$ is a minimizer of $\mathcal{J}_{B_{R+1}}$, then

  $$\mathcal{J}_{B_{R+1}}(u) \le CR^{n-1}, \tag{2.5}$$

  for some $C > 0$.
  In addition, if $u(0) = 0$, then

  $$\begin{gathered}\text{the Lebesgue measures of } \{u > 1/2\} \text{ and } \{u < -1/2\} \text{ in } B_R\\ \text{are both greater than } cR^n,\end{gathered} \tag{2.6}$$

  for some $c > 0$, see [18].
- *Locally uniform convergence of level sets.* If Ω is a smooth domain, $E \subseteq \mathbb{R}^n$ and $u_\varepsilon : \Omega \to [-1,1]$ is a minimizer of $\mathcal{J}_{\Omega,\varepsilon}$ such that (2.4) holds true, then the set $\{|u_\varepsilon| \le 1/2\}$ converges locally uniformly in Ω to ∂E as $\varepsilon \searrow 0$, namely

  $$\lim_{\varepsilon\searrow 0}\sup_{x\in\Omega'} \operatorname{dist}(x,\partial E) = 0$$

  for any $\Omega' \Subset \Omega$, see again [18].

- *Pointwise gradient bounds.* If $u : \mathbb{R}^n \longrightarrow [-1, 1]$ is a critical point for the energy functional in (2.2) for any bounded domain $\Omega \subset \mathbb{R}^n$, then we have the following pointwise gradient bound:

$$|\nabla u(x)|^2 \leq 2W\big(u(x)\big) \quad \text{for all } x \in \mathbb{R}^n, \tag{2.7}$$

see [42].

The inequality in (2.7) can be seen as part of a family of formulas related to Hamiltonian Identities, see [38]. As a matter of fact, we observe that in dimension 1, the bound in (2.7) reduces to the classical Conservation of Energy Principle: indeed, if $u : \mathbb{R} \longrightarrow [-1, 1]$ is a solution of $\ddot{u} = W'(u)$ in $\mathbb{R}$, it follows that

$$\frac{d}{dx}\left(\frac{|\dot{u}(x)|^2}{2} - W(u(x))\right) = \dot{u}(x)\,\ddot{u}(x) - W'(u(x))\,\dot{u}(x) = 0,$$

and therefore

$$\frac{|\dot{u}(x)|^2}{2} - W(u(x)) = \frac{|\dot{u}(y)|^2}{2} - W(u(y)) \leq \frac{|\dot{u}(y)|^2}{2}, \tag{2.8}$$

for any $y \in \mathbb{R}$. Now, two cases occur: either $\dot{u}$ never vanishes, or

$$\dot{u}(y_0) = 0 \tag{2.9}$$

for some $y_0 \in \mathbb{R}$. In the first case, u is monotone, say increasing, and since it is bounded it has a limit at $+\infty$ and, as a consequence,

$$\lim_{y \to +\infty} \dot{u}(y) = 0.$$

Using this information in (2.8), one obtains that

$$\frac{|\dot{u}(x)|^2}{2} - W(u(x)) \leq \lim_{y \to +\infty} \frac{|\dot{u}(y)|^2}{2} = 0,$$

which is (2.7) in this case. If instead (2.9) holds true, it is enough to compute (2.8) at $y := y_0$ and deduce (2.7) in this case as well.

Furthermore, one of the most important problems related to equation (2.1) is the following celebrated conjecture by Ennio De Giorgi:

Conjecture 0.1 (see [22]) Let $u : \mathbb{R}^n \longrightarrow [-1, 1]$ be a solution of (2.1) in the whole of $\mathbb{R}^n$ such that

$$\frac{\partial u}{\partial x_n}(x) > 0 \quad \text{for all } x \in \mathbb{R}^n. \tag{2.10}$$

Is it true that u is 1D – that is, $u(x) = u_0(\omega \cdot x)$ for some $u_0 : \mathbb{R} \longrightarrow \mathbb{R}$ and $\omega \in S^{n-1}$ – at least if $n \leq 8$?

For a survey on Conjecture 0.1, we refer to [29]. See also [27] for a series of related rigidity problems in elliptic equations. Here, we just recall that a positive answer to Conjecture 0.1 (and, in fact, to more general problems) has been provided in dimensions 2 and 3, see [2,4,6,36].

In its full generality and to the best of our knowledge, the problem posed in Conjecture 0.1 is still open in dimensions 4 to 8, see [37], but the claim in Conjecture 0.1 holds true also in dimensions 4 to 8 under the following limit assumption:

$$\lim_{x_n \to \pm\infty} u(x', x_n) = \pm 1, \tag{2.11}$$

see [46–48].

Furthermore, condition (2.11) can be relaxed. For instance, in [30] the behavior of the limit profiles

$$u^{\pm}(x') := \lim_{x_n \to \pm\infty} u(x', x_n)$$

for monotone solutions u is taken into account, and it is proved that if both u^+ and u^- are 2D (i.e. they depend on at most two Euclidean variables), then (2.11) is automatically satisfied and, as a consequence, if $N \leq 8$, then u is 1D. Interestingly, when $N \leq 4$, it is enough to suppose that one between u^+ and u^- is 2D in order to deduce (2.11) and that u is 1D (see in particular Theorems 1.1 and 1.2 in [30]).

These types of results focus on the 1D symmetry "inherited from the profiles $u^{\pm}$ at infinity" of the monotone solutions of (2.1). Other symmetry results can be obtained from suitable asymptotic properties of flatness type of the level sets of the solutions, or from the graph properties of such level sets, see in particular [30, 33]. Results for quasi-minimal solutions are also available in the literature, see [28, 30].

Conjecture 0.1 is also related to a number of problems arising in related fields, such as geometry and dynamical systems, see [32] and the references therein.

We also recall that a variant of Conjecture 0.1, in which the monotonicity property in (2.10) is replaced by a uniform limit assumption at infinity, was proposed independently by Gary William Gibbons:

Conjecture 0.2 (see [19]) Let $u : \mathbb{R}^n \to [-1, 1]$ be a solution of (2.1) in the whole of $\mathbb{R}^n$ such that

$$\lim_{x_n \to \pm\infty} u(x', x_n) = \pm 1 \quad \text{uniformly with respect to } x' := (x_1, \ldots, x_{n-1}) \in \mathbb{R}^{n-1}. \tag{2.12}$$

Is it true that $u(x) = u_0(x_n)$ for some $u_0 : \mathbb{R} \to \mathbb{R}$?

Interestingly, Conjecture 0.2 turns out to be true in any dimension, as proved independently in [5,7,26]. In this sense, the uniform limit condition in (2.12) happens to be significantly stronger and to impose further crucial restrictions when compared with the monotonicity assumption in (2.10). Also, the uniform assumption in (2.12) ends up being dramatically stronger than the simple limit condition in (2.11) and so it provides significantly different types of results, in particularly ruling out the example in [44].

A variational variant of Conjecture 0.1 replaces the monotonicity assumption in (2.10) with a minimality assumption: with respect to this, it is proved in [46] that

$$\begin{gathered}\text{if } u:\mathbb{R}^n\to[-1,1] \text{ is a minimizer for } \mathcal{J}_\Omega\\ \text{in any bounded domain } \Omega\subset\mathbb{R}^n, \text{ and } n\le 7, \text{ then } u \text{ is 1D.}\end{gathered} \tag{2.13}$$

The example in [44] implies that a similar result cannot hold in dimension 9 and higher. The case of dimension 8 has been recently addressed in [41], which established the existence of minimizers for $\mathcal{J}_\Omega$ in dimension 8 and above whose level sets are asymptotic to a non-trivial cone (as a consequence, the result in (2.13) is optimal in terms of dimension). Using a method introduced in [39], the construction in [41] also provides, as a byproduct, several examples in dimension 9 and higher which are structurally different from the one in [44]. See also [14] for a detailed study of the Morse index of the saddle-shaped solutions of the Allen-Cahn equation.

3 The fractional Allen-Cahn equation

Now, we consider a nonlocal analogue of the Allen-Cahn equation in (2.1) and we investigate which of the above classical results remain valid also in this generality. The model that we take into account aims at dealing with long-range interactions which can influence the coexistence of these two phases, and indeed these contributions "coming from far" can (and typically do) produce a number of new phenomena with respect to the classical case.

The problem that we discuss involves the fractional Laplace[1] operator with $s\in(0,1)$, defined as

$$(-\Delta)^s u(x) := c_{n,s}\int_{\mathbb{R}^n}\frac{2u(x)-u(x+y)-u(x-y)}{|y|^{n+2s}}\,dy. \tag{3.1}$$

Here $c_{n,s}$ is a suitable renormalization constant, chosen in such a way that, if u is smooth and rapidly decreasing, then the Fourier Transform of $(-\Delta)^s u$ coincides with $|\xi|^{2s}\hat{u}(\xi)$, being $\hat{u}$ the Fourier Transform of u (equivalently, the fractional Laplacian acts as multiplication by $|\xi|^{2s}$ in Fourier space). See e.g. [1,8,23,35,40,54,56] for the basics on this operator and for several motivations and applications. The fractional counterpart of the Allen-Cahn equation in (2.1) is

$$(-\Delta)^s u = u-u^3 \quad \text{in } \Omega. \tag{3.2}$$

[1] As a matter of fact, in the literature, several notions of fractional Laplacians have been considered, such as a "regional" and a "spectral" fractional version of the Laplacian. These operators are, in general, structurally very different from the integral operator defined in (3.1), see e.g. [53]. In our context, the choice of the operator in (3.1) seemed preferable since it possesses natural scaling properties and therefore it offers a natural compatibility with spatial blow-downs of the solution. In addition, from the physical point of view, the integral kernel models explicitly a point by point interaction, which can be seen as a natural "global" resistance to phase changes in the model.

Akin its classical counterpart, equation (3.2) also comes from a variational principle and it corresponds to the minimization of the energy functional

$$\mathcal{J}_{s,\Omega}(u) := \frac{c_{n,s}}{4} \iint_{Q_\Omega} \frac{\left|u(x) - u(y)\right|^2}{|x - y|^{n+2s}}\, dx\, dy + \int_\Omega W\big(u(x)\big)\, dx, \tag{3.3}$$

up to scaling constants which are omitted for the sake of simplicity. In (3.3) we used the notation

$$\begin{aligned} Q_\Omega &:= \mathbb{R}^{2n} \setminus (\mathbb{R}^n \setminus \Omega)^2 \\ &= \Big(\Omega \times \Omega\Big) \cup \Big(\Omega \times (\mathbb{R}^n \setminus \Omega)\Big) \cup \Big((\mathbb{R}^n \setminus \Omega) \times \Omega\Big). \end{aligned}$$

Comparing the nonlocal energy in (3.3) with its classical counterpart in (2.2), we notice that the fractional model takes into account long-range particle interactions. As a matter of fact, the local interfacial term modeled in (2.2) by the Dirichlet energy is replaced in (3.3) by a Gagliardo-Sobolev-Slobodeckij seminorm. The role of the domain Q_Ω is to collect all the couples $(x, y) \in \mathbb{R}^n \times \mathbb{R}^n$ for which at least one of the points x, y belongs to the container Ω: interestingly, while in (2.2) the interface term takes into account the points in Ω, which can be seen as the complement in $\mathbb{R}^n$ of the "inactive" set $\mathbb{R}^n \setminus \Omega$, the domain Q_Ω describing the long-range interaction in (3.3) consists in the complement in $\mathbb{R}^{2n}$ of the "inactive" couples of points in $(\mathbb{R}^n \setminus \Omega)^2$ (hence, Q_Ω takes into account all the "active" points which interact with points inside the container).

In recent years, a great amount of research was carried out on equation (3.2) and on the energy functional (3.3) (and, in fact, also other types of long-range interactions have been taken into account, see [3]). Our goal here is to describe the results obtained in this fractional framework which are related (and possibly similar in spirit, or significantly different) with the ones described in Section 2.

First of all, we point out that an analogue of the classical Γ-convergence result holds true in the nonlocal setting, with an important variant: namely, the limit in (2.4) remains valid, but the limit set E has different features, according to the fractional parameter s. We will state this result in the forthcoming Theorem 3.1. To this end, to treat the case $s \in (0, 1/2)$, we need to recall the notion of fractional minimal surface, as introduced in [15]. Given $s \in (0, 1/2)$ and two (measurable) disjoint sets $A, B \subseteq \mathbb{R}^n$, one defines the s-interaction between A and B by

$$I_s(A, B) := \iint_{A \times B} \frac{dx\, dy}{|x - y|^{n+2s}}.$$

Then, the s-perimeter of a set E in the domain Ω is defined as

$$\mathrm{Per}_s(E, \Omega) := I_s(E \cap \Omega, E^c \cap \Omega) + I_s(E \cap \Omega, E^c \cap \Omega^c) + I_s(E \cap \Omega^c, E^c \cap \Omega),$$

where we used the complimentary set notation $A^c := \mathbb{R}^n \setminus A$.

To state the Γ-convergence result, it is also convenient to introduce a scaled version of the fractional functional in (3.3). Namely, we set

$$\mathcal{J}_{s,\Omega,\varepsilon}(u) := \begin{cases} \varepsilon^{2s-1} \displaystyle\iint_{Q_\Omega} \frac{|u(x)-u(y)|^2}{|x-y|^{n+2s}}\,dx\,dy + \frac{1}{\varepsilon}\int_\Omega W(u(x))\,dx & \text{if } s \in \left(\frac{1}{2},1\right), \\ \dfrac{1}{|\log\varepsilon|}\left[\displaystyle\iint_{Q_\Omega} \frac{|u(x)-u(y)|^2}{|x-y|^{n+2s}}\,dx\,dy + \frac{1}{\varepsilon}\int_\Omega W\big(u(x)\big)\,dx\right] & \text{if } s = \frac{1}{2}, \\ \displaystyle\iint_{Q_\Omega} \frac{|u(x)-u(y)|^2}{|x-y|^{n+2s}}\,dx\,dy + \frac{1}{\varepsilon^{2s}}\int_\Omega W\big(u(x)\big)\,dx & \text{if } s \in \left(0,\frac{1}{2}\right). \end{cases}$$

This scaled functional is obtained from (3.3), using the blow-down sequence in (2.3), after a multiplication that keeps the energy of the one-dimensional profile bounded uniformly in ε. In this setting, we can state the Γ-convergence result for the fractional Allen-Cahn functional as follows:

Theorem 3.1 (Theorem 1.5 in [51]) *If Ω is a smooth domain and $u_\varepsilon : \Omega \to [-1,1]$ is a sequence of minimizers for $\mathcal{J}_{s,\Omega,\varepsilon}$ such that*

$$\sup_{\varepsilon\in(0,1)} \mathcal{J}_{s,\Omega,\varepsilon}(u_\varepsilon) < +\infty,$$

then, up to a subsequence,

$$\lim_{\varepsilon\searrow 0} u_\varepsilon = u_0 := \chi_E - \chi_{\mathbb{R}^n\setminus E} \quad \text{in } L^1(\Omega),$$

for some set $E \subseteq \mathbb{R}^n$.

If $s \in [1/2, 1)$, the set E minimizes the perimeter in $\overline{\Omega}$ with respect to its boundary datum.

If instead $s \in (0, 1/2)$ and u_ε converges weakly to u_0 in $\mathbb{R}^n \setminus \Omega$, then the set E minimizes the fractional perimeter Per_s *in Ω with respect to its datum in $\mathbb{R}^n \setminus \Omega$.*

It is interesting to remark that the Γ-convergence results in Theorem 3.1 are significantly easier in the case $s \in (0, 1/2)$, since characteristic functions are admissible competitors, having finite energy. Instead, the case $s \in [1/2, 1)$ is much harder to treat, since one has to "reconstruct" a local energy from all the nonlocal contributions in the limit, and therefore a fine measure theoretic analysis of integral contributions is needed in this case.

We now briefly discuss the fractional version of the energy and density estimates. We will see that the estimate in (2.6), which is somehow geometric (stating that, in the large, both phases occupy a non-negligible portion of a large ball centered at the interface), still holds in the fractional case. Conversely, the energy bound in (2.5) is influenced by the fractional parameter s, in the same way as the one presented in Theorem 3.1: indeed, for large values of the parameter s, the estimate in (2.5) remains the same, while for small values of s the energy contributions "coming from infinity" are not anymore negligible and they carry an additional amount of energy in a large ball (though this energy produced by the phase transition remains negligible with respect to the size of the ball). The precise result goes as follows:

Theorem 3.2 (Theorems 1.3 and 1.4 in [52]) *If $R \geq 1$ and $u : B_{R+1} \to [-1,1]$ is a minimizer of $\mathcal{J}_{s,B_{R+1}}$, then*

$$\mathcal{J}_{s,B_{R+1}}(u) \leq \begin{cases} CR^{n-1} & \text{if } s \in \left(\frac{1}{2}, 1\right), \\ CR^{n-1}\log R & \text{if } s = \frac{1}{2}, \\ CR^{n-2s} & \text{if } s \in \left(0, \frac{1}{2}\right), \end{cases}$$

for some $C > 0$.

In addition, if $u(0) = 0$, then

$$\text{the Lebesgue measures of } \{u > 1/2\} \text{ and } \{u < -1/2\} \text{ in } B_R \\ \text{are both greater than } cR^n,$$

for some $c > 0$.

Of course, the constants in Theorem 3.2 depend, in general, on n and s. Though weaker (at least for small s) than in the classical case, the estimates in Theorem 3.2 are sufficient to obtain the locally uniform convergence of the level sets of minimizers, as stated in the following result:

Corollary 3.1 (Corollary 1.7 in [52]) *If Ω is a smooth domain, $E \subseteq \mathbb{R}^n$ and $u_\varepsilon : \Omega \to [-1,1]$ is a minimizer of $\mathcal{J}_{s,\Omega,\varepsilon}$ such that (2.4) holds true, then the set $\{|u_\varepsilon| \leq 1/2\}$ converges locally uniformly in Ω to ∂E as $\varepsilon \searrow 0$.*

Now, we discuss the fractional analogue of (2.7). To this aim, it is convenient to introduce the notion of extension solution of (3.2) (see [16]). Namely, we consider the Poisson Kernel

$$\mathbb{R}^n \times (0, +\infty) =: \mathbb{R}^{n+1}_+ \ni (x,t) \longmapsto P(x,t) := \bar{c}_{n,s} \frac{t^{2s}}{(|x|^2 + t^2)^{\frac{n+2s}{2}}},$$

where $\bar{c}_{n,s} > 0$ is the normalizing constant for which

$$\int_{\mathbb{R}^n} P(x,t)\, dx = 1,$$

for any $t > 0$. Given $u : \mathbb{R}^n \to [-1,1]$, we define

$$\mathbb{R}^{n+1}_+ \ni (x,t) \longmapsto E_u(x,t) := \int_{\mathbb{R}^n} P(x-y,t)\, u(y)\, dy.$$

Then, if u is sufficiently smooth, we have that E_u reconstructs the fractional Laplacian of u as a weighted Neumann term: more precisely, one has that E_u satisfies

$$\begin{cases} \operatorname{div}(t^\alpha \nabla E_u) = 0 & \text{in } \mathbb{R}^{n+1}_+, \\ \tilde{c}_s \lim_{t \searrow 0} t^\alpha \partial_t E_u = -(-\Delta)^s u & \text{in } \mathbb{R}^n, \end{cases}$$

where $\alpha := 1 - 2s \in (-1, 1)$. The constant $\tilde{c}_s > 0$ is needed just for normalization purposes (and it can be explicitly calculated, see e.g. Remark 3.11(a) in [11]). Hence, if u is a solution of (3.2), then E_u is a solution of

$$\begin{cases} \operatorname{div}(t^\alpha \nabla E_u) = 0 & \text{in } \mathbb{R}^{n+1}_+, \\ \tilde{c}_s \lim_{t \searrow 0} t^\alpha \partial_t E_u = u^3 - u & \text{in } \mathbb{R}^n. \end{cases} \tag{3.4}$$

Since, to the best of our knowledge, the fractional counterpart of (2.7) is at the moment understood only when $n = 1$, we will consider in (3.4) the case in which $x \in \mathbb{R}$, namely

$$\begin{cases} \operatorname{div}(t^\alpha \nabla E_u) = 0 & \text{in } \mathbb{R}^2_+, \\ \tilde{c}_s \lim_{t \searrow 0} t^\alpha \partial_t E_u = u^3 - u & \text{in } \mathbb{R}, \end{cases}$$

and look at the related energy functional

$$F(x, y) := (1 - s) \int_0^y t^\alpha \left(|\partial_x E_u(x, t)|^2 - |\partial_t E_u(x, t)|^2 \right) dt.$$

In this setting, the following result holds true:

Theorem 3.3 (Theorem 2.3(i) of [11]) *Let $u : \mathbb{R} \to [-1, 1]$ be a solution of* (3.2) *such that*

$$\partial_x u(x) > 0 \qquad \text{and} \qquad \lim_{x \to \pm\infty} u(x) = \pm 1.$$

Then, for any $x \in \mathbb{R}$ and any $y \geq 0$ we have that

$$F(x, y) \leq W(u(x)) = F(x, +\infty).$$

Interestingly, semilinear fractional equations possess a formal Hamiltonian structure in infinite dimensions (see Section 1.1 in [11]) and Theorem 3.3 recovers the classical Conservation of Energy Principle as $s \nearrow 1$ (see Section 6 in [11]). It is an open problem to understand the possible validity of results as in Theorem 3.3 when $n \geq 2$.

The last part of this note aims at discussing the recent developments of the symmetry results for solutions of equation (3.2), in view of the problems posed in Conjectures 0.1 and 0.2 for the classical case. As a matter of fact, the analogue of Conjecture 0.2 possesses a positive answer also in the fractional setting, for any dimension n and any fractional exponent $s \in (0, 1)$, see Theorem 2 in [31].

As for the analogue of Conjecture 0.1 in the fractional framework, the problem is open in its generality, but it possesses a positive answer for all $n \leq 3$ and $s \in (0, 1)$, and also for $n = 4$ and $s = 1/2$, according to the following result:

Theorem 3.4 *Let* $u : \mathbb{R}^n \longrightarrow [-1, 1]$ *be a solution of* (3.2) *in the whole of* $\mathbb{R}^n$ *such that*

$$\frac{\partial u}{\partial x_n}(x) > 0 \quad \text{for all } x \in \mathbb{R}^n.$$

Suppose that either

$$n \leq 3 \qquad \text{and} \qquad s \in (0, 1),$$

or

$$n = 4 \qquad \text{and} \qquad s = \frac{1}{2}.$$

Then u is 1*D.*

Theorem 3.4 is due to [13] when $n = 2$ and $s = 1/2$, [12, 55] when $n = 2$ and $s \in (0, 1)$, [9] when $n = 3$ and $s = 1/2$, [10] when $n = 3$ and $s \in (1/2, 1)$, [24] (based also on preliminary rigidity results in [25]) when $n = 3$ and $s \in (0, 1/2)$, [34] when $n = 4$ and $s = 1/2$. The cases remained open will surely provide several very interesting and challenging complications. It is also worth to point out that, at the moment, there is no counterexample in the literature to statements as the one in Theorem 3.4 in higher dimensions – nevertheless an important counterexample to the validity of Theorem 3.4 in dimension $n \geq 9$ when $s \in (1/2, 1)$ has been recently announced by H. Chan, J. Dávila, M. del Pino, Y. Liu and J. Wei (see the comments after Theorem 1.3 in [20]).

The validity of Theorem 3.4 in higher dimensions under the additional limit assumption in (2.11) has been also investigated in the recent literature. At the moment, the best result known on this topic can be summarized as follows:

Theorem 3.5 *Let* $n \leq 8$*. Then, there exists* $\varepsilon_0 \in \left(0, \frac{1}{2}\right]$ *such that for any* $s \in \left(\frac{1}{2} - \epsilon_0, 1\right)$ *the following statement holds true.*

Let $u : \mathbb{R}^n \longrightarrow [-1, 1]$ *be a solution of* (3.2) *in the whole of* $\mathbb{R}^n$ *such that*

$$\frac{\partial u}{\partial x_n}(x) > 0 \quad \text{for all } x \in \mathbb{R}^n$$

$$\text{and} \qquad \lim_{x_n \to \pm\infty} u(x', x_n) = \pm 1 \quad \text{for all } x' \in \mathbb{R}^{n-1}.$$

Then, u is 1*D.*

Theorem 3.5 consists in fact of the superposition of three different results, also obtained with a different approach. The result of Theorem 3.5 when s is larger than $1/2$ follows from Theorem 1.3 in [49]. When $s = 1/2$, the result was announced after Theorem 1.1 in [49] and established in Theorem 1.3 of [50]. The case $s \in \left(\frac{1}{2} - \epsilon_0, \frac{1}{2}\right)$ has been established in Theorem 1.6 of [25]. In this latter framework, the quantity ϵ_0 is a universal constant (unfortunately, not explicitly computed by the proof), and the arguments of the proof rely on it in order to deduce the flatness of the corresponding limit interface, which is in this case described by nonlocal minimal surfaces: since such flatness results are only known above the threshold provided by ϵ_0 (see Theorems 2–5 in [17]), also Theorem 3.5 suffers of this restriction. Of course, it is an important open problem to establish whether Theorem 3.5 holds true for a wider range of fractional parameter, as well as it would be very interesting to establish optimal regularity results for nonlocal minimal surfaces.

The fractional counterpart of classical symmetry results under the minimality assumption in (2.13) has also been taken into account, with results similar to Theorem 3.5, which can be summarized as follows:

Theorem 3.6 *Let $n \le 7$. Then, there exists $\varepsilon_0 \in \left(0, \frac{1}{2}\right]$ such that for any $s \in \left(\frac{1}{2} - \epsilon_0, 1\right)$ the following statement holds true.*

Let $u : \mathbb{R}^n \to [-1, 1]$ be a minimizer for $\mathcal{J}_{s,\Omega}$ in any bounded domain $\Omega \subset \mathbb{R}^n$. Then, u is 1D.

Once again, Theorem 3.6 is a collage of different results obtained by different methods and dealing with different parameter ranges. Namely, the statement in Theorem 3.6 when s is larger than $1/2$ has been proved in Theorem 1.2 of [49], and the case $s = 1/2$ has been treated in Theorem 1.2 of [50]. The case $s \in \left(\frac{1}{2} - \epsilon_0, \frac{1}{2}\right)$ has been established in Theorem 1.5 of [25] (once again, in this context, the threshold given by ϵ_0 is used to apply the regularity results for nonlocal minimal surfaces in [17] and it is a very interesting problem to determine the possible validity of Theorem 3.6 when the dimensional and quantitative conditions are violated).

We think that it is important to stress the fact that the differences between the fractional exponent ranges in the previous results do not reflect a series of merely technical difficulties, but instead it reveals fundamental structural differences between the phase transitions when $s \in [1/2, 1)$ and when $s \in (0, 1/2)$. These differences are somehow inherited by the dichotomy provided in Theorem 3.1: indeed, as pointed out in this result, when $s \in [1/2, 1)$ the nonlocal phase transitions end up showing an interface corresponding to a local problem, while when $s \in (0, 1/2)$ the nonlocal features of the problem persist at any scale and produce a limit interface of nonlocal nature. The structural differences between local and nonlocal minimal surfaces may therefore produce significant differences on the phase transitions too: as a matter of fact, it happens that when $s \in (0, 1/2)$ the long-range interactions of points of the interface provide a number of additional rigidity properties which have no counterpart in the classical case. To exhibit a particular phenomenon related to this feature, we recall the forthcoming result in Theorem 3.7. To state this result in a concise way, we introduce the notion of "asymptotically flat" interface, which can be stated as follows. First of all, we say that the interface of u in B_R is trapped in a slab of width $2aR$ in direction $\omega \in S^{n-1}$ if

$$\begin{aligned} &\{x \in B_R \text{ s.t. } \omega \cdot x \le -aR\} \subseteq \left\{x \in B_R \text{ s.t. } u(x) \le -\frac{9}{10}\right\} \\ \text{and } &\left\{x \in B_R \text{ s.t. } u(x) \le \frac{9}{10}\right\} \subseteq \{x \in B_R \text{ s.t. } \omega \cdot x \le aR\}. \end{aligned} \tag{3.5}$$

Of course, when $a \geq 1$, such condition is always satisfied, but the smaller the a is, the flatter the interface is in the ball B_R. We say that the interface of u is asymptotically flat if there exists $R_0 > 0$ such that for any $R \geq R_0$ there exist $\omega(R) \in S^{n-1}$ and $a(R) \geq 0$ such that the interface of u in B_R is trapped in a slab of width $2a(R)\,R$ in direction $\omega(R)$ with

$$\lim_{R\to+\infty} a(R) = 0.$$

Roughly speaking, the interface of u is asymptotically flat if, in large balls, it is trapped into slabs with small ratio between the width of the slab and the radius of the ball (possibly, up to rotations which can vary from one scale to another). In this setting, we have:

Theorem 3.7 (Theorem 1.2 in [25]) *Let $s \in (0, 1/2)$ and u be a solution of* (3.2) *in $\mathbb{R}^n$. Assume that the interface of u is asymptotically flat. Then, u is $1D$.*

We think that Theorem 3.7 reveals several surprising aspects of nonlocal phase transitions in the regime $s \in (0, 1/2)$, where the contributions from infinity happen to be dominant. Indeed, the result in Theorem 3.7 is valid for all solutions, without any monotonicity or energy restrictions. This suggests that if one has a phase coexistence in this regime, plugging additional energy into the system can only produce two alternatives:

- either the interface oscillates significantly at infinity (i.e., the flatness assumption of Theorem 3.7 is not satisfied),
- or the graph of the function u that describes the state parameter of the system can oscillate, but (due to Theorem 3.7) such function is necessarily 1D and therefore the phase separation occurs along parallel hyperplanes, with possible multiplicity.

It is also interesting to stress that a result as the one in Theorem 3.7 does not hold for the classical Allen-Cahn equation (and indeed Theorem 3.7 reveals a purely nonlocal phenomenon). As a matter of fact, in Theorem 1 of [45] a solution of (2.1) in $\mathbb{R}^3$ is constructed whose level sets resemble an appropriate dilation of a catenoid: namely, the level sets of this solution lie in the asymptotically flat region $\{x = (x', x_3) \in \mathbb{R}^3 \text{ s.t. } |x_3| \leq C(1 + \log(1 + |x'|)\}$, for a suitable $C > 0$. In particular, condition (3.5) is satisfied with $\omega(R) := (0, 0, 1)$ and $a(R) := \frac{C(1+\log(1+R)}{R}$, which is infinitesimal as $R \to +\infty$ and, as a byproduct, the interface of this solution is asymptotically flat. Clearly, the solution constructed in [45] is not 1D, since its level sets are modeled on a catenoid rather than on a plane, and therefore this example shows that an analogue of Theorem 3.7 is false in the classical case.

A fractional counterpart of [45] has been recently provided in [20], in the fractional regime $s \in (1/2, 1)$. In particular, Theorem 1.3 of [20] establishes the existence of an entire solution of (3.2) in $\mathbb{R}^3$ vanishing on a rotationally symmetric surface which resembles a catenoid with sublinear growth at infinity. This example shows that an analogue of Theorem 3.7 is false when $s \in (1/2, 1)$.

At the moment, it is an open problem to construct solutions of (3.2) in $\mathbb{R}^3$ with level sets modeled on a catenoid when $s = 1/2$, see Remark 1.4 in [20]: on the one hand, the case $s = 1/2$ relates the large-scale picture of the interfaces to the classical (and not to the nonlocal) minimal surfaces (recall Theorem 3.1), therefore it is still conceptually possible to construct catenoid-like examples in this setting; on the other hand, the infinite dimensional gluing method in [20] deeply relies on the condition $s \in (1/2, 1)$, therefore important modifications would be needed to achieve similar results when $s = 1/2$.

Interestingly, nonlocal catenoids corresponding to the case $s \in (0, 1/2)$ have been constructed in [21] but, remarkably, such surfaces possess linear (rather than sub-linear) growth at infinity (therefore, possible solutions of (3.2) modeled on such catenoids would not possess asymptotically flat interfaces, which is indeed in agreement with Theorem 3.7).

We end this note with a few comments on the proof of Theorem 3.7: the main argument is an "improvement of flatness" which says that if a sufficiently sharp interface is appropriately flat "from the unit ball B_1 towards infinity", then it is even flatter in $B_{1/2}$ (see Theorem 1.1 in [25] for full details). Suitable iterations of this argument give a control of the interface all the way to infinity, showing in particular that (possibly after a rotation) the interface is trapped between a graph that is Lipschitz and sublinear at infinity and its translate. This control of the growth at infinity of the interface in turn allows the use of the sliding method "in a tilted direction". Namely, one fixes $e' \in \mathbb{R}^{n-1}$ with $|e'| = 1$ and $\delta > 0$ and set

$$e_\delta := \frac{(e', \delta)}{\sqrt{1+\delta^2}} \in S^{n-1} \qquad \text{and} \qquad u^{(t)}(x) := u(x - e_\delta t).$$

We point out that $u^{(t)}$ is the translation of the original solution u in the slightly oblique direction e_δ and so the growth control of the interface, combined with a precise estimate of the decay of the solution and the maximum principle, implies that $u^{(t)}$ lies below u for t sufficiently large (say, $t \geq T(e', \delta)$, and we observe that the use of maximum principle here relies on the monotonicity property of the Allen-Cahn nonlinearity outside the interface, namely the function $f(r) := r - r^3$ is decreasing when $|r - 1| \leq \frac{1}{10}$).

Then, one keeps sliding $u^{(t)}$, reducing the value of t, and using again the maximum principle it follows that $u^{(t)} \leq u$ for any $t \geq 0$. As a consequence of this, for any $t \geq 0$, any $x = (x', x_n) \in \mathbb{R}^n$ and any $e' \in S^{n-2}$,

$$u\left((x', x_n) - \frac{(e't, \delta t)}{\sqrt{1+\delta^2}}\right) = u(x - e_\delta t) = u^t(x) \leq u(x)$$

and accordingly, sending $\delta \searrow 0$,

$$u(x' - e't, x_n) \leq u(x). \tag{3.6}$$

Writing (3.6) with e' replaced by $-e'$ (as well as x replaced by y), it follows that

$$u(y' + e't, y_n) \leq u(y), \tag{3.7}$$

for any $y \in \mathbb{R}^n$ and any $e' \in S^{n-2}$. Then, choosing $y := x - (e't, 0)$ in (3.7) and using again (3.6),

$$u(x) = u(x' - e't + e't, x_n) \le u(x' - e't, x_n) \le u(x)$$

and therefore

$$u(x) = u(x' - e't, x_n),$$

for every $x \in \mathbb{R}^n$, every $t \ge 0$ and every $e' \in S^{n-2}$. This shows that, possibly after a rotation, the solution u depends only on x_n and so it completes the proof of Theorem 3.7.

Closing remarks

The recent literature has taken into account a fractional version of the Allen-Cahn equation. From the physical viewpoint, this new equation has the advantage of comprising in the model the long-range interaction which can influence the distribution of phases, by preventing sudden phase changes that would be allowed by the double-well potential by itself.

In mathematical terms, the nonlocal character of this equations poses a number of very challenging questions and creates several new and interesting phenomena.

In particular, the minimal interfaces of the fractional Allen-Cahn equation is influenced by the fractional parameter itself: for values of the fractional parameter close to the integer, the large scale behavior of the interfaces coincides with the classical ones, but for small values of the fractional parameter the interfaces are related to nonlocal minimal surfaces and maintain thereby their nonlocal character at any scale.

A problem of fundamental importance is also to understand under which conditions global solutions are necessary one-dimensional. This is the fractional counterpart of a classical problem posed by Ennio De Giorgi. Positive results in the fractional setting are available in dimension $n \le 3$, and also in dimension $n = 4$ for the square root of the Laplacian. The other cases are still open to the best of our knowledge and they provide some of the most important questions in the research focused on fractional equations.

Acknowledgements

It is a pleasure to thank Xavier Cabré and Joaquim Serra for their very useful comments and for several pleasant discussions. This work has been supported by the Australian Research Council Discovery Project "N.E.W. Nonlocal Equations at Work". The authors are members of GNAMPA/INdAM.

References

1. Abatangelo, N., Valdinoci, E.: Getting acquainted with the fractional Laplacian. Springer INdAM Ser. (2019)
2. Alberti, G., Ambrosio, L., Cabré, X.: On a long-standing conjecture of E. De Giorgi: symmetry in 3D for general nonlinearities and a local minimality property. Acta Appl. Math. **65**(1-3), 9–33 (2001). Special issue dedicated to Antonio Avantaggiati on the occasion of his 70th birthday.
3. Alberti, G., Bellettini, G.: A nonlocal anisotropic model for phase transitions. I. The optimal profile problem. Math. Ann. **310**(3), 527–560 (1998).
4. Ambrosio, L., Cabré, X.: Entire solutions of semilinear elliptic equations in $\mathbf{R}^3$ and a conjecture of De Giorgi. J. Amer. Math. Soc. **13**(4), 725–739 (2000).
5. Barlow, M.T., Bass, R.F., Gui, C.: The Liouville property and a conjecture of De Giorgi. Comm. Pure Appl. Math. **53**(8), 1007–1038 (2000).
6. Berestycki, H., Caffarelli, L., Nirenberg, L.: Further qualitative properties for elliptic equations in unbounded domains. Ann. Scuola Norm. Sup. Pisa Cl. Sci. (4) **25**(1-2), 69–94 (1998) (1997). Dedicated to Ennio De Giorgi
7. Berestycki, H., Hamel, F.c., Monneau, R.: One-dimensional symmetry of bounded entire solutions of some elliptic equations. Duke Math. J. **103**(3), 375–396 (2000).
8. Bucur, C., Valdinoci, E.: Nonlocal diffusion and applications, *Lecture Notes of the Unione Matematica Italiana*, vol. 20. Springer, [Cham]; Unione Matematica Italiana, Bologna (2016).
9. Cabré, X., Cinti, E.: Energy estimates and 1-D symmetry for nonlinear equations involving the half-Laplacian. Discrete Contin. Dyn. Syst. **28**(3), 1179–1206 (2010).
10. Cabré, X., Cinti, E.: Sharp energy estimates for nonlinear fractional diffusion equations. Calc. Var. Partial Differential Equations **49**(1-2), 233–269 (2014).
11. Cabré, X., Sire, Y.: Nonlinear equations for fractional Laplacians, I: Regularity, maximum principles, and Hamiltonian estimates. Ann. Inst. H. Poincaré Anal. Non Linéaire **31**(1), 23–53 (2014).
12. Cabré, X., Sire, Y.: Nonlinear equations for fractional Laplacians II: Existence, uniqueness, and qualitative properties of solutions. Trans. Amer. Math. Soc. **367**(2), 911–941 (2015).
13. Cabré, X., Solà-Morales, J.: Layer solutions in a half-space for boundary reactions. Comm. Pure Appl. Math. **58**(12), 1678–1732 (2005).
14. Cabré, X., Terra, J.: Saddle-shaped solutions of bistable diffusion equations in all of $\mathbb{R}^{2m}$. J. Eur. Math. Soc. (JEMS) **11**(4), 819–843 (2009).
15. Caffarelli, L., Roquejoffre, J.M., Savin, O.: Nonlocal minimal surfaces. Comm. Pure Appl. Math. **63**(9), 1111–1144 (2010).
16. Caffarelli, L., Silvestre, L.: An extension problem related to the fractional Laplacian. Comm. Partial Differential Equations **32**(7-9), 1245–1260 (2007).
17. Caffarelli, L., Valdinoci, E.: Regularity properties of nonlocal minimal surfaces via limiting arguments. Adv. Math. **248**, 843–871 (2013).
18. Caffarelli, L.A., Córdoba, A.: Uniform convergence of a singular perturbation problem. Comm. Pure Appl. Math. **48**(1), 1–12 (1995).
19. Carbou, G.: Unicité et minimalité des solutions d'une équation de Ginzburg-Landau. Ann. Inst. H. Poincaré Anal. Non Linéaire **12**(3), 305–318 (1995).
20. Chan, H., Liu, Y., Wei, J.: A gluing construction for fractional elliptic equations. Part I: a model problem on the catenoid. arXiv e-prints (2017)
21. Dávila, J., del Pino, M., Wei, J.: Nonlocal s-minimal surfaces and Lawson cones. J. Differential Geom. **109**(1), 111–175 (2018).
22. De Giorgi, E.: Convergence problems for functionals and operators. Proceedings of the International Meeting on Recent Methods in Nonlinear Analysis (Rome, 1978). pp. 131–188 (1979)
23. Di Nezza, E., Palatucci, G., Valdinoci, E.: Hitchhiker's guide to the fractional Sobolev spaces. Bull. Sci. Math. **136**(5), 521–573 (2012).

24. Dipierro, S., Farina, A., Valdinoci, E.: A three-dimensional symmetry result for a phase transition equation in the genuinely nonlocal regime. Calc. Var. Partial Differential Equations **57**(1), 57:15 (2018).
25. Dipierro, S., Serra, J., Valdinoci, E.: Improvement of flatness for nonlocal phase transitions. Amer. J. Math., in press
26. Farina, A.: Symmetry for solutions of semilinear elliptic equations in $\mathbf{R}^N$ and related conjectures. Ricerche Mat. **48**(suppl.), 129–154 (1999). Papers in memory of Ennio De Giorgi (Italian). ISSN 0035-5038
27. Farina, A.: Liouville-type theorems for elliptic problems. Handbook of differential equations: stationary partial differential equations. Vol. IV, pp. 61–116, Handb. Differ. Equ., Elsevier/North-Holland, Amsterdam, 2007.
28. Farina, A., Valdinoci, E.: Geometry of quasiminimal phase transitions. Calc. Var. Partial Differential Equations **33**(1), 1–35 (2008).
29. Farina, A., Valdinoci, E.: The state of the art for a conjecture of De Giorgi and related problems. Recent progress on reaction-diffusion systems and viscosity solutions, pp. 74–96, World Sci. Publ., Hackensack, NJ (2009).
30. Farina, A., Valdinoci, E.: 1D symmetry for solutions of semilinear and quasilinear elliptic equations. Trans. Amer. Math. Soc. **363**(2), 579–609 (2011).
31. Farina, A., Valdinoci, E.: Rigidity results for elliptic PDEs with uniform limits: an abstract framework with applications. Indiana Univ. Math. J. **60**(1), 121–141 (2011).
32. Farina, A., Valdinoci, E.: Some results on minimizers and stable solutions of a variational problem. Ergodic Theory Dynam. Systems **32**(4), 1302–1312 (2012).
33. Farina, A., Valdinoci, E.: 1D symmetry for semilinear PDEs from the limit interface of the solution. Comm. Partial Differential Equations **41**(4), 665–682 (2016).
34. Figalli, A., Serra, J.: On stable solutions for boundary reactions: a De Giorgi-type result in dimension 4 + 1. arXiv e-prints (2017)
35. Garofalo, N.: Fractional thoughts. New developments in the analysis of nonlocal operators, pp. 1–135, Contemp. Math., 723, Amer. Math. Soc., Providence, RI (2019).
36. Ghoussoub, N., Gui, C.: On a conjecture of De Giorgi and some related problems. Math. Ann. **311**(3), 481–491 (1998).
37. Ghoussoub, N., Gui, C.: On De Giorgi's conjecture in dimensions 4 and 5. Ann. of Math. (2) **157**(1), 313–334 (2003).
38. Gui, C.: Hamiltonian identities for elliptic partial differential equations. J. Funct. Anal. **254**(4), 904–933 (2008).
39. Jerison, D., Monneau, R.: Towards a counter-example to a conjecture of De Giorgi in high dimensions. Ann. Mat. Pura Appl. (4) **183**(4), 439–467 (2004).
40. Landkof, N.S.: Foundations of modern potential theory. Springer-Verlag, New York-Heidelberg (1972). Translated from the Russian by A. P. Doohovskoy; Die Grundlehren der mathematischen Wissenschaften, Band 180
41. Liu, Y., Wang, K., Wei, J.: Global minimizers of the Allen-Cahn equation in dimension $n \geq 8$. J. Math. Pures Appl. (9) **108**(6), 818–840 (2017).
42. Modica, L.: A gradient bound and a Liouville theorem for nonlinear Poisson equations. Comm. Pure Appl. Math. **38**(5), 679–684 (1985).
43. Modica, L., Mortola, S.: Un esempio di Γ^--convergenza. Boll. Un. Mat. Ital. B (5) **14**(1), 285–299 (1977)
44. del Pino, M., Kowalczyk, M., Wei, J.: On De Giorgi's conjecture in dimension $N \geq 9$. Ann. of Math. (2) **174**(3), 1485–1569 (2011).
45. del Pino, M., Kowalczyk, M., Wei, J.: Entire solutions of the Allen-Cahn equation and complete embedded minimal surfaces of finite total curvature in $\mathbb{R}^3$. J. Differential Geom. **93**(1), 67–131 (2013)
46. Savin, O.: Regularity of flat level sets in phase transitions. Ann. of Math. (2) **169**(1), 41–78 (2009).
47. Savin, O.: Phase transitions, minimal surfaces and a conjecture of De Giorgi Current developments in mathematics, 2009, pp. 59–113 (2010), Int. Press, Somerville, MA

48. Savin, O.: Some remarks on the classification of global solutions with asymptotically flat level sets. Calc. Var. Partial Differential Equations **56**(5), Art. 141, 21 (2017).
49. Savin, O.: Rigidity of minimizers in nonlocal phase transitions. Anal. PDE **11**(8), 1881–1900 (2018).
50. Savin, O.: Rigidity of minimizers in nonlocal phase transitions II. Anal. Theory Appl. **35**(1), 1–27 (2019)
51. Savin, O., Valdinoci, E.: Γ-convergence for nonlocal phase transitions. Ann. Inst. H. Poincaré Anal. Non Linéaire **29**(4), 479–500 (2012).
52. Savin, O., Valdinoci, E.: Density estimates for a variational model driven by the Gagliardo norm. J. Math. Pures Appl. (9) **101**(1), 1–26 (2014).
53. Servadei, R., Valdinoci, E.: On the spectrum of two different fractional operators. Proc. Roy. Soc. Edinburgh Sect. A **144**(4), 831–855 (2014).
54. Silvestre, L.E.: Regularity of the obstacle problem for a fractional power of the Laplace operator. ProQuest LLC, Ann Arbor, MI (2005). Thesis (Ph.D.)–The University of Texas at Austin
55. Sire, Y., Valdinoci, E.: Fractional Laplacian phase transitions and boundary reactions: a geometric inequality and a symmetry result. J. Funct. Anal. **256**(6), 1842–1864 (2009).
56. Stein, E.M.: Singular integrals and differentiability properties of functions. Princeton Mathematical Series, No. 30. Princeton University Press, Princeton, N.J. (1970)

Elements of Statistical Inference in 2-Wasserstein Space

Johannes Ebert, Vladimir Spokoiny and Alexandra Suvorikova *

Abstract This work addresses an issue of statistical inference for the datasets lacking underlying linear structure, which makes impossible the direct application of standard inference techniques and requires a development of a new tool-box taking into account properties of the underlying space. We present an approach based on optimal transportation theory that is a convenient instrument for the analysis of complex data sets. The theory originates from seminal works of a french mathematician Gaspard Monge published at the end of 18th century. This chapter recalls the basics on optimal transportations theory, explains the ideas behind statistical inference on non-linear manifolds, and as an illustrative example presents a novel approach of construction of non asymptotic confidence sets for so called Wasserstein barycenter, a generalized analogous of Euclidean mean to the case of non-linear space endowed with a particular distance belonging to a class of Earth-Mover distances that it is a main object of study in optimal transportation theory. The chapter is based on the paper [18].

1 Introduction

Many applications in modern statistics go beyond the scope of classic setting and deal with data which lie on a certain manifold. For example, this is the case in

Johannes Ebert
Humboldt University of Berlin, e-mail: `johannes.ebert@hotmail.com`

Vladimir Spokoiny
Weierstrass Institute for Applied Analysis and Stochastics, e-mail: `vladimir.spokoiny@wias-berlin.de`

Alexandra Suvorikova
Potsdam University, e-mail: `suvorikova@math.uni-potsdam.de`

* the work was performed while at the Weierstrass Institute for Applied Analysis and Stochastics.

M. Hintermüller, J. F. Rodrigues (eds.), *Topics in Applied Analysis and Optimisation*,
CIM Series in Mathematical Sciences, https://doi.org/10.1007/978-3-030-33116-0_6

computer vision, medical image analysis, bioinformatics, statistics on shape space and so on. Usually one is interested in detection of common underlying structure of a random data set reflected by geometrical properties inherent to the majority of observations. It is a very general concept that describes some hidden properties of the data set which have to be recovers, as they are of interest for further investigation. For instance, the problem of classification of neuro-cognitive states of mind is associated with detection of brain activity patterns in fMRI [32]. Another example comes from bioinformatics, namely from computational epigenetic, that aims to detect common patterns in gene expression regulation [12, 23]. The latter one is supposed to be one of the crucial aspects of morphogenesis. Pattern can also be interpreted in a more specific way as a "typical" geometric shape inherent to all observed items. Following [24] one may define shape as whatever remains after proper normalization of the object (i.e. rotations, dilations, and shifts are factored out). For example, the work [29] estimates "typical" spatial configuration of protein backbones. Basically, this setting appears in problems where the data is subjected to deformations through a random warping procedure. Such problems are also common for image analysis [6, 39] and shape analysis [22].

The obvious complexity of such data sets prevents us from modelling them as vectors in Euclidean space $\mathbb{R}^d$, as this class of models is too poor to capture all the effects of interest. Thus one needs to come up with a richer model class that would also be user-friendly in terms of theoretical analysis and computational tractability. A possible way to treat such data sets is to model them as measures, supported on some Polish space. As an illustration one may refer to [21]. Probably the most natural way of information extraction and filtering out the random noise is averaging. However, in the case of measures assumption of linearity appears to be too restrictive. For example in case of two Gaussian distributions naïve Euclidean average yields a mixture of Gaussians and not another Guassian: one can see that a family of Gaussians is not closed with respect to Euclidean averaging. Thus one needs a suitable analogue of classical Euclidean mean, generalised to an arbitrary metric space (X, ρ). The solution was suggested by a french mathematician Maurice René Fréchet in [19]. Let $\mathbb{P}$ be a data-generating distribution supported on X . A straightforward generalisation of the least-square estimator leads to the concept of the Fréchet mean that is the (set of) global minima of the variance

$$x^* \subseteq \arg\min_{x \in X} \int_X \rho^2(x, y)\mathbb{P}(dy),$$

where x^* is referred to as the population Fréchet mean that is not necessarily unique. However, under certain settings it can be considered as the typical patten or typical representative induced by $\mathbb{P}$. In what follows we assume, that $\mathbb{P}$ and X are such that x^* is unique. The issue is discussed in more details in Section 2. Given a data set $(y_1, ..., y_n)$ where observations are sampled independently in an identical way (an i.i.d. data set) s.t. $y_i \sim \mathbb{P}$, one can build an empirical estimator of x^*

$$x_n \stackrel{\text{def}}{=} \operatorname*{arg\,min}_{x \in X} \frac{1}{n} \sum_{i=1}^{n} \rho^2(x, y_i).$$

There are several works [8–10] that present a detailed study of the asymptotic properties of the empirical Fréchet mean in case X is a finite-dimensional differentiable manifold. The monograph [8] also describes the procedure of asymptotic confidence set construction for x_n. In this work we consider a particular case of (X, ρ), namely $X = \mathcal{P}_2(\mathbb{R}^d)$ is the space of all probability measures with finite second moment supported on $\mathbb{R}^d$. We also set $\rho = W_2$ to be 2-Wasserstein distance. Being an object of study in optimal transportation theory [4, 40, 41], it also appears in statistical data analysis in machine learning, see for example [34] and references therein. The concept of Fréchet mean for 2-Wasserstein space is specified by a seminal paper [1] which also provides conditions of its existence and uniqueness. We postpone the introduction of the distance and related quantities to Section 2 and for the present we keep in mind only the fact that $(\mathcal{P}_2(\mathbb{R}), W_2)$ is just a metric space which provides a convenient toolbox for description and processing objects possessing complex geometrical structure. From now on we refer to Fréchet mean in Wasserstein space as the Wasserstein barycenter. Thus, given a measure $\mathbb{P}$ supported on $\mathcal{P}_2(\mathbb{R}^d)$, its population barycenter μ^* and its empirical counterpart are written as

$$\mu^* \stackrel{\text{def}}{=} \operatorname*{arg\,min}_{\mu \in \mathcal{P}_2(\mathbb{R}^d)} \int_{\mathcal{P}(\mathbb{R}^d)} W_2^2(\mu, \nu) \mathbb{P}(d\nu), \quad \mu_n \stackrel{\text{def}}{=} \operatorname*{arg\,min}_{\mu \in \mathcal{P}_2(\mathbb{R}^d)} \frac{1}{n} \sum_{i=1}^{n} W_2^2(\mu, \nu_i),$$

where the latter is constructed using an i.i.d. sample $\{\nu_1, ..., \nu_n\}$. Wasserstein barycenters appear to be a popular tool for information aggregation in variety domains, such as image processing [35, 37], mathematical economics [13] and other statistical applications [3]. The work [11] provides a characterization of the population barycenter for various parametric classes of random transformations for probability measures with compact support. Recently, [28] established the convergence of the empirical barycenter of an i.i.d. sample of random measures on a locally geodesic metric space towards its population barycenter. The works [2] presents central limit theorem for μ_n. [27] generalises central limit theorem for the case of a barycenter, restricted on some affine sub-space of $\mathcal{P}_2(\mathbb{R}^d)$ and provides its concentration properties. Finally [25] presents law of large numbers for baryceneters constructed using an arbitrary transportation costs. The idea of statistical inference in Wasserstein space is introduced, for example, in [16]. The authors consider a statistical deformation model and obtain the asymptotic distribution and a bootstrap procedure for the Wasserstein barycenter. They use the results to construct a goodness-of-fit test for the deformation model. However, their study is limited to probability measures on the real line. The similar setting is discussed in [36]. Authors study the subspace of Gaussian measures on $\mathbb{R}^d$ and estimate Wasserstein distance $W_2(\nu_1, \nu_2)$ between two Gaussians ν_1 and ν_2, knowing its empirical counterparts $\hat{\nu}_1$, $\hat{\nu}_2$. Empirical measures are estimated using i.i.d. samples $X_1, ..., X_n \sim \nu_1$ and $Y_1, ..., Y_m \sim \nu_2$, with all $X_i, Y_j \in \mathbb{R}^d$.

The current study sets out to generalize the results in [16, 36] to the case where random observed objects are measures on the space $\mathbb{R}^d$. Namely, we consider an i.i.d. sample $\{\nu_1, ..., \nu_n\}$, $\nu_i \sim \mathbb{P}$ and introduce a following non-parametric test

$$T_n = \sqrt{n} W_2(\mu_n, \mu^*).$$

In what follows we present the procedure of construction of non-asymptotic data-driven confidence sets $C\big(z^{\flat}(\alpha)\big)$ around μ_n, s.t.

$$\mathbb{P}\Big(\mu^* \notin C\big(z^{\flat}(\alpha)\big)\Big) \leq \alpha.$$

The procedure exploits the idea of the multiplier bootstrapping technique [14,30,38]. The result is presented in Section 3.

The second part of this work presents a non-parametric two-sample test in 2-Wasserstein space which is further applied to the problem of change point detection. The idea is applicable to any real-world task which aims to detect geometrical changes in observed objects. Further we briefly recall what the problem of change point detection is. Let ν_t be a stochastic process in discrete time. The time moment t^* is supposed to be a change point if at this moment the data stream in hand undergoes some abrupt structural. This can be expressed as follows:

$$\begin{cases} \nu_t \sim \mathbb{P}_1, & t < t^*, \\ \nu_t \sim \mathbb{P}_2, & t \geq t^* \end{cases}$$

The goal is to detect the regime switch as soon as possible under a given false-alarm rate. We recommend [5] and references herein for a survey on existing methods of change point detection. In what follows we focus on a test in a running window. In brief, the idea is to split the window into to consecutive parts and carry out the homogeneity test using these two sub-samples. Let t be a candidate for a change point and let $(\nu_{t-h}, ..., \nu_{t+h-1})$ be observed data in the rolling window of size $2h$. The the above mentioned model of change point detection can be rewritten as

$$\begin{cases} \nu_i \sim \mathbb{P}_1, & i \in \{t-h, ..., t-1\}, \\ \nu_i \sim \mathbb{P}_2, & i \in \{t, ..., t+h-1\}. \end{cases}$$

Then the hypothesis of homogeneity H_0 and its alternative H_1 are

$$H_0 : \mathbb{P}_1 = \mathbb{P}_2, \qquad H_1 : \mathbb{P}_1 \neq \mathbb{P}_2.$$

As a test statistic we use the rescaled 2-Wasserstein distance between averages in the left and right half-windows respectively:

$$T_h(t) \stackrel{\text{def}}{=} \sqrt{h} W_2\big(\mu_l(t), \mu_r(t)\big),$$

where $\mu_l(t)$ and $\mu_r(t)$ are empirical barycenters in the left and right halves of a scrolling window:

$$\mu_l(t) \stackrel{\text{def}}{=} \arg\min_{\mu \in \mathcal{P}_2(\mathbb{R}^d)} \frac{1}{h} \sum_{t-h}^{h-1} W_2^2(\mu, \nu_i), \quad \mu_r(t) \stackrel{\text{def}}{=} \arg\min_{\mu \in \mathcal{P}_2(\mathbb{R}^d)} \frac{1}{h} \sum_{t}^{t+h-1} W_2^2(\mu, \nu_i).$$

A change point is supposed to be detect at time moment t if $T_h(t)$ exceeds some critical level:

$$T_h(t) \geq z_h(\alpha, t).$$

The crucial step of the method is the fully data-driven calibration of critical values $z_h(\alpha, t)$, that is also based on the idea of multiplier bootstrap.

As a starting point, we restrict the discussion to the case of scatter-location family of measures, for which an explicit representation of the Wasserstein distance exists [3]. Section 3 presents the procedure of construction of non-asymptotic confidence sets for empirical Wasserstein barycenters. Section 4 studies its application to detection of structural breaks in data with complex geometry. Theoretical justification is obtained for measures with commuting covariance matrices. Both algorithms are tested under the most general setting on artificial and real data (MNIST database of hand written digits). The results are presented in Section 6. All proofs and conditions can be found in [18].

2 Monge-Kantorovich distance for location-scatter family

In this section we recall some basics on optimal transportation theory and define objects which play key-role in the current study. Let $\mathcal{P}_2(\mathbb{R}^d)$ the set of all probability measures with finite second moment supported on $\mathbb{R}^d$. Optimal transportation metrics are tightly connected to the choice of transportation cost function $c(x, y)$ or, informally speaking, the price one has to pay for a transportation of unit of mass from some location x to some location y, $x, y \in \mathbb{R}^d$. 2-Wasserstein distance corresponds to $c(x, y)$ being l^2 Euclidean norm, $c(x, y) = \|x - y\|^2$. Thus for any μ and ν in $\mathcal{P}_2(\mathbb{R}^d)$ 2-Wasserstein distance is defined as

$$W_2^2(\mu, \nu) \stackrel{\text{def}}{=} \inf_{\pi \in \Pi(\mu, \nu)} \int_{\mathbb{R}^d} \|x - y\|^2 d\pi(x, y), \tag{2.1}$$

where $\Pi(\mu, \nu)$ is the set of all joint probability measures π supported on $\mathbb{R}^d \times \mathbb{R}^d$ with marginals μ and ν respectively:

$$\Pi(\mu, \nu) \stackrel{\text{def}}{=} \{\pi \in \mathcal{P}(\mathbb{R}^d \times \mathbb{R}^d) : \text{ for all } \mathcal{A} \in \mathcal{B}(\mathbb{R}^d) :$$

$$\pi(\mathcal{A} \times \mathbb{R}^d) = \mu(\mathcal{A}), \pi(\mathbb{R}^d \times \mathcal{A}) = \nu(\mathcal{A})\},$$

where $\mathcal{B}(\mathbb{R}^d)$ is Borel σ-algebra on $\mathbb{R}^d$. Following [1], we define a barycenter of any given set $\{\nu_1, ..., \nu_n\}$, $\nu_i \in \mathcal{P}_2(\mathbb{R}^2)$, as any minimizer

$$\mu_n \subseteq \underset{\mu \in \mathcal{P}_2(\mathbb{R}^d)}{\arg\min} \frac{1}{n} \sum_{i=1}^{n} W_2^2(\mu, \nu_i). \tag{2.2}$$

An example is presented at Fig. 1. The upper panel depicts an observed sample of $n = 4$ normalized images of size 100×100 pixels. The left-hand lower box stands for Euclidean averaging of images, whereas the right-hand one for 2-Wasserstein barycenter, computed with Bregman projection algorithm proposed in [7]. This example also illustrates the necessity of taking into account of non-linearity of the manifold in hand. The concept of classical Fréchet mean can be generalized by introducing of reweighting of the summands. Let $(w_1, ..., w_n)$ be a set of normalized weights, *i.e.* $\sum_i w_i = 1$ and $w_i \geq 0$ for all $i = 1, \dots, n$. Then the weighted barycenter is

$$\mu_n \stackrel{\text{def}}{=} \underset{\mu \in \mathcal{P}_2(\mathbb{R}^2)}{\arg\min} \sum_{i=1}^{n} w_i W_2^2(\mu, \nu_i).$$

Its existence, uniqueness and regularity are investigated in [1]. The current study

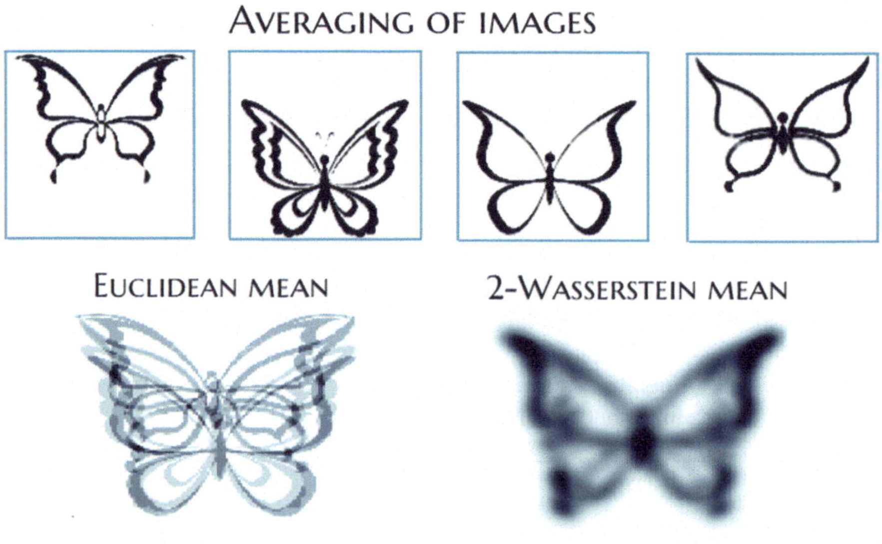

Fig. 1: 2-Wasserstein barycenter of 4 images

investigates the case of a sample $\{\nu_1, \dots, \nu_n\}$ independently generated from some unknown $\mathbb{P}$, supported on $\mathcal{P}_2(\mathbb{R}^d)$. Thus, a measure $\mathbb{P}_n$ induced by the sample

$$\mathbb{P}_n \stackrel{\text{def}}{=} \frac{1}{n} \sum_{i=1}^{n} \delta_{\nu_i}$$

can be considered as an empirical counterpart of $\mathbb{P}$.

After introducing $\mathbb{P}$ one is now able to define a *population* barycenter, that is a set of all $\{\mu^*\}$ which satisfy

$$\mu^* \subseteq \arg\min_{\mu \in \mathcal{P}_2(\mathbb{R}^d)} \int_{\mathcal{P}_2(\mathbb{R}^d)} W_2^2(\mu, \nu) d\mathbb{P}(\nu). \tag{2.3}$$

Further we assume, that $\mathbb{P}$ admits a unique barycenter. The next proposition explains conditions under which the uniqueness assumption holds.

Proposition 2.1 ([28], Proposition 6) *Let $\mathbb{P}$ be such that there exists a set $A \subset \mathcal{P}_2(\mathbb{R}^d)$ of measures such that for all $\mu \in A$,*

$$B \in \mathcal{B}(\mathbb{R}^d), \quad dim(B) \leq d - 1 \longrightarrow \mu(B) = 0,$$

and $\mathbb{P}(A) > 0$, then $\mathbb{P}$ admits a unique barycenter.

The paper [28] shows, that under accepted setting an empirical barycenter is a consistent estimator of the true one:

Proposition 2.2 ([28], Corollary 5) *Suppose that $\mathbb{P} \in \mathcal{P}(\mathcal{P}_2(\mathbb{R}^d))$ has a unique barycenter. Then for any sequence $(\mathbb{P}_n)_{n \geq 1} \subset \mathcal{P}(\mathcal{P}_2(\mathbb{R}^d))$ converging to $\mathbb{P}$: $W_2(\mathbb{P}, \mathbb{P}_n) \to 0$ as $n \to \infty$, any sequence μ_n of their barycenters converges to the barycenter of $\mathbb{P}$:*

$$W_2(\mu^*, \mu_n) \to 0, \quad n \to \infty.$$

In what follows we focus on the measures belonging to the same location-scatter family. It is defined as follows. Let $\mathcal{P}_2^{ac}(\mathbb{R}^d)$ be the set of all absolutely continuous probability measures on $\mathbb{R}^d$. We chose some template measure $\mu_0 \in \mathcal{P}_2^{ac}(\mathbb{R}^d)$ and consider a random variable X induced by the law μ_0: $X \sim \mu_0$. Then a location-scatter family generated from μ_0 is a set of distributions that correspond to a set of possible positive definite affine transformations of X.

($\mathcal{F}$) *Let $\mu_0 \in \mathcal{P}_2^{ac}(\mathbb{R}^d)$ be a template object, s.t. $X \sim \mu_0$,* $\mathrm{e}X = m_0$ *and* $\mathrm{Var}(X) = \mathbb{Q}_0$. *Let the family of transformations be*

$$\mathcal{F}(\mu_0) = \{AX + a, \quad A \in \mathbb{S}_+(d, \mathbb{R}), \quad a \in \mathbb{R}^d\},$$

where $\mathbb{S}_+(d, \mathbb{R})$ is the set of all positive definite symmetric matrices of size $d \times d$ with real entries.

For example, the class of all d-dimensional Gaussians can be considered as a class of all affine transformations of the standard normal distribution:

$$\mathcal{F}(\mathcal{N}(0, I_d)) = \{\mathcal{N}(m, S) : \quad m \in \mathbb{R}^d, \quad S = A^2\}. \tag{2.4}$$

The family plays important role in modern statistical analysis being a powerful modelling tool for many real-world processes and tasks, see e.g. [3] or [31] and references therein. It is worth noting, that 2-Wasserstein distance between any two measures from the class $\mathcal{F}(\mu_0)$ is completely defined by their first and second moments (see e.g. [20, 33]). For instance, let $\mu_0 = \mathcal{N}(0, I)$, and $\mu = \mathcal{N}(m_1, S_1)$, $\nu = \mathcal{N}(m_2, S_2)$. Then the minimum in (2.1) turns into

$$W_2^2(\mu, \nu) = \|m_1 - m_2\|^2 + \operatorname{tr}\left(S_1 + S_2 - 2\left(S_1^{1/2} S_2 S_1^{1/2}\right)^{1/2}\right). \tag{2.5}$$

The paper [3] expands the result obtained in [20] and shows, that 2-Wasserstein distance between any two measures from the same scatter-location family has the same form as (2.5). Furthermore, it also generalizes the result by [1] claiming that an empirical barycenter of any set of Gaussian measures is Gaussian as well. In particular by taking a set $\nu_1 = \mathcal{N}(m_1, S_1), \ldots, \nu_n = \mathcal{N}(m_n, S_n)$ one immediately concludes, that its barycenter μ_n is Gaussian with parameters $\mathcal{N}(r_n, \mathbb{Q}_n)$, where

$$r_n = \frac{1}{n}\sum_{i=1}^{n} m_i, \quad \mathbb{Q}_n = \frac{1}{n}\sum_{i=1}^{n}\left(\mathbb{Q}_n^{1/2} S_i \mathbb{Q}_n^{1/2}\right)^{1/2}.$$

The next proposition shows, that any scatter-location family is closed w.r.t. taking a barycenters as well.

Proposition 2.3 ([3], Theorem 3.11) *Let $\mu_0 \in \mathcal{P}_2^{ac}(\mathbb{R}^d)$ and $\mathbb{P} \in \mathcal{P}(\mathcal{P}_2(\mathbb{R}^d))$. Furthermore, we assume that supp$(\mathbb{P}) = \mathcal{F}(\mu_0)$. In other words each observation $\nu_\omega \sim \mathbb{P}$ and $\nu_\omega \in \mathcal{F}(\mu_0)$, with first and second moments (m_ω, S_ω) respectively. Then μ^* defined in* (2.3) *is the unique barycenter of $\mathbb{P}$ characterized by first and second moments that are defined as*

$$r^* = \int_{\mathbb{R}^d} m_\omega d\mathbb{P}(\omega), \quad \mathbb{Q}^* = \int_{S_+(d,\mathbb{R})}\left(\mathbb{Q}^{*1/2} S_\omega \mathbb{Q}^{*1/2}\right)^{1/2} d\mathbb{P}(\omega).$$

Moreover, $\mu^* \in \mathcal{F}(\mu_0)$.

In the rest of this section we briefly explain data-generating model which plays an important role in what follows. Let $\mu_0 \in \mathcal{P}_2^{ac}(\mathbb{R}^d)$ be a fixed template measure and denote $X \sim \mu_0$, then

$$\mathcal{F}(\mu_0) = \left\{\nu_\omega : \ Y_\omega \sim \nu_\omega \text{ and } Y_\omega = A_\omega X + a_\omega, (A_\omega, a_\omega) \sim \mathbb{P}\right\} \tag{2.6}$$

where supp$(\mathbb{P}) \subset \mathbb{R}^d \times \mathbb{S}_+(d, \mathbb{R})$ and

$$\mathrm{e}_\omega Y_\omega = X, \text{ i.e. } \quad \mathrm{e}_\omega A_\omega = I_d, \quad \mathrm{e}_\omega a_\omega = 0.$$

One can easily derive that μ_0 coincides with μ^* (see Proposition 2.3). Taking into account all aforementioned, one can consider an empirical barycenter μ_n (2.2) as a good candidate for the template estimator.

Remark 1 Unless otherwise noted, from now on we refer to μ_0 as μ^*. It is implicitly assumed, that we always talk about a population barycenter μ^*, that coincides with the template object μ_0 in case of location-scatter family.

The next section provides the procedure of construction confidence sets around μ_n. A possible application to change point detection is presented in Section 4.

3 Bootstrap procedure for confidence sets

This section presents a description of non-asymptotic confidence sets construction and provides its theoretical justification for a particular data model. Let $\{\nu_1, ..., \nu_n\}$ be an observed i.i.d. random sample coming from distribution $\mathbb{P}$. Further we assume, that $\mathbb{P}$ is s.t. μ^* is unique. Let μ_n and μ^* be empirical and population barycenters respectively. We define the following statistic based on the 2-Wasserstein distance between them

$$T_n \overset{\text{def}}{=} \sqrt{n} W_2(\mu_n, \mu^*).$$

A z-confidence set for the population barycenter is defined as

$$C(z) \overset{\text{def}}{=} \{\mu \in \mathcal{P}_2(\mathbb{R}^d) \mid \sqrt{n} W_2(\mu_n, \mu) \le z\}. \tag{3.1}$$

Following classical notations, we define the α-quantile $z_n(\alpha)$ for any $\alpha \in (0, 1)$ as the minimum value that ensures $\mathbb{P}\big(\mu^* \notin C(\mathrm{z})\big) \le \alpha$:

$$z_n(\alpha) \overset{\text{def}}{=} \inf\{z \ge 0 \,|\, \mathbb{P}\big(T_n \ge z\big) \le \alpha\}.$$

The quantile $z_n(\alpha)$ depends on the underlying distribution $\mathbb{P}$, which is generally unknown. We therefore propose to use a weighted bootstrap procedure for the estimation of this quantity. The idea behind the method is as follows. Multiplier bootstrap aims to mimic the distribution of T_n using a newly generated weighted versions of the empirical barycenter $\mu_n^\flat$ by reweighing the summands in (2.2) with random multipliers:

$$\mu_n^\flat \overset{\text{def}}{=} \operatorname*{arg\,min}_{\mu \in \mathcal{P}(\mathbb{R}^d)} \sum_{i=1}^{n} W_2^2(\mu, \nu_i) w_i, \tag{3.2}$$

where the w_i are i.i.d. real-valued variables, s.t. $\mathbb{E}^\flat w_i = \mathrm{Var}(w_i) = 1$ and fulfilling some technical condition $(ED_\flat^W)$ presented in [18]. By construction one can see, that the distribution of $\mu_n^\flat$ is conditional on the observed data $\{\nu_1, ..., \nu_n\}$. In what follows, $\mathbb{P}^\flat$ denotes the distribution of the weights w_i given the sample $\{\nu_1, ..., \nu_n\}$. The counterpart of T_n in the bootstrap world is thus defined as

$$T_n^\flat \overset{\text{def}}{=} \sqrt{n} W_2(\mu_n^\flat, \mu_n). \tag{3.3}$$

Remark 2 (Choice of bootstrap weights) Note, that if the weights are non-negative, e.g. $w_i \sim \mathrm{Po}(1)$ or $w_i \sim \mathrm{Exp}(1)$, the bootstrapped barycenter exists $\mu_n^\flat$. Moreover, if $\mathbb{P}$ is supported on some scatter-location family, then is also unique and belongs to the same scatter-location family. Otherwise if weights are, for instance, normal $w_i \sim \mathcal{N}(1, 1)$, the existence of the solution of (3.2) should be proven.

Introduction of $T_n^{\flat}$ allows definition of bootstrapped counterpart of the quantile $z_n(\alpha)$ which is written as

$$z_n^{\flat}(\alpha) \stackrel{\text{def}}{=} \inf\{z \geq 0 \mid \mathbb{P}^{\flat}(T_n^{\flat} \geq z) \leq \alpha\}. \tag{3.4}$$

Note that this quantity depends on the sample and is therefore random w.r.t. $\mathbb{P}$. Algorithm 1 present a pseudo-code describing the construction of $z_n^{\flat}(\alpha)$. The idea is to replace unknown $z_n(\alpha)$ with known $z_n^{\flat}(\alpha)$ that is computationally tractable. We prove the validity of such a replacement for the case of measures belonging to the same scatter-location family.

Theorem 3.1 (Bootstrap validity for confidence sets) *Let $\mathbb{P}$ be supported on some scale-location family $\mathcal{F}(\mu_0)$, s.t. for any $\mu, \nu \in \mathcal{F}(\mu_0)$ their covariances matrices S_μ, S_ν commute: $S_\mu S_\nu = S_\nu S_\mu$. Let $\nu_1, ..., \nu_n$ be an i.i.d. sample from $\mathbb{P}$. Under some conditions on $\mathbb{P}$ it holds with probability $\mathbb{P}^{\flat} \geq 1 - (e^{-\mathrm{x}} + 2e^{-6\mathrm{x}^2})$, $\mathbb{P} \geq 1 - 6e^{-\mathrm{x}}$*

$$\left|\mathbb{P}\left(T_n \leq z_n^{\flat}(\alpha)\right) - \alpha\right| \leq \Delta_{total}(n),$$

where $z_n^{\flat}(\alpha)$ comes from (3.4) *and $\Delta_{total}(n) \leq C(d)/\sqrt{n}$, where $C(d)$ depends on the dimension d.*

The discussion of details behind the statement can be found in [18].

4 Application to change point detection

In this section we discuss how the procedure of confidence set construction may be extended to change point detection problem. The setting of interest is as follows. Let $\{\mu_t\}_t$ be a flow of measures, sampled independently from some distribution ¶. We aims to check, whether a chosen time moment t^* is a change point or not. To do that, we set $\nu_t \sim \mathbb{P} = \mathbb{P}_1$ if $t < t^*$ and $\nu_t \sim \mathbb{P} = \mathbb{P}_2$ otherwise. Thus the goal is to test

$$H_0 : \mathbb{P}_1 = \mathbb{P}_2, \quad H_1 : \mathbb{P}_1 \neq \mathbb{P}_2.$$

To chose between H_0 and its heterogeneous alternative we suggest to use a classical approach, namely a two-sample test in a scrolling window. We fix $2h$ to be the size of the window. In general, a matter of choice of a proper size h is a non-trivial question, so it is beyond the scope of the current study. We set t to be a candidate for a change point and let the data inside the window be $\nu_{t-h}, ..., \nu_{t+h-1}$. The test statistic is written as

$$T_h(t) = \sqrt{h} W_2(\mu_l(t), \mu_r(t)), \tag{4.1}$$

where $\mu_l(t)$ and $\mu_r(t)$ are barycenters of sub-samples in the left and right halves of the window respectively,

$$\mu_l(t) \stackrel{\text{def}}{=} \arg\min_{\mu \in \mathcal{P}_2(\mathbb{R}^d)} \frac{1}{h} \sum_{i=t-h}^{t-1} W_2^2(\mu, \nu_i), \tag{4.2}$$

$$\mu_r(t) \stackrel{\text{def}}{=} \arg\min_{\mu \in \mathcal{P}_2(\mathbb{R}^d)} \frac{1}{h} \sum_{i=t}^{t+h-1} W_2^2(\mu, \nu_i). \tag{4.3}$$

Change point is supposed to be detected at the moment t if the test exceeds some critical level $z_h(\alpha)$:

$$T_h(t) \geq z_h(\alpha),$$

where $z_h(\alpha)$ is defined as α-quantile of $T_h(t)$ under H_0:

$$z_h(\alpha) \stackrel{\text{def}}{=} \arg\min_z \left\{\mathbb{P}_0\left(T_h(t) \geq z\right) \leq \alpha\right\}.$$

As soon as $z_h(\alpha)$ can not be computed analytically, the core idea of the approach is to replace it with bootstrapped counterpart $z_h^\flat(\alpha)$. While tuning critical values, we assume, that an observed training sample $\{\nu_1, \ldots \nu_M\}$ is i.i.d. Following the already presented framework, we define counterparts of (4.2) and (4.3) in the bootstrap world as

$$\mu_l^\flat(t) \stackrel{\text{def}}{=} \arg\min_{\mu \in \mathcal{P}_2(\mathbb{R}^d)} \sum_{i=t-h}^{t-1} W_2^2(\mu, \nu_i) w_i, \tag{4.4}$$

$$\mu_r^\flat(t) \stackrel{\text{def}}{=} \arg\min_{\mu \in \mathcal{P}_2(\mathbb{R}^d)} \sum_{i=t}^{t+h-1} W_2^2(\mu, \nu_i) w_i, \tag{4.5}$$

where weights w_i follow Condition $(ED_\flat^W)$ and are independent of the observed data set $\{\nu_{t-h}, \ldots, \nu_{t+h-1}\}$. The bootstrapped statistic test is

$$T_h^\flat(t) = \sqrt{h} W_2\left(\mu_l^\flat(t), \mu_r^\flat(t)\right). \tag{4.6}$$

Then its α-quantile $z_h^\flat(\alpha)$ is written as

$$z_h^\flat(\alpha) \stackrel{\text{def}}{=} \arg\min_z \left\{\mathbb{P}^\flat\left(T_h^\flat(t) \geq z\right) \leq \alpha\right\}. \tag{4.7}$$

The procedure of critical value calibration is summarised in Algorithm 2. The below theorem validates such a replacement for the case of scale-location families with commuting covariances.

Theorem 4.2 (Bootstrap validity for change point detection) *Let data generating measure be supported on a scatter-location family presented in Theorem 3.1. Let the size of running window h be fixed. Under H_0 some conditions it holds with probability* $\mathbb{P} \geq 1 - 6e^{-\mathtt{x}} - 2e^{-6\mathtt{x}^2}$, $\mathbb{P}^\flat \geq 1 - (e^{-\mathtt{x}} + 2e^{-6\mathtt{x}^2})$

$$\left|\mathbb{P}\left(\max_{t \in \mathcal{I}} T_h(t) \leq z_h^\flat(\alpha)\right) - \alpha\right| \leq \frac{M}{2h} \Delta_{total,cp}(h),$$

where $z_h^{\flat}(\alpha)$ comes from (3.4) and $\Delta_{total,cp}(h) \le C(d)/\sqrt{h}$, $C(d)$ depends on the dimension d and $\Delta_{total,\,cp}(h)$ is defined in [18].

A more detailed discussion of the theorem can be found in [18].

5 Algorithms

Algorithm 1: Computation of critical value $z_n^{\flat}(\alpha)$

Data: Sample $\{\nu_1, \ldots \nu_n\}$, distribution of w_i, false-alarm rate α
Result: $z_n^{\flat}(\alpha)$
initialize the number of iterations M
for $i \in \{1, \ldots J\}$ **do**
  sample an i.i.d. set $\{w_1, \ldots, w_n\}$
  compute $\mu_n^{\flat}(i)$ using (3.2)
for each $t \ge 0$ compute ecdf $F_{n,M}^{\flat}(t)$:

$$F_{n,M}^{\flat}(t) = \frac{1}{M} \sum_{i=1}^{n} \mathrm{i}\left\{\sqrt{n} W_2\left(\mu_n^{\flat}(i), \mu_n\right) \le t\right\}$$

compute bootstrapped quantile $z_n^{\flat}(\alpha)$:

$$z_n^{\flat}(\alpha) = \inf_{t \ge 0}\left\{F_{n,M}^{\flat}(t) \ge 1 - \alpha\right\}$$

6 Experiments

In this section we illustrate method performance on both real and artificial data sets.

6.1 Coverage probability of the true object μ^*

This section examines the quality of the approximation of $z_n(\alpha)$ by $z_n^{\flat}(\alpha)$ for confidence sets. We assume that each observed object $\nu \sim \mathbb{P}$ is a random transformation of a template μ_0^*, s.t.

$$\mathrm{e}_{\mathbb{P}}\nu = \mu_0^*.$$

As before, denote as μ_n the empirical barycenter of some i.i.d. set $\{\nu_1, \ldots, \nu_n\}$, $\nu_i \sim \mathbb{P}$. For a fixed level rate α the confidence set $C\big(z_n(\alpha)\big)$ is defined as (3.1). Our goal is

Algorithm 2: Computation of critical value $z_h^{\flat}(\alpha)$

Data: Sample $\{\nu_1, \ldots \nu_M\}$, window width h, distribution of w_i, false-alarm rate α
Result: $z_h^{\flat}(\alpha)$
initialize the number of iterations J
initialize vector R, length(R) = J
for $i \in \{1, \ldots J\}$ **do**
  sample an i.i.d. set $\{w_1, \ldots, w_M\}$
  $R(j) \stackrel{\text{def}}{=} 0;$
  for $t \in \{h+1, \ldots, M-h\}$ **do**
    compute $\overline{\mu}_l^{\flat}(t,i), \overline{\mu}_r^{\flat}(t,i), T^{\flat}(t,i)$ using (4.4), (4.5) and (4.6)
    $R(j) \stackrel{\text{def}}{=} \max\big(R(j), T^{\flat}(t,i)\big)$

for x ≥ 0 compute ecdf $F_h^{\flat}(\mathrm{x})$:

$$F_h^{\flat}(\mathrm{x}) = \frac{1}{J}\sum_{j=1}^{J} \mathrm{i}\Big\{R(j) \leq \mathrm{x}\Big\}$$

compute bootstrapped quantile: $z_h^{\flat}(\alpha) = \inf_{\mathrm{x}\geq 0}\Big\{F_h^{\flat}(\mathrm{x}) \geq 1-\alpha\Big\}$

to check the closeness of $z_n^{\flat}(\alpha)$ computed with Algorithm 1 and $z_n^{\flat}(\alpha)$. To do that let's introduce some other "template candidate" μ_1^*: $\mu_1^* \neq \mu_0^*$. We are interested in the estimation of the following probabilities:

$$\mathbb{P}\Big(\mu_0^* \in C\big(z_n^{\flat}(\alpha)\big)\Big), \qquad \mathbb{P}\Big(\mu_1^* \notin C\big(z_n^{\flat}(\alpha)\big)\Big). \tag{6.1}$$

To do that we follow the work by [15] and consider as a template object μ_0^* two concentric circles, that are depicted at the left-hand side of Fig. 2. The middle panel contains four samples of ν. Each ν_i is obtained by random shifts and dilations of each circle. The last box depicts the barycenter μ_n. Naturally, each image can be considered as a uniform measures on $\mathbb{R}^2$. As μ_1^* we consider a single shifted random ellipse presented at Fig.3. To compute optimal transport we use iterative Bregman projections algorithm presented in [7]. For more details on computation issues we refer to [34], [26] or [17].

Remark 3 It is worth noting that the algorithm solves a penalized problem rather then the original one. In other words, instead of minimizing

$$\int_{\mathbb{R}^d} \|x-y\|^2 d\pi(x,y) \to \min_{\pi\in\Pi(\mu,\nu)},$$

it optimizes the following target function

$$\int_{\mathbb{R}^d} \|x-y\|^2 d\pi(x,y) + \gamma \int_{\mathbb{R}^d} \pi(x,y) d\pi(x,y) \to \min_{\pi\in\Pi(\mu,\nu)}.$$

The solution of regularized problem converges to the solution of the original one with the decay of γ. Thus, choosing relatively small regularization parameter $\gamma = .005$, we expect obtained results to be quite close to the solution of the original problem.

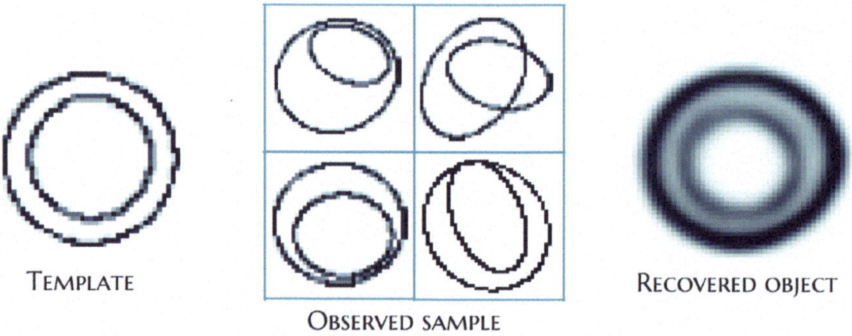

Fig. 2: Recovery of the template object

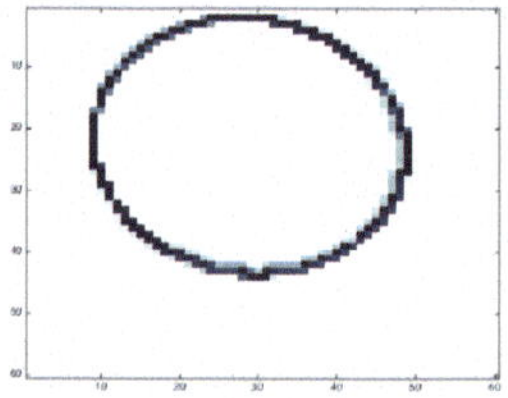

Fig. 3: μ_1^* is a dilated and shifted single ellipse

The experiments are carried for eight sample sizes $n \in \{5, 10, 15, 20, 25, 30, 35, 40\}$. Confidence intervals are estimated for two different false-alarm rates $\alpha \in \{.05, .01\}$. Bootstrap weights follow Poisson distribution $w_i \sim \text{Po}(1)$. Empirical estimators of (6.1) are presented in Tables 1 and 2.

Table 1: $\hat{\mathbb{P}}\left(\mu_0^* \in C\left(z_n^\flat(\alpha)\right)\right)$

	$n = 5$	$n = 10$	$n = 15$	$n = 20$	$n = 25$	$n = 30$	$n = 35$	$n = 40$
$\alpha = .05$	.90	.91	.97	.97	.98	.92	.97	.93
$\alpha = .01$	.91	.96	.99	.99	1	.98	.99	.96

Table 2: $\hat{\mathbb{P}}\left(\mu_1^* \notin C\left(z_n^{\flat}(\alpha)\right)\right)$

	$n=5$	$n=10$	$n=15$	$n=20$	$n=25$	$n=30$	$n=35$	$n=40$
$\alpha = .05$	.85	.84	.94	.94	.94	.98	.98	.97
$\alpha = .01$	.75	.73	.76	.84	.86	.91	.90	.94

6.2 Experiments on the real data

This illustrates performance of Algorithm 1 on the real data. We use MNIST (handwritten digit database) `http://yann.lecun.com/exdb/mnist/`. It contains around 60000 indexed black-and-white images. Each image is a bounding box of 28×28 pixels with a written digit inside. Several examples are presented at Fig. 4. All symbols are approximately of the same size. Given some digit, we define a population barycenter μ^* as a mean over the whole available set of images with this digit. Fig. 5 presents population barycenter for each digit.

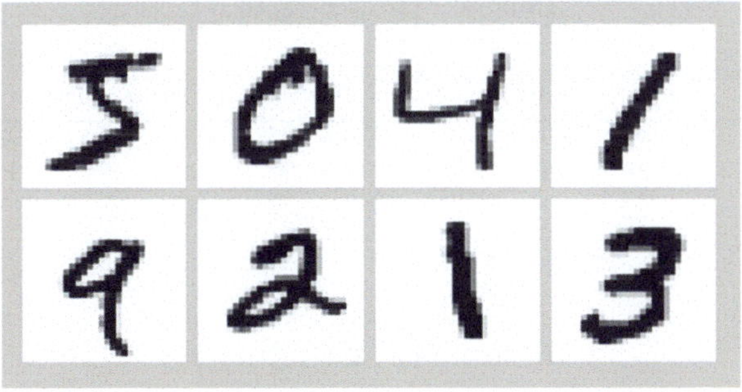

Fig. 4: Random sample from MNIST database

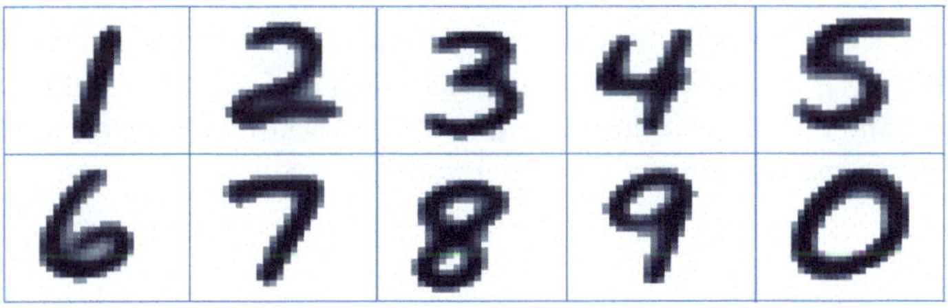

Fig. 5: Barycentric digits, computed over the whole MNIST database

Now we briefly explain the experimental setting. Denote as $\mathcal{S}$ the set of all MNIST images. As before, each image $\nu \in \mathcal{S}$ can be considered as some measure on $\mathbb{R}^2$ with a finite support of size 28 × 28. First fix some reference and test digits and denote them as r and t respectively. Then extract all r- and t-entries from $\mathcal{S}$

$$\mathcal{S}_r = \{\text{set of all } r\text{'s}\}, \qquad \mathcal{S}_t = \{\text{set of all } t\text{'s}\}.$$

From the reference set $\mathcal{S}_r$ we then sample some $\{\nu_1^r, \dots, \nu_n^r\} \subset \mathcal{S}_r$ and denote as μ_n^r its empirical barycenter. Let $C_n(z_\alpha^{\flat})$ be α-confidence set around μ_n^r. The test procedure aims to estimate two following probabilities

$$\mathbb{P}\Big(\mu^r \in C\big(z_n^{\flat}(\alpha)\big)\Big), \qquad \mathbb{P}\Big(\mu^t \notin C\big(z_n^{\flat}(\alpha)\big)\Big),$$

where μ^r and μ^t are empirical barycenters, computed using whole sets $\mathcal{S}_r$ and $\mathcal{S}_t$ respectively (see Fig. 5). As the reference and test digits are used $r = 3$ and $t = 8$ respectively. We consider eight sample sizes $n \in \{5, 10, 15, 20, 25, 30, 35, 40\}$ and two confidence levels $\alpha \in \{.05, .01\}$. Empirical probabilities are estimated using 100 experiments for each n and α. The results are presented in Tables 3 and 4.

Table 3: $\hat{\mathbb{P}}\Big(\mu^3 \in C\big(z_n^{\flat}(\alpha)\big)\Big)$

	$n=5$	$n=10$	$n=15$	$n=20$	$n=25$	$n=30$	$n=35$	$n=40$
$\alpha = .05$	.80	.90	.92	.95	.95	.99	.99	.99
$\alpha = .01$	.91	.94	.97	.97	.95	.99	.99	1

Table 4: $\hat{\mathbb{P}}\Big(\mu^8 \notin C\big(z_n^{\flat}(\alpha)\big)\Big)$

	$n=5$	$n=10$	$n=15$	$n=20$	$n=25$	$n=30$	$n=35$	$n=40$
$\alpha = .05$	.92	1	1	1	1	1	1	1
$\alpha = .01$	.80	.96	.99	1	1	1	1	1

Fig. 6 shows an empirical distribution of 2-Wasserstein distance in two following cases: $\sqrt{n}W_2(\mu_n^r, \mu^r)$ (the left histogram) and $\sqrt{n}W_2(\mu_n^r, \mu^t)$ (the right histogram) respectively. The red vertical line is α-quantile $z_n^{\flat}(\alpha)$ computed with the bootstrap procedure. Fixed parameters are $n = 20$, $\alpha = .01$, $r = 3$, $t = 1$.

6.3 Application to change point detection

This section illustrates the performance of change point detection procedure. We consider the following data generation process. Before the change point observed data is generated from some template μ_0^* and from μ_1^* afterwords. Two examples are

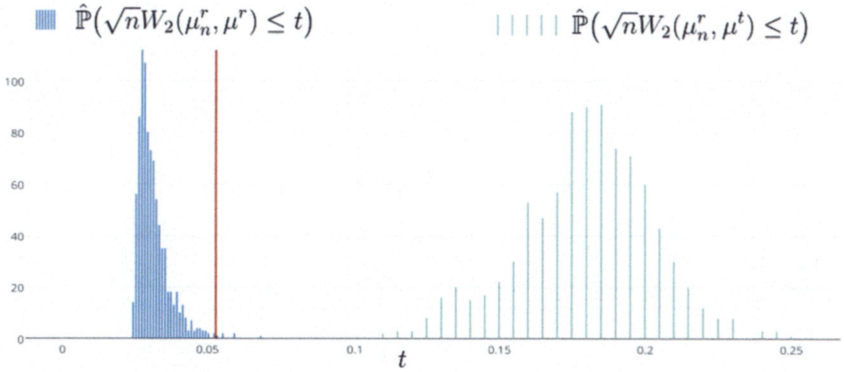

Fig. 6: Distribution of $\sqrt{n}W_2(\mu_n^r,\mu^r)$ and $\sqrt{n}W_2(\mu_n^r,\mu^t)$

presented at Fig. 7. The upper panel shows a switch from nested ellipses to curved triangles. The bottom panel refers to switch from handwritten digits "three" to "five". The second data set is randomly sampled from MNIST databases. Let $T_h(t)$ be the test statistics

$$T_h(t) = \sqrt{h}W_2\left(\mu_h^l(t), \mu_h^r(t)\right),$$

where $\mu_h^l(t)$ and $\mu_h^r(t)$ stand for the barycenter of $\{\nu_{t-h},...,\nu_{t-1}\}$ and $\{\nu_t,...,\nu_{t+h-1}\}$ respectively.

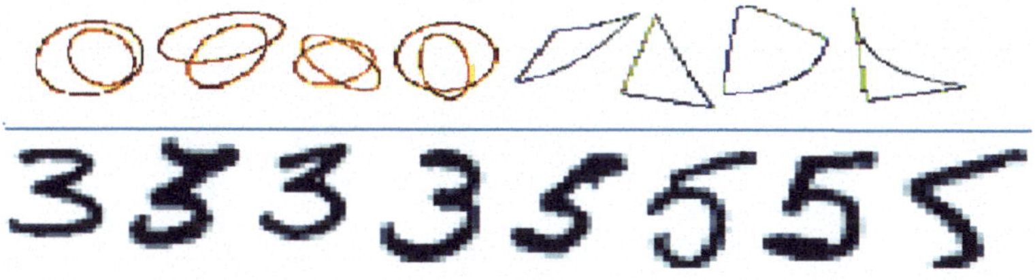

Fig. 7: Change point in a flow of random images

Fig. 8 illustrates work of the algorithm on artificial data, namely on the stream that switches from random nested ellipses to curved triangles. Critical value $z_h^b(\alpha)$ for $\alpha = .01$ is computed with Algorithm 2 and depicted with horizontal line. We use the width of scrolling window $2h = 10$. Switch occurs at $t = 40$ and is marked with the black vertical line. The panel below shows the data before and after the change point, namely ν_t for $t \in \{36, 37, 38, 39, 40, 41, 42, 43\}$. Fig. 9 shows the behaviour of $T_h(t)$ on real data set. Switch occurs at $t = 200$, the width of scrolling window is $2h = 100$. Horizontal lines, that refer to critical levels $\alpha = .01$ and $\alpha = .05$ respectively are computed with Algorithm 2. As before, the panel below depicts observations in the vicinity of the change point: ν_t for $t \in \{196, 197, 198, 199, 200, 201, 202, 203\}$.

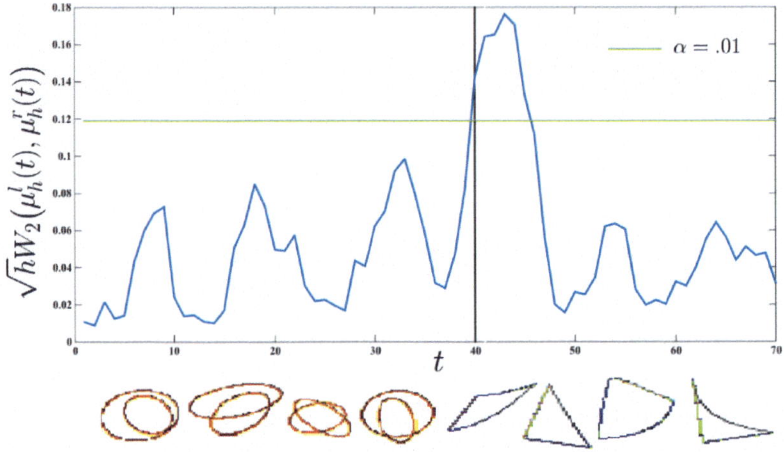

Fig. 8: Distribution of $\sqrt{h}W_2(\mu_h^l(t), \mu_h^r(t))$

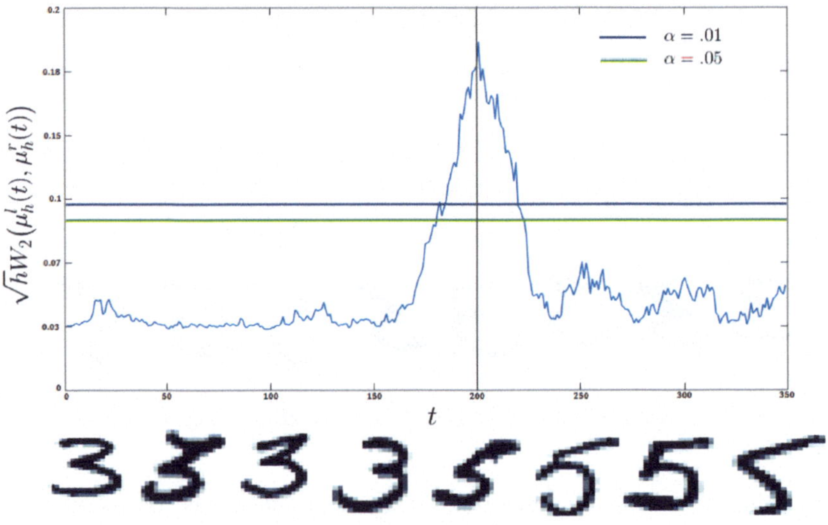

Fig. 9: Distribution of $\sqrt{h}W_2(\mu_h^l(t), \mu_h^r(t))$

References

1. Agueh, M., Carlier, G.: Barycenters in the wasserstein space. SIAM Journal on Mathematical Analysis **43**(2), 904–924 (2011)
2. Ahidar-Coutrix, A., Gouic, T.L., Paris, Q.: On the rate of convergence of empirical barycentres in metric spaces: curvature, convexity and extendible geodesics. arXiv:1806.02740 (2018)
3. Álvarez-Esteban, P.C., del Barrio, E., Cuesta-Albertos, J.A., Matrán, C.: Wide Consensus for Parallelized Inference. arXiv e-prints (2015)
4. Ambrosio, L., Gigli, N.: A user's guide to optimal transport. In: Modelling and optimisation of flows on networks, pp. 1–155. Springer (2013)
5. Aminikhanghahi, S., Cook, D.J.: A survey of methods for time series change point detection. Knowledge and information systems **51**(2), 339–367 (2017)
6. Amit, Y., Grenander, U., Piccioni, M.: Structural image restoration through deformable templates. Journal of the American Statistical Association **86**(414), 376–387 (1991)
7. Benamou, J.D., Carlier, G., Cuturi, M., Nenna, L., Peyré, G.: Iterative bregman projections for regularized transportation problems. SIAM Journal on Scientific Computing **37**(2), A1111–A1138 (2015)
8. Bhattacharya, A.: Nonparametric statistics on manifolds with applications to shape spaces. ProQuest (2008)
9. Bhattacharya, R., Patrangenaru, V.: Large sample theory of intrinsic and extrinsic sample means on manifolds. i. Annals of statistics pp. 1–29 (2003)
10. Bhattacharya, R., Patrangenaru, V.: Large sample theory of intrinsic and extrinsic sample means on manifolds: Ii. Annals of statistics pp. 1225–1259 (2005)
11. Bigot, J., Klein, T.: Characterization of barycenters in the wasserstein space by averaging optimal transport maps. ESAIM: Probability and Statistics **22**, 35–57 (2018)
12. Bock, C., Lengauer, T.: Computational epigenetics. Bioinformatics **24**(1), 1–10 (2008)
13. Carlier, G., Ekeland, I.: Matching for teams. Economic Theory **42**(2), 397–418 (2008)
14. Chernozhukov, V., Chetverikov, D., Kato, K., et al.: Central limit theorems and bootstrap in high dimensions. The Annals of Probability **45**(4), 2309–2352 (2017)
15. Cuturi, M., Doucet, A.: Fast computation of wasserstein barycenters. In: International Conference on Machine Learning, pp. 685–693 (2014)
16. Del Barrio, E., Lescornel, H., Loubes, J.M.: A statistical analysis of a deformation model with wasserstein barycenters : estimation procedure and goodness of fit test. arXiv:1508.06465 [math, stat] (2015)
17. Dvurechensky, P., Dvinskikh, D., Gasnikov, A., Uribe, C.A., Nedić, A.: Decentralize and randomize: Faster algorithm for Wasserstein barycenters. In: S. Bengio, H. Wallach, H. Larochelle, K. Grauman, N. Cesa-Bianchi, R. Garnett (eds.) Advances in Neural Information Processing Systems 31, NeurIPS 2018, pp. 10783–10793. Curran Associates, Inc. (2018). arXiv:1802.04367
18. Ebert, J., Spokoiny, V., Suvorikova, A.: Construction of non-asymptotic confidence sets in 2-wasserstein space. arXiv:1703.03658 (2017)
19. Fréchet, M.: Les éléments aléatoires de nature quelconque dans un espace distancié. In: Annales de l'institut Henri Poincaré, vol. 10, pp. 215–310 (1948)
20. Gelbrich, M.: On a formula for the l2 wasserstein metric between measures on euclidean and hilbert spaces. Mathematische Nachrichten **147**(1), 185–203 (1990)
21. Gramfort, A., Peyré, G., Cuturi, M.: Fast optimal transport averaging of neuroimaging data. In: International Conference on Information Processing in Medical Imaging, pp. 261–272. Springer (2015)
22. Huckemann, S., Hotz, T., Munk, A.: Intrinsic shape analysis: Geodesic principal component analysis for riemannian manifolds modulo lie group actions. discussion paper with rejoinder. Statistica Sinica (2010)
23. Jaenisch, R., Bird, A.: Epigenetic regulation of gene expression: how the genome integrates intrinsic and environmental signals. Nature genetics **33**, 245–254 (2003)

24. Kendall, D.G.: The diffusion of shape. Advances in applied probability **9**(3), 428–430 (1977)
25. Kroshnin, A.: Fréchet barycenters in the monge-kantorovich spaces. arXiv:1702.05740 (2017)
26. Kroshnin, A., Dvinskikh, D., Dvurechensky, P., Gasnikov, A., Tupitsa, N., Uribe, C.: On the complexity of approximating wasserstein barycenter. arXiv:1901.08686 (2019)
27. Kroshnin, A., Spokoiny, V., Suvorikova, A.: Statistical inference for bures-wasserstein barycenters. arXiv:1901.00226 (2019)
28. Le Gouic, T., Loubes, J.M.: Existence and consistency of wasserstein barycenters. Probability Theory and Related Fields **168**(3-4), 901–917 (2017)
29. Liu, W., Srivastava, A., Zhang, J.: Protein structure alignment using elastic shape analysis. In: Proceedings of the First ACM International Conference on Bioinformatics and Computational Biology, pp. 62–70. ACM (2010)
30. Mammen, E., et al.: Bootstrap and wild bootstrap for high dimensional linear models. The annals of statistics **21**(1), 255–285 (1993)
31. Muzellec, B., Cuturi, M.: Generalizing point embeddings using the wasserstein space of elliptical distributions. In: Advances in Neural Information Processing Systems, pp. 10258–10269 (2018)
32. Norman, K.A., Polyn, S.M., Detre, G.J., Haxby, J.V.: Beyond mind-reading: multi-voxel pattern analysis of fmri data. Trends in cognitive sciences **10**(9), 424–430 (2006)
33. Olkin, I., Pukelsheim, F.: The distance between two random vectors with given dispersion matrices. Linear Algebra and its Applications **48**, 257–263 (1982)
34. Peyré, G., Cuturi, M., et al.: Computational optimal transport. Foundations and Trends® in Machine Learning **11**(5-6), 355–607 (2019)
35. Rabin, J., Peyré, G., Delon, J., Bernot, M.: Wasserstein barycenter and its application to texture mixing. In: Scale Space and Variational Methods in Computer Vision, pp. 435–446. Springer (2011)
36. Rippl, T., Munk, A., Sturm, A.: Limit laws of the empirical wasserstein distance: Gaussian distributions. Journal of Multivariate Analysis **151**, 90–109 (2016)
37. Solomon, J., de Goes, F., Peyré, G., Cuturi, M., Butscher, A., Nguyen, A., Du, T., Guibas, L.: Convolutional wasserstein distances: Efficient optimal transportation on geometric domains. ACM Trans. Graph. **34**(4), 66:1–66:11 (2015).
38. Spokoiny, V., Zhilova, M.: Bootstrap confidence sets under model misspecification. The Annals of Statistics **43**(6), 2653–2675 (2015)
39. Trouvé, A., Younes, L.: Metamorphoses through lie group action. Foundations of Computational Mathematics **5**(2), 173–198 (2005)
40. Villani, C.: Topics in optimal transportation. 58. American Mathematical Soc. (2003)
41. Villani, C.: Optimal Transport, *Grundlehren der mathematischen Wissenschaften*, vol. 338. Springer Berlin Heidelberg (2009)

On the Use of ADMM for Imaging Inverse Problems: the Pros and Cons of Matrix Inversions

Mário A. T. Figueiredo

Abstract This paper overviews a line of work on the use of the ADMM (*alternating direction method of multipliers*, a member of the augmented Lagrangian family of methods) to solve regularization formulations of some classical imaging inverse problems. At the core of this line of work is a way of using ADMM to tackle optimization problems where the objective function is the sum of two or more convex functions, each of which having a proximity operator that can be efficiently computed. The approach is illustrated on a variety of well-known problems, namely: image restoration and reconstruction with linear observations (for example, compressive imaging, image deblurring, image inpainting), which may be contaminated with Gaussian or Poisson noise, using synthesis, analysis, or hybrid regularization, and unconstrained or constrained regularization/variational formulations. In all these cases, the proposed approach inherits the convergence properties of ADMM. The main computational bottleneck of the proposed approach is a matrix inversion, which has been often criticized as a hurdle that should be avoided; in contrast, we show that in all the above mentioned problems, this inversion can be tackled very efficiently and we conjecture that it actually underlies the good empirical performance which has been reported for several instances of this class of methods.

1 Introduction

At the core of any digital imaging system, there is an algorithm solving the underlying inverse problem: it takes the data acquired by the sensing apparatus (be it the image sensor in a photographic camera, a magnetic resonance imaging scanner, or an earth observation satellite) and produces/reconstructs images that are adequate for human viewing or further processing or analysis. It is thus clear that imaging inverse

Mário A. T. Figueiredo
Instituto de Telecomunicações, and Instituto Superior Técnico, University of Lisbon, Portugal;
e-mail: mario.figueiredo@tecnico.ulisboa.pt

M. Hintermüller, J. F. Rodrigues (eds.), *Topics in Applied Analysis and Optimisation*,
CIM Series in Mathematical Sciences, https://doi.org/10.1007/978-3-030-33116-0_7

problems, and the ability to solve them, play an important role in modern science, technology, and society in general. Medical imaging, remote sensing, seismography, digital photography and video, astronomic imaging (optical or radio), any other imaging technologies have at their core the solution of inverse problems: they produce visual representations (images) of some underlying reality (*e.g.*, an organ in the body of a patient) from indirect/imperfect observations. Inverse problems are almost always *"ill-posed"*, meaning that even with a perfectly known observation model, the observed data does not uniquely and stably determine the solution (the underlying unknown image).

One approach to deal with the ill-posed nature of imaging inverse problems is to formulate an optimization problem (usually, but not necessarily, convex), typically involving, at least, the following two terms: the *data (fidelity) term*, which encourages the solutions/estimates to be as compatible as possible with the available observations, and the *regularizer*, which penalizes/enforces solutions considered undesirable/desirable. This class of approaches to inverse problems is often termed variational, due to its reliance on an optimization formulation.

Many state-of-the-art regularizers encourage or enforce sparseness of the representation of the underlying image with respect to some redundant frame or dictionary, a feature known to characterize natural noiseless images (see [30] and the many references therein). This sparseness may be expressed in the analysis or synthesis formulations [31, 60], usually via the standard ℓ_1 norm or, more recently and with better performance, by taking into account dependency structures among the representation coefficients via group norms [56]. Another very popular, classical regularizer is the well-known *total variation* (TV), which, simply put, seeks reconstructions with a sparse set of discontinuities [17, 58].

In regularization/variational approaches, optimization naturally takes center stage. However, off-the-shelf methods are usually unable to cope with the very high-dimensionality and non-smoothness of these problems, which has channelled a significant amount of research towards the development of special-purpose fast algorithms that are able to exploit the specific features of these problems [16]. Of course, the literature on this research is vast and cannot be comprehensively covered in this paper. Instead, we will focus on a line of work that we have been developing in recent years and which has produced a series of algorithms [1–3, 10, 34, 35], all of which will be here presented as instances of a common algorithmic template based on the famous *alternating direction method of multipliers* (ADMM) [11, 27, 38, 39].

This paper has no new theoretic results on ADMM or variants thereof, neither is it focused on theoretical aspects of this class of algorithms. Instead, the main goal is to highligh the two following aspects: our way of applying ADMM leads to a flexible and modular toolbox that allows addressing a variety of regularization/variational formulations of imaging inverse problems; although each of the resulting algorithms typically involves a matrix inversion, a fact that has been often seen as an obstacle to be avoided at all costs, we show that, in many problems of interest, this inversion can be tackled very efficiently and we conjecture that it actually underlies the good empirical performance that has been observed empirically in instances of this class of methods.

2 General Problem Formulation

In this paper, we consider only a finite-dimensional setting, where both the underlying image (or some representation thereof) and the observed data belong to finite-dimensional Euclidean spaces, denoted $\mathcal{X}$ and $\mathcal{Y}$, respectively. The variational formulation is based on two main building blocks:

- The *data-fidelity* function, $\Psi : \mathcal{Y} \times \mathcal{X} \to \bar{\mathbb{R}} = \mathbb{R} \cup \{+\infty\}$, such that $\Psi(\mathbf{y}, \mathbf{x})$ expresses how much a given candidate estimate $\mathbf{x}$ deviates from explaining the data observed $\mathbf{y}$. This function is usually derived by modeling how the observations are generated; *e.g.*, in a probabilistic setting, it would correspond to the negative log-likelihood, $\Psi(\mathbf{y}, \mathbf{x}) = -\log p(\mathbf{y}|\mathbf{x})$, but other interpretations are possible.
- The *regularization* function, $\Phi : \mathcal{X} \to \bar{\mathbb{R}}$, such that that $\Phi(\mathbf{x})$ measures how undesirable a candidate estimate $\mathbf{x}$ is. In a Bayesian framework, we have $\Phi(\mathbf{x}) = -\log p_{\mathbf{X}}(\mathbf{x})$, where $p_{\mathbf{X}}$ is the *prior* probability density function, with $\mathbf{x}$ seen as a sample of a random variable $\mathbf{X} \in \mathcal{X}$, but other semantics are valid.

There are three standard ways of combining the two functions defined above, Ψ and Φ, into an optimization problem in such a way that the corresponding solution strikes a balance between the two goals embodied in these functions [42, 45]:

- Tikhonov regularization: $\widehat{\mathbf{x}} \in \operatorname{Arg\,min}_{\mathbf{x}} \Psi(\mathbf{y}, \mathbf{x}) + \alpha\, \Phi(\mathbf{x})$;
- Morozov regularization: $\widehat{\mathbf{x}} \in \operatorname{Arg\,min}_{\mathbf{x}} \Phi(\mathbf{x})$ subject to $\Psi(\mathbf{y}, \mathbf{x}) \leq \delta$;
- Ivanov regularization: $\widehat{\mathbf{x}} \in \operatorname{Arg\,min}_{\mathbf{x}} \Psi(\mathbf{y}, \mathbf{x})$ subject to $\Phi(\mathbf{x}) \leq \tau$,

where Arg min denotes the set of minimizers. If both Ψ and Φ are convex functions of $\mathbf{x}$, these formulations are equivalent, under mild conditions, in the following sense: for any choice of the parameter defining one of the problems (α, δ, or τ), there is a choice of the other two parameters for which all the problems have a common solution [45]. However, in practice it is necessary to choose/adjust these parameters, which is sometimes more conveniently done in one of the formulations than the others. In this paper, we will consider only Tikhonov and Morozov regularization, since most of the derivations for Morozov regularization apply with minor changes to the Ivanov counterpart.

We will confine the discussion to frame-based regularization, which is essentially based on the following observation: the coefficients of the representation of a natural noise-free image on certain *frames*[1] (namely wavelet frames) are *sparse*. Roughly, by sparsity one means that the representation is dominated by a minority of large

[1] In a vector space, say $\mathbb{R}^n$, a frame is a collection of vectors $\{\mathbf{w}_1, \ldots, \mathbf{w}_k\}$, such that there exist constants $0 < A \leq B < \infty$ satisfying $A\|\mathbf{x}\|^2 \leq \sum_j |\langle \mathbf{x}, \mathbf{w}_j \rangle|^2 \leq B\|\mathbf{x}\|^2$, for any $\mathbf{x} \in \mathbb{R}^n$. If $A = B$, the frame is called tight; if $A = B = 1$ it is called a Parseval frame (we will only use Parseval frames). Collecting the frame vectors into a matrix $\mathbf{W} \in \mathbb{R}^{n\times k}$, we have $\mathbf{W}\mathbf{W}^T = \mathbf{I}$, with $\mathbf{W}^T\mathbf{W} = \mathbf{I}$ also holding only if the frame is an orthonormal basis (of course with $k = n$). If $n \geq k$, the frame is called redundant. For a Parseval frame, $\mathbf{W}$ is called the synthesis matrix and $\mathbf{W}^T \in \mathbb{R}^{k\times n}$ is the analysis matrix.

coefficients, with most others being either very small or exactly zero. A detailed discussion of what sparseness exactly means and how it applies to images is beyond the scope of this paper (see [29, 30], for details and pointers to a vast literature); we simply use the classical ℓ_1 norm ($\|\mathbf{v}\|_1 = \sum_i |v_i|$) of the frame coefficients as a measure of (non-)sparseness. There are two main formulations for frame-based, ℓ_1-based regularization [31, 60]:

- In the *analysis* formulation, $\mathbf{x} \in \mathcal{X} = \mathbb{R}^n$ represents[2] an image itself, and the regularizer is applied to its frame-analysis coefficients, thus it has the form $\Phi(\mathbf{x}) = \|\mathbf{W}^T\mathbf{x}\|_1$. The data term has the form $\Psi(\mathbf{y}, \mathbf{x}) = \Upsilon(\mathbf{y}, \mathbf{x})$, where $\Upsilon(\mathbf{y}, \mathbf{x})$ is a function that measures the degree of discrepancy between **image** $\mathbf{x}$ and the data $\mathbf{y}$.
- In the *synthesis* formulation, rather than an image, the variable $\mathbf{x} \in \mathcal{X} = \mathbb{R}^k$ denotes the vector of coefficients of the frame-based representation of an image $\mathbf{W}\mathbf{x}$. The regularizer thus has the form $\Phi(\mathbf{x}) = \|\mathbf{x}\|_1$, while the data term has the form $\Psi(\mathbf{y}, \mathbf{x}) = \Upsilon(\mathbf{y}, \mathbf{W}\mathbf{x})$, where Υ is as defined in the previous paragraph.

If $\mathbf{W}$ is an orthonormal basis, the synthesis and analysis formulations are equivalent; however, for redundant frames, the two formulations yield different results [31]. Hybrid analysis-synthesis formulations are also possible [3] (see Section 6).

3 The Alternating Direction Method of Multipliers

3.1 The Standard ADMM

Although ADMM can be presented for a slightly more general problem formulation [11], we here consider an unconstrained problem of the form

$$\min_{\mathbf{z}\in\mathbb{R}^d} f_1(\mathbf{z}) + f_2(\mathbf{G}\,\mathbf{z}), \tag{3.1}$$

where $f_1 : \mathbb{R}^d \to \bar{\mathbb{R}}$ and $f_2 : \mathbb{R}^p \to \bar{\mathbb{R}}$ are closed (lower semi-continuous), proper (not equal to $+\infty$ everywhere), convex functions, while and $\mathbf{G} \in \mathbb{R}^{p\times d}$ is a matrix of appropriate dimensions. This form is sufficient to deal with all the problems considered below.

The ADMM for problem problem (3.1) is defined as follows:

[2] As is commonly done, $\mathbf{x}$ is the vector representation of an image, obtained by stacking its pixels in lexicographical order.

Algorithm 3: ADMM

Initialization: $k = 0$, choose $\mu > 0$, $\mathbf{u}_0$, and $\mathbf{d}_0$;
repeat
  $\mathbf{z}_{k+1} \leftarrow \arg\min_{\mathbf{z}} f_1(\mathbf{z}) + \frac{\mu}{2}\|\mathbf{G}\,\mathbf{z} - \mathbf{u}_k - \mathbf{d}_k\|_2^2$
  $\mathbf{u}_{k+1} \leftarrow \arg\min_{\mathbf{u}} f_2(\mathbf{u}) + \frac{\mu}{2}\|\mathbf{G}\,\mathbf{z}_{k+1} - \mathbf{u} - \mathbf{d}_k\|_2^2$
  $\mathbf{d}_{k+1} \leftarrow \mathbf{d}_k - (\mathbf{G}\,\mathbf{z}_{k+1} - \mathbf{u}_{k+1})$
  $k \leftarrow k + 1$
until *stopping criterion is satisfied*;

Convergence of ADMM was first shown by Eckstein and Bertsekas [27] (via its connection to the Douglas-Rachford algorithm), and later by other authors under weaker conditions [11]. The result by Eckstein and Bertsekas allows for inexact minimizations, but demands that $\mathbf{G}$ has full column rank, which guarantees that the minimization needed to compute $\mathbf{z}_{k+1}$ has a unique solution; notice that the minimization yielding $\mathbf{u}_{k+1}$ has a unique solution, because the function being minimized is strictly convex. Moreover, for convergence to hold, problem (3.1) must have a solution; in this paper, we will assume that this is always the case.

Theorem 3.1 (Eckstein and Bertsekas [27]) *Consider problem* (3.1), *where* $\mathbf{G} \in \mathbb{R}^{p\times d}$ *has full column rank and* $f_1 : \mathbb{R}^d \to \bar{\mathbb{R}}$ *and* $f_2 : \mathbb{R}^p \to \bar{\mathbb{R}}$ *are closed, proper, and convex functions. Consider arbitrary* $\mu > 0, \mathbf{u}_0, \mathbf{d}_0 \in \mathbb{R}^p$. *Let* $\eta_k \geq 0,\ k = 0, 1, \ldots,$ *and* $\rho_k \geq 0,\ k = 0, 1, \ldots,$ *be two sequences such that* $\sum_{k=0}^{\infty} \eta_k < \infty$ *and* $\sum_{k=0}^{\infty} \rho_k < \infty$. *Consider three sequences* $\mathbf{z}_k \in \mathbb{R}^d, \mathbf{u}_k \in \mathbb{R}^p$, *and* $\mathbf{d}_k \in \mathbb{R}^p$, *for* $k = 0, 1, \ldots,$ *satisfying*

$$\Big\|\mathbf{z}_{k+1} - \arg\min_{\mathbf{z}} f_1(\mathbf{z}) + \frac{\mu}{2}\|\mathbf{G}\mathbf{z}-\mathbf{u}_k-\mathbf{d}_k\|_2^2\Big\| \leq \eta_k$$
$$\Big\|\mathbf{u}_{k+1} - \arg\min_{\mathbf{u}} f_2(\mathbf{u}) + \frac{\mu}{2}\|\mathbf{G}\mathbf{z}_{k+1}-\mathbf{u}-\mathbf{d}_k\|_2^2\Big\| \leq \rho_k$$
$$\mathbf{d}_{k+1} = \mathbf{d}_k - (\mathbf{G}\,\mathbf{z}_{k+1} - \mathbf{u}_{k+1}).$$

Then, if (3.1) *has a solution, that is, if Arg* $\min_{\mathbf{z}\in\mathbb{R}^d} f_1(\mathbf{z}) + f_2(\mathbf{G}\,\mathbf{z}) \neq \emptyset$, *the sequence* $\{\mathbf{z}_k\}$ *converges to some* $\mathbf{z}^* \in$ *Arg* $\min_{\mathbf{z}\in\mathbb{R}^d} f_1(\mathbf{z}) + f_2(\mathbf{G}\,\mathbf{z})$. *If* (3.1) *does not have a solution, then at least one of the sequences* $\{\mathbf{u}_k\}$ *or* $\{\mathbf{d}_k\}$ *diverges.*

Notice that the sequences $\mathbf{z}_k$, $\mathbf{u}_k$ and $\mathbf{d}_k$ defined in the ADMM algorithm satisfy the conditions in Theorem 1 with $\eta_k = \rho_k = 0$. However, the theorem shows that even if the minimizations in lines 3–4 of the ADMM algorithm are inexactly solved, convergence still holds if the error sequences are absolutely summable. This fact is quite relevant in designing instances of ADMM, in cases where these minimizations lack closed form solutions [34]. For recent and comprehensive reviews of ADMM and its relationship with Bregman methods [68], see [11, 32].

A very important issue to which some research has been devoted is the choice of the ADMM penalty parameter μ. Although Theorem 1 guarantees convergence (if its conditions are satisfied), regardless of the value of μ, in practice its choice has a great impact on the speed of convergence. A well-known heuristic to adjust μ along the algorithm that works well in most problems is described in [11]. More recently, a version of ADMM that adaptively adjusts μ (based on the Barzilai-Borwein spectral method for gradient descent [6]) to achieve fast convergence has been proposed and shown to be very effective, yielding fast empirical convergence and robustness with respect to the initial stepsize and problem scaling.

3.2 Using ADMM for More than Two Functions

In many relevant problems, the objective function involves more than just the two terms in (3.1). Consider a generalization of (3.1) where instead of two functions, we have J functions, *i.e.*,

$$\min_{\mathbf{z}\in\mathbb{R}^d} \sum_{j=1}^{J} g_j(\mathbf{H}^{(j)}\,\mathbf{z}), \tag{3.2}$$

with $g_j : \mathbb{R}^{p_j} \to \bar{\mathbb{R}}$ being closed, proper, and convex, and $\mathbf{H}^{(j)} \in \mathbb{R}^{p_j\times d}$ arbitrary matrices of appropriate sizes.

Extending ADMM to handle the more general formulation (3.2) is a quest that has stimulated considerable research efforts (see [19, 41] and the several references therein). In our line of work (which started in [2, 33, 34]), we have sidestepped this issue by reformulating (3.2) into the form (3.1), which can be done in a number of different ways, and then simply apply standard ADMM, inheriting its convergence properties. Similar approaches were independently proposed in different forms by Setzer, Steidl, and Teuber [61], Combettes and Pesquet [22], and Eckstein and Yao [28].

Our approach is based on casting the minimization problem (3.2) into the form (3.1) by using the following correspondences:

- $f_1 = 0$;
- $f_2 : \mathbb{R}^{p_1} \times \cdots \times \mathbb{R}^{p_J} \to \bar{\mathbb{R}}$ is defined as

$$f_2(\mathbf{u}) = \sum_{j=1}^{J} g_j(\mathbf{u}^{(j)}), \tag{3.3}$$

  where $\mathbf{u}^{(j)} \in \mathbb{R}^{p_j}$ and $\mathbf{u} = [(\mathbf{u}^{(1)})^T, \ldots, (\mathbf{u}^{(J)})^T]^T \in \mathbb{R}^p$;
- matrix $\mathbf{G} \in \mathbb{R}^{p\times d}$ is obtained by piling $\mathbf{H}^{(1)}, \ldots, \mathbf{H}^{(J)}$,

$$\mathbf{G} = \left[(\mathbf{H}^{(1)})^T \ \cdots \ (\mathbf{H}^{(J)})^T\right]^T \in \mathbb{R}^{p\times d}, \tag{3.4}$$

  where $p = p_1 + \cdots + p_J$.

In applying ADMM to the resulting problem, it is convenient to define the following partitions (where $\mathbf{d}_k^{(j)}, \mathbf{u}_k^{(j)} \in \mathbb{R}^{p_j}$):

$$\mathbf{d}_k = \left[(\mathbf{d}_k^{(1)})^T \cdots (\mathbf{d}_k^{(J)})^T\right]^T, \tag{3.5}$$

$$\mathbf{u}_k = \left[(\mathbf{u}_k^{(1)})^T \cdots (\mathbf{u}_k^{(J)})^T\right]^T. \tag{3.6}$$

The fact that $f_1 = 0$ transforms line 3 of ADMM into an unconstrained quadratic problem, which has a unique solution, if $\mathbf{G}$ has full column-rank. Given the block structure of $\mathbf{G}$ in (3.4), the corresponding solution can be written as

$$\arg\min_{\mathbf{z}} \|\mathbf{G}\,\mathbf{z} - \boldsymbol{\zeta}_k\|_2^2 = (\mathbf{G}^T\mathbf{G})^{-1}\mathbf{G}^T\boldsymbol{\zeta}_k \tag{3.7}$$

$$= \left[\sum_{j=1}^{J}(\mathbf{H}^{(j)})^T\mathbf{H}^{(j)}\right]^{-1}\sum_{j=1}^{J}(\mathbf{H}^{(j)})^T\boldsymbol{\zeta}_k^{(j)}, \tag{3.8}$$

where $\boldsymbol{\zeta}_k^{(j)} = \mathbf{u}_k^{(j)} + \mathbf{d}_k^{(j)}$ and $\boldsymbol{\zeta}_k = [(\boldsymbol{\zeta}_k^{(1)})^T \cdots (\boldsymbol{\zeta}_k^{(J)})^T]^T$.

Due to the form of f_2 in (3.3), line 4 of the ADMM algorithm becomes,

$$\begin{bmatrix}\mathbf{u}_{k+1}^{(1)}\\ \vdots \\ \mathbf{u}_{k+1}^{(J)}\end{bmatrix} \leftarrow \arg\min_{\begin{bmatrix}\mathbf{u}^{(1)}\\ \vdots \\ \mathbf{u}^{(J)}\end{bmatrix}\in\mathbb{R}^p} \sum_{j=1}^{J} g_j(\mathbf{u}^{(j)}) + \|\mathbf{u}^{(j)} - \mathbf{s}_k^{(j)}\|_2^2, \tag{3.9}$$

where $\mathbf{s}_k^{(j)} = \mathbf{H}^{(j)}\mathbf{z}_{k+1} - \mathbf{d}_k^{(j)}$. Due to the obvious separability of this problem, the minimization in (3.9) can be decoupled into a set of J independent minimizations,

$$\mathbf{u}_{k+1}^{(j)} \leftarrow \arg\min_{\mathbf{v}\in\mathbb{R}^{p_j}} g_j(\mathbf{v}) + \frac{\mu}{2}\|\mathbf{v} - \mathbf{s}_k^{(j)}\|_2^2, \tag{3.10}$$

for $j = 1, \ldots, J$.

The minimization problem in the right hand side of (3.10) defines the so-called *Moreau proximity operator* of g_j/μ (denoted as $\mathrm{prox}_{g_j/\mu}$) [7,50], applied to $\mathbf{s}_k^{(j)}$, thus

$$\mathbf{u}_{k+1}^{(j)} \leftarrow \mathrm{prox}_{g_j/\mu}(\mathbf{s}_k^{(j)}) \equiv \arg\min_{\mathbf{x}} \frac{\mu}{2}\|\mathbf{x} - \mathbf{s}_k^{(j)}\|_2^2 + g_j(\mathbf{x}). \tag{3.11}$$

For several functions, the corresponding Moreau proximity operators can be computed exactly in closed form [7,21]. An important and well-known case is the ℓ_1 norm ($\|\mathbf{x}\|_1 = \sum_i |x_i|$), for which the corresponding proximity operator is the famous soft thresholding function,

$$\mathrm{prox}_{\gamma\|\cdot\|_1}(\mathbf{v}) = \mathrm{soft}(\mathbf{v},\gamma) = \mathrm{sign}(\mathbf{v}) \odot \max\{|\mathbf{v}| - \gamma, 0\}, \tag{3.12}$$

where $\mathrm{sign}(\cdot)$ is the component-wise application of the sign function, $\odot$ is the component-wise product, $|\mathbf{v}|$ denotes the vector of absolute values of the elements of $\mathbf{v}$, and the maximum is computed in a component-wise fashion, *i.e.*, $(\max\{\mathbf{a}, \mathbf{b}\})_i = \max\{a_i, b_i\}$.

The computational bottleneck of this instance of ADMM is the matrix inversion in (3.8), which has often been seen as an obstacle that should be avoided at all costs. However, we will see below how this inversion can be very efficiently computed in a variety of imaging inverse problems of interest.

In the following sections, we will describe several instantiations of the formulation in (3.2) that we have used in recent years to address several imaging inverse problems, focusing in each case on how the matrix inversion in (3.8) and the proximity operators in (3.10) can be efficiently computed. We will not present any numerical/experimental results, referring the interested reader to the original publications where each of these algorithms was first presented.

4 Linear Observations with Gaussian Noise

4.1 Observation Model

The most studied imaging inverse problem involves linear observations additively contaminated by independent (a.k.a. white) Gaussian noise. The formal probabilistic model for the observed data $\mathbf{y}$ is

$$\mathbf{y} \sim \mathcal{N}(\mathbf{Bx}, \mathbf{I}), \tag{4.1}$$

where $\mathbf{B}$ is the matrix representation of the direct/observation operator and $\mathcal{N}(\boldsymbol{\mu}, \mathbf{C})$ denotes a Gaussian distribution of mean vector $\boldsymbol{\mu}$ and covariance $\mathbf{C}$ (there is no loss of generality in taking unit variance, since we are assuming that the noise variance is known). In the case of image deconvolution, under periodic boundary conditions, $\mathbf{B}$ is a block-circulant matrix with circulant blocks. Matrix $\mathbf{B}$ can also represent other linear operators, such as tomographic (Radon) projections, or the loss of image pixels (in image inpainting problems). Given (4.1), the natural choice for the data term is the negative log-likelihood

$$\Upsilon(\mathbf{y}, \mathbf{x}) = \frac{1}{2}\|\mathbf{y} - \mathbf{Bx}\|_2^2 + K, \tag{4.2}$$

where K is an irrelevant additive constant that we will set to zero.

4.2 Tikhonov Analysis Regularization

The analysis formulation of Tikhonov regularization yields the following unconstrained optimization problem:

$$\min_{\mathbf{x}} \frac{1}{2}\|\mathbf{y} - \mathbf{B}\mathbf{x}\|_2^2 + \alpha\|\mathbf{W}^T\mathbf{x}\|_1, \tag{4.3}$$

where $\mathbf{W}^T$ is the analysis matrix of some Parseval frame contained in the columns of $\mathbf{W}$.

Problem (4.3) has the canonical form (3.2), with $J = 2$, $\mathbf{H}^{(1)} = \mathbf{B}$, $\mathbf{H}^{(2)} = \mathbf{W}^T$, $g_1(\mathbf{u}) = \frac{1}{2}\|\mathbf{y} - \mathbf{u}\|_2^2$, and $g_2(\mathbf{u}) = \alpha\|\mathbf{u}\|_1$. To implement the ADMM instance introduced in Subsection 3.2, the necessary building blocks are the proximity operators of g_1 and g_2 and the matrix inversion in (3.8). The proximity operators in this case are simple:

$$\begin{aligned} \operatorname{prox}_{g_1/\mu}(\mathbf{s}) &= \arg\min_{\mathbf{x}} \frac{\mu}{2}\|\mathbf{s} - \mathbf{x}\|_2^2 + \frac{1}{2}\|\mathbf{y} - \mathbf{x}\|_2^2 \\ &= \frac{\mathbf{y} + \mu\mathbf{s}}{1 + \mu}; \end{aligned} \tag{4.4}$$

$$\operatorname{prox}_{g_2/\mu}(\mathbf{s}) = \operatorname{soft}(\mathbf{s}, \alpha/\mu); \tag{4.5}$$

The matrix inversion in (3.8), with $\mathbf{H}^{(1)} = \mathbf{B}$, $\mathbf{H}^{(2)} = \mathbf{W}^T$, becomes

$$\left[\mathbf{B}^T\mathbf{B} + \mathbf{W}\mathbf{W}^T\right]^{-1} = \left[\mathbf{B}^T\mathbf{B} + \mathbf{I}\right]^{-1}, \tag{4.6}$$

because $\mathbf{W}^T$ is the analysis matrix of a Parseval frame, thus $\mathbf{W}\mathbf{W}^T = \mathbf{I}$. The cost of computing this inverse depends critically on the structure of matrix $\mathbf{B}$. In the following paragraphs, we will review how in several problems of interest, this matrix inversion can be computed with low cost.

The algorithm also involves matrix-vector products with $\mathbf{W}$ and $\mathbf{W}^T$, that is, frame synthesis and analysis operations; we only consider frames for which fast $O(n \log n)$ implementations of these operations exist [47]. Examples of such frames include undecimated wavelets, complex wavelets, curvelets, and shearlets [29].

The resulting algorithm was proposed in [1] and was called SALSA (*split augmented Lagrangian shrinkage algorithm*). Notice that convergence of SALSA is guaranteed by Theorem 1, since matrix $\mathbf{G}$ (see (3.4)), which in this case is $\mathbf{G} = [\mathbf{B}^T\ \mathbf{W}]^T$, has full column rank, because the analysis matrix $\mathbf{W}^T$ of a Parseval frame has a trivial null-space ($\mathbf{W}^T\mathbf{v} = 0 \Rightarrow \mathbf{v} = 0$). More details and experimental results can be found in [1].

4.2.1 Periodic Deconvolution

If $\mathbf{B}$ is the matrix representation of a two-dimensional periodic convolution, it is a block-circulant matrix with circulant blocks that can be factorized as

$$\mathbf{B} = \mathbf{U}^H\mathbf{D}\mathbf{U}, \tag{4.7}$$

where $\mathbf{U}$ is the matrix representing the two-dimensional discrete Fourier transform (DFT), $\mathbf{U}^H = \mathbf{U}^{-1}$ is its inverse[3] ($\mathbf{U}$ is a *unitary* matrix, meaning that $\mathbf{U}\mathbf{U}^H = \mathbf{U}^H\mathbf{U} = \mathbf{I}$), and $\mathbf{D}$ is a diagonal matrix with the DFT coefficients of the convolution kernel represented by the multiplication by $\mathbf{B}$. Consequently (with $\mathbf{B}^T = \mathbf{B}^H$, since $\mathbf{B}$ is a real matrix),

$$\left[\mathbf{B}^T\mathbf{B} + \mathbf{I}\right]^{-1} = \mathbf{U}^H\left[|\mathbf{D}|^2 + \mathbf{I}\right]^{-1}\mathbf{U}, \tag{4.8}$$

where $|\mathbf{D}|^2$ is the matrix with the squared absolute values of the entries of $\mathbf{D}$. Since $|\mathbf{D}|^2 + \mathbf{I}$ is diagonal, its inversion has $O(n)$ cost. Finally, the required products by $\mathbf{U}$ and $\mathbf{U}^H$ can be carried out with $O(n \log n)$ cost by resorting to the *fast Fourier transform* (FFT) algorithm.

4.2.2 Deconvolution with Unknown Boundaries

In deconvolution, the pixels located near the boundary of the observed image depend on pixels (of the unknown image) located outside of its domain. The typical way to formalize this issue is to adopt a so-called *boundary condition* (BC). In the context of image restoration, the used of periodic BC as in the previous paragraph dates back at least to the 1970s [5]. Other important and well-studied alternatives are the following:

- The zero BC assumes that the external pixels have zero value, thus the matrix representing the convolution is block-Toeplitz, with Toeplitz blocks [51]. By analogy with the BC for ordinary or partial differential equations that assumes fixed values at the domain boundary, this is commonly referred to as Dirichlet BC [51].
- In the reflexive and anti-reflexive BC, the pixels outside the image domain are a reflection of those near the boundary, using even or odd symmetry, respectively [25]. Because in the reflexive BC, the discrete derivative at the boundary is fixed, by analogy with the BC for differential equations with fixed derivative at the domain boundary, the reflexive BC is often referred to as Neumann BC [51,52].

As shown in [4], these classical BCs are unnatural, and do not correspond to any realistic imaging system; their use is essentially motivated by computational convenience. Namely, periodic BCs allow a very fast implementation of the convolution as a point-wise multiplication in the DFT domain, efficiently implementable using the FFT. However, real imaging systems, there is no reason why the external (unobserved) pixels should obey a periodic, or any other, BC. A well-known consequence of this mismatch is the degradation of the deconvolved images, namely with the appearance of ringing artifacts near the boundaries. There are several techniques aiming at reducing these boundary artifacts [44]. Convolutions under Dirichlet BC can be efficiently computed in the DFT domain by using zero-padding [51]. However, not only the above-mentioned mismatch problem remains (*i.e.*, it is unrealistic

[3] The notation $\mathbf{A}^H$ denotes the conjugate transpose of matrix $\mathbf{A}$.

to assume that the boundary pixels are zero), but also the matrix inversion in (3.8) can non longer be effciently computed.

A realistic BC for practical imaging systems assumes that the external pixels are unknown. The corresponding observation model is the composition of a spatial mask with a convolution under arbitrary BC (the most convenient one being periodic, due to the resulting computational efficiency) [18, 48, 49, 57, 62]. This assumption corresponds to setting

$$\Upsilon(\mathbf{y}, \mathbf{x}) = \frac{1}{2}\|\mathbf{y} - \mathbf{M}\,\mathbf{A}\,\mathbf{x}\|_2^2, \tag{4.9}$$

where $\mathbf{M} \in \{0, 1\}^{m\times n}$ (with $m < n$) is a masking matrix, *i.e.*, a matrix whose rows are a subset of the rows of an identity matrix. Assuming that $\mathbf{A}$ models the convolution with a blurring filter with a limited support of size $(1 + 2\,l) \times (1 + 2\,l)$, and that $\mathbf{x}$ and $\mathbf{Ax}$ represent square images of dimensions $\sqrt{n} \times \sqrt{n}$, then matrix $\mathbf{M} \in \mathbb{R}^{m\times n}$, with $m = (\sqrt{n} - 2\,l)^2$, represents the removal of a band of width l of the outermost pixels of the full convolved image $\mathbf{Ax}$. The observation model in (4.9) can be seen as hybrid deconvolution-inpainting problem [18], with the missing pixels constituting the unknown boundary. For $\mathbf{M} = \mathbf{I}$, (4.9) becomes a standard periodic deconvolution problem. Conversely, for $\mathbf{A} = \mathbf{I}$, (4.9) becomes a pure inpainting problem. Naturally, (4.9) can be used to model problems where not only the boundary, but also other pixels, are missing [4].

Combining (4.9) with the Tikhonov analysis regularization yields

$$\min_{\mathbf{x}} \frac{1}{2}\|\mathbf{y} - \mathbf{M}\,\mathbf{A}\,\mathbf{x}\|_2^2 + \alpha\|\mathbf{W}^T\mathbf{x}\|_1, \tag{4.10}$$

which corresponds to (4.3) with $\mathbf{B} = \mathbf{M}\,\mathbf{A}$. A naïve approach would be to simply replace $\mathbf{B}$ by $\mathbf{M}\,\mathbf{A}$ everywhere and proceed as in subsection 4.2.1. However, the resulting matrix $\mathbf{B}^T\mathbf{B} + \mathbf{I} = \mathbf{A}^T\mathbf{M}^T\mathbf{M}\mathbf{A} + \mathbf{I}$ becomes harder to invert, even with $\mathbf{A}$ circulant with circulant blocks: due to the presence of $\mathbf{M}$, (4.8) is no longer valid.

The approach proposed in [4] is to use a different way of mapping (4.10) into (3.2): $\mathbf{H}^{(1)} = \mathbf{A}$, $\mathbf{H}^{(2)} = \mathbf{W}^T$, $g_1(\mathbf{u}) = \frac{1}{2}\|\mathbf{y} - \mathbf{M}\,\mathbf{u}\|_2^2$, and $g_2(\mathbf{u}) = \alpha\|\mathbf{u}\|_1$. With this choice, the matrix to be inverted is $\mathbf{A}^T\mathbf{M}^T\mathbf{M}\mathbf{A} + \mathbf{I}$, which can be done again using the FFT, since $\mathbf{A}$ is circulant with circulant blocks. The other difference resides in function g_1, but it turns out that the corresponding proximity operator can still be computed efficiently:

$$\begin{aligned} \mathrm{prox}_{g_1/\mu}(\mathbf{s}) &= \arg\min_{\mathbf{x}} \|\mathbf{M}\,\mathbf{x} - \mathbf{y}\|_2^2 + \mu\|\mathbf{x} - \mathbf{s}\|_2^2 && (4.11)\\ &= \left(\mathbf{M}^T\mathbf{M} + \mu\mathbf{I}\right)^{-1}\left(\mathbf{M}^T\mathbf{y} + \mu\mathbf{s}\right); && (4.12) \end{aligned}$$

given the structure of $\mathbf{M}$, matrix $\mathbf{M}^T\mathbf{M}$ is diagonal, thus the inversion in (4.12) has $O(n)$ cost, the same being true about the product $\mathbf{M}^T\mathbf{y}$, which corresponds to extending the observed image $\mathbf{y}$ to the size of $\mathbf{x}$, by creating a boundary of zeros around it. In fact, (4.12) is a particular case of the result in Proposition 24.14 in [7], because $\mathbf{M}\,\mathbf{M}^T = \mathbf{I}$, but (4.11)–(4.12) shows that it is trivial to obtain in this case.

In summary, deconvolution under unknown boundary conditions can be addressed by this version of ADMM as efficiently as with periodic boundary conditions. For more details and experimental results, the interested reader is referred to [4].

4.2.3 Image Inpainting

In image inpainting problems, the observed image $\mathbf{y}$ results from not observing some elements of $\mathbf{x}$. The corresponding observation matrix $\mathbf{B}$ has dimension $m \times n$ (where $m < n$ is the number of observed pixels) and it is composed of m rows of an $n \times n$ identity matrix. In this case, $\mathbf{B}^T\mathbf{B}$ is a diagonal matrix of dimension $n \times n$ with ones and zeros in the diagonal (with the zeros corresponding to the non-observed image pixels). Consequently, $\mathbf{B}^T\mathbf{B} + \mathbf{I}$ is a diagonal matrix and its inversion has $O(n)$ cost, as do matrix-vector products by $\mathbf{B}$ and $\mathbf{B}^T$.

4.2.4 Compressive Fourier Imaging

The third and final observation model herein considered is that of partial Fourier observations, which is used as a basic model of magnetic resonance imaging (MRI) [46], and has been the focus of much recent interest due to its connection to compressed sensing [13,26]. In this imaging model, the observation matrix is given by $\mathbf{B} = \mathbf{M}\,\mathbf{U}$, where $\mathbf{M}$ is again $m \times n$ (with $m < n$) masking binary matrix, with similar properties as the the observation matrix in inpainting problems, and $\mathbf{U}$ is the DFT matrix. In this case,

$$\begin{aligned}\left[\mathbf{B}^T\mathbf{B} + \mathbf{I}\right]^{-1} &= \left[\mathbf{U}^H\mathbf{M}^T\mathbf{M}\mathbf{U} + \mathbf{I}\right]^{-1}\\ &= \mathbf{I} - \mathbf{U}^H\mathbf{M}^T\left[\mathbf{M}\,\mathbf{U}\,\mathbf{U}^H\mathbf{M}^T + \mathbf{I}\right]^{-1}\mathbf{M}\,\mathbf{U}\\ &= \mathbf{I} - \frac{1}{2}\,\mathbf{U}^H\mathbf{M}^T\mathbf{M}\,\mathbf{U},\end{aligned} \tag{4.13}$$

where the second equality is a consequence of the application of the famous Sherman-Morrison-Woodbury (SMW) matrix inversion formula, and the third equality results from the fact that $\mathbf{U}\mathbf{U}^H = \mathbf{I}$ and $\mathbf{M}\mathbf{M}^T = \mathbf{I}$ (due to its structure). Again, the cost of computing and applying this matrix is dominated by the $O(n \log n)$ cost of the FFT implementations of the products by $\mathbf{U}$ and $\mathbf{U}^H$.

4.3 Tikhonov Synthesis Regularization

The synthesis formulation of Tikhonov regularization corresponds to the optimization problem

$$\min_{\mathbf{x}} \frac{1}{2}\|\mathbf{y} - \mathbf{B}\mathbf{W}\mathbf{x}\|_2^2 + \alpha\|\mathbf{x}\|_1, \tag{4.14}$$

where $\mathbf{x}$ is no longer the image to be recovered, but the coefficients of its representation on the Parseval frame that constitutes the columns of $\mathbf{W}$. The standard approach for solving (4.14) is the so-called proximal-gradient (or iterative shrinkage/thresholding–IST) algorithm, which has been rediscovered several times under different perspectives and in several communities [12, 24, 36, 43, 55] (see also [7,23,54]). However, IST applied to the problem in (4.14) is known to be quite slow, specially when $\mathbf{B}\,\mathbf{W}$ is poorly conditioned, a fact that has stimulated much research aimed at developing faster variants thereof [8,9,66].

Problem (4.14) has the canonical form (3.2), with $J = 2$, $\mathbf{H}^{(1)} = \mathbf{BW}$, and $\mathbf{H}^{(2)} = \mathbf{I}$, $g_1(\mathbf{u}) = \frac{1}{2}\|\mathbf{y}-\mathbf{u}\|_2^2$, and $g_2(\mathbf{u}) = \alpha\|\mathbf{u}\|_1$. Notice that convergence of ADMM/SALSA in this case is also guaranteed by Theorem 1, since matrix $\mathbf{G}$ (see (3.4)), in this case is $\mathbf{G} = [(\mathbf{BW})^T\ \ \mathbf{I}]^T$, which has full column rank regardless of $\mathbf{BW}$.

The proximity operators of g_1 and g_2 are as in (4.4) and (4.5), while the matrix inversion in (3.8) now takes the form

$$\begin{aligned}\left[\mathbf{W}^T\mathbf{B}^T\mathbf{B}\mathbf{W}+\mathbf{I}\right]^{-1} &= \mathbf{I}-\mathbf{W}^T\mathbf{B}^T\left[\mathbf{B}\mathbf{W}\mathbf{W}^T\mathbf{B}^T+\mathbf{I}\right]^{-1}\mathbf{B}\mathbf{W}\\ &= \mathbf{I}-\mathbf{W}^T\mathbf{B}^T\left[\mathbf{B}\mathbf{B}^T+\mathbf{I}\right]^{-1}\mathbf{B}\mathbf{W},\end{aligned} \tag{4.15}$$

where the first equality results from the application of the SMW matrix inversion formula and the second one from the fact that $\mathbf{W}$ contains a Parseval frame, thus $\mathbf{W}\mathbf{W}^T = \mathbf{I}$. We are thus left with the problem of inverting matrix $\mathbf{B}\mathbf{B}^T + \mathbf{I}$, which again depends of the particular problem at hand.

4.3.1 Periodic Deconvolution

If $\mathbf{B}$ represents a periodic convolution, $\mathbf{B} = \mathbf{U}^H\mathbf{D}\mathbf{U}$, we have $\left[\mathbf{B}\mathbf{B}^T + \mathbf{I}\right]^{-1} = \mathbf{U}^H\left[|\mathbf{D}|^2+\mathbf{I}\right]^{-1}\mathbf{U}$, exactly as in (4.8). Inserting this equality in (4.15) finally yields

$$\left[\mathbf{W}^T\mathbf{B}^T\mathbf{B}\mathbf{W}+\mathbf{I}\right]^{-1} = \mathbf{I}-\mathbf{W}^T\mathbf{U}^H\mathbf{D}\left[|\mathbf{D}|^2+\mathbf{I}\right]^{-1}\mathbf{D}\mathbf{U}\mathbf{W}. \tag{4.16}$$

As above, because matrix $\mathbf{D}\left[|\mathbf{D}|^2+\mathbf{I}\right]^{-1}\mathbf{D}$ is diagonal, the cost of matrix-vector products by the matrix in (4.16) is $O(n\log n)$, corresponding to FFT implementations of the products by $\mathbf{U}$ and $\mathbf{U}^H$ and of the fast frame analysis ($\mathbf{W}^T$) and synthesis ($\mathbf{W}$).

4.3.2 Deconvolution with Unknown Boundaries

In this case, $\mathbf{B} = \mathbf{M}\mathbf{A}$ and, as in subsection 4.2.2, we proceed by redefining g_1 as $g_1(\mathbf{u}) = \frac{1}{2}\|\mathbf{y} - \mathbf{M}\mathbf{u}\|_2^2$, of which the proximity operator was given in (4.12). Everything else is as in the periodic deconvolution case, with matrix $\mathbf{A}$ replacing $\mathbf{B}$ in the expressions of the previous subsection.

4.3.3 Image Inpainting

In the image inpainting problem, $\mathbf{B}\mathbf{B}^T = \mathbf{I}$, thus $\left(\mathbf{B}\mathbf{B}^T + \mathbf{I}\right)^{-1} = \frac{1}{2}\mathbf{I}$. Inserting this equality into (4.15), we obtain

$$\left[\mathbf{W}^T\mathbf{B}^T\mathbf{B}\mathbf{W} + \mathbf{I}\right]^{-1} = \mathbf{I} - \frac{1}{2}\mathbf{W}^T\mathbf{B}^T\mathbf{B}\mathbf{W}, \tag{4.17}$$

Since matrix $\mathbf{B}^T\mathbf{B}$ is diagonal, the cost of products by the matrix in (4.17) is $O(n \log n)$, corresponding to fast frame analysis ($\mathbf{W}^T$) and synthesis ($\mathbf{W}$) operations.

4.3.4 Compressive Fourier Imaging

For partial Fourier observations, we have $\mathbf{B} = \mathbf{M}\mathbf{U}$, where, as above, $\mathbf{U}$ is the DFT matrix and $\mathbf{M}$ contains a subset of the rows of an identity. In this case,

$$\left[\mathbf{B}\mathbf{B}^H + \mathbf{I}\right]^{-1} = \left[\mathbf{M}\mathbf{U}\mathbf{U}^H\mathbf{M}^T + \mathbf{I}\right]^{-1} = \frac{1}{2}\mathbf{I}, \tag{4.18}$$

again because $\mathbf{U}\mathbf{U}^H = \mathbf{I}$ and $\mathbf{M}\mathbf{M}^T = \mathbf{I}$. Inserting this equality and $\mathbf{B} = \mathbf{M}\mathbf{U}$ into (4.15) yields

$$\left[\mathbf{W}^T\mathbf{B}^T\mathbf{B}\mathbf{W} + \mathbf{I}\right]^{-1} = \mathbf{I} - \frac{1}{2}\,\mathbf{W}^T\mathbf{U}^H\mathbf{M}^T\mathbf{M}\,\mathbf{U}\,\mathbf{W}. \tag{4.19}$$

Since $\mathbf{M}^T\mathbf{M}$ is diagonal, the cost of products by the matrix in (4.19) is $O(n \log n)$, corresponding to fast frame analysis ($\mathbf{W}^T$) and synthesis ($\mathbf{W}$) operations and the FFT implementations of the products by $\mathbf{U}$ and $\mathbf{U}^H$.

4.4 Morozov Analysis Regularization

The frame-based analysis formulation of Morozov regularization (see Section 2, for the definition) consists in tackling the inverse problem via the following constrained optimization problem:

$$\min_{\mathbf{x}} \|\mathbf{W}^T\mathbf{x}\|_1 \quad \text{subject to} \quad \|\mathbf{y} - \mathbf{B}\mathbf{x}\|_2 \leq \delta. \tag{4.20}$$

In [2], we proposed addressing problem (4.20) by rewriting it in (apparently) unconstrained form

$$\min_{\mathbf{x}} \iota_{\mathcal{B}_\delta(\mathbf{y})}(\mathbf{B}\mathbf{x}) + \|\mathbf{W}^T\mathbf{x}\|_1, \tag{4.21}$$

where $\iota_{\mathcal{S}}$ is the indicator function of set $\mathcal{S}$, which is defined as

$$\iota_{\mathcal{S}}(\mathbf{x}) = \begin{cases} 0 & \Leftarrow \mathbf{x} \in \mathcal{S} \\ \infty & \Leftarrow \mathbf{x} \notin \mathcal{S}, \end{cases}$$

and $\mathcal{B}_\delta(\mathbf{y}) = \{\mathbf{x} : \|\mathbf{x} - \mathbf{y}\|_2 \leq \delta\}$ denotes an Euclidean ball of radius δ centered at $\mathbf{y}$.

Clearly, problem (4.21) can be mapped into the canonical form (3.2), with $J = 2$, $g_1(\mathbf{u}) = \iota_{\mathcal{B}_\delta(\mathbf{y})}(\mathbf{u})$, $g_2(\mathbf{u}) = \|\mathbf{u}\|_1$, $\mathbf{H}^{(1)} = \mathbf{B}$, and $\mathbf{H}^{(2)} = \mathbf{W}^T$. The proximity operator of this g_1 is

$$\begin{aligned} \text{prox}_{g_1/\mu}(\mathbf{s}) &= \arg\min_{\mathbf{x}} \frac{\mu}{2}\|\mathbf{s} - \mathbf{x}\|_2^2 + \iota_{\mathcal{B}_\delta(\mathbf{y})}(\mathbf{x}) \\ &= \arg\min_{\mathbf{x}\in\mathcal{B}_\delta(\mathbf{y})} \|\mathbf{s} - \mathbf{x}\|_2^2 \qquad (4.22) \\ &= \mathbb{P}_{\mathcal{B}_\delta(\mathbf{y})}(\mathbf{s}), \qquad (4.23) \end{aligned}$$

where $\mathbb{P}_{\mathcal{S}}$ denotes the Euclidean projection on a (convex) set $\mathcal{S}$. In the case of the Euclidean ball $\mathcal{B}_\delta(\mathbf{y})$, this is simply

$$\mathbb{P}_{\mathcal{B}_\delta(\mathbf{y})}(\mathbf{s}) = \begin{cases} \mathbf{s} & \Leftarrow \|\mathbf{s} - \mathbf{y}\|_2 \leq \delta \\ \mathbf{y} + \delta\,\frac{\mathbf{s}-\mathbf{y}}{\|\mathbf{s}-\mathbf{y}\|_2} & \Leftarrow \|\mathbf{s} - \mathbf{y}\|_2 > \delta. \end{cases} \qquad (4.24)$$

Finally, as above, $\text{prox}_{g_2/\mu}(\mathbf{s}) = \text{soft}(\mathbf{s}, 1/\mu)$.

The matrix inversion in (3.8), with $\mathbf{H}^{(1)} = \mathbf{B}$, and $\mathbf{H}^{(2)} = \mathbf{W}^T$, has the exact same form as in (4.6), and all the derivations (for the analysis and synthesis formulations of periodic deconvolution, inpainting, and compressive Fourier imaging) carried out for the Tikhonov regularization also apply in this case. The resulting algorithm was proposed in [2] and was termed CSALSA (*constrained split augmented Lagrangian shrinkage algorithm*). Convergence of CSALSA results from the same arguments used to show convergence of SALSA. Finally, notice that the relationship between the Morozov analysis and synthesis formulations is exactly the same as that between the Tikhonov counterparts (the only difference being the replacement of the linear proximity operator (4.4) by the projection (4.23)), so we will abstain from studying it in detail here.

The only case that requires some additional thought is that of unknown boundaries. Following the same strategy as in the Tikhonov regularization case, we redefine g_1 as $g_1(\mathbf{u}) = \iota_{\mathcal{B}_\delta(\mathbf{y})}(\mathbf{M}\,\mathbf{u})$; consequently, the corresponding proximity operator is no longer simply an Euclidean projection on an Euclidean ball as in (4.24), although it can still be computed efficiently and in close-form, as shown next. Using Proposition 24.14 in [7],

$$\text{prox}_{g_1/\mu}(\mathbf{s}) = \mathbf{s} + \mathbf{M}^T\left(\mathbb{P}_{\mathcal{B}_\delta(\mathbf{y})}(\mathbf{M}\,\mathbf{s}) - \mathbf{M}\,\mathbf{s}\right), \qquad (4.25)$$

which, as above, can be computed with $O(n)$ cost.

5 Poissonian Observations

5.1 Observation Model

In many imaging modalities (namely if the number of photos being detected by the imaging device is low, a scenario often known as *photon-limited imaging* [64, 65]), additive Gaussian noise is not an adequate model, and the observed data is better modeled by using a Poisson distribution,

$$\mathbf{y} \sim \mathcal{P}(\mathbf{Bx}), \tag{5.1}$$

where $\mathbf{B}$ is the matrix representation of the linear observation model and $\mathcal{P}(\boldsymbol{\lambda})$ denotes the distribution of a Poisson process of intensity vector $\boldsymbol{\lambda}$, *i.e.*,

$$\mathbb{P}[\mathbf{Y} = \mathbf{y}|\mathbf{x}] = \prod_i \frac{(\mathbf{B}\,\mathbf{x})_i^{y_i}\, e^{-(\mathbf{B}\,\mathbf{x})_i}}{y_i!} \tag{5.2}$$

where $\lambda_i \geq 0$, for all i.

Poissonian models are highly relevant in fields such as astronomical imaging [63], biomedical imaging (such as *positron-emission tomography* – PET [53]) [59, 65], and photographic imaging [37]. Given (5.1), the natural choice for the data term is the negative log-likelihood,

$$\Upsilon(\mathbf{y}, \mathbf{x}) = \sum_i \xi((\mathbf{Bx})_i, y_i), \tag{5.3}$$

where

$$\xi(z, y) = z + \iota_{\mathbb{R}_+}(z) - y\ \log(z_+), \tag{5.4}$$

where $z_+ = \max\{0, z\}$ and $0 \log(0) \equiv 0$. The inclusion of the indicator function $\iota_{\mathbb{R}_+}$ forces the non-negativity of the elements $\mathbf{B}\,\mathbf{x}$, since these can also be understood as Poisson intensities (see [34], for a more detailed derivation). This function is sometimes referred to as the Kullback-Leibler model, although only because it is formally similar to a Kullback-leibler divergence (apart from the indicator function) [61].

5.2 Tikhonov Analysis and Synthesis Regularization

The analysis formulation of the Tikhonov regularization approach to the image reconstruction/restoration problem with Poissonian observations leads the following unconstrained optimization problem:

$$\min_{\mathbf{x}} \sum_i \xi((\mathbf{Bx})_i, y_i) + \alpha \|\mathbf{W}^T \mathbf{x}\|_1 + \iota_{\mathbb{R}_+^n}(\mathbf{x}), \tag{5.5}$$

where the purpose of the indicator function of the first orthant, $\iota_{\mathbb{R}_+^n}$, is to enforce non-negativity of the solution, because the elements of $\mathbf{x}$ are Poisson intensities.

Problem (5.5) clearly follows the canonical form in (3.2), with $J = 3$, $g_1(\mathbf{u}) = \sum_i \xi(u_i, y_i)$, $g_2(\mathbf{u}) = \alpha\|\mathbf{u}\|_1$, $g_3(\mathbf{u}) = \iota_{\mathbb{R}_+^n}(\mathbf{u})$, $\mathbf{H}^{(1)} = \mathbf{B}$, $\mathbf{H}^{(2)} = \mathbf{W}^T$, and $\mathbf{H}^{(3)} = \mathbf{I}$. The building blocks of the ADMM instance introduced in Subsection 3.2 are the proximity operators of g_1, g_2, and g_3 and the matrix inversion in (3.8). The proximity operator of g_2 is as above: $\text{prox}_{g_2/\mu}(\mathbf{s}) = \text{soft}(\mathbf{s}, \alpha/\mu)$. The proximity operator of g_3 is simply the projection on the first orthant:

$$\text{prox}_{g_3/\mu}(\mathbf{s}) = \max\{\mathbf{s}, 0\}. \tag{5.6}$$

Concerning the proximity operator of g_1, it can be shown that it is given (component-wise) by

$$\left(\text{prox}_{g_1/\mu}(\mathbf{s})\right)_i = \frac{1}{2}\left(s_i - \frac{1}{\mu} + \sqrt{\left(s_i - \frac{1}{\mu}\right)^2 + \frac{4\,y_i}{\mu}}\right), \tag{5.7}$$

Notice that $\left(\text{prox}_{g_1/\mu}(\mathbf{s})\right)_i$ is always necessarily non-negative, regardless of its argument $\mathbf{s}$.

All the derivations made above concerning the matrix inversion in (3.8) apply unchanged to this case. The resulting family of algorithms was introduced in [34] and was therein called PIDAL (*Poisson image deconvolution via augmented Lagrangian*). Convergence of PIDAL is guaranteed by Theorem 1, since g_1, g_2, and g_3 are closed, proper, convex functions, and matrix $\mathbf{G}$ (see (3.4)), which is here equal to $\mathbf{G} = [\mathbf{B}^T\ \mathbf{W}\ \mathbf{I}]^T$, has full column rank due to the presence of $\mathbf{I}$. For the proof of existence of solutions of (5.5), the reader is referred to [34].

The PIDAL algorithm for the synthesis formulation of Tikhonov regularization for linear-Poisson observations,

$$\min_{\mathbf{x}} \sum_i \xi((\mathbf{BWx})_i, y_i) + \alpha\|\mathbf{x}\|_1 + \iota_{\mathbb{R}_+^n}(\mathbf{Wx}), \tag{5.8}$$

is obtained by using the same g_1, g_2, and g_3 functions, but a different set of $\mathbf{H}^{(j)}$ matrices: $\mathbf{H}^{(1)} = \mathbf{BW}$, $\mathbf{H}^{(2)} = \mathbf{I}$, and $\mathbf{H}^{(3)} = \mathbf{W}$. Concerning the matrix inversion in (3.8), we now have

$$\begin{aligned}\left(\mathbf{W}^T\mathbf{B}^T\mathbf{B}\mathbf{W} + \mathbf{I} + \mathbf{W}^T\mathbf{W}\right)^{-1} &= \left(\mathbf{W}^T(\mathbf{B}^T\mathbf{B} + \mathbf{I})\mathbf{W} + \mathbf{I}\right)^{-1}\\ &= \mathbf{I} - \mathbf{W}^T\left(\mathbf{W}\mathbf{W}^T + (\mathbf{B}^T\mathbf{B} + \mathbf{I})^{-1}\right)^{-1}\mathbf{W}\\ &= \mathbf{I} - \mathbf{W}^T\left(\mathbf{I} + (\mathbf{B}^T\mathbf{B} + \mathbf{I})^{-1}\right)^{-1}\mathbf{W},\end{aligned} \tag{5.9}$$

where the second equality results from the SMW matrix inversion identity and the third one from the assumption that $\mathbf{W}$ is a Parseval frame. We are thus left with the problem of inverting $(\mathbf{B}^T\mathbf{B} + \mathbf{I})$, which we have already addressed above for the case of periodic deconvolution, deconvolution with unknown boundaries, image inpainting, and partial Fourier observations. For more details and the use of other regularizers (namely, total variation), as well as numerical experimental results, the reader is referred to [34].

The Morozov formulation for the linear-Poisson case is not as straightforward as in the Gaussian case. In fact, the required projection (that takes the place of (4.23)) doesn't have a simple closed-form solution, and has to be computed numerically [14].

Finally, we should mention that we have also proposed a closely-related ADMM-based algorithm for the recovery of images observed under multiplicative noise, which is the fundamental model in coherent imaging systems (such as synthetic aperture radar and sonar, ultrasound imaging, laser imaging). The algorithm, called MIDAL (*multiplicative image denoising by augmented Lagrangian*), was introduced in [10].

6 Hybrid Analysis-Synthesis Regularization

Although some research has been devoted to comparing the analysis and synthesis formulations [31, 60], and their relative merits, there is no clear consensus on which of the two is to be preferred for a given problem. This choice can be avoided by combining the two formulations into a hybrid synthesis-analysis criterion, as we have proposed in [3]. Considering the linear-Gaussian observation model (the adaptation to Poissonian observation case is straightforward), we proposed the following (Tikhonov-type) analysis-synthesis hybrid formulation:

$$\min_{\mathbf{x}} \frac{1}{2}\|\mathbf{y} - \mathbf{B}\,\mathbf{W}_1\,\mathbf{x}\|_2^2 + \alpha\|\mathbf{x}\|_1 + \beta\|\mathbf{W}_2^T\,\mathbf{W}_1\,\mathbf{x}\|_1; \tag{6.1}$$

in (6.1), $\mathbf{W}_1$ and $\mathbf{W}_2$ are the synthesis matrices of two Parseval frames (the same one, or two different ones).

Clearly, the optimization problem in (6.1) can be written in the canonical form (3.2), by letting $J = 3$, $g_1(\mathbf{u}) = \frac{1}{2}\|\mathbf{y} - \mathbf{u}\|_2^2$, $g_2(\mathbf{u}) = \alpha\|\mathbf{u}\|_1$, $g_3(\mathbf{u}) = \beta\|\mathbf{u}\|_1$, $\mathbf{H}^{(1)} = \mathbf{B}\mathbf{W}_1$, $\mathbf{H}^{(2)} = \mathbf{I}$, and $\mathbf{H}^{(3)} = \mathbf{W}_2^T\,\mathbf{W}_1$. The proximity operator of g_1 is the linear shrinkage in (4.4), while those of g_2 and g_3 are soft thresholding functions (4.5). The final component needed to build the ADMM-based method is the matrix inversion in (3.8). Invoking the Parseval nature of both frames and applying the SMW formula, we obtain

$$\left[\mathbf{W}_1^T\mathbf{B}^T\mathbf{B}\mathbf{W}_1 + \mathbf{W}_1^T\mathbf{W}_2\mathbf{W}_2^T\mathbf{W}_1 + \mathbf{I}\right]^{-1} = \mathbf{I} - \mathbf{W}_1^T\left[(\mathbf{B}^T\mathbf{B} + \mathbf{I})^{-1} + \mathbf{I}\right]^{-1}\mathbf{W}_1, \tag{6.2}$$

meaning that we are, once again, left with the problem of inverting matrix $(\mathbf{B}^T\mathbf{B} + \mathbf{I})$, which we have already addressed above for the case of periodic deconvolution, deconvolution with unknown boundaries, image inpainting, and partial Fourier observations. Finally, notice that convergence of the resulting ADMM algorithm (assuming the solution set is not empty) is guaranteed by the presence of an identity block in matrix $\mathbf{G} = \left[(\mathbf{B}\mathbf{W}_1)^T \;\; \mathbf{I} \;\; (\mathbf{W}_2^T\,\mathbf{W}_1)^T\right]^T$, which ensures that it has full column-rank.

7 Conclusions and Discussion

We have overviewed a line of work that we have pursued in the past several years, by exploiting a particular way of applying the *alternating direction method of multipliers* (ADMM) to address a variety of convex optimization problems arising in imaging inverse problems. We provided an integrated view of several formulations for different problems: Tikhonov (unconstrained optimization) and Morozov (constrained optimization) regularization, analysis, synthesis, and analysis-synthesis hybrid formulations, deconvolution under periodic or unknown boundary conditions, and Gaussian or Poissonian noise models. We have abstained from presenting any experimental results, since detailed experimental assessment of this class of approaches can be found in the papers were these algorithms were originally introduced and analyzed [1], [2], [3], [4], [34].

We stress that, at this point, we make no claims that each of the reviewed algorithms is a state-of-the-art method for the corresponding problem; when they were originally proposed in [1], [2], [3], [4], [34], they were, empirically, indeed state-of-the-art methods in terms of speed, although of course such a claim is a weak one, as experimental assessments of algorithm speed is a delicate issue in itself. Our goal here was twofold: (a) to show how our approach to using ADMM to tackle problems where the objective function has two or more terms (see Subsection 3.2) constitutes a flexible and modular toolbox that allows addressing a variety of regularization formulations of imaging inverse problems; (b) to give evidence that the matrix inversion that is required to implement the resulting algorithms can be efficiently computed in a variety of problems and formulations, thus dispensing practitioners (addressing such problems) from having to look for alternative algorithmic frameworks that avoid matrix inversions (such as the *Chambolle-Pock algorithm* [15], or linearized versions of ADMM [40]).

Finally, we end the paper with the conjecture that the matrix inversion mentioned in the previous paragraph is indeed responsible for the good numerical performance that has been empirically observed for the class of methods herein reviewed. This conjecture has connections with very recent observations by Combettes and Glaudin [20]. To focus the discussion, consider what is arguably the simplest problem herein considered: the analysis formulation of Tikhonov regularization for linear observations with Gaussian noise in (4.3). Although this objective function includes a smooth term, $\|\mathbf{Bx}-\mathbf{y}\|_2^2/2$ with a gradient of known Lipschitz constant, the proposed algorithms never invoke its gradient, $\mathbf{B}^T(\mathbf{Bx}-\mathbf{b})$ (as is commonly done for smooth terms), but the inverse of its regularized Hessian $\left(\mathbf{B}^T\mathbf{B}+\mathbf{I}\right)^{-1}$. This inverse regularized Hessian is closely related to the proximity operator of $\|\mathbf{Bx}-\mathbf{y}\|_2^2/2$, which is given by $\operatorname{prox}_{\|\mathbf{B}\cdot-\mathbf{b}\|_2^2/2}(\mathbf{s}) = \left(\mathbf{B}^T\mathbf{B}+\mathbf{I}\right)^{-1}(\mathbf{B}^T\mathbf{b}+\mathbf{s})$. As empirically observed in [20], algorithms in which smooth functions are activated via their proximity operator (*i.e.*, involving the inverse of its regularized Hessian) perform better (in the problems considered in [20]) than algorithms where this functions are activated via their gradient. Clarifying these connections, as well as connections with regularized Newton methods, is a topic for current and future research.

Acknowledgments

The work herein described was partially supported by *Fundação para a Ciência e Tecnologia*, grant UID/EEA/50008/2013. The author acknowledges the long-term collaboration and the many insightful conversations with his colleague José Bioucas-Dias, on the topics of this paper and many others.

References

1. M. Afonso, J. Bioucas-Dias, M. Figueiredo, "Fast image recovery using variable splitting and constrained optimization," *IEEE Transactions Image Processing,* vol. 19(9), pp. 2345–2356, 2010.
2. M. Afonso, J. Bioucas-Dias, M. Figueiredo, "An augmented Lagrangian approach to the constrained optimization formulation of imaging inverse problems," *IEEE Transactions Image Processing,*, vol. 20(3), pp. 681–695, 2011.
3. M. Afonso, J. Bioucas-Dias, M. Figueiredo, "Hybrid synthesis-analysis frame-based regularization: a criterion and an algorithm," *Workshop on Signal Processing with Adaptive Sparse Structured Representations – SPARS'2011*, Edinburgh, UK, 2011.
4. M. Almeida, M. Figueiredo, "Deconvolving images with unknown boundaries using the alternating direction method of multipliers," *IEEE Transactions on Image processing*, vol. 22(8), pp. 3074–3086, 2013.
5. H. Andrews, B. Hunt, *Digital Image Restoration*, Prentice-Hall, 1977.
6. J. Barzilai, J. Borwein, "Two-point step size gradient methods," *IMA Journal of Numerical Analysis*, vol. 8, pp. 141–148, 1988.
7. H. Bauschke, P. Combettes, *Convex Analysis and Monotone Operator Theory in Hilbert Spaces*, Springer, 2017.
8. A. Beck, M. Teboulle, "A fast iterative shrinkage-thresholding algorithm for linear inverse problems," *SIAM Journal on Imaging Sciences*, vol. 2(1), pp. 183–202, 2009.
9. J. Bioucas-Dias, M. Figueiredo, "A new TwIST: two-step iterative shrinkage/thresholding algorithms for image restoration," *IEEE Transactions Image Processing*, vol. 16(12), pp. 2992–3004, 2007.
10. J. Bioucas-Dias, M. Figueiredo, "Multiplicative noise removal using variable splitting and constrained optimization," *IEEE Transactions Image Processing,* vol. 19(7), pp. 1720–1730, 2010.
11. S. Boyd, N. Parikh, E. Chu, B. Peleato, J. Eckstein, "Distributed optimization and statistical learning via the alternating direction method of multipliers," *Foundations and Trends in Machine Learning*, vol. 3, pp. 1–122, 2011.
12. R. Bruck, "'An iterative solution of a variational inequality for certain monotone operator in a Hilbert space,"' *Bulletin of the American Mathematical Society*, vol. 81(5), pp. 890–892, 1975.
13. E. Candès, J. Romberg, T. Tao. "Robust uncertainty principles: Exact signal reconstruction from highly incomplete frequency information," *IEEE Transactions on Information Theory,* vol. 52(2), pp. 489–509, 2006.
14. M. Carlavan, L. Blanc–Féraud, "Two constrained formulations for deblurring Poisson noisy images," *Proceedings of the IEEE International Conference on Image Processing – ICIP'2011*, Brussels, Belgium, 2011.
15. A. Chambolle and T. Pock, "A first-order primal-dual algorithm for convex problems with applications to imaging," *Journal of Mathematical Imaging and Vision*, vol. 40(1), pp. 120–145, 2011.

16. A. Chambolle and T. Pock, “An introduction to continuous optimization for imaging," *Acta Numerica*, vol. 25, pp. 161–319, 2016.
17. T. Chan, S. Esedoglu, F. Park, and A. Yip, “Recent developments in total variation image restoration," in *Handbook of Mathematical Models in Computer Vision*, N. Paragios, Y. Chen, O. Faugeras (Editors), Springer, 2005.
18. T. Chan, A. M. Yip, and F. E. Park, “Simultaneous total variation image inpainting and blind deconvolution," *International Journal of Imaging Systems and Technology*, vol. 15(1), pp. 92–102, 2005.
19. C. Chen, B. He, Y. Ye, X. Yuan “The direct extension of ADMM for multi-block convex minimization problems is not necessarily convergent," *Mathematical Programming*, vol. 155(1–2), pp. 57–79, 2016.
20. P. Combettes and L. Glaudin, “Proximal activation of smooth functions in splitting algorithms for convex minimization," arXiv:1803.02919, 2018.
21. P. Combettes, J.-C. Pesquet, “Proximal thresholding algorithm for minimization over orthonormal bases," *SIAM Journal on Optimization*, vol. 18(4), pp.1351–1376, 2007.
22. P. Combettes, J.-C.Pesquet, “Proximal splitting methods in signal processing," in *Fixed-point algorithms for inverse problems in science and engineering*, Springer, pp. 185–212, 2011.
23. P. Combettes, V. Wajs, “Signal recovery by proximal forward-backward splitting," *SIAM Journal on Multiscale Modeling & Simulation*, vol. 4(4), pp. 1168–1200, 2005.
24. I. Daubechies, M. De Friese, C. De Mol. “An iterative thresholding algorithm for linear inverse problems with a sparsity constraint," *Communication in Pure and Applied Mathematics*, vol. 57, pp. 1413–1457, 2004.
25. M. Donatelli, C. Estatico, A. Martinelli, and S. Serra-Capizzano, “Improved image deblurring with anti-reflective boundary conditions and re-blurring," *Inverse Problems*, vol. 22(6), pp. 2035–2053, 2006.
26. D. Donoho. “Compressed sensing," *IEEE Transactions on Information Theory*, vol. 52(4), pp. 1289–1306, 2006.
27. J. Eckstein, D. Bertsekas, “On the Douglas-Rachford splitting method and the proximal point algorithm for maximal monotone operators," *Mathematical Programming*, vol. 55(1–3), pp. 293–318, 1992.
28. J. Eckstein, W. Yao, “Understanding the convergence of the alternating direction method of multipliers: Theoretical and computational perspectives," *Pacific Journal of Optimization*, vol. 11(4), pp. 619-644, 2015.
29. M. Elad, *Sparse and Redundant Representations: From Theory to Applications in Signal and Image Processing*, Springer, 2010.
30. M. Elad, M. Figueiredo, Y. Ma, “On the role of sparse and redundant representations in image processing," *Proceedings of the IEEE*, vol. 98(6), pp. 972–982, 2010.
31. M. Elad, P. Milanfar, and R. Rubinstein, “Analysis versus synthesis in signal priors," *Inverse Problems*, vol. 23, pp. 947–968, 2007.
32. E. Esser, “Applications of Lagrangian-based alternating direction methods and connections to split Bregman", Tech. Rep. 09-31, Comp. and Applied Math., UCLA, 2009.
33. M. Figueiredo, J. Bioucas-Dias, “Deconvolution of Poissonian images using variable splitting and augmented Lagrangian optimization", *Proceedings of the IEEE Workshop on Statistical Signal Processing*, pp. 733–736, Cardiff, UK, 2009.
34. M. Figueiredo, J. Bioucas-Dias, “Restoration of Poissonian images using alternating direction optimization," *IEEE Transactions on Image Processing*, vol. 19(12), pp. 3133–3145, 2010.
35. M. Figueiredo, J. Bioucas-Dias, “An alternating direction algorithm for (overlapping) group regularization," *Workshop on Signal Processing with Adaptive Sparse Structured Representations – SPARS'2011*, Edinburgh, 2011.
36. M. Figueiredo, R. Nowak, “An EM algorithm for wavelet-based image restoration," *IEEE Transactions on Image Processing*, vol. 12(8), pp. 906–916, 2003.
37. A. Foi, S. Alenius, M. Trimeche, V. Katkovnik, K. Egiazarian, “A spatially adaptive Poissonian image deblurring," *Proceedings of the IEEE International Conference on Image Processing – ICIP'2005*, Genova, Italy, 2005.

38. D. Gabay, B. Mercier, "A dual algorithm for the solution of nonlinear variational problems via finite-element approximations," *Computers and Mathematics with Applications*, vol. 2(1), pp. 17–40, 1976.
39. R. Glowinski, A. Marroco, "Sur l'approximation, par elements finis d'ordre un, et la resolution, par penalisation-dualité d'une classe de problemes de Dirichlet non lineares," *Revue Française d'Automatique, Informatique et Recherche Opérationelle*, vol. 9(2), pp. 41–76, 1975.
40. D. Goldfarb, S. Ma, K. Scheinberg, "Fast alternating linearization methods for minimizing the sum of two convex functions," *Mathematical Programming*, vol. 141(1–2), pp. 349–382, 2013.
41. M. Hong, Z.-Q. Luo, "On the linear convergence of the alternating direction method of multipliers," *Mathematical Programming*, vol. 162(1), pp. 165–199, 2017.
42. V. Ivanov, V. Vasin, V. Tanana, *Theory of Linear Ill-posed Problems and Its Applications*, Walter de Gruyter & Co., 2002. (Original Russian edition: 1978.)
43. P. Lions, B. Mercier, "'Splitting algorithms for the sum of two nonlinear operators,"' *SIAM Journal on Numerical Analysis*, vol. 16, pp. 964–979, 1979.
44. R. Liu and J. Jia, "Reducing boundary artifacts in image deconvolution," *Proceedings of the IEEE International Conference on Image Processing – ICIP'2008*, San Diego, USA, 2008.
45. D. Lorenz, N. Worliczek, "'Necessary conditions for variational regularization schemes,"' *Inverse Problems*, vol. 29(7), p. 075016, 2013.
46. M. Lustig, D. Donoho, J. Pauly, "'Sparse MRI: the application of compressed sensing for rapid MR imaging,"' *Magnetic Resonance in Medicine,* vol. 58(6), pp. 1182–1195, 2007.
47. S. Mallat, *A Wavelet Tour of Signal Processing* (3rd Edition), Academic Press, 2008.
48. A. Matakos, S. Ramani, and J. Fessler, "'Image restoration using non-circulant shift-invariant system models,"' in *Proceedings of the IEEE International Conference on Image Processing – ICIP'2012*, Orlando, USA, 2012.
49. A. Matakos, S. Ramani, and J. Fessler, "'Accelerated edge-preserving image restoration without boundary artifacts,'"' *IEEE Transactions on Image Processing*, vol. 22(5), pp. 2019–2029, 2013.
50. J. J. Moreau, "'Proximité et dualité dans un espace Hilbertien"', *Bulletin de la Société Mathematique de France*, vol. 93, pp. 273–299, 1965.
51. M. Ng, *Iterative methods for Toeplitz systems*, Oxford University Press, 2004.
52. M. Ng, R. Chan, and W.-C. Tang, "'A fast algorithm for deblurring models with Neumann boundary conditions,"' *SIAM Journal on Scientific Computing*, vol. 21(3), pp. 851–866, 1999.
53. J. Ollinger, J. Fessler, "'Positron-emission tomography"', *IEEE Signal Processing Magazine*, vol. 14(1), pp. 43–55, 1997.
54. N. Parikh, S. Boyd, "'Proximal algorithms"', *Foundations and Trends in Optimization*, vol. 1(3), pp. 123–231, 2014.
55. G. Passty, "'Ergodic convergence to a zero of the sum of monotone operators in Hilbert space,"' *Journal of Mathematical Analysis and Applications*, vol. 72(2), pp. 383–390, 1979.
56. N. Rao, N. Kingsbury, S. Wright, R. Nowak, "Convex approaches to model wavelet sparsity patterns," *Proceedings of the IEEE International Conference on Image Processing – ICIP'2011*, Brussels, Belgium, 2011.
57. S. J. Reeves, "Fast image restoration without boundary artifacts," *IEEE Transactions on Image Processing*, vol. 14(10), pp. 1448–1453, 2005.
58. L. Rudin, S. Osher, and E. Fatemi, "Nonlinear total variation based noise removal algorithms," *Physica D*, vol. 60, pp. 259–268, 1992.
59. P. Sarder and A. Nehorai. "Deconvolution method for 3-D fluorescence microscopy images," *IEEE Signal Processing Magazine*, vol. 23(3), pp. 32–45, 2006.
60. I. Selesnick, M. Figueiredo, "Signal restoration with overcomplete wavelet transforms: comparison of analysis and synthesis priors," *Proceedings of SPIE*, vol. 7446, 2009.
61. S. Setzer, G. Steidl, and T. Teuber, "Deblurring Poissonian images by split Bregman techniques," *Journal of Visual Communication and Image Representation,* vol. 21(3), pp. 193–199, 2010.
62. M. Šorel, "Removing boundary artifacts for real-time iterated shrinkage deconvolution," *IEEE Transactions on Image Processing*, vol. 21(4), pp. 2329–2334, 2012.

63. J.-L. Starck, F. Murtagh, *Astronomical Image and Data Analysis,* Springer, 2006.
64. R. Willett, "'The dark side of image reconstruction: Emerging methods for photon-limited imaging,"' *SIAM News*, 2014.
65. R. Willett and R. Nowak. "Platelets: a multiscale approach for recovering edges and surfaces in photon-limited medical imaging," *IEEE Transactions Medical Imaging,* vol. 22(3), pp. 332–350, 2003.
66. S. Wright, R. Nowak, M. Figueiredo, "Sparse reconstruction by separable approximation", *IEEE Transactions on Signal Proc.*, vol. 57(7), pp. 2479-2493, 2009.
67. Z. Xu, M. Figueiredo, T. Goldstein, "Adaptive ADMM with spectral penalty parameter selection," *Proceedings of the 20th International Conference on Artificial Intelligence and Statistics – AISTATS'2017,* PMLR vol. 54, pp. 718–727, 2017.
68. W. Yin, S. Osher, D. Goldfarb, J. Darbon, "Bregman iterative algorithms for ℓ_1-minimization with applications to compressed sensing," *SIAM Journal on Imaging Sciences*, vol. 1(1), pp. 143–168, 2008.

Models and Numerical Methods for Electrolyte Flows

Jürgen Fuhrmann, Clemens Guhlke, Alexander Linke, Christian Merdon and Rüdiger Müller

1 Introduction

Liquid electrolytes are fluidic mixtures containing electrically charged ions. Electrochemical energy conversion systems like fuel cells and batteries contain liquid electrolytes. In biological tissues, nanoscale pores in the cell membranes separate different types of ions inside the cell from those in the intercellular space. Nanopores between electrolyte reservoirs can be used for analytical applications in medicine. Water purification technologies like electrodialysis rely on the electrolytic flow properties. This short and by far not exhaustive list of occurrences of electrolytic flow processes shows the importance of correct modeling of electrolyte flows. Due to the complex physical interactions present in this type of flows, in many case numerical simulation techniques are required to facilitate a deeper understanding of the flow behavior..

This contribution introduces a coupled modeling and simulation approach which has several new aspects. Section 2 reviews a modeling approach which uses recently obtained formulations [1] based on first principles of nonequilibrium thermodynamics. It allows to include ion-solvent interactions, finite ion size and solvation effects in a consistent manner. At the end of this section, a short overview over existing analytical results is given. Section 3 starts with an overview on previous results on numerical methods. It introduces a finite volume discretization approach for ion transport in a self-consistent electric field which is motivated by results from semiconductor device simulation and has been adapted to the improved ion transport models [2]. For fluid flow, the recently developed [3] pressure robust mixed finite element method is introduced. The section is finalized by a short description of the fix point approach for coupling flow and charge transport. Section 4 provides the results of a number of numerical examples which verify the presented approach and exhibit its potential for further research.

Weierstrass Institute for Applied Analysis and Stochastics, Mohrenstr. 39, 10117 Berlin, Germany
e-mail: juergen.fuhrmann@wias-berlin.de

M. Hintermüller, J. F. Rodrigues (eds.), *Topics in Applied Analysis and Optimisation*,
CIM Series in Mathematical Sciences, https://doi.org/10.1007/978-3-030-33116-0_8

2 Continuum models

Electroosmotic flows are characterized by the presence of an electric field that exerts a net force on the fluid molecules in regions where the local net charge due to the ions present therein is nonzero. Being part of the momentum balance for the barycentric velocity of the fluid, this net force induces fluid motion. Correspondingly, the dissolved ionic species molecules in the fluid are advected by the barycentric velocity field. Motion of ions relative to the barycentric velocity is induced by the gradients of their chemical potential and the electrostatic potential. A counterforce to the motion of dissolved molecules is due to elastic interactions between the ions and the solvent.

In addition to these processes, the spatial distribution of the net charge of ions in a self-consistent way contributes to the electric field.

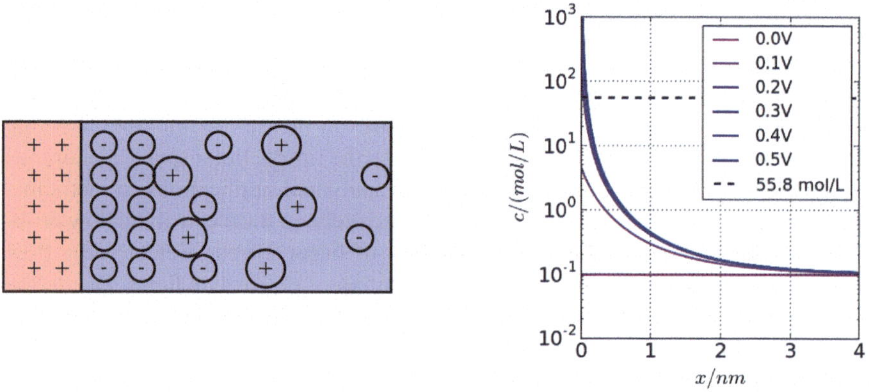

Fig. 1: Left: accumulation of negative ions at positively charged electrode surface. Right: already for moderate applied voltages, the classical Nernst-Planck model, which ignores the ion-solvent interaction, predicts ion concentrations at an ideally polarizable electrode which are significantly larger than the molar density of the solvent – in this case water with a molar density of $55.8\,\mathrm{mol}\cdot\mathrm{dm}^{-3}$.

Classical models for electrolytes [4,5] rely on a *dilute solution* assumption. In this case the ion-solvent interaction is neglected. As a result, there is no mechanism to limit the accumulation of ions inside narrow boundary layers that screen the electric field at electrodes or charged walls, see Fig. 1.

2.1 Bulk equations

The limitations of classical models are well known and several remedies for these shortcomings have been suggested early on [6]. Here a model for the flow of electrolytes is considered which has been introduced in [1, 7, 8] based on consistent application of the principles of nonequilibrium thermodynamics [9]. It includes ion volume and solvation effects and consistently couples the transport equations to the momentum balance and generalizes previous approaches to include steric (ion size) effects [10–14], see also [15].

In a given bounded space-time domain $\Omega \times (0, t^\sharp) \subset \mathbb{R}^d \times (0, \infty)$, and with appropriate initial and boundary conditions, the system (2.1a)– (2.5c) describes the isothermal evolution of the molar concentration of N charged species $c_1 \dots c_N$ with charge numbers $z_1 \dots z_N$ dissolved in a solvent of concentration c_0. Species are further characterized by their molar volumes v_i and molar masses M_i. The electric field is described as the negative gradient of the electrostatic potential ψ. The barycentric velocity of the mixture is denoted by $\mathbf{u}$, and p is the pressure. The following equations are considered:

$$\partial_t(\rho\mathbf{u}) - \nu\Delta\mathbf{u} + \rho(\mathbf{u}\cdot\nabla)\mathbf{u} + \nabla p = -q\nabla\psi \tag{2.1a}$$

$$\nabla\cdot\rho\mathbf{u} = 0 \tag{2.1b}$$

$$\partial_t c_i + \nabla\cdot(\mathbf{N}_i + c_i\mathbf{u}) = 0 \qquad (i = 1 \dots N) \tag{2.1c}$$

$$-\nabla\cdot(\varepsilon\nabla\psi) = F\sum_{i=1}^{N} z_i c_i = q. \tag{2.1d}$$

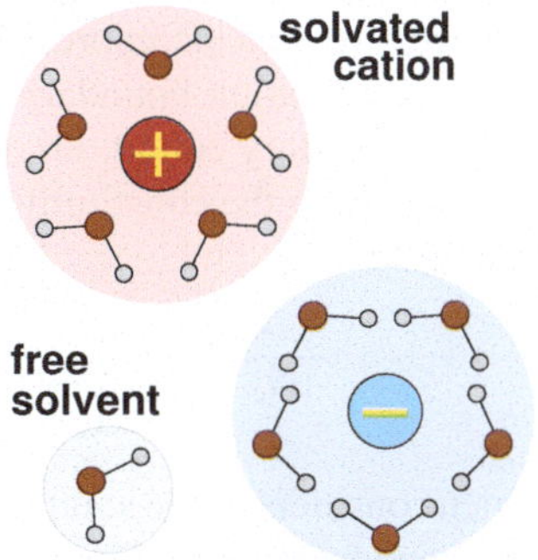

Fig. 2 Constituents of the liquid electrolyte are the free solvent molecules and solvated ions, i.e. larger complexes that are build from a center ion and a solvation shell of bounded polar solvent molecules.

Equation (2.1a) together with (2.1b) comprises the incompressible Navier–Stokes equations for a fluid of viscosity ν and constant density ρ. In the general case, where molar volumes and molar masses are not equal, ρ would depend on the local composition of the electrolyte.

In regions where the space charge $q = F\sum_{i=1}^{N} z_i c_i$ (F being the Faraday constant) is nonzero, the electric field $-\nabla\psi$ becomes a driving force of the flow.

The partial mass balance equations (2.1c) describe the redistribution of species concentrations due to advection in the velocity field $\mathbf{u}$ and molar diffusion fluxes $\mathbf{N}_i$. The Poisson equation (2.1d) describes the distribution of the electrostatic potential ψ under a given configuration of the space charge. The constant ε is the dielectric permittivity of the medium.

The fluxes $\mathbf{N}_i$, the molar chemical potentials μ_i and the incompressibility constraint for a liquid electrolyte are given by

$$\mathbf{N}_i = -\frac{D_i}{RT} c_i \left(\nabla\mu_i - \frac{\kappa_i M_0 + M_i}{M_0} \nabla\mu_0 + z_i F \nabla\psi \right) \qquad (i = 1 \dots N) \tag{2.2a}$$

$$\mu_i = (\kappa_i v_0 + v_i)(p - p^\circ) + RT \ln \frac{c_i}{\overline{c}} \qquad (i = 0 \dots N) \tag{2.2b}$$

$$1 = v_0 c_0 + \sum_{i=1}^{N} (\kappa_i v_0 + v_i) c_i. \tag{2.2c}$$

The generalized Nernst-Planck flux (2.2a) combines the gradients of the species chemical potentials $\nabla\mu_i$, the gradient of the solvent chemical potential $\nabla\mu_0$ and the electric field $-\nabla\psi$ as driving forces for the motion of ions of species i relative to the barycentric velocity $\mathbf{u}$. In this equation, D_i are the diffusion coefficients, R is the molar gas constant, and T is the temperature. Equation (2.2b) is a constitutive relation for the chemical potential μ_i depending on the local pressure and concentration. Here, p° is a reference pressure, and $\overline{c} = \sum_{i=0}^{N} c_i$ is the summary species concentration. In (2.2c) a simple model for solvated ions is applied, see [2,7,16]. In polar solvents like water, ions carry a shell of electrically attracted solvent molecules, see Fig. 2. As a result, the mass and volume of a solvated ion are given by $\kappa_i M_0 + M_i$ and $\kappa_i v_0 + v_i$, respectively. The incompressibility constraint (2.2c) limits the accumulation of ions in the polarization boundary layer to physically reasonable values, see Fig. 3. For typical boundary conditions, the reader is referred to numerical example section 4.

Comparing the constitutive equations (2.5a)-(2.5c) to the classical Nernst-Planck flux [4,5]

$$\mathbf{N}_i = -D_i \left(\nabla c_i + z_i c_i \frac{F}{RT} \nabla\psi \right) \qquad (i = 1 \dots N), \tag{2.3}$$

which considers dilute solutions, one observes that in (2.3) the ion-solvent interaction described by the term $\nabla\mu_0$ is ignored. Moreover in (2.3) implicitly a material model is assumed that neglects the pressure dependence of μ_i, which is inappropriate in charged boundary layers, see Fig. 1, right.

The mass density of the mixture is

$$\rho = M_0 c_0 + \sum_{i=1}^{N} (\kappa_i M_0 + M_i) c_i = \frac{M_0}{v_0} + \sum_{i=1}^{N} \left(M_i - M_0 \frac{v_i}{v_0} \right) c_i. \tag{2.4}$$

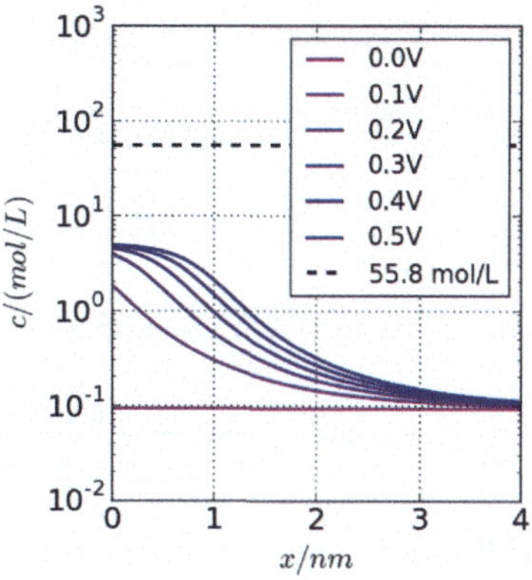

Fig. 3 Physically reasonable ion concentrations at ideally polarizable electrode in equilibrium for the generalized Nernst-Planck flux (2.5a).

As for reasonable solvation numbers in the range of $\kappa_i \approx 10$, the ionic concentrations are necessarily small, i.e. $v_0\, c_i \ll 1$, the density ρ is dominated by the density of the solvent $\rho_0 = \frac{M_0}{v_0}$. For simplicity, and due to the fact that the pressure robust Navier-Stokes solver described in Section 3.3 currently is available only for constant density ρ, in the sequel it is assumed that all species molar masses and molar volumes are equal: $v_i = v_0$, $M_i = M_0$ $(i = 1 \dots N)$, leading to

$$\mathbf{N}_i = -\frac{D_i}{RT} c_i \left(\nabla\mu_i - (\kappa_i + 1)\nabla\mu_0 + z_i F \nabla\psi\right) \qquad (i = 1 \dots N) \tag{2.5a}$$

$$\mu_i = v_0(\kappa_i + 1)(p - p^\circ) + RT \ln \frac{c_i}{\bar{c}} \qquad (i = 0 \dots N) \tag{2.5b}$$

$$1 = v_0 c_0 + \sum_{i=1}^{N} (\kappa_i + 1) v_0 c_i. \tag{2.5c}$$

2.2 Reformulation in species activities

In order to develop a space discretization approach for the generalized Nernst-Planck fluxes (2.5a), after [17], the system is re-formulated in terms of (effective) species activities $a_i = \exp\left(\frac{\mu_i - (\kappa_i+1)\mu_0}{RT}\right)$. The quantity $\mu_i - (\kappa_i + 1)\mu_0$ is sometimes denoted as entropy variable [18]. Introducing the activity coefficients $\gamma_i = \frac{a_i}{c_i}$ and its inverse (reciprocal) $\beta_i = \frac{1}{\gamma_i} = \frac{c_i}{a_i}$ allows to transform the Nernst-Planck-Poisson system consisting of (2.1c), (2.1d), (2.5a) to

$$-\nabla \cdot (\varepsilon \nabla \psi) = F \sum_{i=1}^{N} z_i \beta_i a_i = q. \tag{2.6a}$$

$$0 = \partial_t(\beta_i a_i) + \nabla \cdot (\mathbf{N}_i + \beta_i a_i \mathbf{u}) \qquad (i = 1 \dots N) \tag{2.6b}$$

$$\mathbf{N}_i = -D_i \beta_i \left(\nabla a_i + a_i z_i \frac{F}{RT} \nabla \psi \right) \qquad i = (1 \dots N). \tag{2.6c}$$

From (2.5b) and (2.5c) one obtains

$$a_i = \frac{v_0 \beta_i a_i}{1 - v_0 \sum_{j=1}^{N} \beta_j a_j (\kappa_j + 1)} \qquad (i = 1 \dots N)$$

which is a linear system of equations which allows to express $\beta_1 \dots \beta_n$ through $a_1 \dots a_n$. It has the unique solution [17]

$$\beta_i = \beta = \frac{1}{v_0 + v_0 \sum_{j=1}^{N} a_j (\kappa_j + 1)} \qquad (i = 1 \dots N). \tag{2.7}$$

It follows immediately that for any nonnegative solution $a_1 \dots a_n$ of system (2.6), the resulting concentrations are bounded in a physically meaningful way:

$$0 \le c_i = \beta_i a_i \le \frac{1}{v_0}. \tag{2.8}$$

A similar observation in the context of cross diffusion systems has been described in [18].

In the general case with different molar volumes and molar masses, system (2.7) becomes nonlinear, the quantities β_i differ between species and in addition depend on the pressure p [2, 17], leading to a nonlinear system of equations

$$\beta_i = B_i(a_1 \dots a_n, \beta_1 \dots \beta_n, p) \qquad (i = 1 \dots N). \tag{2.9}$$

2.3 Analytical treatment

Long before the advent of computers, the need to understand mechanisms of electrokinetic phenomena like electroosmosis and electrophoresis led to the development of various asymptotic and analytical tools to handle these complex effects, mostly relying on the classical Nernst-Planck fluxes. Fundamental in this context is the Helmholtz-Smoluchowski theory [19] which quantitatively explains the electroosmotic flow phenomenon. For comprehensive treatment see e.g. [20–22]. See also section 4.1 of this contribution for a short overview.

A particular intriguing phenomenon from the mathematical and application point of view is the development of electroconvective instabilities in electrodialysis cells [23, 24].

Apparently, mathematical existence and uniqueness theory started much later. In the case of of the classical Nernst-Planck flux, existence of a local solution has been established in [25]. Existence, and in some cases uniqueness of solutions has been proven in [26–28]. In [29], the existence of unique local strong solutions in bounded n-dimensional domains as well as the the existence of unique global strong solutions and exponential convergence to uniquely determined steady states in two space dimensions has been proven. The authors of [30] prove global existence and stability results for large data in two space dimension. In [31], global weak solutions in three space dimensions are constructed. Recently, existence theory for the improved model in the case of a compressible flow has been developed in [32–34].

An analytical solution of the Poisson–Nernst–Planck–Stokes equations in a cylindrical channel has been derived in [35] in the context of fuel cell membranes.

3 Numerical methods

3.1 Previous work

A number of contributions is devoted to the discretization and numerical solution for the case of the classical Nernst-Planck flux. The authors of [36] consider a finite element discretization of the coupled system. A mixed finite element method for the 2D Stokes-Nernst-Planck-Poisson system is considered in [37]. A similar approach is used in [38]. In [39, 40], a Galerkin pseudospectral method is used to perform simulations of electrokinetic instabilities over permselective membranes in a rectangular domain. For a similar problem, the authors of [41] apply a finite volume method for both the Navier-Stokes and the Nernst-Planck-Poisson subsystems, coupling them via a fixed point iteration method. The authors of [42] apply a finite difference method. In [43], a rather recent overview on the state of the art for this problem class is given. In [44], a mixed finite element method is considered and analyzed. Finite difference methods are used e.g. in [45], [46] and in [47] for a micro/nanofluidic applications. Extension to the nonisothermal case is considered in [48] using a finite volume scheme on a regular grid in cylindrical coordinates. A three-dimensional model including thermal and mechanical effects based on an edge averaged exponentially fitted finite element method with applications in membrane biology is described in [49].

An interesting comparison between a custom developed and a commercially available simulation tool one finds in [50]. The commercially available code COMSOL Multiphysics is used to obtain numerical results on electroosmotic flows in [51] and [52]. The authors of [53] modified the Coulomb force term in (2.1a) by adding the sum of the concentration gradients in order to minimize spurious flows due to large pressure gradients and implemented this approach into COMSOL Multiphysics, see also section 4.4 of this contribution.

Coupling between fluid flow and modified Poisson-Nernst-Planck models taking into account steric effects up to now has been considered only by very few authors.

In [54], a constant flow situation for a modified Nernst-Planck flux along a pore is assumed, allowing to decouple the problem. In order to model the behavior of ionic liquids, a simple upwind finite volume method on a rectangular grid to discretize the modified Poisson-Nernst-Planck equations according to [15] has been coupled to the lattice Boltzmann method for the Navier-Stokes equations in [55]. The authors of [56] implemented a 2D numerical model of a nanopore with reservoirs into ANSYS Fluent.

In the sequel, a coupling approach is presented which combines novel, pressure robust mixed finite element methods [3, 57] with a thermodynamically consistent two point flux finite volume method designed for the discretization of modified Nernst-Plank-Poisson equations using ideas from semiconductor device simulation.

3.2 Thermodynamically consistent finite volume methods for species transport

A two point flux finite volume method on boundary conforming Delaunay meshes is used to approximate the Nernst-Planck-Poisson part of the problem. It has been inspired by the successful Scharfetter-Gummel box method for the solution of charge transport problems in semiconductors [58,59]. For a recent overview on this method see [60]. It was initially developed for drift-diffusion problems in non-degenerate semiconductors exhibiting Boltzmann statistics for charge carrier densities whose fluxes are equivalent to the classical Nernst-Planck flux (2.3).

The simulation domain Ω is partitioned into a finite number of closed convex polyhedral control volumes $K \in \mathcal{K}$ such that $K \cap L = \partial K \cap \partial L$ and $\overline{\Omega} = \bigcup_{K \in \mathcal{K}} K$. With each control volume a node $\mathbf{x}_K \in K$ is associated. If the control volume intersects with the boundary $\partial\Omega$, its corresponding node shall be situated on the boundary: $\mathbf{x}_K \in \partial\Omega \cap K$. The partition shall be *admissible* [61], that is for two neighboring control volumes K, L, the *edge* $\overline{\mathbf{x}_K \mathbf{x}_L}$ is orthogonal to the interface between the control volumes $\partial K \cap \partial L$. Let $\mathbf{h}_{KL} = \mathbf{x}_L - \mathbf{x}_K$ and $h_{KL} = |\mathbf{h}_{KL}|$. Then, the normal vectors to ∂K can be calculated as $\mathbf{n}_{KL} = \frac{1}{h_{KL}} \mathbf{h}_{KL}$.

A constructive way to obtain such a partition is based on the creation of a boundary conforming Delaunay triangulation resp. tetrahedralization of the domain and the subsequent construction of its dual, the Voronoi tessellation intersected with the domain, see e.g. [59, 60, 62], see also Fig. 4.

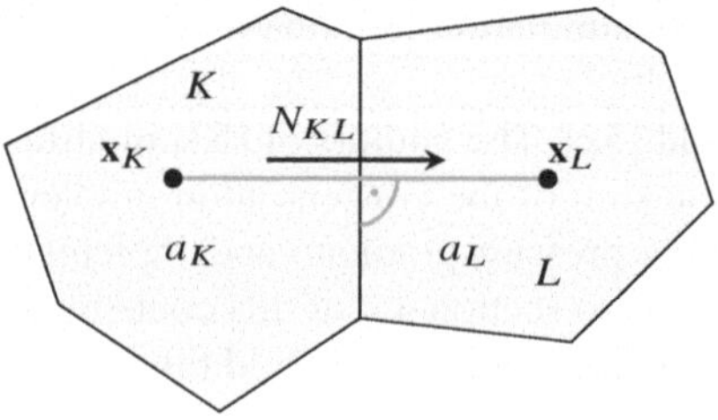

Fig. 4 Two neighboring control volumes K and L with collocation points $\mathbf{x}_K, \mathbf{x}_L$ stored activities a_K, a_L and flux N_{KL}.

The time axis is subdivided into intervals

$$0 = t^0 < t^1 < \cdots < t^{N_t} = t^\sharp.$$

Denote by $\mathbf{J}_i = c_i\mathbf{u} + \mathbf{N}_i = \beta_i a_i \mathbf{u} + \mathbf{N}_i$ the convection diffusion flux of the model under consideration. The general approach to derive a two point flux finite volume scheme for a conservation law (index i omitted)

$$\partial_t c + \nabla \cdot \mathbf{J} = 0$$

consists in integrating the equation over a space-time control volume $K \times [t^{n-1}, t^n]$:

$$0 = \int_{t^{n-1}}^{t^n} \int_K (\partial_t c + \nabla \cdot \mathbf{J})\, d\mathbf{x}\, dt = \int_{t^{n-1}}^{t^n} \int_K \partial_t c\, d\mathbf{x}\, dt + \int_{t^{n-1}}^{t^n} \int_{\partial K} \mathbf{J} \cdot \mathbf{n}\, ds\, dt$$

$$= \int_K (c^n - c^{n-1})\, d\mathbf{x} + \sum_{\substack{L \text{ neighbor} \\ \text{of } K}} \int_{t^{n-1}}^{t^n} \int_{\partial K \cap \partial L} \mathbf{J} \cdot \mathbf{n}_{KL}\, ds\, dt$$

This is approximated via

$$|K| \frac{c_K^n - c_K^{n-1}}{t^n - t^{n-1}} + \sum_{L \text{ neighbor of } K} |\partial K \cap \partial L| J_{KL}^n = 0,$$

and it remains to define the numerical fluxes J_{KL}^n which should approximate the continuous fluxes between two neighboring control volumes and depend on the unknown values in the two collocation points $\mathbf{x}_K$ and $\mathbf{x}_L$ at moment t^n in order to obtain a fully implicit in time scheme.

The modification of the Scharfetter-Gummel scheme [58] proposed in [17] is based on the similarity of the expressions (2.6c) and (2.3) up to the pre-factor β. The later is the same as the drift-diffusion flux in non-degenerate semiconductors, for which this discretization scheme was initially derived. Let $B(\xi) = \frac{\xi}{e^\xi - 1}$ be the Bernoulli function. Set

$$J_{KL} = D\beta_{KL} \cdot \frac{B\left(-\delta_{KL} - \frac{u_{KL}}{D}\right) a_K - B\left(\delta_{KL} + \frac{u_{KL}}{D}\right) a_L}{h_{KL}}$$

where β_{KL} is an average of the inverse activity coefficients β_K and β_L, $\delta_{KL} = \frac{zF}{RT}(\psi_K - \psi_L)$ is proportional to the local electric force, and

$$u_{KL} = \int_{\partial K \cap \partial L} \mathbf{u} \cdot \mathbf{h}_{KL}\, ds \tag{3.1}$$

is the normal integral over the interface $\partial K \cap \partial L$ of the convective flux scaled by h_{KL}. If the continuous flux is divergence free, i.e. it fulfills equation (2.1b), the flux projections u_{KL} fulfill the discrete divergence condition

$$\sum_{L \text{ neighbor of } K} |\partial K \cap \partial L| u_{KL} = 0 \tag{3.2}$$

which in the absence of charges and coupling through the activity coefficients guarantees a discrete maximum principle of the approximate solution [63].

The resulting time discrete implicit Euler finite volume upwind scheme guarantees nonnegativity of discrete activities and exact zero fluxes under thermodynamic equilibrium conditions. Moreover it guarantees the bounds (2.8) [17].

Existence and convergence theory for the discrete problem for generalized Nernst-Planck fluxes is still open. For cases similar to the classical Nernst-Planck fluxes, in one space dimension, an independent derivation, and second order convergence in the discrete maximum norm for the Scharfetter-Gummel scheme has been shown in [64]. Under the assumption that second derivatives of the continuous solution exist, in [65] for moderately sized drift terms and two-dimensional, square grids, second order convergence for the scheme in the L_2-norm has been shown. Reinterpretations of the finite volume Scharfetter-Gummel scheme as a nonstandard finite element method allowed to obtain convergence estimates schemes on Delaunay grids [66, 67]. For a general approach to the convergence theory of finite volume schemes, see [61]. In [68], weak convergence (no order estimate) for a generalization of the Scharfetter-Gummel scheme to nonlinear convection-diffusion problems has been shown, however, this proof does not cover the case of the generalized Nernst-Planck flux.

The discretizaton ansatz leads to a large nonlinear discrete system in the unknowns $\psi, a_1 \dots a_n$ which is solved by Newton's method in every time step. For the general model, the nonlinear equations for the inverse activity coefficients $\beta_1 \dots \beta_n$ according to (2.9) and a Laplace equation for the pressure have to be added to the overall system [2].

3.3 Pressure robust, divergence free finite elements for fluid flow.

For a recent and comprehensive introduction into the field of finite element methods for incompressible flow, see e.g. [69]. Mixed finite element methods approximate the Stokes resp. Navier–Stokes equations based on an inf-sup stable pair of velocity ansatz space V_h and pressure ansatz space Q_h. A fundamental property of the Stokes and Navier–Stokes equations consists in the fact that — under appropriate boundary conditions — the addition of a gradient force to the body force on the right-hand side of the momentum balance (2.1a) leaves the velocity unchanged, as it just can be compensated by a change in the pressure. Most classical mixed finite element methods for the Navier–Stokes equations do not preserve this property [70]. As a consequence, the corresponding error estimates for the velocity depend on the pressure [69]. Moreover, the divergence constraint of the discrete solution $\mathbf{u}_h$ is fulfilled only in a weak finite element sense:

$$\int_\Omega q_h \mathrm{div}(\mathbf{u}_h)\, dx = 0 \quad \text{for all } q_h \in Q_h. \tag{3.3}$$

This raises problems when coupling the flow simulation to a transport simulation using finite volume methods, because the maximum principle for the species concentration is directly linked to the divergence constraint in the finite volume sense (3.2) [63].

Pressure robust mixed methods, first introduced in [3], are based on the introduction of a divergence free velocity reconstruction Π into the discrete weak formulation of the flow problem. The resulting discretization of the stationary Stokes equation (provided here for simplicity) reads as: find $(\mathbf{u_h}, p_h) \in V_h \times Q_h$ such that

$$\int_\Omega \nu \nabla \mathbf{u_h} : \nabla \mathbf{v_h} dx + \int_\Omega p \nabla \cdot \mathbf{v_h} dx = \int_\Omega \mathbf{f} \cdot (\Pi \mathbf{v_h}) dx \quad \text{for all } \mathbf{v_h} \in V_h$$
$$\int_\Omega q_h \nabla \cdot \mathbf{u_h}\, dx = 0 \quad \text{for all } q_h \in Q_h.$$

This formulation differs from that of the classical mixed methods only in the introduction of a *reconstruction operator* Π with the following properties:

(i) If $\mathbf{u}_h$ is divergence free in the weak finite element sense (3.3), then the reconstruction $\Pi\mathbf{u}_h$ is pointwise divergence free in the continuous sense:

$$\nabla \cdot (\Pi \mathbf{u}_h) = 0.$$

(ii) The change of the test function by the reconstruction operator causes a consistency error that should have the same asymptotic convergence rate of the original method and should not depend on the pressure.

Under these conditions, the resulting velocity error estimate is independent of the pressure [57]. Furthermore, using the reconstruction $\Pi\mathbf{u}_h$ in the coupling to the discretization of the ion flux guarantees the divergence condition (3.2) for the projections obtained via (3.1). Using this method, even for a complicated structure of the pressure as in the case of electrolyte flows, a good velocity approximation can be obtained without the need to resort to high order pressure approximations. This leads to a significant reduction of degrees of freedom numbers necessary to obtain a given accuracy of the velocity. The action of Π on a discrete velocity field can be calculated locally, on elements or element patches. Therefore its implementation leads to low overhead in calculations.

For a comprehensive overview on this method, and the role of the divergence constraint in flow discretizations, see the survey article [57].

3.4 Coupling strategy.

The coupling approach between the Navier–Stokes solver and the Nernst-Planck-Poisson solver is currently based on a fixed point iteration strategy:

Set $\mathbf{u_h}, p_h$ to zero, calculate initial solution for (2.1d)–(2.5c);
while *not converged* **do**
  Provide ψ_h, q_h to Navier–Stokes solver;
  Solve (2.1a)–(2.1b) for $\mathbf{u_h}, p_h$;
  Project $\Pi\mathbf{u_h}, p_h$ to the Poisson-Nernst-Planck solver;
  Solve (2.6a)–(2.6c);
end

The projection of $\Pi\mathbf{u}$ to the finite volume solver according to (3.1) includes the integration of the reconstructed finite element solution $\Pi\mathbf{u}_h$ over interfaces between neighboring control volumes of the finite volume method. For a detailed explanation of this algorithmically challenging step see [63]. Sufficient accuracy of this step guarantees that the projected velocity is divergence free in the sense (3.2). In the implementation, the integrals are calculated by quadrature rules, and for a given discretization grid, the projection operator is assembled into a sparse matrix, allowing for computationally efficient repeated application of the projection operator during the fixed point iteration. As a consequence, in the case of electroneutral, inert transported species, the maximum principle for species concentrations is guaranteed [63]. In combination with pressure robust finite element methods, this coupling approach was first applied to modeling of thin layer flow cells [71].

4 Numerical examples

In this section, first stationary simulation results based on the coupled method are presented which are mainly intended to verify the correctness of the method and its implementation.

The discretization methods and the coupling strategy introduced above are implemented in the framework of the toolbox pdelib [72] that is developed at WIAS. In the following examples the solution of the Nernst-Planck-Poisson system is performed using Newton's method with full analytical Jacobians combined with parameter embedding to tackle strong nonlinearities starting with the equilibrium solution, for details, see e.g. [60]. For the flow part of the problem, the stationary Stokes solution was solved using a second order finite element method. Its velocity space consists of piecewise quadratic continuous vector fields enhanced with cell bubble functions and its pressure space consists of piecewise linear and discontinuous scalar fields [73]. This method allows for an easy divergence free reconstruction operator into the Raviart-Thomas finite element space of first order by standard interpolation [57,74]. Linear systems were solved using the sparse direct solver Pardiso [75,76].

4.1 Infinite pore with charged walls

For an extensive treatment of this case for the classical Nernst-Planck-Poisson flux, see [22]. As similar treatment one can find in [54]. Consider a stationary, laminar electroosmotic Stokes flow in an infinite domain under a constant in space applied electric field. Suppose that velocity and concentrations do not depend on the longitudinal coordinate z. This problem can be seen as a model of a pore of infinite length.

Let $\Omega_\oslash \subset \mathbb{R}^{d-1}$ be a convex cross-sectional domain and $\Omega = \Omega_\oslash \times \mathbb{R} \subset \mathbb{R}^d$. Assume a constant flow along the z-Axis such that $u_\oslash = 0$, and $\partial_z u_z = 0$. Set $\psi = \psi_\oslash - (z - z_0)E_z$ where E_z is the constant z-component of the electric field. Similarly, assume $p = p_\oslash + (z - z_0)\Pi_z$ where Π_z is the constant z component of the pressure gradient. These functions are linear in z and $\nabla \cdot (\varepsilon \nabla \psi) = \nabla_\oslash \cdot (\varepsilon \nabla_\oslash \psi)$. Further, consider zero ionic current in the cross sectional direction: $N_{i,\oslash} = 0$.

From the momentum equation (2.1a) one obtains in $\Omega_\oslash$:

$$\nabla_\oslash p = -q \nabla_\oslash \psi \tag{4.1a}$$

$$-\eta \Delta_\oslash u_z + \Pi_z = q E_z. \tag{4.1b}$$

Due to the assumptions on $\mathbf{u}$, (2.1b) is fulfilled, and, moreover, $\nabla \cdot (\beta_i a_i \mathbf{u}) = 0$. Then, together with the zero lateral current condition, and the fact that $\mathbf{N}_z$ must be z-independent, the continuity equation (2.1c) is fulfilled. As its right hand side is independent on z, the Poisson equation (2.1d) gives

$$-\nabla_\oslash \varepsilon \nabla_\oslash \psi = q \tag{4.2a}$$

Finally, the zero lateral current condition reduces the Nernst-Planck equation (2.5a) to the equilibrium condition

$$a_i = \exp\left(\frac{z_i F}{RT}(\phi_i - \psi)\right) \tag{4.2b}$$

where ϕ_i are constant electrochemical (quasi-Fermi) potentials which can be obtained from a bulk concentration condition [17]. As in [17], turn (4.1a) into the second order equation

$$-\Delta_\oslash p(x, y) = \nabla_\oslash \cdot (q \nabla_\oslash \psi). \tag{4.2c}$$

With appropriate boundary conditions on $\partial\Omega_\oslash$

$$\varepsilon \partial_n \psi = \sigma \qquad\qquad \partial_n p = -q \partial_n \psi \tag{4.3}$$

system (4.2a)-(4.2c) together with (2.7) corresponds to the equilibrium case described in [17] and can be generalized to the full model from [2].

Given the lateral charge distribution from the solution of (4.2a), equation (4.1b) together with the no-slip boundary condition

$$u_z|_{\partial\Omega_\oslash} = 0 \tag{4.4}$$

allows to calculate the lateral distribution of the velocity component u_z via the solution of an elliptic equation.

4.2 Slit between two infinite plates

For the case of flow in the slit of width $2w$ between two infinite plates, set $d = 2$, $\Omega_\oslash = (0, w)$, and set symmetry boundary conditions at $x = 0$ and a fixed pressure value:

$$\varepsilon\partial_x\psi(0) = 0, \qquad p(0) = 0, \qquad \eta\partial_x u_z(0) = 0 \tag{4.5}$$

From (4.2a) and (4.1b) follows

$$\eta\partial_{xx}u_z = -\varepsilon E_z\partial_{xx}\psi - \Pi_z. \tag{4.6}$$

Integrating once gives

$$\eta\partial_x u_z = -\varepsilon E_z\partial_x\psi - \Pi_z x + C. \tag{4.7}$$

with $C = 0$ due the boundary condition at $x = 0$. Let x_ζ be such that $u_z(x_\zeta) = 0$. A second integration gives

$$\int_0^{x_\zeta} \eta\partial_x u_z dx = \int_0^{x_\zeta} -\varepsilon E_z\partial_x\psi - \Pi_z x \, dx$$
$$-\eta u_z(0) = -\varepsilon E_z \left(\psi(x_\zeta) - \psi(0)\right) - \Pi_z \frac{1}{2} x_\zeta^2$$

Assuming $\Pi_z = 0$, i.e. that the pressure gradient is absent as the driving force, one can define the electroosmotic velocity $v_{eo} = u_z(0)$. Assuming that the pore width is sufficiently large to see a deviation from electroneutrality only in a small boundary layer close to x_ζ, this is the velocity of the plug flow initiated by electroosmotic forces in the boundary layer. One obtains the famous Helmholtz-Smoluchowski formula [19]

$$v_{eo} = \frac{\varepsilon E_z}{\eta}\zeta \tag{4.8}$$

where $\zeta = \psi(x_\zeta) - \psi(0)$ is the *zeta potential*. With the definition of the electrochemical potentials ϕ_i from molarity and electroneutrality one can assume $\psi(0) = 0$. Moreover, for the flow model discussed in this example, with constant viscosity and no-slip boundary condition at the pore wall, $x_\zeta = w$, and $\zeta = \psi|_{x=w}$ is the potential at the pore wall which is induced by the surface charge σ. Note that this derivation of the zeta potential does not depend on the particular variant of the Nernst-Planck flux.

Assuming the case of flow of a binary electrolyte in the slit between two infinite parallel plates, for the classical Nernst-Planck flux the zeta potential can be calculated explicitly and expressed by the Grahame equation [22].

$$\zeta = \frac{2RT}{F}\operatorname{asinh}\left(\frac{\sigma}{\sqrt{8\varepsilon RT c_0}}\right), \tag{4.9}$$

where $c_0 = c(0)$ is the bulk electrolyte concentration. In [77, 78], asymptotic theory is used to approximate the lateral distribution of the electrostatic potential and the flow velocity:

$$\psi(x) = \zeta \frac{\cosh \frac{x}{\lambda}}{\cosh \frac{w}{\lambda}}, \qquad v_z(x) = v_{eo}\left(1 - \frac{\cosh \frac{x}{\lambda}}{\cosh \frac{w}{\lambda}}\right), \tag{4.10}$$

where $\lambda = \sqrt{\frac{\varepsilon RT}{2F^2 c_0}}$ is the Debye length. These expressions can be used to benchmark the implementation of numerical methods.

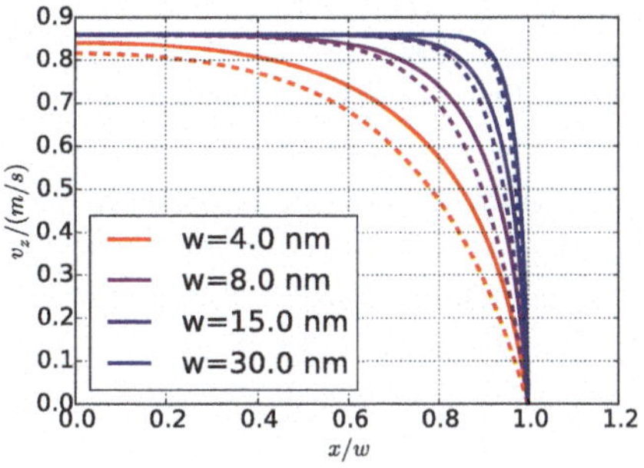

Fig. 5 Comparison of simulation results with the classical asymptotic Helmholtz-Smoluchowski theory for different pore widths w. Depicted is the velocity component u_z. Lines: numerical solution of the cross-sectional problem. Dashed: approximation from [77, 78].

Using the finite volume method referenced in [2, 17] to solve the modified Poisson-Boltzmann part, and a similar method to solve the equation for u_z, the slit problem is solved on a boundary layer grid such that $h_{min} = 0.1\,\text{nm}/2.0^r$ close to $x = w$ $h_{max} = 0.2w/2.0^r$ close to $x = 0$ where $r = 0, 1\ldots$ denotes the refinement level and such that subsequent intervals follow a geometric progression. Figure 5 shows the development of the plug flow profile for an increasing pore width w, and at the same time an increasing coincidence of the numerical solution with the asymptotic expression. Fig. 6 supports the verification of the accuracy of the numerical calculation of the zeta potential for the Gouy-Chapman model. One observes second order convergence to the value given by the Grahame equation (4.9) with respect to the number of grid points.

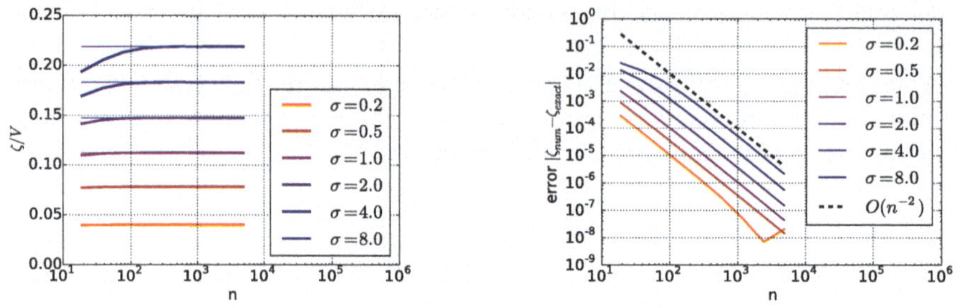

Fig. 6: Result and accuracy of numerical zeta potential calculation for dilute solution model for a slit with half width 20 nm vs. number of grid points. Left: Thin lines: value of zeta potential according to the Grahame equation (4.9), thick lines: numerically calculated zeta potential. Right: error vs. grid size.

4.3 Finite pore with charged walls

Now, consider a slit of width $2w$ between two parallel plates of infinite width and finite length l. The geometry of this problem is represented by a rectangular domain $(0,l)\times(0,w)$. Let $\Gamma_{in} = l\times(0,w)$, $\Gamma_{out} = 0\times(0,w)$, $\Gamma_{wall} = (0,l)\times w$, $\Gamma_{sym} = (0,l)\times 0$

In addition to the charged wall boundary conditions from Section 4.1, it is necessary to introduce boundary conditions at the inlet and the outlet.

Consider the coupled system consisting of (2.1a)–(2.1b) and (2.6a)–(2.6c) with two ionic species of opposite charges $z_1 = 1$, $z_2 = -1$. Let $v_0 = 1.0/M_w$, where $M_w = 55.8\,\mathrm{mol}\cdot\mathrm{dm}^{-3}$ is the molarity of liquid water at room temperature. Choose an activity value a_{res} such that $c_{res} = \beta a_{res} = 1\,\mathrm{mol}\cdot\mathrm{dm}^{-3}$. Then, set the following boundary conditions:

$$
\begin{aligned}
a_i|_{\Gamma_{in,out}} &= a_{res} && (i=1,2) && \text{(Reservoir)}\\
N_i\cdot\mathbf{n}|_{\Gamma_{wall}} &= 0 && (i=1,2) && \text{(Impermeable wall)}\\
\phi|_{\Gamma_{out}} &= 0V && && \\
\phi|_{\Gamma_{in}} &= 0.5V && && \text{(Applied electric field)}\\
\varepsilon\nabla\phi\cdot\mathbf{n}|_{\Gamma_{wall}} &= \sigma = 10\mu Ascm^{-2} && && \text{(Charged wall)}\\
\mathbf{u}|_{\Gamma_{wall}} &= 0 && && \text{(No-Slip)}\\
\eta\nabla\mathbf{u}\cdot\mathbf{n}|_{\Gamma_{in,out}} &= p\mathbf{n} && && \text{("Do nothing")}
\end{aligned}
$$

These inlet and outlet boundary conditions impose an electric field along the pore. They assume ion reservoirs of fixed concentration at both ends of the pore. Further, unhindered electrolyte flow into and out of the pore is assumed. At Γ_{sym}, set symmetry boundary conditions for all variables.

The solution for the classical Nernst-Planck model is depicted in Fig. 7 and 8 (left). It shows the onset of the typical plug flow behavior to be expected for electroosmotic

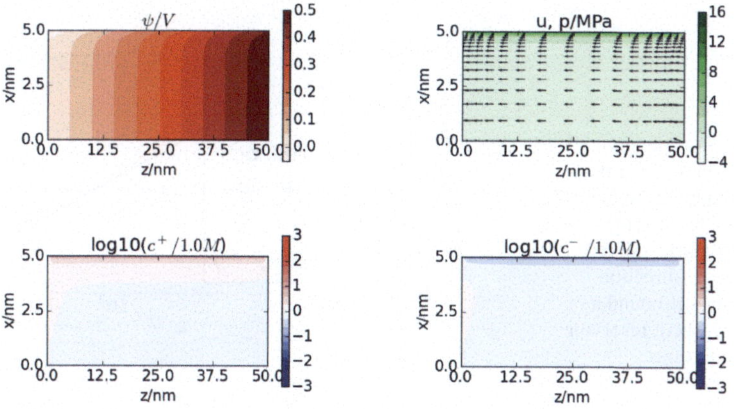

Fig. 7: Electro-osmotic flow of an electrolyte with concentration 1 mol·dm^{-3} through a straight nanopore of width 10 nm with charged walls for an imposed potential difference of 0.5 V in longitudinal direction. Top left: distribution of the electrostatic potential. Top right: velocity field (arrows) and pressure (color). Bottom row: positive resp. negative ion concentration.

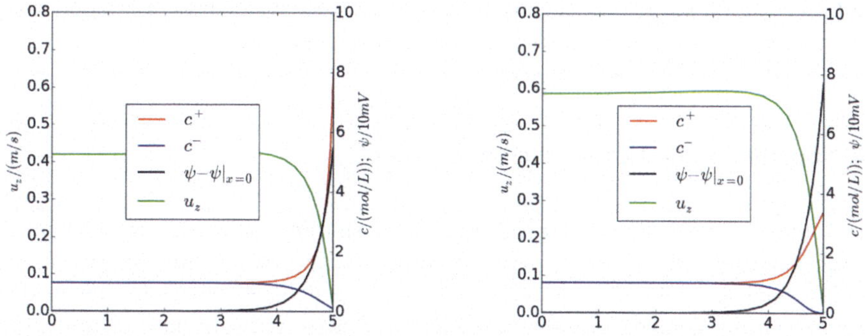

Fig. 8: Velocity, concentration and electrostatic potential profiles in a 10 nm pore. Left: Classical Nernst-Planck model. Right: Improved model with solvation number $\kappa = 10$.

flow in wider pores. Fig 8 at the same time demonstrates the influence of the model improvements discussed in this paper. Including the solvation effect increases the potential and decreases the concentrations at the charged wall boundary. As a consequence, the electroosmotic velocity increases.

In order to verify the coupling approach with the pressure robust Navier–Stokes solver, regard a second version of this problem, where for the velocity, periodic

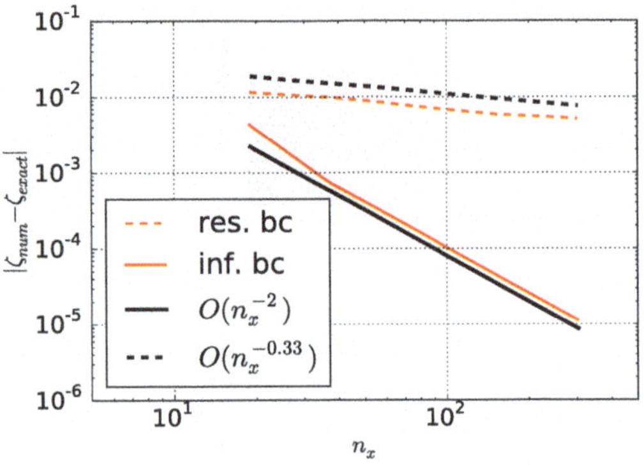

Fig. 9 Error in zeta potential at the half length of the pore for Gouy-Chapman model for a pore of finite length vs. number of grid points in cross section direction. Lines: infinite slit boundary conditions. Dashed: reservoir boundary conditions.

boundary conditions are considered. In order to avoid the edge effect for the species activities and the electrostatic potential at the reservoir boundaries, the solution of the 1D cross sectional Poisson-Boltzmann problem is taken as boundary value at the inlet resp. outlet. The problem is solved by the iterative procedure described in Algorithm 4. For this case of infinite slit boundary conditions Fig. 9 demonstrates a similar second order convergence rate of the zeta potential at point $(w, \frac{l}{2})$ as for the one-dimensional case, cf. Fig. 6. Conversely, in the case of reservoir boundary conditions, the edge effect caused by the mismatch between the boundary conditions and the infinite slit situation leads to a lower convergence rate.

4.4 Spurious velocities in electrophoresis

The authors of [53] discuss the occurrence of spurious velocities in mechanical equilibrium due to the pressure gradient failing to cancel the Coulomb force in the finite element approximation. As a consequence, a straightforward finite element discretization using standard mixed finite element methods results in spurious velocities in an equilibrium situation where both $\mathbf{u} = 0$ and $\mathbf{N}_i = 0$, $(i = 1 \dots n)$. In order to remedy this situation, using the fundamental property of the Navier-Stokes equation mentioned at the beginning of section 3.3, they add the gradient force $RT \sum_{i=1}^{N} \nabla c_i$ to the right hand side of the momentum balance (2.1a). In mechanical equilibrium, this indeed cancels out the pressure and removes the main source of spurious velocities.

While this approach appears to be a clever way to improve simulation results with existing software implementations (like COMSOL Multiphysics in [53]), in the sequel it will be demonstrated that the coupled method presented in this paper without this modification delivers solutions with a similar or lower magnitude of spurious velocities.

In order to discuss this situation, regard a charged $(1e/\text{nm}^2)$ circle of radius 10 nm in a two-dimensional box of side length 60 nm Assume symmetry boundary

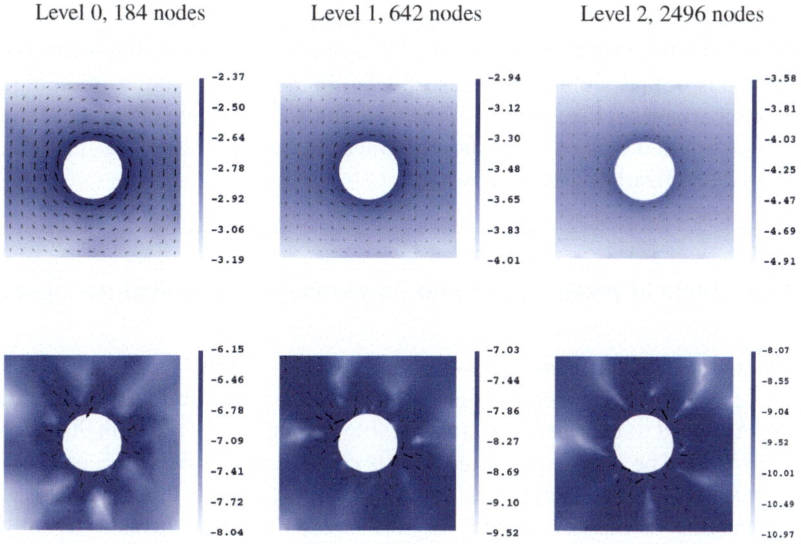

Fig. 10: Absolute value (color, scale is $\log_{10}(u/(ms^{-1}))$ and arrow plots of velocities in equilibrium on coarse, middle and fine grids. Top row: classical mixed FEM. Bottom row: pressure robust FEM.

Fig. 11: Left: coarsest, boundary adapted grid (184 nodes). Right: absolute value and streamlines of solution with applied bias of 0.1 V on level 2 grid.

conditions at the side walls, charged wall and no-slip boundary conditions at the circle, reservoir, zero potential, reservoir and periodic flow boundary conditions at the top and bottom walls.

Fig. 10, bottom row shows directions and the absolute value of spurious velocities in equilibrium for three subsequent grid refinement levels, with logarithms of maximum values of the velocity -6.15, -7.02 and -8.07, respectively. In the top row, results are given for the case of flow simulation without the pressure robust correction. This corresponds to the results obtained in [53] (see Fig. 5 therein), however an exact comparison is not possible due to lack of detail in the problem description in [53].

For comparison, in Fig. 11 (right) the absolute value of the velocity and streamlines for an applied bias of 0.1 V are presented. For reference, in Fig. 11 (left), the level zero discretization grid is shown.

4.5 Ionic current rectification and flow vortex in a conical nanopore

Consider the artificial conical nanopore described in [79, 80]. The length of the pore is $12\mu m$, the width of the small opening is 3 nm, the width of the large opening is 300 nm. Attached to the openings are reservoir regions which are both 900 nm wide and 6 μm long. The outer wall of the pore is charged with $1\,\mathrm{As}\cdot\mathrm{m}^{-2}$, and the bulk concentration of the electrolyte is $0.1\,\mathrm{mol}\cdot\mathrm{dm}^{-3}$ resp. $1\,\mathrm{mol}\cdot\mathrm{dm}^{-3}$.

Experimental results documented in [81] (see e.g. Fig. 2 therein) show a significant rectification effect for the ionic current induced by potential differences of different signs applied between the reservoirs attached to the pore. Fig. 12 demonstrates this effect, the influence of coupling to the flow in the model, and the influence of the solvation on the simulation results. Fig. 12 (left) shows a significant rectification effect in the case of electrolyte concentration $0.1\,\mathrm{mol}\cdot\mathrm{dm}^{-3}$. As expected, due to lower concentrations in the boundary layer, ionic currents for the full model with solvation are smaller than in the case of the classical Nernst-Planck flux. For electrolyte concentration $1\,\mathrm{mol}\cdot\mathrm{dm}^{-3}$, Fig. 12 (right) shows higher ionic currents, a diminished rectification effect.

Numerical simulations reported in [52] suggest the existence of a vortex stretched along the pore. For practical reasons, these have been performed on a "shortened" pore of length 1 μm. Fig. 13 demonstrates similar results for the pore geometry discussed here in the case of classical Nernst-Planck flux. One observes a vortex in

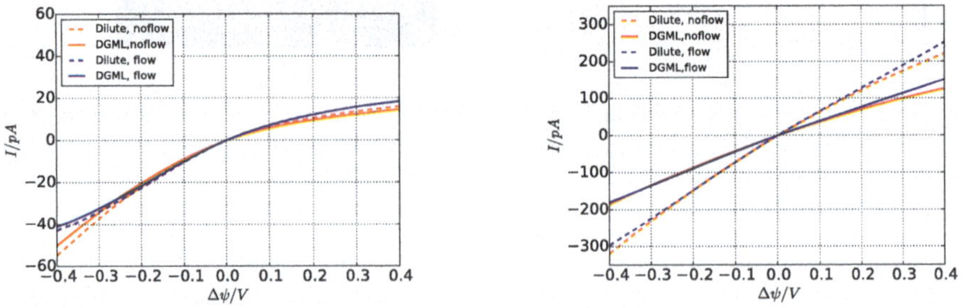

Fig. 12: Simulated current rectification in a conical nanopore. Left: electrolyte concentration $0.1\,\mathrm{mol}\cdot\mathrm{dm}^{-3}$. Right: electrolyte concentration $1\,\mathrm{mol}\cdot\mathrm{dm}^{-3}$. DGML: model according to (2.2). Dilute: dilute solution model (2.3).

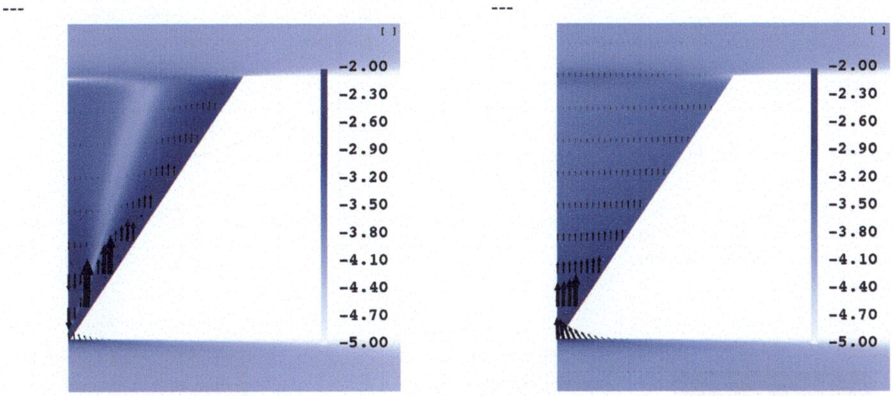

Fig. 13: Simulated velocity field in a conical nanopore for classical Nernst-Planck flux. Color scale: $\log ||\mathbf{u}||$. y axis compressed by factor of 27.

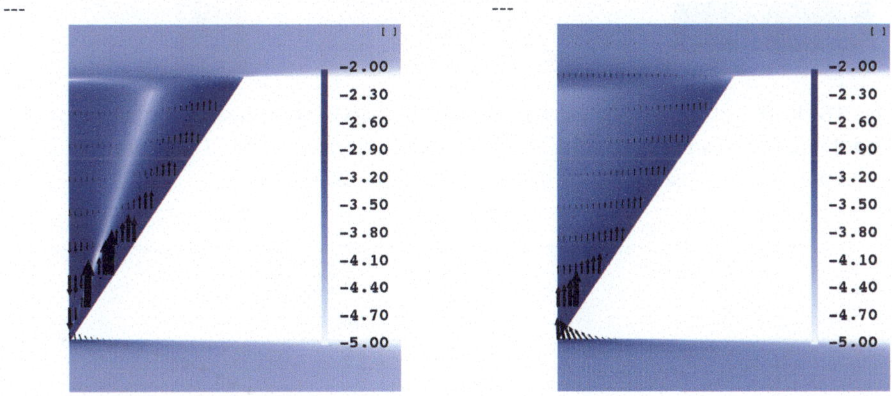

Fig. 14: Simulated velocity field in a conical nanopore for Nernst-Planck flux with solvation effect. Color scale: $\log ||\mathbf{u}||$. y axis compressed by factor of 27. Arrow scale is the same as in Fig. 13.

the case of electrolyte concentration $0.1\,\mathrm{mol}\cdot\mathrm{dm}^{-3}$. According to Fig. 14, similar behavior is observed for the model with solvation, though in this case, the velocities are slightly larger than in the case without solvation. An explanation for this behavior may be the fact that according to (4.9), the zeta potential is inversely proportional to the square root of the concentration, therefore, for lower concentrations, there is a tendency to increase the electroosmotic flow velocity close to the wall. Fig 15 demonstrates this prediction. In order to maintain mass conservation, a high velocity at the wall need to be compensated by a flow in inverse direction in the center of the pore.

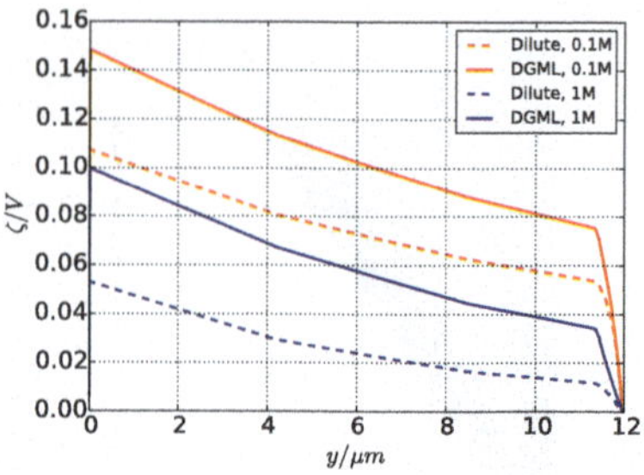

Fig. 15 Simulated zeta potential for different models and electrolyte concentrations. DGML: model according to (2.2). Dilute: dilute solution model (2.3).

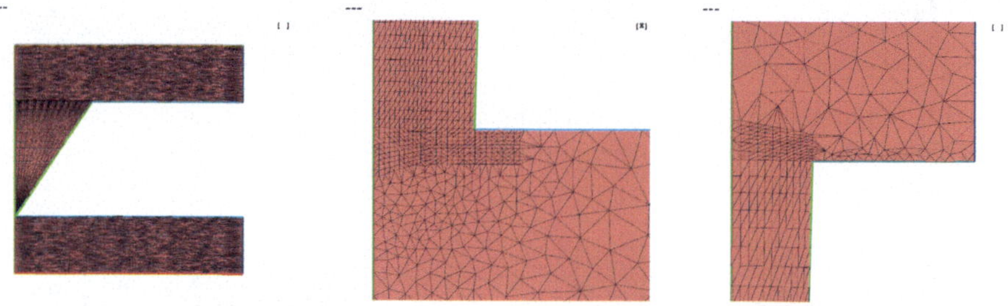

Fig. 16: Left: full view of discretization grid, y axis compressed by factor of 27. Middle: detail at small opening. Right: detail at wide opening. The boundary layer at the pore wall is resolved by a fine grid along the whole pore.

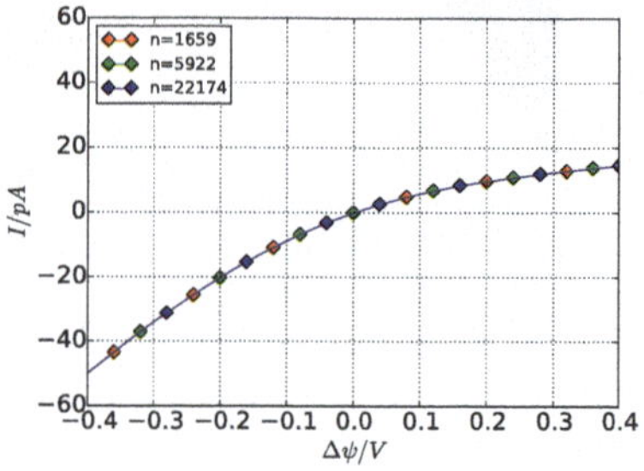

Fig. 17 IV-curve for electrolyte concentration $0.1\,\text{mol}\cdot\text{dm}^{-3}$ with solvation effect on three subsequent grids

Calculations have been performed on a tailored boundary conforming Delaunay grid [62] consisting of 5922 discretization nodes. It has been combined from a tapered, topologically rectangular grid for the inner part of the pore, from a rectangular grid in a boundary layer of width 1.5 nm, and from triangular grids for the lower and upper reservoir regions created using Triangle [82] with a number of a priori given grid points. Fig. 16 exhibits some features of this grid. Notable is the resolution of the polarization boundary layer with strongly anisotropic elements. Even on coarser grids, if they are well tailored, the presented method allows to obtain qualitatively meaningful results. Fig. 17 demonstrates the results for the calculation of the IV curve on three subsequent grids.

5 Conclusions

In this contribution, first results on a novel approach to the numerical solution of the Nernst-Planck-Poisson-Navier-Stokes system have been presented. The underlying model is based on first principles of nonequilibrium thermodynamics and includes ion-solvent interaction and solvation effects.

The discretization methods used are designed to preserve qualitative physical properties of the continuous model independent of the mesh size like mass conservation, pressure robustness, consistency to the thermodynamic equilibrium and maximum principle. A number of directions for future work arise:

- Investigation and improvement of the fixed point coupled solution approach.
- Incorporation of non-constant density into the pressure robust mixed finite element approach.
- Automatization of boundary layer adapted mesh generation for the finite volume method for general geometries in two and three space dimensions.
- Investigation of the existence and convergence of discrete solutions.
- Applications in nanofluidics, cell biology and other fields.

Acknowledgement

The research described in this paper has been supported by the German Federal Ministry of Education and Research Grant 03EK3027D (Network "Perspectives for Rechargeable Magnesium-Air batteries") and the Einstein Center of Mathematics, Berlin, project CH11 "Sensing with Nanopores".

References

1. W. Dreyer, C. Guhlke, and R. Müller. Overcoming the shortcomings of the Nernst–Planck model. *Phys. Chem. Chem. Phys.*, 15(19):7075–7086, 2013.
2. J. Fuhrmann. A numerical strategy for Nernst–Planck systems with solvation effect. *Fuel Cells*, 16(6):704–714, 2016.
3. A. Linke. On the role of the Helmholtz decomposition in mixed methods for incompressible flows and a new variational crime. *Comput. Method. Appl. Mech. Eng.*, 268:782–800, 2014.
4. W. Nernst. Zur Kinetik der in Lösung befindlichen Körper. *Z. Phys. Chemie*, 2(1):613–637, 1888.
5. M. Planck. Über die Erregung von Electricität und Wärme in Electrolyten. *Ann. Phys.*, 275(2):161–186, 1890.
6. O. Stern. Zur Theorie der elektrolytischen Doppelschicht. *Z. f. Electrochemie*, 30:508, 1924.
7. W. Dreyer, C. Guhlke, and M. Landstorfer. A mixture theory of electrolytes containing solvation effects. *Electrochem. Commun.*, 43:75–78, 2014.
8. M. Landstorfer, C. Guhlke, and W. Dreyer. Theory and structure of the metal-electrolyte interface incorporating adsorption and solvation effects. *Electrochim. Acta*, 201:187–219, 2016.

9. S. R. de Groot and P. O. Mazur. *Non-Equilibrium Thermodynamics*. Dover Publications, 1962.
10. J. J. Bikerman. Structure and capacity of electrical double layer. *Philos. Mag.*, 33(220):384–397, 1942.
11. V. Freise. Zur Theorie der diffusen Doppelschicht. *Z. Elektrochem.*, 56:822–827, 1952.
12. N. F. Carnahan and K. E. Starling. Equation of state for nonattracting rigid spheres. *J. Chem. Phys.*, 51:635, 1969.
13. G. A. Mansoori, N. F. Carnahan, K. E. Starling, and T. W. Leland Jr. Equilibrium thermodynamic properties of the mixture of hard spheres. *J. Chem. Phys.*, 54:1523, 1971.
14. A. A. Kornyshev and M. A. Vorotyntsev. Conductivity and space charge phenomena in solid electrolytes with one mobile charge carrier species, a review with original material. *Electrochim. Acta*, 26(3):303–323, 1981.
15. M. S. Kilic, M. Z. Bazant, and A. Ajdari. Steric effects in the dynamics of electrolytes at large applied voltages. I. Double-layer charging. *Phys. Rev. E*, 75(2):021502, 2007.
16. W. Dreyer, C. Guhlke, M. Landstorfer, and R. Müller. New insights on the interfacial tension of electrochemical interfaces and the Lippmann equation. *Eur. J. Appl. Math.*, 29(4):708–753, 2018.
17. J. Fuhrmann. Comparison and numerical treatment of generalised Nernst–Planck models. *Comp. Phys. Comm.*, 196:166 – 178, 2015.
18. A. Jüngel. The boundedness-by-entropy method for cross-diffusion systems. *Nonlinearity*, 28(6):1963, 2015.
19. M von Smoluchowski. Elektrische Endoosmose und Strömungströme. In L. Graetz, editor, *Handbuch der Electrizität und des Magnetismus Band II*, pages 366–427. Barth, Leipzig, 1921.
20. S.S. Dukhin and B.V. Deryagin. *Surface and Colloid Science: Electrokinetic Phenomena*. Plenum Press, 1974.
21. J. Lyklema. *Fundamentals of interface and colloid science: soft colloids*, volume 5. Elsevier, 2005.
22. R. J. Hunter. *Zeta potential in colloid science: principles and applications*, volume 2. Academic Press, 2013.
23. I. Rubinstein and B. Zaltzman. Electro-osmotically induced convection at a permselective membrane. *Phys. Rev. E*, 62(2):2238, 2000.
24. E.V. Dydek, B. Zaltzman, I. Rubinstein, D.S. Deng, A. Mani, and M.Z. Bazant. Overlimiting current in a microchannel. *Phys. Rev. Lett.*, 107(11):118301, 2011.
25. J. W. Jerome. Analytical approaches to charge transport in a moving medium. *Transport Theor. Stat.*, 31(4-6):333–366, 2002.
26. R.J. Ryham. *An energetic variational approach to mathematical modeling of charged fluids: charge phases, simulation and well posedness*. PhD thesis, Pennsylvania State Univ., 2006.
27. R. J. Ryham. Existence, uniqueness, regularity and long-term behavior for dissipative systems modeling electrohydrodynamics. arXiv:0910.4973, 2009.
28. M. Schmuck. Analysis of the Navier–Stokes–Nernst–Planck–Poisson system. *Math. Models Methods Appl. Sci.*, 19(06):993–1014, 2009.
29. D. Bothe, A. Fischer, and J. Saal. Global well-posedness and stability of electrokinetic flows. *SIAM J. Math. Anal.*, 46(2):1263–1316, 2014.
30. P. Constantin and M. Ignatova. On the Nernst–Planck–Navier–Stokes system. *Arch. Ration. Mech. Anal.*, 232(3):1379–1428, 2019.
31. A. Fischer and J. Saal. Global weak solutions in three space dimensions for electrokinetic flow processes. *J. Evol. Equ.*, 17(1):309–333, 2017.
32. W. Dreyer, P. E. Druet, P. Gajewski, and C. Guhlke. Analysis of improved Nernst–Planck–Poisson models of compressible isothermal electrolytes. Part I: Derivation of the model and survey of the results, 2017. WIAS Berlin, preprint 2395.
33. W. Dreyer, P. E. Druet, P. Gajewski, and C. Guhlke. Analysis of improved Nernst–Planck–Poisson models of compressible isothermal electrolytes. Part II: Approximation and a priori estimates, 2017. WIAS Berlin, preprint 2396.

34. W. Dreyer, P. E. Druet, P. Gajewski, and C. Guhlke. Analysis of improved Nernst–Planck–Poisson models of compressible isothermal electrolytes. Part III: Compactness and convergence, 2017. WIAS Berlin, preprint 2397.
35. P. Berg and J. Findlay. Analytical solution of the Poisson–Nernst–Planck–Stokes equations in a cylindrical channel. *Proc. R. Soc. A*, 467(2135):3157–3169, 2011.
36. A. Prohl and M. Schmuck. Convergent finite element discretizations of the Navier–Stokes–Nernst–Planck–Poisson system. *ESAIM: Math. Modelling and Num. Analysis*, 44(3):531–571, 2010.
37. F. Frank, N. Ray, and P. Knabner. Numerical investigation of homogenized Stokes–Nernst–Planck–Poisson systems. *Comp. Vis. Sci.*, 14(8):385–400, 2011.
38. F. Keller, M. Feist, H. Nirschl, and W. Dörfler. Investigation of the nonlinear effects during the sedimentation process of a charged colloidal particle by direct numerical simulation. *J. Colloid Interf. Sci.*, 344(1):228–236, 2010.
39. E.A. Demekhin, V.S. Shelistov, and S.V. Polyanskikh. Linear and nonlinear evolution and diffusion layer selection in electrokinetic instability. *Phys. Rev. E*, 84(3):036318, 2011.
40. H.C. Chang, E.A. Demekhin, and V.S. Shelistov. Competition between Dukhin's and Rubinstein's electrokinetic modes. *Phys. Rev. E*, 86(4):046319, 2012.
41. V. S. Pham, Z. Li, K. M. Lim, J. K. White, and J. Han. Direct numerical simulation of electroconvective instability and hysteretic current-voltage response of a permselective membrane. *Phys. Rev. E*, 86(4):046310, 2012.
42. C.L. Druzgalski, M.B. Andersen, and A. Mani. Direct numerical simulation of electroconvective instability and hydrodynamic chaos near an ion-selective surface. *Phys. Fluids*, 25(11):110804, 2013.
43. V.V. Nikonenko, A.V. Kovalenko, M. K. Urtenov, N.D. Pismenskaya, J. Han, Ph. Sistat, and G. Pourcelly. Desalination at overlimiting currents: State-of-the-art and perspectives. *Desalination*, 342:85–106, 2014.
44. M. He and P. Sun. Mixed finite element analysis for the Poisson–Nernst–Planck/Stokes coupling. *J. Comp. Appl. Math.*, 341:61–79, 2018.
45. J.Y. Lin, L.M. Fu, and R.J. Yang. Numerical simulation of electrokinetic focusing in microfluidic chips. *J. Micromech. Microeng.*, 12(6):955, 2002.
46. H. Daiguji, Y. Oka, and K. Shirono. Nanofluidic diode and bipolar transistor. *Nano Lett.*, 5(11):2274–2280, 2005.
47. Y. Wang, K. Pant, Zh. Chen, G. Wang, W. F. Diffey, P. Ashley, and Sh. Sundaram. Numerical analysis of electrokinetic transport in micro-nanofluidic interconnect preconcentrator in hydrodynamic flow. *Microfluid. Nanofluid.*, 7(5):683, 2009.
48. G.Y. Tang, C. Yang, C.J. Chai, and H.Q. Gong. Modeling of electroosmotic flow and capillary electrophoresis with the joule heating effect: The Nernst- Planck equation versus the Boltzmann distribution. *Langmuir*, 19(26):10975–10984, 2003.
49. R. Sacco, P. Airoldi, A.G. Mauri, and J.W. Jerome. Three-dimensional simulation of biological ion channels under mechanical, thermal and fluid forces. *Appl. Math. Model.*, 43:221–251, 2017.
50. E. Karatay, C. L. Druzgalski, and A. Mani. Simulation of chaotic electrokinetic transport: Performance of commercial software versus custom-built direct numerical simulation codes. *J. Colloid Interf. Sci.*, 446:67–76, 2015.
51. Y. Ai, M. Zhang, S.W. Joo, M.A. Cheney, and Sh. Qian. Effects of electroosmotic flow on ionic current rectification in conical nanopores. *J. Phys. Chem. C*, 114(9):3883–3890, 2010.
52. M.R. Powell, N. Sa, M. Davenport, K. Healy, I. Vlassiouk, S.E. Letant, L.A. Baker, and Z.S. Siwy. Noise properties of rectifying nanopores. *J. Phys. Chem. C*, 115(17):8775–8783, 2011.
53. G. Rempfer, G. B. Davies, Ch. Holm, and J. de Graaf. Reducing spurious flow in simulations of electrokinetic phenomena. *J. Chem. Phys.*, 145(4):044901, 2016.
54. P Berg and BE Benjaminsen. Effects of finite-size ions and relative permittivity in a nanopore model of a polymer electrolyte membrane. *Electrochim. Acta*, 120:429–438, 2014.
55. Ch. Wang, J. Bao, W. Pan, and X. Sun. Modeling electrokinetics in ionic liquids. *Electrophoresis*, 38(13-14):1693–1705, 2017.

56. I. I. Ryzhkov and A.V. Minakov. Finite ion size effects on electrolyte transport in nanofiltration membranes. *J. Sib. Fed. Univ. Mathematics & Physics*, 10(2):186, 2017.
57. V. John, A. Linke, C. Merdon, M. Neilan, and L. G. Rebholz. On the divergence constraint in mixed finite element methods for incompressible flows. *SIAM Review*, 59(3):492–544, 2017.
58. D.L. Scharfetter and H.K. Gummel. Large-signal analysis of a silicon Read diode oscillator. *IEEE Trans. Electron. Dev.*, 16(1):64–77, 1969.
59. R.E. Bank, D.J. Rose, and W. Fichtner. Numerical methods for semiconductor device simulation. *SIAM J. Sci. Stat. Comp.*, 4(3):416–435, 1983.
60. P. Farrell, N. Rotundo, D.H. Doan, M. Kantner, J. Fuhrmann, and Th. Koprucki. Numerical methods for drift-diffusion models. In J. Piprek, editor, *Handbook of Optoelectronic Device Modeling and Simulation: Lasers, Modulators, Photodetectors, Solar Cells, and Numerical Methods*, volume 2, chapter 50, pages 733–771. CRC Press, Boca Raton, 2017.
61. R. Eymard, Th. Gallouët, and R. Herbin. Finite volume methods. *Handbook of numerical analysis*, 7:713–1018, 2000.
62. H. Si, K. Gärtner, and J. Fuhrmann. Boundary conforming Delaunay mesh generation. *Comput. Math. Math. Phys.*, 50:38–53, 2010.
63. J. Fuhrmann, A. Linke, and H. Langmach. A numerical method for mass conservative coupling between fluid flow and solute transport. *Appl. Numer. Math.*, 61(4):530–553, 2011.
64. A. M. Il'in. A difference scheme for a differential equation with a small parameter multiplying the second derivative. *Mat. zametki*, 6:237–248, 1969.
65. R. D. Lazarov, I.D. Mishev, and P. S. Vassilevski. Finite volume methods for convection-diffusion problems. *SIAM J. Numer. Anal.*, 33(1):31–55, 1996.
66. J. J. H. Miller and S. Wang. An analysis of the Scharfetter-Gummel box method for the stationary semiconductor device equations. *RAIRO-Math. Model. Num.*, 28(2):123–140, 1994.
67. J. Xu and L. Zikatanov. A monotone finite element scheme for convection-diffusion equations. *Math. Comp.*, 68(228):1429–1446, 1999.
68. R. Eymard, J. Fuhrmann, and K. Gärtner. A finite volume scheme for nonlinear parabolic equations derived from one-dimensional local Dirichlet problems. *Numer. Math.*, 102(3):463–495, 2006.
69. V. John. *Finite element methods for incompressible flow problems*. Springer, 2016.
70. V. Girault and P.-A. Raviart. *Finite element methods for Navier–Stokes equations: theory and algorithms*, volume 5. Springer, 2012.
71. C. Merdon, J. Fuhrmann, A. Linke, T. Streckenbach, F. Neumann, M. Khodayari, and H. Baltruschat. Inverse modeling of thin layer flow cells for detection of solubility, transport and reaction coefficients from experimental data. *Electrochim. Acta*, 211:1–10, 2016.
72. J. Fuhrmann, T. Streckenbach, et al. pdelib. `http://pdelib.org`, 2018.
73. Ch. Bernardi and G. Raugel. Analysis of some finite elements for the Stokes problem. *Math. Comp.*, 44(169):71–79, 1985.
74. A. Linke and Ch. Merdon. Pressure-robustness and discrete Helmholtz projectors in mixed finite element methods for the incompressible Navier–Stokes equations. *Comput. Method. Appl. Mech. Eng.*, 311:304–326, 2016.
75. O. Schenk and K. Gärtner. Solving unsymmetric sparse systems of linear equations with PARDISO. *Future Gener. Comput. Syst.*, 20(3):475–487, 2004.
76. O. Schenk, K. Gärtner, et al. PARDISO solver project. URL: http://www.pardiso-project.org, 2017. Accessed 2017-01-01.
77. J.Th. Overbeek. Electrokinetic phenomena. In H. R. Kruyt, editor, *Colloid science*, volume 1, pages 194–244. Elsevier, Amsterdam, 1952.
78. D. Burgreen and F.R. Nakache. Electrokinetic flow in ultrafine capillary slits. *J. Phys. Chem.*, 68(5):1084–1091, 1964.
79. Z. Siwy and A. Fuliński. Fabrication of a synthetic nanopore ion pump. *Phys. Rev. Lett.*, 89(19):198103, 2002.
80. M.T. Wolfram, M. Burger, and Z.S. Siwy. Mathematical modeling and simulation of nanopore blocking by precipitation. *J. Phys.-Condens. Mat.*, 22(45):454101, 2010.

81. Z. Siwy, Y. Gu, H.A. Spohr, D. Baur, A. Wolf-Reber, R. Spohr, P. Apel, and Y.E. Korchev. Rectification and voltage gating of ion currents in a nanofabricated pore. *Europhys. Lett.*, 60(3):349, 2002.
82. J. Shewchuk. Triangle: A two-dimensional quality mesh generator and Delaunay triangulator. URL: http://www.cs.cmu.edu/~quake/triangle.html. Accessed 2017-01-01.

Consequences of Uncertain Friction for the Transport of Natural Gas through Passive Networks of Pipelines

Holger Heitsch and Nikolai Strogies

Abstract Assuming a pipe-wise constant structure of the friction coefficient in the modeling of natural gas transport through a passive network of pipes via semilinear systems of balance laws with associated linear coupling and boundary conditions, uncertainty in this parameter is quantified by a Markov chain Monte Carlo method. Information on the prior distribution is obtained from practitioners. The results are applied to the problem of validating technical feasibility under random exit demand in gas transport networks. The impact of quantified uncertainty to the probability level of technical feasible exit demand situations is studied by two example networks of small and medium size. The gas transport of the network is modeled by stationary solutions that are steady states of the time dependent semilinear problems.

Key words: uncertainty quantification, Markov chain Monte Carlo, reliability of gas networks, nomination validation, spheric-radial decomposition

1 Introduction

The transport of natural gas through a single pipe can be modeled by a simplification of the full Euler equations, describing the conservation of mass as well as balance of momentum and energy in fluid dynamics. An overview on existing models for transport of natural gas can be found in [6] and we employ the notation of this work. Assuming a heat flux through the pipe walls compensating discontinuities of temperature in case of shock- and rarefaction waves, energy is no longer a balanced quantity (see [20, Section 14.6]). Working under such a regime, an associated system approximately describing the underlying physics is given by the fully nonlienar system of balance laws

Holger Heitsch · Nikolai Strogies
Weierstrass Institute, Mohrenstr. 39, 10117 Berlin, Germany
e-mail: holger.heitsch@wias-berlin.de

M. Hintermüller, J. F. Rodrigues (eds.), *Topics in Applied Analysis and Optimisation*, CIM Series in Mathematical Sciences, https://doi.org/10.1007/978-3-030-33116-0_9

$$\begin{aligned}\rho_t + q_x &= 0,\\ q_t + (p(\rho) + \tfrac{q^2}{\rho})_x &= \lambda \tfrac{q|q|}{\rho} - g\rho h',\end{aligned} \tag{ISO 1}$$

which is a well known model for gas transport, see, e.g., [5, 12, 17]. Here, ρ, q, g, h' denote density, volume flow, gravitational constant and slope of the pipe, respectively. Further, $p(\rho)$ represents the pressure depending on the density of the natural gas, usually described by an equation of state, and λ is the friction coefficient, also known as Darcy friction factor, quantifying the influence of friction at the pipe wall on the flow behavior. A priory, the friction coefficient is assumed to be a function of the spatial position accounting for local effects in the internal coating of the pipe or local changes in the diameter caused by pollution.

Assuming additional simplifications in (ISO 1), like considering only planar networks with $h' \equiv 0$, neglecting the influence of $\frac{q^2}{\rho}$ in the flux term and utilizing the simplified pressure law $p(\rho) = a^2\rho$, where $a > 0$ denotes the constant speed of sound, we obtain a semilinear system of first-order partial differential equations. Defined on a pipe which is represented by the interval (x_L, x_R) with $x_L < x_R$, it is given by

$$\begin{aligned}\rho_t + q_x &= 0,\\ q_t + a^2\rho_x &= -\lambda\frac{q|q|}{\rho},\end{aligned} \qquad \text{on } (0,T) \times (x_L, x_R), \tag{ISO 2}$$

where $T > 0$ represents the time horizon. A brief discussion of the above mentioned simplifications can be found in [12,22]. To obtain a well posed forward problem, we in addition consider the initial conditions

$$\rho(0,\cdot) = \rho_0(\cdot), \quad q(0,\cdot) = q_0(\cdot) \text{ for } x \in (x_L, x_R), \tag{IC}$$

and boundary conditions. System (ISO 2) is strictly hyperbolic with one strictly negative and one strictly positive eigenvalue. Consequently, conditions on linear combinations of the state variables ρ and q are required at both ends of the pipe. In other words, there have to exist certain vectors $c_L, c_R \in \mathbb{R}^2$ and functions $d_L, d_R \in L^\infty(0,T)$ such that

$$c_L^\top(\rho(t,x_L), q(t,x_L)) = d_L(t), \quad c_R^\top(\rho(t,x_R), q(t,x_R)) = d_R(t). \tag{BC}$$

Besides the time-dependent models, a stationary model is considered as well, where the boundary data are constant. It is obtained by neglecting the time derivatives in system (ISO 2). The resulting ordinary differential equations

$$q_x = 0, \quad \rho_x = -\frac{\lambda}{a^2}\frac{q|q|}{\rho},$$

with associated initial conditions $q(x_L) = q$, $\rho(x_L) = \rho$ that are obtained from the original boundary conditions, can be solved explicitly, providing

$$q(x) \equiv q, \quad \rho(x) = \sqrt{\rho(x_L)^2 - 2\bar{\lambda}(x)\frac{q|q|}{a^2}} \tag{ISO-ALG}$$

with $\bar{\lambda}(x) = \int_{x_L}^{x} \lambda(\tau)d\tau$. Note that the initial conditions can be imposed at arbitrary spatial positions along the pipe. In particular, data for volume flow and density do not necessarily have to be imposed at the same position.

In the context of obtaining information on the friction coefficient out of measurements, for example, measurements of pressure at certain points in the network, the Bayesian framework is based on the uncertainty-to-observation operator $\mathcal{G}$ which maps the underlying unknown (the friction coefficient) onto the measurement data y. It represents a composition of the solution operator associated to the underlying PDE problem (ISO 2), extended to a passive network as discussed below, applied to the coefficient λ and a data formation operator. In case of density being measured at finitely many points in time $t_j \in [0, T], j = 1, \ldots, K$, and at a position $\bar{x}$ within the network, we consider

$$\mathcal{G}(\lambda) = (\rho(t_1, \bar{x}), \ldots, \rho(t_K, \bar{x}))^{\top}. \tag{1.1}$$

Since the measuring process introduces errors to the observation, we find

$$y = \mathcal{G}(\lambda) + \eta \Leftrightarrow \eta = y - \mathcal{G}(\lambda), \tag{1.2}$$

with $\eta \in \mathbb{R}^K$ denoting the measurement error. In the work at hand, we assume this error to be Gaussian noise with mean zero, associated covariance matrix $\Gamma \in R^{K \times K}$ and probability density function (PDF) denoted by π_{DL}. Given a friction coefficient λ, the probability of obtaining measurement data y is given according to

$$P(y|\lambda) = \pi_{DL}(y - \mathcal{G}(\lambda)), \tag{1.3}$$

also called likelihood of the data. Now, using Bayes' theorem, we can incorporate available "prior" knowledge on λ and provide its probability, given the measured data y, as

$$P(\lambda|y) \propto P(y|\lambda)P(\lambda). \tag{1.4}$$

Our knowledge on the distribution of λ, i.e., $P(\lambda)$, is called prior and the probability of λ given the data y, i.e., $P(\lambda|y)$, is called posterior. Both probability distributions are equipped with associated PDF's π_{PR} and π_{PO}.

The Bayesian approach to inverse problems involving a partial differential equation has been intensively studied in the last years. For an overview we refer to [2] and the references therein. In particular we mention [4], where a Multi-Level Markov chain Monte Carlo method has been investigated that allows for incorporating mesh-refinement strategies for the partial differential equation. To the best of our knowledge, besides the investigations in [13, 18], where the first one merely considers a single pipe while allowing for a spatially distributed friction coefficient and the latter one not employing the Bayesian approach, obtaining information on distributions of the friction coefficient has not been subject of investigation so far.

2 Results for the state equations

This section is dedicated to discussions on notions of solutions for the underlying system in both the time independent and dependent setting, and the extension of the respective ordinary and partial differential equations to passive networks of pipelines.

2.1 Steady states

As outlined above, the time independent model of gas flow results in an algebraic solution representing the connection between density and flow along a pipe as described in (ISO-ALG). As consequence, in a passive gas network there exists a explicit characterization of gas flow feasibility that is based on the algebraic formulation of mass and momentum conservation, respectively, the Kirchhoff's first and second law. Feasibility of gas load (or nomination) is equivalent to the existence of a pressure-flow profile fulfilling that Kirchhoff's laws and meeting nodal bounds on the pressure. For a characterization of the set of all capacities that can be realized, functional relations in the nomination space are sought, that hold if and only if the nomination is feasible. These functional relations become closer and closer coupled among each other the more intertwined cycles there are in the network. In what follows, a general characterization is derived that still contains as many implicit indeterminates as there are fundamental cycles in the network.

The gas transportation network is considered as a connected directed graph $G = (\mathcal{V}, \mathcal{E})$, with $|\mathcal{V}| = n + 1$ nodes and $|\mathcal{E}| = m \geq n$ edges. Assume the network is in steady state let be $q \in \mathbb{R}^m$ the flow along the edges of G and $(p_0, p) \in \mathbb{R}^{n+1}$ the pressure at nodes in $\mathcal{V}$. The network topology let be given by $\mathrm{i} \in \mathbb{R}^{n \times m}$, a reduced node-arc incidence matrix of G. Inflows and outflow are described by a balanced load vector $(b_0, b) \in \mathbb{R}^{n+1}$, i.e., it holds $-\mathbb{1}^\top b = b_0$, where $\mathbb{1}$ denotes the vector of all ones in suitable dimension, here n. Moreover, we make the sign convention that $b_i \leq 0$ at injection points (entries) and $b_i \geq 0$ at withdrawal points (exits). Mass, or mass flow, conservation at each node in $\mathcal{V}$ (Kirchhoff's first law) now reads

$$\mathrm{i}q = b. \tag{2.1}$$

Denoting with $o(e) \in \mathcal{V}$ and $\pi(e) \in \mathcal{V}$ the origin and head of some edge $e \in \mathcal{E}$, respectively, then the pressure drop between the ends of pipe $e \in \mathcal{E}$ causes a constant flow along the pipe due to the condition (Kirchhoff's second law)

$$(p_k)^2 - (p_\ell)^2 = \Lambda_e |q_e| q_e, \tag{2.2}$$

where $k = o(e)$ and $\ell = \pi(e)$. The latter equation is equivalent to (ISO-ALG), but formulated in terms of pressure values rather than density right now. Here, the the so-called *roughness* coefficient Λ_e combines constant parameters and the integral friction coefficient λ_e of some pipe $e \in \mathcal{E}$. In particular, with the law $p = a^2 \rho$ we have

$$\Lambda_e = \Lambda_e(\lambda_e) = 2a^2 l_e \lambda_e, \tag{2.3}$$

where l_e denotes the length of pipe e, and a is the speed of sound again. With technical lower and upper pressure limits p^{min}, p^{max} we are led to introduce the following set M_{feas} of *feasible load (nomination) vectors*.

Definition 1 A load vector (b_0, b) is feasible load vector, if and only if (b_0, b) is contained in the feasibility set M_{feas} defined as

$$M_{\text{feas}} := \left\{(b_0, b) \;\middle|\; -\mathbb{1}^\top b = b_0;\ \exists (q, p) \text{ with } p \in \left[p^{min}, p^{max}\right] \text{ and } (2.1), (2.2)\right\}. \tag{2.4}$$

The following provides a characterization of the set M_{feas}, where all pressure variables and "most of" the flow variables are eliminated. With the notation $\Lambda := \text{diag}\{\Lambda_e \mid e \in \mathcal{E}\}$ for roughness, from [11] we take the following result.

Theorem 2.1 ([11, Theorem 1]) *Let* $\mathrm{i} = (\mathrm{i}_B, \mathrm{i}_N)$ *be a partition into basis and nonbasis submatrices of the incidence matrix* i. *Let* Λ_B, Λ_N *and* q_B, q_N *be according partitions of* Λ *and* q. *Define*

$$h : \mathbb{R}^n \times \mathbb{R}^{|N|} \to \mathbb{R}^n,\ h(u, v) := \left(\mathrm{i}_B^\top\right)^{-1} \Lambda_B \left|\mathrm{i}_B^{-1}(u - \mathrm{i}_N v)\right| \left(\mathrm{i}_B^{-1}(u - \mathrm{i}_N v)\right). \tag{2.5}$$

Then, M_{feas} *consists of all* (b_0, b) *with* $-\mathbb{1}^\top b = b_0$ *for which there is a* $z \in \mathbb{R}^{|N|}$ *such that*

$$\mathrm{i}_N^\top h(b, z) = \Lambda_N |z| z \tag{2.6}$$

$$(p_0^{min})^2 \leq \min_{k=1,\ldots,n} \left[(p_k^{max})^2 + h_k(b, z)\right] \tag{2.7}$$

$$(p_0^{max})^2 \geq \max_{\ell=1,\ldots,n} \left[(p_\ell^{min})^2 + h_\ell(b, z)\right] \tag{2.8}$$

$$\min_{k=1,\ldots,n} \left[(p_k^{max})^2 + h_k(b, z)\right] \geq \max_{\ell=1,\ldots,n} \left[(p_\ell^{min})^2 + h_\ell(b, z)\right]. \tag{2.9}$$

Up to finding an auxiliary variable z satisfying (2.6), Theorem 2.1 identifies fully explicit feasibility conditions with respect to the load vector (b_0, b) and the side constraints (pressure bounds). Therefore, the feasibility test for balanced (b_0, b) reduces to determining the unique z solving (2.6) and then checking the inequality system (2.7), (2.8), (2.9). Observe that the dimension of z corresponds to the number of columns of the nonbasis part i_N of the reduced incidence matrix i, hence to the number of fundamental cycles in the network. Obviously, the situation should be particularly comfortable for networks without cycles, as is illustrated now.

Special case of tree networks

Suppose $G = (\mathcal{V}, \mathcal{E})$ is a tree (trivially a spanning tree of itself). Fix an arbitrary leaf node as root and number it by 0. Direct all edges in $\mathcal{E}$ away from the root. The incidence matrix i of G already is the basis matrix i_B so that there is no nonbasis portion i_N. Using depth-first search, number the nodes so that numbers increase along any path from the root to one of the leaves. For $k, \ell \in \mathcal{V}$, denote $k \succeq \ell$ if, in G, the unique directed path from the root to k, denoted $\Pi(k)$, passes through ℓ. Since i_N vacuous and $\Lambda = \Lambda_B$, one obtains for h as defined in (2.5)

$$h(b,z) \;=\; h(b) \;=\; \left(\mathrm{i}^{-1}\right)^\top \Lambda \left|\mathrm{i}^{-1} b\right| \left(\mathrm{i}^{-1} b\right)$$

and componentwise, for $k = 1, \ldots, n$, and emphasizing the implicit dependency of the friction coefficients λ_e according to (2.3),

$$h_k(b,\lambda) = \sum_{e \in \Pi(k)} 2a^2 l_e \lambda_e \left| \sum_{t \in \mathcal{V}, t \succeq \pi(e)} b_t \right| \left(\sum_{t \in \mathcal{V}, t \succeq \pi(e)} b_t \right), \quad k = 0, \ldots, n, \tag{2.10}$$

as shown in [11]. To reduce technicality we assume that the network has the node 0 as the only entry and all remaining nodes as exits, again with all edges directed away from 0. Then in (2.10) flow and edge directions conform, leading to

$$h_k(b,\lambda) = \sum_{e \in \Pi(k)} 2a^2 l_e \lambda_e \left(\sum_{t \in \mathcal{V}, t \succeq \pi(e)} b_t \right)^2, \qquad k = 0, \ldots, n. \tag{2.11}$$

Now, Theorem 2.1 specializes as follows:

Corollary 2.1 *If the network is a tree with a single entry as its root, then the set of feasible load vectors is given by*

$$M_{\text{feas}} = \left\{ (-\mathbb{1}^\top b, b) \,\middle|\, 0 \le \min_{k=0,\ldots,n} \left\{ (p_k^{max})^2 + h_k(b,\lambda) \right\} - \max_{\ell=0,\ldots,n} \left\{ (p_\ell^{min})^2 + h_\ell(b,\lambda) \right\} \right\} \tag{2.12}$$

with $h_k(b,\lambda)$ is as in (2.11), $k = 0, \ldots, n$. Note that $h_0(\cdot,\cdot) \equiv 0$ here.

2.2 Time dependent problems

The state system (ISO 2) can be written as

$$y_t + A y_x = g(y), \tag{2.13}$$

with $y := (\rho, q)^\top$ denoting the state vector. Here $g(y)$ and the A are defined by

$$g(y) = (0, -\lambda \tfrac{q|q|}{\rho})^\top \text{ and } A = \begin{pmatrix} 0 & 1 \\ a^2 & 0 \end{pmatrix}.$$

The eigenvalues of A, given by $\sigma_1 = -a < a = \sigma_2$, define *characteristic lines.* Indeed, given a position $x \in (x_L, x_R)$ and a point in time $\tau \in (0, T)$, the characteristics passing through (τ, x) are defined as solutions to the ordinary differential equations

$$\dot{s}_i(t; \tau, t) = \sigma_i$$

with $s_i(\cdot; \tau, x) : \mathbb{R} \to \mathbb{R}$ satisfying $s_i(\tau; \tau, x) = (\tau, x)$. The index i relates characteristic and eigenvalue. Since the domain Q is bounded, we define the times $\underline{t}_i(\tau, x) \in [0, \tau]$ and $\overline{t}_i(\tau, x) \in [\tau, T]$, specifying the time, the i-th characteristic passing through (τ, x) satisfies $(t, s_i(t; \tau, x)) \in Q$, i.e., we either have $s_i(\underline{t}_i(\tau, x); \tau, x) = x, x \in [x_L, x_R]$, in case the characteristic intersects with $\{0\} \times [x_L, x_R]$ or $s_1(\underline{t}_1(\tau, x); \tau, x) = x_R$ or $s_2(\underline{t}_2(\tau, x); \tau, x) = x_L$ if the characteristic intersects with the boundary $(0, T) \times \{x_R\}$ or $(0, T) \times \{x_L\}$, respectively. The time $\overline{t}_i(\tau, x)$ is defined correspondingly. Once more, the value a represents the speed of sound in the model, i.e., the speed information is propagated with through the spatial domain.

Based on a transformation of the representation (2.13), we consider *broad solutions* for (ISO 2) as follows. Due to strict hyperbolicity of (ISO 2), there exist a matrix $L \in \mathbb{R}^{n\times n}$ such that $A = LDL^{-1}$ where $D = \text{diag}(\sigma_i) \in \mathbb{R}^{2\times 2}$. Multiplying (2.13) from the left by L^{-1}, using the linearity of differentiation and setting $\mathcal{T}(y) := L^{-1}y$ and $f(y) := L^{-1}g(y)$, we obtain

$$(\mathcal{T}(y))_t + D(\mathcal{T}(y))_x = f(y),$$

a system of scalar, linear transport equations, merely coupled by the source term on the right hand side. Given (ISO 2), the transformation matrices are

$$L = c_a \begin{pmatrix} 1 & 1 \\ -a & a \end{pmatrix}, \quad L^{-1} = (2c_a)^{-1} \begin{pmatrix} 1 & -a^{-1} \\ 1 & a^{-1} \end{pmatrix}.$$

Also note that the linear transport equations even have constant coefficients. In case of a single equation of this type, it is well known that solutions are described by ordinary differential equations along the characteristic line, defined by the constant coefficient. The concept of broad solutions extends this property to systems in that the transformed components of $\mathcal{T}(y)$ are absolutely continuous functions along the corresponding characteristic lines. Broad solutions to semilinear systems of balance laws for unbounded domains have been studied for example in [1,21], while in [14,15] bounded domains have been considered. We recall the definition from [18] as follows.

Definition 2 A *broad solution* of (ISO 2) is a function $y = L\mathcal{T}(y) : Q \to \mathbb{R}^2$ such that, for almost every $(\tau, x) \in Q$, the map $t \mapsto \mathcal{T}_i(y)[t, s_i(t; \tau, x)]$ is an absolutely continuous function satisfying

1. at almost every $(\tau, x) \in Q$ the ordinary differential equations

$$\frac{d}{dt}\mathcal{T}_i(y)[t, s_i(t; \tau, x)] = f_i(t, s_i(t; \tau, x), y(t, s_i(t; \tau, x))) \tag{2.14}$$

almost everywhere on $(\underline{t}_i(\tau, x), \overline{t}_i(\tau, x))$ for $i = 1, 2$,

2. the initial condition at $x \in (x_L, x_R)$

$$\mathcal{T}_i(y)[0,x] = c_i y_0(x),$$

3. the boundary condition at x_L in the sense that for almost every $t \in (0,T)$ we have

$$\mathcal{T}_2(y)[t,x_L] = c_2 C_L^{-1} \begin{pmatrix} \mathcal{T}_1(y)[t,x_L] \\ d_L(t) \end{pmatrix},$$

4. the boundary condition at x_R in the sense that for almost every $t \in (0,T)$ we have

$$\mathcal{T}_1(y)[t,x_R] = c_1 C_R^{-1} \begin{pmatrix} \mathcal{T}_2(y)[t,x_R] \\ d_R(t) \end{pmatrix},$$

5. in case of a network, 1. and 2. hold true on all pipes of the network, 3. and 4. are satisfied at entry- and exit nodes (cf. Sect. 2.1), and in addition, coupling conditions (2.16) and (2.17) are satisfied at interior nodes (nodes with no consumption) of the network.

The lateral boundaries of Q are approached by exactly one of the characteristics for every $t \in (0,T)$. To be able to reconstruct the original variables at the boundary, the matrices

$$C_L = \begin{pmatrix} c_1 \\ c_L \end{pmatrix}^\top, \quad C_R = \begin{pmatrix} c_2 \\ c_R \end{pmatrix}^\top$$

have to be invertible, restricting the possible linear combinations of y that can be prescribed. Here, $c_1, c_2 \in \mathbb{R}^2$ denote the rows of L^{-1}. As it can be seen in [18, Proposition 2], solutions as in (ISO-ALG) (corresponding to Definition 1 in Sect. 2.1) form steady states for the time dependent problem (ISO 2) in the sense of Definition 2.

The coupling conditions (2.16) and (2.17) (see below) allow for reconstructing the original state variables from the transformed variables $\mathcal{T}^{(i)}$ as the following example demonstrates.

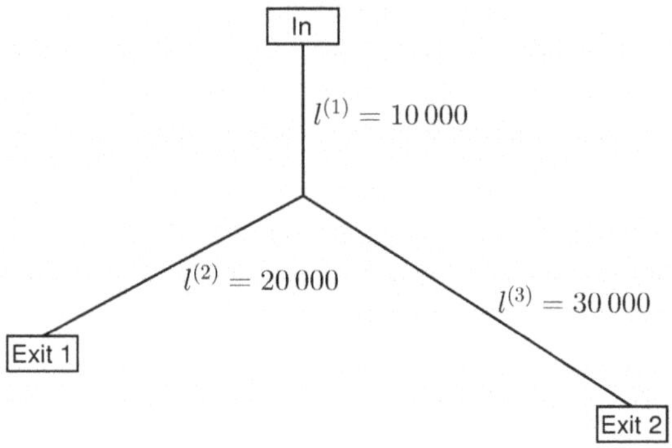

Fig. 1 Sketch of a basic passive network involving three pipes, one entry node, one interior node and two exit nodes

Example Consider the internal node of the Y-shaped network depicted in Fig. 1. For the interior node we obtain $n_L = \{1\}$, $n_R = \{2,3\}$ (sets of incoming/outgoing pipes) and consequently, the second component of $\mathcal{T}$ from pipe 1, $\mathcal{T}_2^{(1)}$, and the first component of $\mathcal{T}$ from pipe 2 and 3, $\mathcal{T}_1^{(2)}$ and $\mathcal{T}_1^{(3)}$, respectively, approach the junction. As a consequence, the original state variables $\{(\rho^{(i)}, q^{(i)})\}_{i=1}^3$ have to satisfy the linear system

$$\begin{pmatrix} (2c_a)^{-1} & (2c_a a)^{-1} & 0 & 0 & 0 & 0 \\ 0 & 0 & (2c_a)^{-1} & (2c_a a)^{-1} & 0 & 0 \\ 0 & 0 & 0 & 0 & (2c_a)^{-1} & (2c_a a)^{-1} \\ 1 & 0 & -1 & 0 & 0 & 0 \\ 1 & 0 & 0 & 0 & -1 & 0 \\ 0 & 1 & 0 & -1 & 0 & -1 \end{pmatrix} \begin{pmatrix} \rho^{(1)} \\ q^{(1)} \\ \rho^{(2)} \\ q^{(2)} \\ \rho^{(3)} \\ q^{(3)} \end{pmatrix} = \begin{pmatrix} \mathcal{T}_2^{(1)} \\ \mathcal{T}_1^{(2)} \\ \mathcal{T}_1^{(3)} \\ 0 \\ 0 \\ 0 \end{pmatrix} \tag{2.15}$$

at the junction, and $\mathcal{T}_2^{(1)}, \mathcal{T}_1^{(2)}, \mathcal{T}_1^{(3)}$ are obtained as linear combinations of the solution according to the transformation matrix L. □

Concerning the existence of broad solutions of (ISO 2) and their properties, we restrict ourselves to briefly recalling the results from [18]. Under certain conditions, we assume to be satisfied in the cases at hand, existence is established in Proposition 4 and Remark 1 of [18]. Further, the Lipschitz continuous dependency of the trace evaluation of ρ on the friction coefficient λ is proven in Proposition 6 of [18] where the latter result depends on Proposition 5 and Proposition 3 of the given reference. In particular, all of the given results also apply for passive networks of pipelines as discussed next.

2.3 Extension to passive networks

So far, the physical process of natural gas being transported is merely introduced on a single pipe. An extension to passive networks, represented as directed graph $G = (\mathcal{V}, \mathcal{E})$ with the notation of Sect. 2.1, is obtained as follows. We restrict ourselves to tree networks with single entry (root of the tree), interior nodes (no consumption), and exit nodes assuming they are all leaves of the tree. Every edge $e \in \mathcal{E}$ models a pipe on its associated domain $(x_L^{(e)}, x_R^{(e)})$, where the transport equation (ISO 2) has to hold. In addition, for every node $k \in \mathcal{V}$, there exist index sets n_L^k, n_R^k denoting its incoming and outgoing edges, respectively. Besides entry and exit nodes with assumed only one adjacent pipeline, i.e., n_L^k is empty set, n_R^k is singleton and vice versa, respectively, internal nodes connect at least two pipes, and thus, they correspond to points of pipe interaction. The type of such interactions is limited to junctions only, rendering the network passive in that no active elements like, e.g., compressors or valves are considered. As in the single-pipe scenario, entries and exits of the network require the definition of boundary conditions while, in order to obtain a well posed system of partial differential equations, initial conditions have to be introduced as well. In case of junctions, additional coupling conditions have to be imposed, characterizing the interplay of solutions to (ISO 2) on each of the adjacent pipes. On the one hand, the volume flow has to be balanced such that Kirchhoff's circuit law

$$\sum_{e \in n_L^k} q^{(e)}(x_R^{(e)}) + dq_k = \sum_{e \in n_R^k} q^{(e)}(x_L^{(e)}) \tag{2.16}$$

holds true. Here, dq_k denotes possible injection or extraction of gas at the corresponding node. On the other hand, the pressure has to be conserved, i.e., we require

$$p^{(e)}(x_R^{(e)}) = p_k = p^{(e')}(x_L^{(e')}) \text{ for all } e \in n_L^k \text{ and } e' \in n_R^k. \tag{2.17}$$

As a consequence of the particular form (ISO-ALG) for steady states on single pipes, the representation of steady states for networks reduces to real numbers for every node and pipe where the numbers associated to pipes describe the constant volume flow along this element. This is due to the fact, that the data fixing the solution can be imposed at arbitrary position, and the numbers for the junctions describe pressure or density. The validation of feasibility of the solution now consist of checking the coupling conditions, and verifying, that the drop of density between two nodes can be described according to (ISO-ALG) along each pipe, i.e., if

$$\rho(x_L^{(e)})^2 - \rho(x_R^{(e)})^2 = 2\bar{\lambda}(x_R^{(e)})\frac{q^{(e)}|q^{(e)}|}{a^2}$$

holds for $e \in \mathcal{E}$. We immediately observe that in case of stationary solutions, the drop of density merely depends on $\bar{\lambda}(x_R^{(e)})$. Thus, instead of considering distributed frictions coefficients as in [13, 18], we restrict the discussion on pipe-wise constant coefficients $\lambda_e \in \mathbb{R}$, and thus, $\bar{\lambda}(x_R^{(e)}) = \lambda_e(x_R^{(e)} - x_L^{(e)})$ (cf. formula (2.3) Sect. 2.1). This is sufficient as we concentrate on a particular representative of all functions with the same integral value.

When considering networks of pipelines for gas transport in the time dependent setting, an important fact lies in the lack of observeability of distributed information along the pipes for a fixed point in time. On the one hand, this introduces the requirement of taking measurements of the state at a fixed spatial position like entry- or exit nodes for several times. On the other hand, it implies a lack of knowledge about the initial state within the pipes which is a more severe drawback. The structure of broad solutions as solution of ordinary differential equations with an initial condition depending on distributed information on the initial state of the system strictly requires corresponding knowledge. As in [7, 18], we assume the initial conditions of the time dependent PDE-problems to be given as steady states that can be computed efficiently (see, e.g. [19]).

3 Uncertainty quantification for the semilinear model

This section introduces the Bayesian approach to inverse problems, clarifies the involved probability distributions and provides an algorithm for sampling from the posterior distribution.

The Bayesian approach to inverse problems as, e.g., discussed in [2], is based on the following theorem.

Theorem 3.2 (Bayes' Theorem) *Assume that*

$$Z := \int_{\mathbb{R}^n} \pi_{DL}(y - \mathcal{G}(\lambda))\pi_{PR}(\lambda)d\lambda > 0. \tag{3.1}$$

Then $\lambda|y$ *is a random variable with Lebegue density* $\pi_{PO}(\lambda)$ *given by*

$$\pi_{PO}(\lambda) = \frac{1}{Z}\pi_{DL}(y - \mathcal{G}(\lambda))\pi_{PR}(\lambda).$$

The result is suited to finite dimensional problems, but similar principles hold in case of a function space setting, where merely a Radon-Nikodym density of measures associated to posterior and prior density is given (see [2] for details). In case of pipewise constant friction coefficients and a finite number of density measurements, the inverse problem renders finite dimensional in that input as well as observations are finite dimensional objects. However, we still have to consider solutions to the state system that are given in their associated function spaces.

Concerning the data likelihood we assume the measurement errors at each spatial position and time to be independent and identically distributed according to a normal distribution with mean zero and a variance of 0.001.

Based on consultations with industry partners, the friction coefficients within a newly produced pipes can be assumed to be distributed according to a truncated normal distribution, i.e., a distribution that is derived from that of a normally distributed random variable by bounding it from above or below and a probability density function given as

$$\pi_{TN}(\lambda) = \frac{\phi\left(\frac{\lambda-\mu}{\sigma}\right)}{\sigma\left(\Phi\left(\frac{b-\mu}{\sigma}\right) - \Phi\left(\frac{a-\mu}{\sigma}\right)\right)}. \tag{3.2}$$

Here, ϕ and Φ denote the probability and cumulative density functions of the standard normal distribution, respectively, where $\mu = 0.018172$ and $\sigma = 0.0005$ are mean and variance. The truncation is set up by values $a^{new} = 0.01813$ and $b^{new} = 0.018206$ denoting the respective upper and lower bounds. While operating, the flow-performance of pipes worsens in that the friction coefficient increases. In order to incorporate this knowledge into the prior-modeling, we allow for an increased upper bound truncating the normal distribution in case of aged pipes, i.e., setting $b = 0.021$. The lower bound remains as before in this scenario as we do not expect the friction coefficient to drop in the aging process. Fig. 2 provides rescaled probability density function for both cases and demonstrates the increased domains of the probability density function, and thus, possible choices of λ.

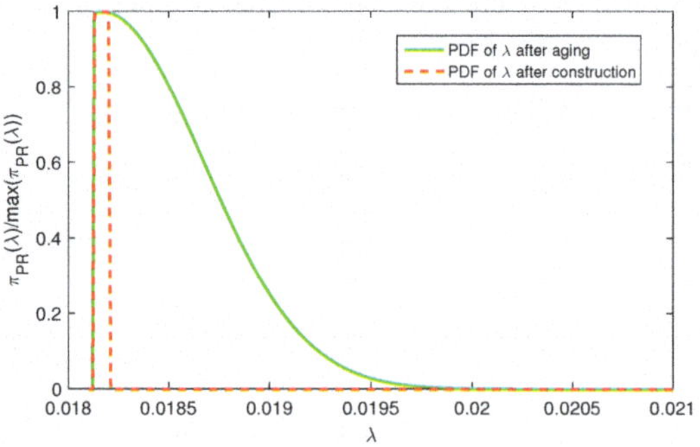

Fig. 2 Shape of the normalized PDF for the truncated normal distribution

The following result enables the usage of Theorem 3.2.

Proposition 3.1 *The factor Z as defined in* (3.1) *is strictly positive.*

Proof As outlined above, the prior is assumed to be a truncated normal distribution on each of the pipes. Moreover, they are independently distributed providing

$$\pi_{PR}(\lambda) = \prod_{i=1}^{|\mathcal{E}|} \pi_{TN}(\lambda_i). \tag{3.3}$$

Here, π_{TN} denotes the pdf of the probability distribution introduced in (3.2), λ_i denotes the i-th component of the vector of friction coefficients and $|\mathcal{E}|$ the total number of pipes. In other words, the support of π_{PR} and consequently π_{PO} is compact. According to the results from Sect. 2.2, the observation operator $\mathcal{G}(\lambda)$ depends Lipschitz continuously on λ, and in addition, the probability density function π_{DL} is a continuous function rendering the concatenation

$$\lambda \mapsto \pi_{DL}(y - \mathcal{G}(\lambda)) \tag{3.4}$$

continuous. The celebrated Weierstrass' Theorem now guarantees the existence of a minimum for $\pi_{DL}(y-\mathcal{G}(\lambda))$ on the compact support of π_{PR}, denoted by $\underline{\pi}$. Moreover, since the support is bounded, this minimum has to be strictly positive. Consequently, we estimate

$$Z \geq \underline{\pi} \int_{\mathbb{R}^n} \pi_{PR}(\lambda) d\lambda > 0, \tag{3.5}$$

where the latter inequality holds due to the properties of the truncated normal distribution. □

The Markov chain Monte Carlo method is designed to generate a Markov chain with associated stationary distribution that equals $P(\lambda|y)$. The main advantage lies in avoiding the computation of Z given in (3.1). This quantity is expensive to compute as it usually requires the application of Monte Carlo methods as well. Thus, we apply a method merely working with ratios of different posterior distribution, ensuring this factor to be canceled out. The respective algorithm is called Metropolis Hastings algorithm and defined next.

Data: Proposal density $\pi(\lambda'|\lambda)$, measured data y.
Initialization, i.e., choose $\lambda^{(0)} \in \mathbb{R}^{N_P}$, set $i = 0$.
Compute $\bar{\pi}_{PO}(\lambda^{(0)}) := \pi_{DL}(y - \mathcal{G}(\lambda^{(0)}))\pi_{PR}(\lambda^{(0)})$.
for $i \geq 0$ **do**

Draw proposed friction coefficient λ' from proposal density $\pi(\cdot|\lambda_{(i)})$.
Compute

$$\alpha = \min\left\{1, \frac{\bar{\pi}_{PO}(\lambda')\pi(\lambda'|\lambda^{(i)})}{\bar{\pi}_{PO}(\lambda^{(i)})\pi(\lambda_{(i)}|\lambda')}\right\}. \tag{3.6}$$

Set $\lambda^{(i+1)} = \lambda'$ with probability α and $\lambda^{(i+1)} = \lambda^{(i)}$ with probability $1 - \alpha$.

Algorithm 1 Metropolis Hastings algorithm

---

4 Numerical Realization

Within this section, two major points are pursued. On the one hand, the numerical realization of solving the coupled systems of time dependent partial differential equations is described. On the other hand, a set of examples for the method is provided and put into context with different approaches for gaining statistics and choices for the measurement positions.

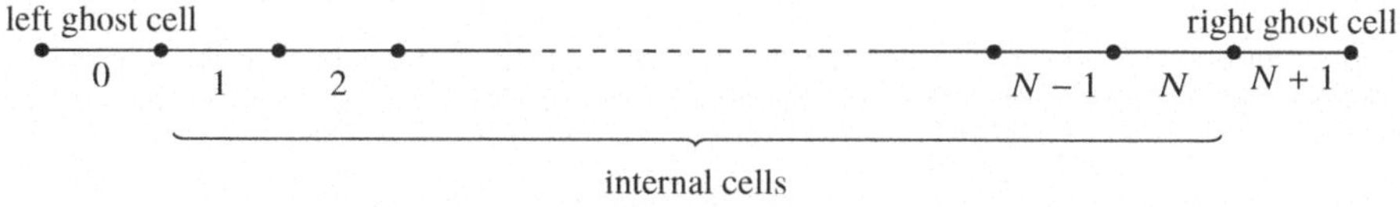

Fig. 3: Spatial discretization of a pipe whit ghost cells

Concerning the discretization of the state systems, we employ a numerical scheme that is based on piecewise constant averages of the state functions and also utilized in [13, 18]. These cell averages are computed on a uniform grid dividing pipes into N cells of width Δx, which are referred to as internal cells. The boundary and coupling conditions at entry and exit nodes and junctions are realized by ghost cells, i.e., each pipe has a ghost cell on both ends as depicted in Fig. 3.

The scheme is inspired by the particle method from [8] that is consistent with entropy solutions of scalar conservation laws and that was already used for semilinear systems of conservation laws in [14,15]. Note that the special structure of Eigenvalues for the differential operator under consideration renders the method more structures than initially intended, as it can be interpreted on a uniform grid. At time step n an explicit Euler method is used to approximate the solutions to (2.14) for each component of the transformed variables, i.e., for the first characteristic,

$$\mathcal{T}_1(y_{i-1}^{n+1}) = \mathcal{T}_1(y_i^n) + \tfrac{\Delta t}{c_a a}\lambda_i \frac{q_i^n |q_i^n|}{\rho_i^n},$$

and for $\mathcal{T}_2$, respectively. For equidistantly distributed particles with distance Δx, a time step $\Delta t = a^{-1}\Delta x$ and integral averages y_i^n and λ_i, the original state variables can be reconstructed by $L\mathcal{T}(y)$, yielding the discretization scheme

$$\begin{aligned}
\rho_i^{n+1} &= \tfrac{1}{2}(\rho_{i+1}^n + \rho_{i-1}^n) - \tfrac{c_{CFL}}{2}(q_{i+1}^n - q_{i-1}^n) - \tfrac{\Delta t}{2a}\left(\lambda_{i-1}\frac{q_{i-1}^n|q_{i-1}^n|}{\rho_{i-1}^n} - \lambda_{i+1}\frac{q_{i+1}^n|q_{i+1}^n|}{\rho_{i+1}^n}\right),\\
q_i^{n+1} &= \tfrac{1}{2}(q_{i+1}^n + q_{i-1}^n) - \tfrac{a^2 c_{CFL}}{2}(\rho_{i+1}^n - \rho_{i-1}^n) - \tfrac{\Delta t}{2}\left(\lambda_{i-1}\frac{q_{i-1}^n|q_{i-1}^n|}{\rho_{i-1}^n} + \lambda_{i+1}\frac{q_{i+1}^n|q_{i+1}^n|}{\rho_{i+1}^n}\right),
\end{aligned} \tag{4.1}$$

$i = 1, \ldots, N$, corresponding to the Lax-Friedrichs scheme, where c_{CFL} refers to the Courant number.

The similarity of the particle method to a classical TVD discretization scheme, mirroring the nature of broad solutions by integrating along characteristic lines, suggests a procedure for obtaining the values of the state at the ghost cells by employing the concept of broad solutions in the following way. At entry or exit nodes of the network, the updated value for the state is obtained by evaluating the ingoing transformed variable and solving the linear systems defined by the matrices C_L or C_R, respectively, i.e, to solve

$$C_L^{-1}\begin{pmatrix}\mathcal{T}_1(y_0^{n+1})\\ d_0(t^{n+1})\end{pmatrix} \text{ and } C_R^{-1}\begin{pmatrix}\mathcal{T}_2(y_{N+1}^{n+1})\\ d_{N+1}(t^{n+1})\end{pmatrix}$$

for boundary data d_0, d_{N+1} on the left and right ghost cell, respectively. Explicitly, this provides

$$\begin{aligned}
\rho_{0/N+1}^{n+1} &= \rho_{1/N}^n \mp \tfrac{1}{a} q_{1/N}^n \pm \tfrac{1}{a} d_{0/N+1}(t^{n+1}) \pm \tfrac{\Delta t}{a}\lambda_{1/N}\frac{q_{1/N}^n|q_{1/N}^n|}{\rho_{1/N}^n},\\
q_{0/N+1}^{n+1} &= d_{0/N+1}(t^{n+1}).
\end{aligned} \tag{4.2}$$

At junctions, all ingoing transformed variables and the coupling conditions form linear systems that have to be solved for the values of the state variables on the corresponding ghost cells. In case of a junction with three pipes, the system is given as (2.15).

Summarizing, we have established an algebraic expression that provides the updated state at all ghost cells g of the network, based on the current iterate and the boundary conditions at the next time step $d(t^{n+1})$, i.e.,

$$(\rho_g^{n+1}, q_g^{n+1}) = f_g(\rho^n, q^n, \lambda, d(t^{n+1})).$$

In general, the linear systems that have to be solved to determine the values of the states at the ghost cells are very small and constant over time. As a consequence, the most effective way to compute the associated solutions is given by precomputing the inverse matrices once and using them to compute the values on the ghost cells directly. Since we consider steady states in the initial conditions of the systems, we employ the closed form solutions of them, depending merely on the current value of the friction coefficient. The associated cell averages can be computed exactly from this closed forms.

In the following numerical experiments we utilized a mesh width of $\Delta x = 10$ and fixed the speed of sound to $a = 300$. Moreover, we considered a time horizon of $T = 200$ when solving the underlying system of partial differential equations and utilized a varying volume flow at the entry node of the corresponding networks given by

$$q(t, x_{In}) = 300 + 20 * sin((2\pi/50)t),$$

while the volume flow at the exit nodes is set constant, equaling the initial value of the volume flow in the adjacent pipe.

4.1 Discussion of numerical examples

In this section we are going to discuss two different network examples in order to illustrate the results of Algorithm 5 in particular situations. The first example is taken from Sect. 2.2, while the second example considers a slightly larger network.

Example 1

This example compares results from the Markov chain Monte Carlo approach presented in the work at hand and the statistical approaches for the example considered in [18]. The experimental setup consists of a plain network depicted in Fig. 1 with three pipelines, connected via a single junction. Artificial data have been generated by solving the underlying problem for the friction coefficient

$$\lambda_N = (0.018172, 0.018195, 0.018145)^\top$$

on a fine discretization with associated initial data, providing a reference solution $(\bar{\rho}, \bar{q})$. Here, the pipes are numbered counter-clock wise, beginning with the pipe at the entry node. In [18], an inverse problem for identifying the friction coefficient from noisy measurements was studied. To obtain statistical information on the results, the noise in the measurements was generated several times and the corresponding identification problems of the form

$$
\begin{aligned}
&\text{minimize} \\
&\tfrac{1}{2}\Big(\sum_{i=0}^{20}(\rho(10\cdot i, x_{E_1}) - \rho^d_{E_1}(10\cdot i))^2 + (\rho(10\cdot i, x_{E_2}) - \rho^d_{E_2}(10\cdot i))^2\Big) + \tfrac{\alpha}{2}\|\lambda - \lambda_M\|^2_{l^2} \\
&\text{subject to} \\
&(\rho^{(i)}, q^{(i)}) = \mathcal{S}(\lambda^{(i)}) \text{ on } (0,T)\times(x_L^{(i)}, x_R^{(i)}) \\
&(\rho^{(i)}(0,x), q^{(i)}(0,x)) \text{ solves (ISO-ALG) with given } q_0^{(i)} \text{ for } \lambda^{(i)} \\
&\rho(t, x_R^{(1)}) = \rho(t, x_L^{(2)}) = \rho_L^{(3)},\ \ q^{(1)}(t, x_R^{(1)}) = q^{(2)}(t, x_L^{(2)}) + q^{(3)}(t, x_L^{(3)}) \\
&q^{(1)}(t, x_L^{(1)}) = q^d_{In}(t), q^{(2)}(t, x_R^{(2)}) = q^d_{Exit_1}(t), q^{(3)}(t, x_R^{(3)}) = q^d_{Exit_2}(t) \\
&\rho^{(1)}(t, x_L^{(1)}) = 52.3 \\
&10^{-9} \le \lambda^{(i)} \le \overline{\lambda} \\
&\text{with} \\
&\rho^d_{E_1}(10\cdot i) = \bar{\rho}(10\cdot i, x_{E_1}) + \eta_i, \rho^d_{E_2}(10\cdot i) = \bar{\rho}(10\cdot i, x_{E_2}) + \mu_i
\end{aligned}
$$

were solved for variables $\eta, \mu \in \mathbb{R}^{20}$ that are component wise i.i.d. with respect to $\mathcal{N}(0, 10^{-3})$. This corresponds to repeating the same experiment several times and resulted in

$$E^{Inverse}[\lambda] = (0.018105, 0.018288, 0.018284)^\top$$

with

$$C^{Inverse} = cov(\lambda_i, \lambda_j) = \begin{pmatrix} 0.30719 & -0.42635 & -0.63932 \\ -0.42635 & 0.59173 & 0.88732 \\ -0.63932 & 0.88732 & 1.33057 \end{pmatrix} \cdot 10^{-6}$$

for expected value and covariance matrix, respectively.

For the Bayesian approach, we utilized the same physical set up, i.e., we considered the Y-shaped network presented in Fig. 1 with the same initial values, in particular, $q = (300, 120, 180)^\top$ and $\rho(x_{Entry}) = 52.3$. To demonstrate the influence of certain parameters that determine the Bayesian approach for solving the inverse problem, we investigate different scenarios in the latter case. First, we considered data that are, similar to the identification problem, based on λ_N, i.e., we are considering 'new' pipes in that their friction coefficient is take from the interval [0.01813, 0.018206]. Here, we compare the influence of the chosen prior distribution and present results for both possibilities sketched in Fig. 2 referred to as scenario 1 and 2, respectively. In the considered scenarios, the expected values are given by

$$E^{Bayes_1}[\lambda] = (0.018171, 0.018196, 0.018149)^\top,$$
$$E^{Bayes_2}[\lambda] = (0.018160, 0.018221, 0.018212)^\top,$$

where the associated covariance matrices is of order 10^{-10} and 10^{-8}, respectively. We refrain from presenting the full covariance matrices since in both examples the order of them is to small to influence the outcome in the numerical experiments of Sect. 5.2.

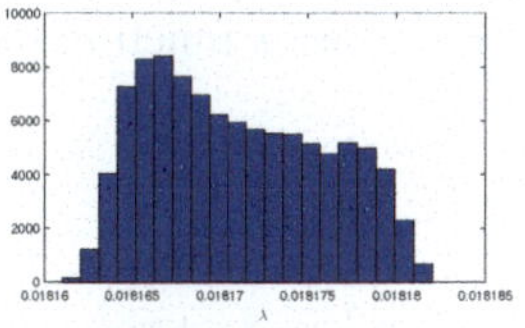

Pipe 1, Scenario 1 | Pipe 2, Scenario 1 | Pipe 3, Scenario 1

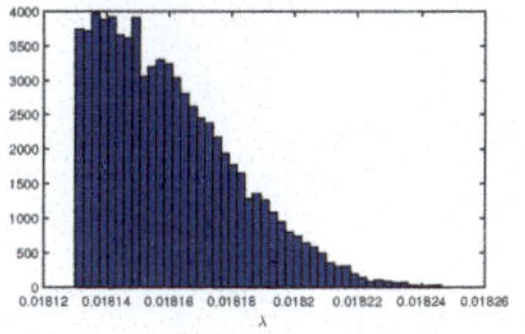

Pipe 1, Scenario 2 | Pipe 2, Scenario 2 | Pipe 3, Scenario 2

Fig. 4: Histograms of the resulting trajectories of Algorithm 1 for scenarios 1 and 2

Fig. 4 provides histograms of the components of the Markov chain generated by Algorithm 5, and thus, it gives an idea of the associated probability density functions. Moreover, we provide the Markov chain for scenario 1 in Fig. 5.

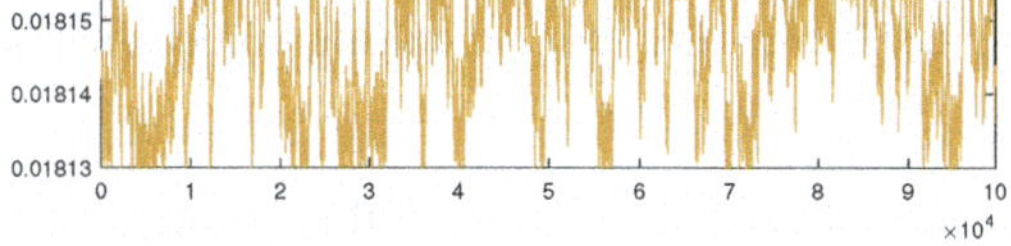

Fig. 5 Trajectory generated in Example 1 and scenario 1

Recall that the Bayesian approach is designed to gain information on the posterior probability density function, while the statistics based on the inverse problem from

[18] can at most be considered as fair approximation. Thus, the difference in the results of both approaches is not suspicious as the latter method, e.g. does not incorporate prior knowledge on the distribution of λ to the process. As it can be seen in the orders of the covariance matrices associated with scenario 1 and 2, the stochastic effect of the friction coefficient in freshly produced pipes is very small and has almost no influence on the problem of validating technical feasibility of stationary solutions for the underlying system under random demand.

However, to demonstrate, even when utilizing the Bayesian approach, that operating a network of pipelines introduces stochastic effects into problems as the validation problem, we in addition consider scenario 3 with the prior function for aged pipes and data that are generated with friction coefficients

$$\lambda_O = (0.02, 0.019, 0.0195)^\top,$$

i.e., we consider aged pipes, where λ became larger. In that situation, the results computed by Algorithm 5 turn out as

$$E^{Bayes_3}[\lambda] = (0.020244, 0.018660, 0.018977)^\top$$

and

$$C^{Bayes_3} = cov(\lambda_i, \lambda_j) = \begin{pmatrix} 0.30359 & -0.41560 & -0.61995 \\ -0.41560 & 0.58673 & 0.84921 \\ -0.61995 & 0.84921 & 1.30697 \end{pmatrix} \cdot 10^{-7}$$

for expected value and covariance matrix, respectively. In contrast to the previous scenarios, the order of the covariance matrix became larger, and thus, it also introduces effects to the probability level of feasibility sets (see Sect. 5.2). The histograms of the components of the Markov chain are depicted in Fig. 6. Here, we observe larger supports in all of them, compared with the previous histograms shown in Fig. 4.

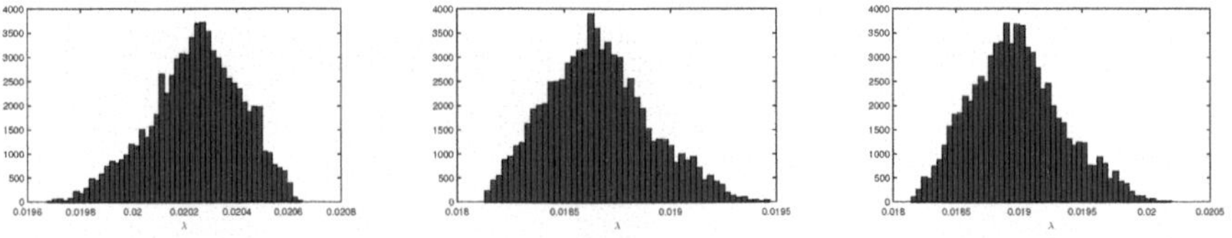

Pipe 1, Scenario 3 Pipe 2, Scenario 3 Pipe 3, Scenario 3

Fig. 6: Histograms of the resulting trajectories of Algorithm 1 for scenario 3

In all considered cases, we observed a characteristic sign-structure in the covariance matrix that is implied by the network topology. The coupling and boundary conditions enforce low values of λ in the pipes that are connected with the exit nodes, if the friction coefficient in the pipe connected to the entry node is large and vice versa, to compensate for density drops in the first pipe that are to large or to low, compared to the true solution.

Example 2

The second example aims on a larger network that is obtained by extending the network structure from Exampele 1 by further pipes as displayed in Fig. 7.

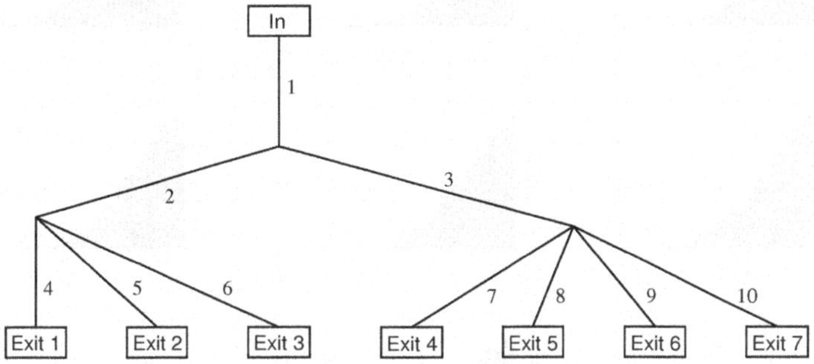

Fig. 7: Sketch of the network from Example 2

The network consist of 10 pipes of length 7500 with three interior, one entry and seven exit nodes. The initial state is based on the pipewise constant volume flow

$$q = (300, 120, 180, 30, 40, 50, 30, 40, 50, 60)^{\top}$$

and $\rho(x_{Entry}) = 52.3$. Again, measurements of the density are taken at the exit nodes at merely 10 points in time and compared to a perturbed reference solution generated on a fine mesh for the friction coefficients

$$\lambda_O = 10^{-2} \cdot (1.8172, 1.8195, 1.8145, 1.82, 1.816, 1.818, 1.817, 1.8152, 1.8175, 1.8141)^{\top} .$$

Compared to the setting of the previous example, we restrict ourselves only on scenario 2, i.e., obtaining information on newly produced pipes, but, with the prior density that also covers aged pipes. The results of Algorithm 5 are given by the expected values

$$E[\lambda] = 10^{-2} \cdot (1.8547, 1.8539, 1.8546, 1.8546, 1.8562, 1.8559, 1.8564, 1.8553, 1.8550, 1.8520)^{\top}$$

after 75000 iterations. Here, the expected value of the estimation differs slightly from λ_O.

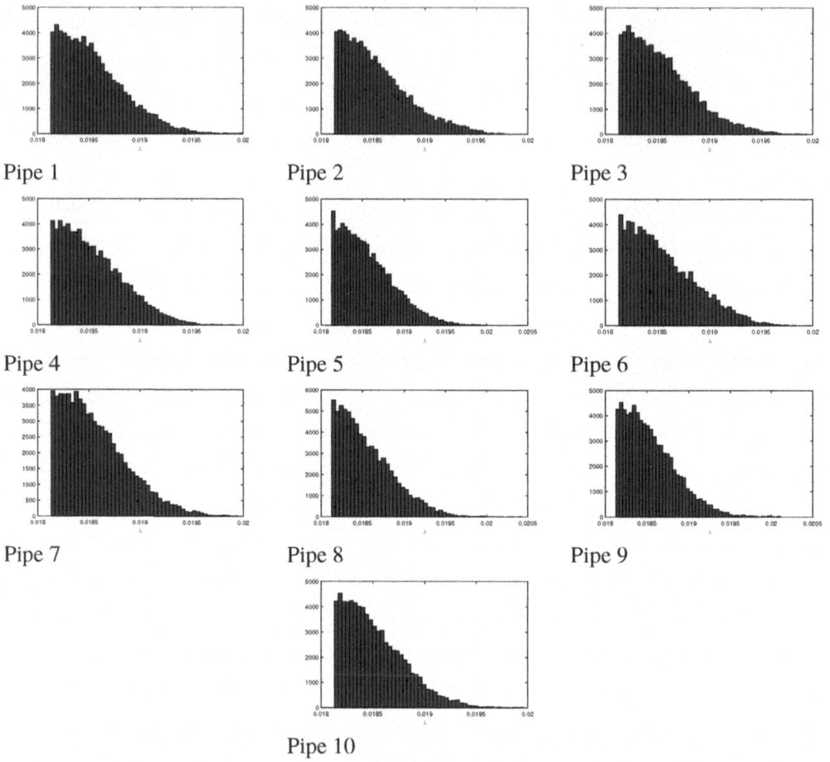

Fig. 8: Histograms of the resulting trajectories from Algorithm 1 for Example 2

Moreover, due to the obviously increased network complexity, we observe that the structure of the network can no longer be derived from the sign structure of the covariance matrix. The computed covariance matrix reads

$$cov(\lambda_i, \lambda_j) = 10^{-7} \cdot$$

$$\begin{pmatrix} 0.934 & -0.014 & 0.001 & -0.039 & -0.005 & 0.005 & 0.015 & -0.001 & 0.027 & -0.012 \\ -0.014 & 0.976 & 0.018 & 0.044 & 0.013 & 0.068 & 0.067 & 0.024 & -0.028 & -0.009 \\ 0.001 & 0.018 & 0.948 & -0.036 & -0.046 & -0.009 & -0.030 & 0.000 & 0.042 & -0.003 \\ -0.039 & 0.044 & -0.036 & 0.910 & 0.015 & -0.002 & -0.001 & 0.038 & 0.001 & -0.030 \\ -0.005 & 0.013 & -0.046 & 0.015 & 1.026 & -0.013 & 0.034 & -0.040 & 0.018 & -0.023 \\ 0.005 & 0.068 & -0.009 & -0.002 & -0.013 & 1.008 & -0.022 & -0.041 & -0.024 & 0.034 \\ 0.015 & 0.067 & -0.030 & -0.001 & 0.034 & -0.022 & 1.007 & -0.043 & -0.022 & -0.021 \\ -0.001 & 0.024 & 0.000 & 0.038 & -0.040 & -0.041 & -0.043 & 1.043 & 0.000 & -0.017 \\ 0.027 & -0.028 & 0.042 & 0.001 & 0.018 & -0.024 & -0.022 & 0.000 & 0.974 & 0.036 \\ -0.012 & -0.009 & -0.003 & -0.030 & -0.023 & 0.034 & -0.021 & -0.017 & 0.036 & 0.830 \end{pmatrix}$$

right now. The order of the covariance matrix is quite larger than before, even in scenario 2, where we look at newer pipes with less uncertainty. In Fig. 8 we provide the histograms for the considered scenario.

5 Determining probabilities of feasibility sets

Next, the random nature of the exit load vector as well as the uncertainty of friction along the pipes is taken into account. For simplification, as before we want to restrict to a tree shaped gas transport network involving a single entry node (labeled with 0) and a number of n exit nodes. Further we assume that the network is in steady state (see Sect. 2.1). The aim of this section is the computation of the probability of the event that a random load (or demand) vector is technical feasible under uncertain friction in the sense of (2.12). Since the demand vector (b_0, b) must be balanced, in the following we assume that $b_0 = -\mathbb{1}^\top b$, that is, the total exit demand can always be satisfied by the corresponding supply at the single entry node. By doing so, the following set of feasible pairs of exit load vectors and friction coefficients becomes relevant

$$\tilde{M}_{\text{feas}} := \left\{(b, \lambda) \in \mathbb{R}^{n \times m} \,\middle|\, g_{k,l}(b, \lambda) \geq 0;\ k, l = 0, \ldots, n;\ k \neq l\right\}, \tag{5.1}$$

where technical feasibility can be formulated by a set of constraint mappings $g_{k,l}(\cdot, \cdot)$ arriving in a natural way from (2.12). In particular, we have that

$$g_{k,l}(b, \lambda) := (p_k^{max})^2 + h_k(b, \lambda) - (p_l^{min})^2 - h_l(b, \lambda), \tag{5.2}$$

where $h_k(\cdot, \cdot)$ taken from (2.11), $k, l = 0, \ldots, n$ and $k \neq l$. More precisely, if (b, λ) is identified with some random vector $\big(b(\omega), \lambda(\omega)\big)$ on a probability space $(\Omega, \mathcal{A}, \mathbb{P})$, then

$$\mathbb{P}\left\{\omega \,\middle|\, \big(b(\omega), \lambda(\omega)\big) \in \tilde{M}_{\text{feas}}\right\} \tag{5.3}$$

marks the probability of exit demand vectors to be feasible in the context of uncertain friction. The main variation of exit load data is temperature driven. However, even at fixed temperature, considerable random variation remains. That is why exit loads can be understood as a stochastic process depending on temperature and may be characterized by a finite family of multivariate distributions, each of them referring to some (rather narrow) range of temperature and reflecting the joint distribution of loads at the given set of exit points, see [19, Chapter 13].As recorded in the same reference [19, Table 13.3], these distributions are most likely to be Gaussian (possibly truncated) or lognormal. Our assumption to consider a multivariate Gaussian distribution for b can therefore be seen as a prototype setting which maybe adapted without much effort to more realistic settings (multivariate log-normal distributions etc.). As well as the demand vector the uncertainty of friction coefficients approximately follows a multivariate Gaussian distribution whose parameter can be estimated as seen before. Thus, we assume that

$$b \sim \mathcal{N}(\mu_1, \Sigma_1) \qquad \text{and} \qquad \lambda \sim \mathcal{N}(\mu_2, \Sigma_2), \tag{5.4}$$

where μ_1, μ_2 and Σ_1, Σ_2 denote mean values and covariance matrices of the demand and friction random vectors, respectively. Clearly, by formula (5.1) we could use the final inequality system in order to test feasibility of simulated outcomes of the pairs (b,λ) according to the given Gaussian distributions. The average number of feasible simulations would yield the Monte Carlo estimate for the desired probability in (5.3). Such Monte Carlo approach has two drawbacks: first it may come up with a comparatively large variance for the obtained probability estimation and, second, it does not provide us with information about the sensitivity of this probability with respect to changes of external parameters, that could subject of optimization. This sensitivity (derivative) information is crucial, however, in order to set up any efficient algorithm of nonlinear optimization in order to solve optimization problems in the context of gas transmission, e.g. the maximization of booking capacities [16].

5.1 Spheric-radial decomposition

Instead of crude Monte Carlo sampling we rather propose here the so-called spheric-radial decomposition of Gaussian random vectors (see, e.g. [3,9]). This alternative not only may significantly reduce the variance of probability estimations but, moreover, it offers the possibility of efficiently approximating gradients of (5.3) with respect to external network parameters. This last feature is of supreme importance for optimization problems under probabilistic constraints [23].

Theorem 5.3 *Let (b,λ) be a Gaussian random vector distributed according to* (5.4). *Then for the probability of random load and random friction being technical feasible it holds that*

$$\mathbb{P}\left((b,\lambda)\in\tilde{M}_{\text{feas}}\right)=\int\limits_{(v_1,v_2)\in\mathbb{S}^{n+m-1}}\mu_\chi\left\{r\geq 0\,\middle|\,(rL_1v_1+\mu_1,rL_2v_2+\mu_2)\in\tilde{M}_{\text{feas}}\right\}d\mu_\eta,$$

where matrices L_i are such that $\Sigma_i=L_iL_i^\top$ (e.g., Cholesky decomposition), $i=1,2$, and, μ_χ is the law of chi-distribution with $n+m$ degrees of freedom and μ_η is the law of uniform distribution at the Euclidean unit sphere $\mathbb{S}^{n+m-1}$.

In order to evaluate the integrand in the spheric integral above, for any fixed direction $(v_1,v_2)\in\mathbb{S}^{n+m-1}$, one has to compute the χ-probability of the one-dimensional set

$$\{r\geq 0\,|\,(rL_1v_1+\mu_1,rL_2v_2+\mu_2)\in\tilde{M}_{\text{feas}}\}.$$

Thus, using (5.1) computing the probability of feasibility amounts to characterizing the set

$$\{r\geq 0\,|\,g(rL_1v_1+\mu_1,rL_2v_2+\mu_2)\geq 0\}\qquad(v\in\mathbb{S}^{n+m-1}),\tag{5.5}$$

where we define

$$g(b,\lambda):=\min_{\substack{k,l=0,\ldots,m\\k\neq l}}\left\{(p_k^{\max})^2+h_k(b,\lambda)-(p_l^{\min})^2-h_l(b,\lambda)\right\}.\tag{5.6}$$

Applying the idea of spheric-radial decomposition presented in Theorem 5.3, we propose the following algorithm to computing the probability $\mathbb{P}\big((b,\lambda) \in \tilde{M}_{\text{feas}}\big)$.

Data: Let be (b, λ) random vector according to (5.4).
Set $S = 0$ and sample N points $\{v^1, \ldots, v^N\}$ uniformly distributed on the sphere $\mathbb{S}^{n+m-1}$.
for $i = 1, \ldots, N$ **do**
  Find the zero's of the one dimensional function
  $\theta(r) := g(rL_1 v_1^i + \mu_1, rL_2 v_2^i + \mu_2)$
  with g defined in (5.6) and represent the set
  $M^i := \{r \geq 0 \mid \theta(r) \geq 0\}$
  corresponding to (5.5) as a disjoint union of intervals
  $M^i = \cup_{j=1}^{s} [\alpha_j, \beta_j]$,
  where α_j, β_j are the zero's obtained before and ordered appropriately.
  Compute the χ-probability of M^i according to
  $\mu_\chi(M^i) = \sum_j F_\chi(\beta_j) - F_\chi(\alpha_j)$,
  where F_χ refers to the cumulative distribution function of the one-dimensional χ-distribution with $n + m$ degrees of freedom.
  Put $S := S + \mu_\chi(M^i)$.
Set $\mathbb{P}\big((b,\lambda) \in \tilde{M}_{\text{feas}}\big) := S/N$.

Algorithm 2 Spheric-radial decomposition

A few words on this algorithm are in order at this place. The algorithm clearly provides an approximation to the spheric integral in Theorem 5.3 by means of a finite sum based on sampling of the sphere, and then, averaging the values of the integrand over all samples. Of course, this approximation will improve with the sampling size which may be large depending on the dimension $n + m$ of the problem (i.e., exit nodes and edges in the network) and on the desired precision for the probability.

We recall that the uniform distribution on the sphere $\mathbb{S}^{n+m-1}$ can be represented as the distribution of $\eta/\|\eta\|$ (Euclidean norm), where η has a standard Gaussian distribution in $\mathbb{R}^{n+m}$, i.e., $\eta \sim \mathcal{N}(0, I)$. Then, the simplest idea to sample points v^i on the sphere would be to independently sample $m + n$ values w_j of a one-dimensional standard normal distribution by using standard random generators and then putting $v^i := w/\|w\|$ for $w := (w_1, \ldots, w_{n+m})$. When replacing such Monte Carlo sampling of the normal distribution (based on random number generators) by Quasi-Monte Carlo sampling (based on deterministic low discrepancy sequences), one observes a dramatic improvement in the precision of the result. For the problem of nomination validation in gas networks (with fixed friction coefficients), this was revealed in [11].

5.2 Preliminary numerical results

A considerably numerical study with respect to computations of probabilities of technical feasibility in stationary networks under stochastic exit demand can be found in [11]. However, the impact of uncertainty of friction along the pipes to the probability level of feasible exit load situation has not been investigated so far. Therefore, the aim of the numerical study in this paper is to incorporate the effect of random friction into the consideration in [11]. For our numerical tests we proceed with the two example networks already discussed in Sect. 4.1. Even if we consider tree shaped networks only, the results can easily transferred to more complex situation involving cycles within the network topology.

Example 1

For the purpose of illustration we start with the small network serving as example for computing the probability of feasibility as in (5.3). The network consists of one entry node, one passive (interior) node and two exit nodes, where we consider a stochastic gas demand. The shape of the network is given by Fig. 1. Observe, that the passive node can be formally modeled as an exit node, but, with zero consumption. As there exist 3 arcs joining the nodes, we have three frictional driven roughness coefficients $\Lambda_e = 2a^2 l_e \lambda_e$, where $e = 1, 2, 3$ (cf. (2.2)).

We like to mention here that this example network already was considered in [10], where a numerical study of the influence of uncertain friction (roughness) is provided. But, compared to the propose of this paper, no distribution information for friction is used. Instead, a complete robust approach for describing uncertainty with respect to friction along the pipes is applied.

Table 1: Distribution for the exit demand of Example 1

Mean μ_1	Covariance Σ_1
$\begin{pmatrix} 120.0 \\ 180.0 \end{pmatrix}$	$\begin{pmatrix} 1\,600.0 & 800.0 \\ 800.0 & 4\,000.0 \end{pmatrix}$

Table 2: Fixed network parameter in Example 1

Node	Pressure p^{min}	Pressure p^{max}
Entry	40.0	52.0
Interior	40.0	52.0
Exit 1	40.0	52.0
Exit 2	40.0	52.0

In this paper we are going to base the computations directly on the distribution information of friction estimated by the Markov chain Monte Carlo method in Sect. 4.1. Because the variance with respect to the friction coefficients in case of new pipes turn out to be too small, we provide the computations for Example 1 only for the results of Algorithm 5 when considering aged pipes. In particular, for the multivariate Gaussian friction coefficient $\lambda \sim (\mu_2, \Sigma_2)$ the mean value is set to $\mu_2 = E^{Bayes_3}[\lambda]$ with corresponding covariance matrix $\Sigma_2 = C^{Bayes_3}$ (see Sect. 4.1, Example 1). The length of the pipes is set to $l = (10\,000, 20\,000, 30\,000)$. In addition, the parameter for the bivariate Gaussian distributed exit demand vector $b \sim \mathcal{N}(\mu_1, \Sigma_1)$ as well as the fixed network parameters are displayed in Table 1 and Table 2. The pressure limits are given in bar, whereas the exit demand is given in volume flow (as before). Note that exit demand is frequently given in thermal power P instead of mass or volume flow q by network owners. However, the relation between thermal power and mass flow is given by the equation $P = qH_c$, where H_c refers to the calorific value of the gas.

For our study how the uncertainty of friction affects technical feasibility, we performed a test series to compute the probability of feasibility with respect to stochastic exit demand. In particular, we compare different random friction situation by randomly chosen samples simulated from the distribution. In addition, we compare the results with the computations that we obtain when using the friction mean value μ_2, the friction mean value plus standard deviation $\mu_2 + \sigma_2$, and the friction mean value minus standard deviation $\mu_2 - \sigma_2$, in order to incorporate the friction part according to (5.6). Finally, we determine the probability of technical feasibility in a fully stochastic environment, i.e., where we assume that both friction and exit load follow the stated multivariate Gaussian distributions.

Table 3: Impact of uncertain friction to the technical feasibility in Example 1. The table compares the probabilities of the obtained feasibility sets in different situations with respect to friction, computed via spheric-radial decomposition.

	Random scenarios			Reference scenarios			Stochastic
Samples	$\lambda_{[1]}$	$\lambda_{[2]}$	$\lambda_{[3]}$	μ_2	$\mu_2 + \sigma_2$	$\mu_2 - \sigma_2$	$\mathcal{N}(\mu_2, \Sigma_2)$
1 000	70.417	68.917	69.227	69.699	68.787	70.620	69.694
5 000	70.320	68.816	69.126	69.600	68.685	70.524	69.521
10 000	70.319	68.818	69.127	69.601	68.688	70.522	69.549
50 000	70.301	68.792	69.103	69.579	68.662	70.504	69.580
100 000	70.297	68.789	69.100	69.575	68.659	70.500	69.580

Table 3 summarizes the numerical results. All computations were performed by Algorithm 6. The number of samples on the unique sphere varies between 1 000 and 100 000 samples. Clearly, the accuracy increases with a higher number of samples on the sphere used for the approximation. However, we obtain a notable effect of using different random outcomes for the friction coefficients (column 2 to 4). We obtain a variation by the computed probability levels of about 2 percent compared to the average value of nearly 70 percent (column 5). On the other hand, in that example problem we realize that a fully stochastic approach for both friction and load distribution (last column) does not return a significant difference in contrast to proceeding with stochastic exit demand and the mean value for friction. Considering the small network, we conclude that a accurate approximation of the mean value for the friction part seems to be sufficient for a quite fair approximation of the probability level of feasibility sets with respect to random exit demand.

Example 2

Now we discuss Example 2 from Sect. 4.1. The underlying network can be viewed as enlargement of the network in Example 1 by adding additional pipes end nodes to the net as illustrated in Fig. 7. We base our computations directly on the estimated data from Sect. 4.1 (Example 2) for the friction coefficients. In a different way from Example 1 the following numerical results are related to non-aged pipes. However, for the computations we proceed in almost the same manner as before. The length of pipes we slightly enlarge to $l_e = 10\,000$ for all involved pipes $e = 1, \dots, 10$. The Gaussian distribution for the friction comes up with the Markov chain Monte Carlo estimation by mean $\mu_2 = E[\lambda]$ and by covariance $\Sigma_2 = cov(\lambda_i, \lambda_j)$ (see Sect. 4.1, Example 2). All additional parameters, in particular, the assumed Gaussian distribution for the exit demand are displayed by Table 4 and Table 5.

Table 4: Distribution parameter for the exit demand for Example 2

Mean μ_1	Covariance Σ_1
$\begin{pmatrix} 30.0 \\ 40.0 \\ 50.0 \\ 30.0 \\ 40.0 \\ 50.0 \\ 60.0 \end{pmatrix}$	$\begin{pmatrix} 64.0 & 8.0 & 32.0 & -16.0 & 8.0 & -16.0 & 16.0 \\ 8.0 & 82.0 & 31.0 & 7.0 & 19.0 & -29.0 & -7.0 \\ 32.0 & 31.0 & 146.0 & -16.0 & -23.0 & 5.0 & 16.0 \\ -16.0 & 7.0 & -16.0 & 42.0 & 15.0 & -7.0 & 0.0 \\ 8.0 & 19.0 & -23.0 & 15.0 & 99.0 & 20.0 & 17.0 \\ -16.0 & -29.0 & 5.0 & -7.0 & 20.0 & 98.0 & 0.0 \\ 16.0 & -7.0 & 16.0 & 0.0 & 17.0 & 0.0 & 112.0 \end{pmatrix}$

Table 5: Fixed network parameter in Example 2

Node	Pressure p^{min}	Pressure p^{max}
Entry	45.0	52.0
Interior 1–3	45.0	52.0
Exit 1–7	45.0	52.0

As in Example 1, the considered test series aim to approximate the probability of feasibility under stochastic exit demand in different situation with respect to friction.

Again, the computations include three randomly selected friction samples simulated from the distribution, results that are obtained when using the friction mean value μ_λ, the friction mean value plus standard deviation $\mu_\lambda + \sigma_\lambda$, and the friction mean value minus standard deviation $\mu_\lambda - \sigma_\lambda$. As well as considering these fixed friction vectors we determine the probability of technical feasibility in a fully stochastic environment once more, i.e., where we assume that both friction and exit load follow the given multivariate Gaussian distributions.

Table 6: Numerical results, related to Example 1. The table compares the probabilities of the obtained feasibility sets in different situations with respect to friction, computed via spheric-radial decomposition.

	Random scenarios			Reference scenarios			Stochastic
Samples	$\lambda_{[1]}$	$\lambda_{[2]}$	$\lambda_{[3]}$	μ_2	$\mu_2 + \sigma_2$	$\mu_2 - \sigma_2$	$\mathcal{N}(\mu_2, \Sigma_2)$
1 000	93.140	95.015	91.574	93.376	91.950	94.629	93.429
5 000	93.309	95.121	91.780	93.534	92.146	94.748	93.513
10 000	93.211	95.053	91.667	93.439	92.036	94.673	93.449
50 000	93.268	95.061	91.768	93.489	92.126	94.690	93.430
100 000	93.252	95.049	91.749	93.473	92.107	94.677	93.417

Table 6 shows the very similar results of all computations, but, for Example 2 now. As before, applying randomly selected friction coefficients obviously guides to a falsified estimation of the probability of feasibility. Only a proper estimation of the average friction can serve as a good base for accurate computation. Nevertheless, in the environment of the enlarged network topology also the calculation wit the mean value differs slightly from the fully stochastic result. Using the mean value seems to overestimate the probability level compared with the stochastic approach. The latter observation gives a hint to what happens when taking more involved networks with a much higher number of pipes and nodes into the consideration. In such a case we expect that the usage of expected friction only causes a to optimistic overestimation of the probability level for the feasibility of random exit demand.

Acknowledgements The authors wish to thank the German Research Foundation (DFG) for their support within projects B02, B04 of TRR 154 of the Collaborative Research Centres (SFB).

References

1. A. Bressan. *Hyperbolic systems of conservation laws*, volume 20 of *Oxford Lecture Series in Mathematics and its Applications*. Oxford University Press, Oxford, 2000. The one-dimensional Cauchy problem.
2. M. Dashti and A. M. Stuart. *The Bayesian Approach to Inverse Problems*, pages 311–428. Springer International Publishing, 2017.

3. I. Déak. Subroutines for computing normal probabilities of sets-computer experiences. *Annals of Operations Research*, 100:103–122, 2000.
4. T. J. Dodwell, C. Ketelsen, R. Scheichl, and A. L. Teckentrup. A hierarchical multilevel Markov chain Monte Carlo algorithm with applications to uncertainty quantification in subsurface flow. *SIAM/ASA J. Uncertain. Quantif.*, 3(1):1075–1108, 2015.
5. P. Domschke. *Adjoint-Based Control of Model and Discretization Errors for Gas Transport in Networked Pipelines*. PhD thesis, TU Darmstadt, 2011.
6. P. Domschke, B. Hiller, J. Lang, and C. Tischendorf. Modellierung von Gasnetzwerken: Eine Übersicht. Technical Report 2717, Technische Universität Darmstadt, 2017.
7. H. Egger, T. Kugler, and N. Strogies. Parameter identification in a semilinear hyperbolic system. *Inverse Problems*, 33(5):055022, 25, 2017.
8. Y Farjoun and B. Seibold. Solving one dimensional scalar conservation laws by particle management. *Meshfree methods for partial differential equations IV*, 65:95, 2008.
9. A. Genz and F. Bretz. *Computation of multivariate normal and t-probabilities*, volume 195 of *Lecture Notes in Statistics*. Springer, Heidelberg, 2009.
10. T. González Grandón, H. Heitsch, and R. Henrion. A joint model of probabilistic/robust constraints for gas transport management in stationary networks. *Computational Management Science*, 14:443–460, 2017.
11. C. Gotzes, H. Heitsch, R. Henrion, and R. Schultz. Feasibility of nominations in stationary gas networks with random load. *Mathematical Methods of Operations Research*, 84:427–457, 2016.
12. M. Gugat, M. Herty, and V. Schleper. Flow control in gas networks: exact controllability to a given demand. *Math. Methods Appl. Sci.*, 34(7):745–757, 2011.
13. S. Hajian, M. Hintermüller, C. Schillings, and N. Strogies. A Bayesian approach for parameter identification in gas networks. Technical report, WIAS Berlin, 2018. Preprint.
14. F. M. Hante. *Hybrid Dynamics Comprising Modes Governed by Partial Differential Equations: Modeling, Analysis and Control for Semilinear Hyperbolic Systems in One Space Dimension*. PhD thesis, University Erlangen-Nuremberg, 2010.
15. F. M. Hante and G. Leugering. Optimal boundary control of convention-reaction transport systems with binary control functions. In *HSCC*, pages 209–222. Springer, 2009.
16. H. Heitsch. On probabilistic capacity maximization in a stationary gas network. Technical report, WIAS Berlin, 2018. Preprint.
17. M. Herty, J. Mohring, and V. Sachers. A new model for gas flow in pipe networks. *Math. Methods Appl. Sci.*, 33(7):845–855, 2010.
18. M. Hintermüller and N. Strogies. On the identification of the friction coefficient in a semilinear system for gas transport through a network. Technical report, WIAS Berlin, 2018. Preprint, Submitted.
19. T. Koch, B. Hiller, M. E. Pfetsch, and L. Schewe, editors. *Evaluating gas network capacities*, volume 21 of *MOS-SIAM Series on Optimization*. Society for Industrial and Applied Mathematics (SIAM), Philadelphia, PA, 2015.
20. R. J. LeVeque. *Finite volume methods for hyperbolic problems*. Cambridge Texts in Applied Mathematics. Cambridge University Press, Cambridge, 2002.
21. B. L. Roždestvenskiĭ and N. N. Janenko. *Systems of quasilinear equations and their applications to gas dynamics*, volume 55 of *Translations of Mathematical Monographs*. American Mathematical Society, Providence, RI, 1983. Translated from the second Russian edition by J. R. Schulenberger.
22. J. Stolwijk and V. Mehrmann. Error analysis and model adaptivity for flows in gas networks. *Analele Stiintifice Univ. Ovidius Constanta. Seria Matematica*, 2017. Accepted for publication.
23. W. v. Ackooij and R. Henrion. (Sub-) gradient formulae for probability functions of random inequality systems under Gaussian distribution. *SIAM/ASA Journal on Uncertainty Quantification*, 5:63–87, 2017.

Probabilistic Methods for Spatial Multihop Communication Systems

Benedikt Jahnel and Wolfgang König

Abstract We present and comment some recent modeling and results from stochastic geometry about the functionality of spatial communication networks with a multihop system of message transmissions. Our novel approaches concern connectivity on random street systems, frustration probabilities for service quality under constraints with regard to interference and capacity, and a new model of random message routing. Our main focus is on the description of the influence of spatial aspects, predominantly the locations of all the users. As a leading mathematical tool, we introduce the probabilistic theory of large deviations to the study of such systems in a high-density situation.

1 Introduction

Stochastic geometry is applied to the modeling of spatial ad hoc communication networks for decades. It is even fair to say that such networks triggered the development of one of the fundamental and ubiquitous mathematical models, the *Boolean model*, and connectivity questions about random point configurations in space, the *continuum percolation theory*. The obvious interpretation is that the points are the locations of the users, and a message can make a hop from one of these points to another one if their distance is below a certain threshold. In this way, a geometric random graph appears. The main characteristic of an ad hoc network is that the hops of the message are iterated, such that it may travel potentially a long way from node to node. The connectivity structure of this graph is decisive for many aspects of the service quality of the communication network.

Benedikt Jahnel
Weierstrass Institute, Mohrenstr. 39 in 10117 Berlin, e-mail: jahnel@wias-berlin.de

Wolfgang König
Weierstrass Institute, Mohrenstr. 39 in 10117 Berlin, e-mail: koenig@wias-berlin.de, and
TU Berlin, Str. des 17. Juni 136 in 10623 Berlin

M. Hintermüller, J. F. Rodrigues (eds.), *Topics in Applied Analysis and Optimisation*,
CIM Series in Mathematical Sciences, https://doi.org/10.1007/978-3-030-33116-0_10

In such a system, the users also act as relays and there are *a priori* no base stations necessary, i.e., there is no node that all the messages have to travel to or from. Therefore, the system is decentralized, and its functionality does not depend on a small subset of components. Furthermore, it can potentially handle a larger amount of information than cellular systems that are based on base stations. Since base stations are very expensive, ad hoc systems are also significantly cheeper. These three advantages of ad hoc systems are known since decades. However, as a matter of fact, ad hoc systems have hardly ever been installed in reality on a large scale and there are multiple possible reasons for that. One may be that, due in part to technical restrictions, there is an upper bound for the number of hops that a message can make in order to avoid too much loss of service quality such as transmission errors or time delay. This is a serious restriction that indirectly also puts an upper bound to the spatial size of an ad hoc system. Therefore, one ansatz is to consider a hybrid system of thinly distributed base stations with a local ad hoc system around each base station. Some of the models that we discuss later start from this idea and consider only one of these ad hoc cells with just one base station in the centre.

Another reason that ad hoc systems are not operative on a large scale may be theoretical problems in the details of the technical realization of the functionality. This in turn may be due to a lack of reliable mathematical assertions about their properties, e.g., concerning the two main aspects of interference and capacity, but also optimal-routing questions, security aspects or compatibility with legal rules, or even more refined characteristics such as user-friendliness. Hence, there is a lot of mathematical research activity on many aspects of ad hoc communication systems since decades and this will certainly continue in the future.

From a mathematical perspective, many relevant questions about the functionality of ad hoc networks have been raised and answered, and a number of variants and more structured versions of the above model have been introduced and studied. A rich list of mathematical models and tools has been created over the decades, and important properties have been stated and proved, and there is no end to this development. In particular, the rich discipline of probability theory is employed in many ways and is considered to be very successful. It is the purpose of this article to report on some of the latest developments in these directions, which center around the following questions:

- How can one describe connectivity properties in random environments, e.g., on a random street system?
- How small is the probability for bad quality of service of the type that too large a portion of users do not have a connection in the presence of constraints with respect to interference and capacity?
- How can one describe the global message flow in the system, if the message trajectories are randomly controlled just by aiming at a low interference?

Each of these questions is innovative for its own sake and is born out of its own motivation. The first one aims at a more realistic modeling of the topology of the communication area, the second studies the probabilities of frustration events, i.e., unwanted events that are likely to occur very rarely, and the third one discusses a new ansatz for random message routing, controlled by geometric relevant properties. These three kinds of questions concern three of the most important aspects of the functionality of an ad hoc system: The theoretical possibility to send messages through the system, i.e., connectivity; interference and capacity; and finally routing of many messages at the same time. In the treatment of the latter two, we will introduce a mathematical theory to the study of telecommunication systems that has hardly been used there yet, see the end of this section.

In the research that we describe here, it is not the aim to deliver complete solutions and optimal numerical values for any kind of relevant quantity, but to reveal fundamental mechanisms, interrelations and critical effects that arise naturally in such systems. Therefore, we want to keep the amount of parameters as low as possible, such that we can concentrate on a certain aspect and can rigorously explain and quantify its influence on the system. For the sake of this, we decided to keep the aspect of temporal evolution low and assumed in most of the models that all the messages are emitted at one fixed time and finish the entire trajectory during this time unit. Further simplifications are dealt with, e.g., the existence of just one channel between any two nodes.

On the other hand, the spatial aspect and the description of the entire ensemble of user locations is important to us and taken great care of; all the final descriptions of the models decisively depend on these locations; they are important ingredients of our analysis. Since we aim at the description of large systems, we consider limits of large user numbers. Here one has two fundamental settings: The high-density limit, in which the diverging number of users are located in a fixed, compact area and form a cloud profile, and the thermodynamic limit, where the area grows with the user number such that the average number of users per unit area is kept fixed. In this way, we will obtain macroscopic pictures from microscopic models.

A purpose of this text is to demonstrate the use and the value of the mathematical *theory of large deviations* for the investigation of some important questions about spatial random networks. This theory is a branch of asymptotic probability theory and investigates the exponential decay rates of the probabilities of rare events. The prototype of such events are events of a large deviation of a convergent random quantity (e.g., in a law of large numbers) away from its limit, which explains the name of the theory. One strong point of this theory is that it does not only give analytic formulas that express the decay rate, but also explain the 'typical', the most likely way to realize the rare event. In connection with spatial mathematical models for telecommunication networks, there is a variety of ways to employ this theory, e.g., for the analysis of events of a bad quality of service in the two above mentioned limiting settings of a large user number. The converging random quantity there is the point cloud of locations of all the users.

The remaining part of this text is organized as follows. In Section 2, we give a brief account on the basics of the mathematical modeling of ad hoc systems, and in Sections 3, 4, and 5, we describe the three respective questions formulated above and the answers in terms of mathematical results.

2 Basics of the mathematical modeling

This section is to set down notations and to introduce to the thinking and to some basic mathematical methods of stochastic geometry. See [2, 3, 8] for general introductions to this subject. We also explain the basics of the theory of large deviations. We tried to make the paper self-contained in the sense that the notation is used consistently in all the sections throughout the paper.

2.1 Location of users: The Poisson point process

We fix a communication area, $W \subset \mathbb{R}^d$, which may be also unbounded, e.g., $W = \mathbb{R}^d$. In W, we consider a collection of points $X = (X_i)_{i \in I}$, which is meant to model the locations of the users. We assume that I is at most countable, for bounded W even finite. Furthermore, we would like that the points accumulate nowhere and that $X_i \neq X_j$ for $i \neq j$. It is our goal to consider mainly large systems, i.e., large I, and we cannot assume that we know all the details of all the locations of the users. Therefore the approach of the choice is to pick X randomly, and all the X_i are assumed to be independently distributed. That is, we will assume that X is a *Poisson point process* with a σ-finite Borel measure μ on W as its intensity measure. This means that, for any choice of mutually disjoint bounded sets $A_1, \dots, A_m \subset W$, the number N_k of the points $X_i \in A_k$, $k = 1, \dots, m$, forms an independent collection of random variables, and each N_k is Poisson-distributed with parameter $\mu(A_k)$.

A particular role is played by μ equal to λLeb, where Leb is the Lebesgue measure (restricted to W), which makes the spatial distribution of X stochastically homogeneous. The scalar quantity $\lambda > 0$ is referred to as the intensity. One needs to choose more individual intensity measures μ if one wants to model rural or urban areas or lakes, forests, highly frequented places and so on. In Section 3 we will choose μ randomly and obtain by definition a *Cox point process*.

This high degree of spatial independence makes the model mathematically more easily tractable. The Poisson point process has been established as the standard basic model for modeling countably many independent random locations of points without accumulation points. It is easy to perform simulations of such a process, as it has the property that, once the Poisson number I of points is fixed, all the X_i are independent sites with distribution $\mu/\mu(W)$, if $\mu(W) < \infty$.

2.2 Connectivity: Continuum percolation

Here we give a short introduction to the cornerstone theory for the spatial modeling of telecommunication systems. One of the standard references is [19].

Every user X_i can transmit a message, and another user X_j can receive it if the distance is not larger than a given threshold $R > 0$. In order to account for this, we draw a straight line from X_i to X_j in this case for any i and j and obtain a geometric graph, called the *Gilbert graph*, with vertex set X, which describes the connectivity structure of the network. Now we can talk in a natural way in terms of the notion of connectedness. The connected components of the geometric graph are sometimes also called just components or clusters. The same notion applies to the random subset $Z_R = \bigcup_{i\in I} B_R(X_i)$, where $B_R(x)$ is the open ball with radius R centered at x. Then Z_R is the *communication area*, the set of space points that can be reached by at least one of the users with a signal. The random set Z_R is the most elementary representative of the *Boolean model*, one fundamental object of stochastic geometry.

For the important special case of $W = \mathbb{R}^d$ and $\mu = \lambda\mathrm{Leb}$ for some $\lambda \in (0, \infty)$, one introduces the quantity $\theta(\lambda, R)$, the *percolation probability*, the probability that the area $Z_R \cup B_R(o)$ has an unboundedly large connected component containing the origin o. If $\theta(\lambda, R) > 0$, then a user placed at the origin can transmit a message over an unbounded distance with positive probability. (There is a similar notion of percolation requiring only that the origin lies in an unbounded component of Z_R, but we do not refer to it.) By simple scaling arguments, it is clear that we may put $R = 1$ and consider $\theta(\lambda) = \theta(\lambda, 1)$. It is clear that θ is an increasing function of λ. Furthermore, there is a critical parameter, the *percolation threshold* λ_c, such that $\theta(\lambda) = 0$ for $\lambda < \lambda_c$ and $\theta(\lambda) > 0$ for $\lambda > \lambda_c$. It is one of the fundamental, non-trivial results of continuum percolation theory that the critical threshold is positive and finite in dimensions $d \geq 2$, see [19], i.e., that both regimes of connectivity exist (depending on the intensity): The supercritical regime where messages can in principle travel unboundedly far, and the subcritical one in which each user can communicate only with finitely many others.

There is no simple expression available for the value of λ_c, but there exist highly accurate approximations, see for example [1, Table 2]. The percolation probability can be further used to approximate a number of important network characteristics. For example, the probability that two users can communicate over long distances is asymptotically given by $\theta(\lambda)^2$, see [8, Corollary 4.2.4]. Similarly, the limiting expected number of connected users in a growing volume is also given by $\theta(\lambda)^2$. A detailed understanding of the function $\lambda \mapsto \theta(\lambda)$, especially near the critical intensity, is one of the longstanding open problems in percolation theory, see Figure 1 for a sketch of the function. We remark, that not even the continuity of this map at the critical value has been rigorously settled yet!

2.3 Palm calculus

For the analysis of a stationary point process X in $\mathbb{R}^d$, it is often desirable to view the point cloud from the perspective of one of the points, which should be a 'typical' one, i.e., one that is picked uniformly at random. Equivalently, one would like to shift the process such that the typical point is located at the origin, i.e., one would like to condition on the event $\{0 \in X\}$. None of these wishes can be satisfied, since there is no uniform distribution on infinitely many points, and the event $\{0 \in X\}$

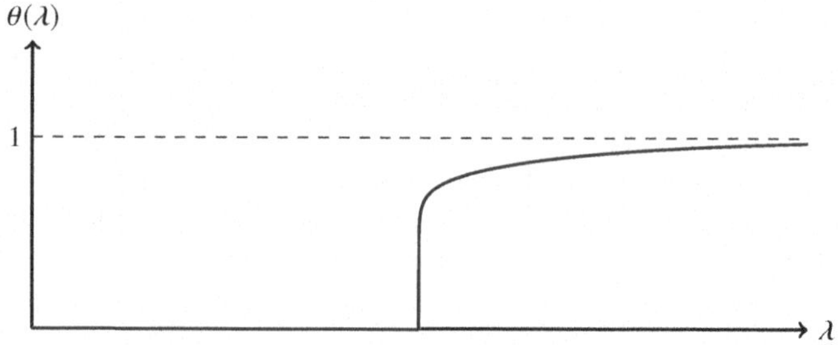

Fig. 1: Approximative form of the percolation probability

has probability zero. The *Palm calculus* resolves this technical problem by providing a version X^*, the *Palm version*, of X with the desired statistical properties. This process is formally defined via the identity

$$\mathbb{E}[f(X^*)] = \frac{1}{\lambda}\mathbb{E}\Big[\sum_{X_i \in X\cap[0,1]^d} f(X - X_i)\Big], \tag{2.1}$$

for all continuous and bounded test functions f. Instead of the unit box, every measurable set $A \subset \mathbb{R}^d$ with positive and finite Lebesgue measure can be taken, and the pre-factor $1/\lambda$ must then be adapted by $1/\mu(A)$ with $\mu = \lambda\text{Leb}$ the intensity measure. Then, in any reasonable way how one could try to approach the above two wishes rigorously, X^* is a rigorous fulfillment. In the case that X is a homogeneous Poisson point process, it is one of the fundamental results of point process theory called the Slivnyak-Mecke formula, that X^* has the same distribution as the process $X \cup \{0\}$, see for example [18]. In this case, note that the percolation probability (defined above as the probability that o lies in an unbounded component of $\bigcup_{x\in X\cup\{0\}} B_R(x)$) can also be written as the probability that o lies in an unbounded component of $\bigcup_{x\in X^*} B_R(x)$.

2.4 Interference

One of the most important aspects of the quality of message transmission in a network is the *interference*, i.e., the question whether or not the transmitted signal is safely received by the intended receiver, but not suppressed by a multitude of signals emitted from other users. For this, the intended signal's strength must be strong enough at the location of the intended recipient in comparison to the sum of all the other signal strengths that are floating around at the time of the transmission. These strengths are small if the sources are far away. This fact induces the necessity to handle the spatial details of the system, i.e., the locations of all the users. Also a general background noise, created by other types of sources, may influence the reception negatively. The problem of interference is present as soon as many users transmit at the same time, and it becomes serious if they have unfavorable locations with respect to each other. It can be tackled only by controlling the locations and/or the time instances of message transmission, which is the subject of another extensive field of research in mathematics and engineering. Taking interference into account, the model renders much more complicated than the Boolean model of Section 2.2, since a multitude of long-distance correlations appear.

Mathematically, one often models interference by use of the *signal-to-noise-and-interference ratio (SINR)* for a transmission of a message from X_i to $x \in W$ at a time instant at which messages are sent out from any of the users in Y, where $Y \subset X$ is some subset of the users:

$$\mathrm{SINR}_{N,\gamma}(X_i, x, Y) = \frac{\ell(|X_i - x|)}{N + \gamma \sum_{X_k \in Y \setminus \{X_i, x\}} \ell(|X_k - x|)}. \tag{2.2}$$

Here, $\ell \colon [0,\infty) \to [0,\infty)$ is the *path-loss function*, which describes the loss of the signal strength over distance. Typical choices are $\ell(r) = (1+r)^{-\alpha}$ or $\ell(r) = \max\{1, r^{-\alpha}\}$ or the perfect-scaling function $\ell(r) = r^{-\alpha}$ with some $\alpha \in (0,\infty)$. In order to ensure that the total expected signal strength in space is finite, one often chooses $\alpha \in (d,\infty)$, which implies that $\int_{\mathbb{R}^d} \ell(|x|)\,\mathrm{d}x < \infty$. This in turn implies that, almost surely, the *interference* $\sum_{i \in I} \ell(|X_i - x|)$ is a finite random variable, in the case $W = \mathbb{R}^d$ and μ (a constant times) the Lebesgue measure. The parameters $N \in [0,\infty)$ and $\gamma \in (0,\infty)$ represent the background noise and the strength of the influence of the interference. If $N = 0$, one speaks of a *signal-to-interference ratio, SIR*. The signal strengths can individually be modeled by replacing $\ell(|X_i - x|)$ with $F_i \ell(|X_i - x|)$ with some positive random variable F_i, but we do not do that here. The parameter α models, in a rather simple way, the strength of the hampering effect coming from environmental objects like walls, trees, houses etc. A comprehensive overview on SIR in stochastic geometry is given in [2, 3].

Now, a transmission from X_i to x, when all the users in Y transmit at the same time, is successful if $\mathrm{SINR}_{N,\gamma}(X_i, x, Y) \geq \tau$, where $\tau \in (0,\infty)$ is a technology-dependent constant often called the *SINR threshold*. It describes the ability to filter out the wanted signal from X_i from the noise and the interference at x.

Let us now draw an edge between the users X_i and X_j if $\mathrm{SINR}_{N,\gamma}(X_i, X_j, X) \geq \tau$, and note that the arising geometric graph is a directed one and that it has now a lot of far-reaching stochastic dependencies, even though the underlying point process X is assumed a Poisson point process. The lack of symmetry can easily be removed, but the dependencies will never disappear. The graph is often called the *SINR-graph*, respectively the SIR-graph if $N = 0$. Here we assumed that, at any considered time, every user transmits a message and therefore contributes to the interference.

Actually, the assumption that the interference is taken with respect to all the users, is a great simplification and is not justified in realistic situations. In fact, in the asymptotic situation of a high-density limit $\lambda \to \infty$ that we will consider several times, the large amount of interference would kill all the functionality of the system immediately, if every user transmits at the same time. Mathematically, we make up for this problem by picking the parameter γ as γ/λ, and we obtain very satisfactory mathematical results. However, a much more realistic way to cope with this problem is to divide all the message transmissions over many time instances and to allow only very particular subsets of users to transmit at the same time. This procedure requires algorithms for the choices of these subsets and renders the model much more complex. Also new questions arise, e.g., about the throughput that one can achieve, but we do not follow this idea here. See the seminal paper [10] for a first and fundamental treatment of this.

2.5 Large deviations

Let us explain what large-deviations theory is and what it can achieve (see [6, 7] for general accounts and [21] for an introduction to the theory with statistical physics flavor). Roughly speaking, it provides tools for expressing, for random events $A_1, A_2, A_3, \dots$ whose probabilities converge to zero, the exponential rate of this decay $\lim_{n\to\infty} \frac{1}{n} \log \mathbb{P}(A_n)$, if it exists. The prototypical example is in terms of a random walk $S_n = X_1 + \dots + X_n$, the partial sum of independent and identically distributed (i.i.d.) random variables X_i. For definiteness, assume that the X_i have expectation equal to zero and additionally very strong integrability properties, more precisely, they should have finite exponential moments of all orders. Then, according to the law of large numbers, S_n/n converges to zero in probability. The event $A_n = \{S_n \geq \varepsilon n\}$ is an event of a large deviation for any fixed $\varepsilon > 0$. With the help of the Markov inequality, one derives an upper bound of the form

$$\limsup_{n\to\infty} \frac{1}{n} \log \mathbb{P}(S_n \geq \varepsilon n) \leq -\mathrm{I}(\varepsilon), \qquad \varepsilon \in (0, \infty), \tag{2.3}$$

where the *rate function* I is given as $\mathrm{I}(\varepsilon) = \sup_{y>0}[y\varepsilon - \log \mathbb{E}(\mathrm{e}^{yX_1})]$. Somewhat deeper techniques show that in (2.3) also the opposite inequality holds and that a version for negative ε holds as well. This may be summarized by saying that $\mathbb{P}(S_n \approx xn) \approx \mathrm{e}^{-n\mathrm{I}(x)}$ for large n and any $x \in \mathbb{R}$. One says that $(S_n/n)_{n\in\mathbb{N}}$ *satisfies a large-deviations principle (LDP)* with rate function I. A proper formulation is in terms of the weak convergence of the set function $\frac{1}{n} \log \mathbb{P}(S_n/n \in \cdot)$ towards the set function $-\inf\{\mathrm{I}(x) \colon x \in \cdot\}$, i.e., in terms of upper bounds for closed sets and a lower bound for open sets. Hence, topology plays an important role in an LDP.

It is important to note that the formula for the exponential rate is explicit and amenable to further analysis; it contains useful and characteristic information about the way how the large-deviations are typically realized. The theory of variational calculus is helpful here.

One useful tool, called the *contraction principle*, is that, for any continuous function F, also the sequence $(F(S_n/n))_{n\in\mathbb{N}}$ satisfies an LDP, and the rate function is $y \mapsto \inf\{\mathrm{I}(x) \colon F(x) = y\}$. Another cornerstones of the theory, *Varadhan's lemma*, states that expectations of the form $\mathbb{E}(\mathrm{e}^{nf(S_n/n)})$ with continuous and bounded f behave like $\exp\{n \sup_{x\in\mathbb{R}}(f(x) - \mathrm{I}(x))\}$, which is a substantial extension of the well-known Laplace principle.

Now, let us consider one of the two limiting situations of importance for the modeling of large networks that we briefly mentioned in Section 1, the high-density limit. Let the communication area W be bounded, and a Poisson point process $X^\lambda = (X_i)_{i\in I_\lambda}$ in W with intensity measure $\lambda\mu$ be given. Here μ is a measure on W, which we want to assume as absolutely continuous with respect to the Lebesgue measure, and λ is a positive parameter, which will be sent to ∞. It is no problem to see that the *normalized empirical measure* $L_\lambda = \lambda^{-1}\sum_{i\in I_\lambda}\delta_{X_i}$, where δ denotes the Dirac measure in x, converges towards the intensity measure. That is, $L_\lambda \Rightarrow \mu$ as $\lambda \to \infty$ in the weak sense, i.e., when testing against continuous and bounded functions. In other words, the dense cloud of users in W approaches the density of the intensity measure μ, i.e., a multitude of microscopic information (every single user location) is approximated by some much simpler macroscopic object, a density, for which there are good perspectives for further analysis. (One might argue that such a limiting setting is useless for describing human beings, since they cannot be squeezed infinitely strongly, but we are heading for approximate formulas, and many of the situations are in reality much better than the approximation by asymptotic formulas.)

Now let us consider the large deviations of the empirical measure L_λ, e.g., we consider probabilities of the form $A_\lambda(\varepsilon) = \{F(L_\lambda) - F(\mu) \ge \varepsilon\}$ for some $\varepsilon \in (0,\infty)$ with some continuous and bounded function F on densities on W. See Section 4.1 for an example of F that expresses service quality, where then $A_\lambda(\varepsilon)$ is an event of a very bad quality of service, which we sometimes call a frustration event for this reason. It is not difficult to get convinced that L_λ should satisfy an LDP with rate function equal to

$$\mathrm{I}(m) = H(m|\mu) = \int f(x)\log f(x)\mu(\mathrm{d}x) - m(W) + \mu(W), \tag{2.4}$$

if the density $\mathrm{d}m/\mathrm{d}\mu = f$ exists, and $H(m|\mu) = \infty$ otherwise. The functional H is called the *Kullback-Leibler divergence* of m with respect to the reference measure μ. (Unfortunately, we did not find this statement in the literature, but this is negligible, since our function F is not continuous anyway and had to be approximated with different techniques.) Using this, one can guess that a statement like

$$\mathbb{P}(A_\lambda(\varepsilon)) \approx \mathrm{e}^{-\lambda\chi_\varepsilon}, \qquad \text{where} \quad \chi_\varepsilon = \inf\{H(m \mid \mu)\colon F(m) - F(\mu) \ge \varepsilon\}, \tag{2.5}$$

should be valid asymptotically as λ tends to infinity. Remarkably, a further analysis of the characteristic variational formula χ_ε can be used to derive deep insights into the distribution of configurations that are responsible for the most likely realization of the event $A_\lambda(\varepsilon)$ of a bad service. This further analysis is indeed sometimes possible since the formula is sensitive to the spatial distribution of m, and the rate function is sufficiently explicit.

3 Continuum percolation in random environments

The analysis of percolation properties of an ad hoc system can be used to derive first rough estimates on its connectivity. For example, it can provide an approximation for the critical number of users that have to participate in the network in order to allow for long-distance communication. Without any knowledge of the distribution of users in space, their means of communications, or the environment, it is reasonable to start with an analysis of the elementary Boolean model as described in Section 2.2, as this gives already a first good understanding. In this section we extend the Boolean model with respect to the way in which users are distributed in space by assuming that they prefer to be located on certain additional structures in the environment. Mathematically, this amounts to a random choice of the intensity measure μ of the Poisson points, and we end up with what is called a *Cox point process*. However, we keep the assumption that connections are established only if users are at distance $\leq R$ with some reach parameter $R \in (0, \infty)$, and thus still neglect interference and further environment constraints. A refined analysis of Cox percolation in the SINR graph can be found in the recent manuscript [23].

3.1 Main example: Voronoi tessellation

Our guiding example is that users are situated in two dimensions on a random street system modeling a city in central Europe, see Figure 2. Here, the streets are represented by a realization of a random tessellation process. For example, as has been understood statistically, see [9], central European cities can be well approximated by *Poisson-Voronoi tessellations*. This object is defined as follows. Based on an underlying Poisson point process, the *Voronoi cell* of a Poisson point contains all sites in W closer to that point than to any other of the Poisson points. Then, we consider the separation lines between neighboring such cells, which are straight line segments, and consider them as the streets. There is a natural (random) measure Λ, which is supported precisely on the union S of the streets. Λ will be often referred to as the *random environment*. We write it as $\Lambda(A) = |A \cap S|$ for measurable sets A in $\mathbb{R}^2$ where $|\cdot|$ here denotes the one-dimensional Hausdorff measure. The *street intensity* is then defined as $\gamma = \mathbb{E}[\Lambda([0,1]^2)]$, the expected street length in a unit area, and can be derived from the intensity of the Poisson point process underlying the Voronoi tessellation, see [20]. This parameter can be used to distinguish for example rural areas from city centers.

Given the street system S, users are now placed on S as a Poisson point process restricted to S, more precisely, the users form a Poisson point process with intensity measure equal to $\lambda\Lambda$, where we introduced an additional scalar user intensity $\lambda \in (0, \infty)$. This is an example of a Cox point process, i.e., a Poisson point process with random intensity measure. Now the percolation probability is denoted $\theta(\lambda, R, \gamma)$ in this situation, and we note that the probability is taken both with respect to the street system and the user process. Note that elementary scaling arguments give that

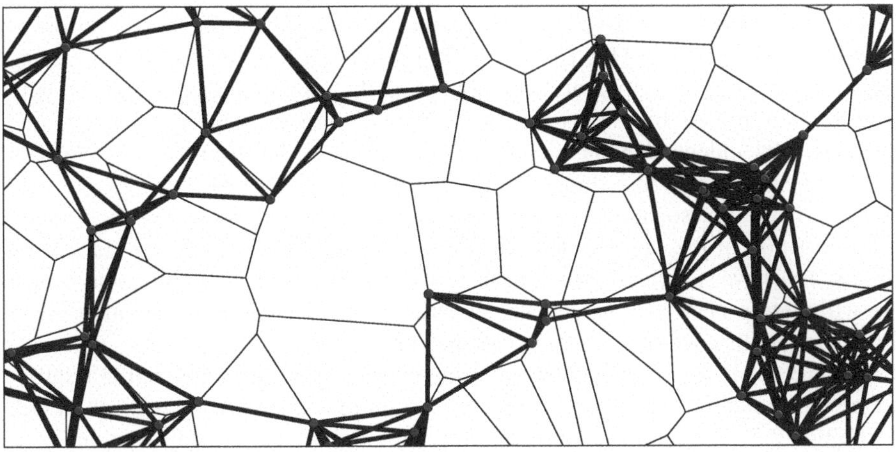

Fig. 2: Realization of the Gilbert graph of users confined to a street system given by a Poisson-Voronoi tessellation

$$\theta(a^{-1}\lambda, aR, a^{-1}\gamma) = \theta(\lambda, R, \gamma), \qquad a \in (0, \infty), \tag{3.1}$$

and hence as in case of the Boolean model, one of the system parameters can be eliminated.

We will refer to the above example briefly as to the *Voronoi environment*. In the following, we will present results for much more general random measures Λ on $\mathbb{R}^2$, where θ in general does not enjoy a scaling property as in (3.1).

3.2 The critical user intensity for percolation

Fix a general random environment measure Λ with intensity $\gamma = \mathbb{E}[\Lambda([0,1]^2)]$. We write the percolation probability for the Poisson point process with intensity measure $\lambda\Lambda$ and radius R as $\theta(\lambda, R, \gamma)$, even though it is not (as in the Voronoi environment) a function of the parameter γ only, but of many other properties of Λ. As in Section 2.2, we define the critical percolation threshold $\lambda_c(R, \gamma)$ as the point at which $\lambda \mapsto \theta(\lambda, R, \gamma)$ switches from zero to some positive value. This is also called the *critical user intensity*.

Let us discuss non-triviality of $\lambda_c(R, \gamma)$, i.e., existence of non-trivial sub- and supercritical regimes with respect to the user intensity. This is not true in general for Cox point processes; indeed, there are random environments of streets that never allow to drive to infinity. In this case, no matter how many users are placed on the street system, there will never be an infinite cluster and the system stays subcritical. Also the converse situation of an environment that guarantees percolation for arbitrarily thinly placed users can be easily constructed or found in the literature, see [4].

For the existence of a subcritical regime of users, the sufficient criterion of *stabilization* is given in [12, Definition 2.3]. This is a kind of qualitative mixing property, which is in simple terms the following: In a stabilizing environment, with high probability, environments in distant regions in space behave asymptotically stochastically independent in a certain dependence of the distance. The Voronoi environment is stabilizing. Indeed, although it can happen that, due to large void spaces without Poisson points of cell centroids, the local street system changes under a perturbation of the street system far away, this is still highly unlikely since large void spaces have sufficiently small probability. A strong form of stabilization, which we will refer to later, is *b-dependence*, where the environment behaves independent in regions of distance $\geq b$ for some parameter $b \in (0, \infty)$.

For the existence of a supercritical regime, apart from the stabilization requirement, the sufficient criterion of *asymptotic essential connectedness* is given in [12, Definition 2.5]. Again, this guarantees that, with high probability, the random environment supports the placement of users and provides appropriate long-term connectivity. The Voronoi environment is asymptoticly essentially connected.

Now we present and discuss the heuristics of some approximations for the critical user intensity in the Voronoi environment in some limiting regimes. A first approximation can be derived by replacing the Cox point process with intensity $\lambda|\cdot \cap S|$ by a Poisson point process with intensity $\lambda\gamma\,\mathrm{Leb}$. Then, writing λ_c for the critical intensity of the Boolean model with reach parameter 1, we can approximate for dense street systems

$$\lambda_c(R, \gamma) \approx \lambda_c \frac{1}{\gamma} \frac{1}{R^2}. \tag{3.2}$$

More precisely, it can be proved using the scaling invariance (3.1), that this approximation becomes exact in the limit of dense streets $\gamma \longrightarrow \infty$ with inversely proportional user intensity.

The above approximation is based on the assumption that the process of users can be approximated by a two-dimensional percolation system. However, this assumption is severely violated in rural areas, where the street system is sparse in relation to the radius R. Then, the Cox percolation becomes essentially one-dimensional and the principal obstructions for percolation are formed by device gaps of length R preventing communication along that edge. However, for $\gamma \longrightarrow 0$ and $\lambda \longrightarrow \infty$ such that $c\gamma = \exp(-\lambda R)\lambda$ for some $c > 0$ we obtain convergence to an inhomogeneous Bernoulli bond percolation model on the street system S in the Voronoi environment. Here, any edge of length ℓ is open with probability $\exp(-\ell c)$ independent from all other edges. Writing c_{crit} for the critical parameter of this Bernoulli bond percolation model we have the approximation,

$$\lambda_c(R, \gamma) \exp(-\lambda_c(R, \gamma) R) \approx \gamma c_{\mathrm{crit}} \tag{3.3}$$

which can be proved to become exact in the limit $\gamma \longrightarrow 0$ with λ scaled as described above.

3.3 Asymptotics for the percolation probability

In this section we formulate and discuss some rigorous assertions about the percolation probability for a Cox process with a general random environment Λ with intensity γ. We continue to write this quantity as $\theta(\lambda, R, \gamma)$, even though it is not (as in the Voronoi environment) a function of the parameter γ only, but of many other properties of Λ.

Before we have a closer look at the percolation probability, we want to discuss a characterization of it, which is used in the proofs and in heuristics. As we explained in Section 2.3, in the special case of $\mu = \lambda\mathrm{Leb}$, the percolation probability can be represented as the Palm probability that the origin o lies in an unbounded component of $\bigcup_{x \in X^*} B_R(x)$, where X^* is the Palm version of the point process X. This definition is fine, as soon as we can make sense out of the Palm version X^* of the Cox process. This is not immediate since adding a user at the origin violates the street constraint if there is no street crossing the origin. In order to resolve this, we define, in the case of a translation-invariant Cox point process X, its Palm version X^* via

$$\mathbb{E}[f(X^*)] = \frac{1}{\lambda \mathbb{E}[\Lambda([0,1]^2)]} \mathbb{E}\Big[\sum_{X_i \in X \cap [0,1]^2} f(X - X_i) \Big], \tag{3.4}$$

for continuous and bounded test functions f.

In the remainder of this section we collect results about the behavior of θ in a number of asymptotic regimes. General results about Cox point processes are proved in [12] and we will here complement them with some associated results for the specific case of the Voronoi environment.

3.3.1 Large communication radius

Let us start by considering the limiting behavior of θ for large interaction radii R. For this it will be necessary to also define the Palm version Λ^* of the environment process Λ, i.e.,

$$\mathbb{E}[f(\Lambda^*)] = \mathbb{E}\Big[\int_{[0,1]^2} \Lambda(\mathrm{d}x) f(\vartheta_x(\Lambda)) \Big], \tag{3.5}$$

where ϑ_x is the shift by x. Then, the main statement is that if Λ is b-dependent and the random variable $\Lambda([0,1]^2)$ has all exponential moments, then the limiting logarithmic moment generating function

$$I^*(t) = -\lim_{R \uparrow \infty} \frac{1}{R^2} \log \mathbb{E}[\exp(-t\Lambda([-R/2, R/2]^2))], \qquad t \in \mathbb{R}, \tag{3.6}$$

exists and

$$\lim_{R \uparrow \infty} \frac{1}{R^2} \log(1 - \theta(\lambda, R, \gamma)) = -\pi I^*(\lambda), \qquad \lambda \in (0, \infty). \tag{3.7}$$

In words, the percolation probability tends to one exponentially fast, and the rate of convergence is given by some averaging characteristic of the environment. The lower bound in (3.7) holds under much weaker conditions, which allow to consider also a Voronoi environment. In this case, by an application of Jensen's inequality, we have that

$$\lim_{R\uparrow\infty}\frac{1}{R^2}\log(1-\theta(\lambda,R,\gamma))\geq -\pi\lambda\gamma. \tag{3.8}$$

Using similar arguments we can derive the corresponding result for the limit of dense streets, i.e.,

$$\lim_{\gamma\uparrow\infty}\frac{1}{\gamma}\log(1-\theta(\lambda,R,\gamma))\geq -\pi\lambda R^2. \tag{3.9}$$

3.3.2 Large user intensity

In the limit $\lambda \to \infty$ of many users, the behavior of θ depends crucially on the structure of the support of Λ. In particular, there is no averaging of the environment involved as in the previous case. Now, the asymptotic rate of convergence to one for the percolation probability turns out to be given by the exponentially cheapest way to isolate the origin. For this, we define C_R to be the family of all compact sets $A \subset \mathbb{R}^d$ that contain the origin and are *R-connected* in the sense that the set $\overline{A}_{R/2} = \{x \in \mathbb{R}^2\colon \mathrm{dist}(x,A) < R/2\}$ is connected. Moreover, let $\partial_R A = \overline{A}_R \setminus A$ denote the *R-boundary* of A. Then, if Λ is b-dependent and ess-inf $\Lambda([-\delta,\delta]^2) > 0$ for every $\delta > 0$, we have that

$$\limsup_{\lambda\uparrow\infty}\frac{1}{\lambda}\log(1-\theta(\lambda,R,\gamma))\leq -\lim_{\epsilon\downarrow 0}\inf_{A\in\mathrm{C}_{R+\epsilon}}\text{ess-inf}\,\Lambda^*(\partial_{R-\epsilon}A), \tag{3.10}$$

and

$$\liminf_{\lambda\uparrow\infty}\frac{1}{\lambda}\log(1-\theta(\lambda,R,\gamma))\geq -\inf_{A\in\mathrm{C}_R}\text{ess-inf}\,\Lambda^*(\partial_R A). \tag{3.11}$$

Note that in general the lower bound is not given by the isolation probability of the origin. Consider for example a situation where the support of Λ does not percolate, which implies that $\sup\{\theta(\lambda,R,\gamma)\colon \lambda > 0\} = 0$. Nevertheless, for the Voronoi environment, for the lower bound, the right-hand side of (3.11) equals $-2R$, which is the least street length that can be realized by $\Lambda^*(\partial_R\{o\})$ with positive probability.

In the sense of large deviations, the asymptotically unlikely event of the origin being not connected to infinity is realized in the least unlikely of all the unlikely ways.

3.3.3 Coupled limits

Let us finally address the behavior of θ under certain coupled limits. First, for any stabilizing environment Λ, under the scaling which corresponds to the scale invariance of the Boolean model with vanishing user intensity, the percolation probability converges to the percolation probability of the Boolean model and the dependence on the environment disappears. More precisely,

$$\lim_{\substack{R\uparrow\infty,\,\lambda\downarrow 0\\ \lambda R^2=\rho}} \theta(\lambda, R, \gamma) = \theta(\rho), \qquad \rho \in (0, \infty), \tag{3.12}$$

where we recall that $\theta(\rho)$ denotes the percolation probability of the Boolean model with $R = 1$. This in particular applies to Voronoi environments.

Conversely, under a coupled limit of small interaction radii and large user intensity, if Λ is given by a tessellation process, it is not sufficient to increase the user intensity by setting $\rho = \lambda R^2$. The reason for this is that percolation on tessellation edges is essentially one-dimensional and hence there is no percolation possible. Using a similar reasoning as in the paragraph prior to (3.3), we can still compare our model to a Bernoulli bond percolation model with percolation probability $\theta_{\mathrm{Ber}}(c)$, based on the same tessellation process. The result is the following. If Λ is stabilizing and essentially asymptotically connected and c is sufficiently small, then

$$\lim_{\substack{\lambda\to\infty,\,R\downarrow 0\\ \lambda\exp(-\lambda R)=c}} \theta(\lambda, R, \gamma) = \theta_{\mathrm{Ber}}(c). \tag{3.13}$$

This also applies to Voronoi environments.

3.4 Summary

To conclude, we have derived a number of approximating formulas for the percolation probability as well as for the critical user intensity for percolation in a refined and more general setting of users positioned in some random environment, see [5]. This covers the important special situation where users are assumed to be located on some street system modeled via Poisson-Voronoi tessellation processes. Depending on the kind of environment, for example rural versus urban city geometries, this should help practitioners to better assess the feasibility, functionality and quality of large ad hoc networks with respect to their connectivity. In Figure 3 we exhibit simulated graphs of the percolation probability as a function of the user intensity for several values of R and a Voronoi environment of an intensity that can statistically be observed in central European cities.

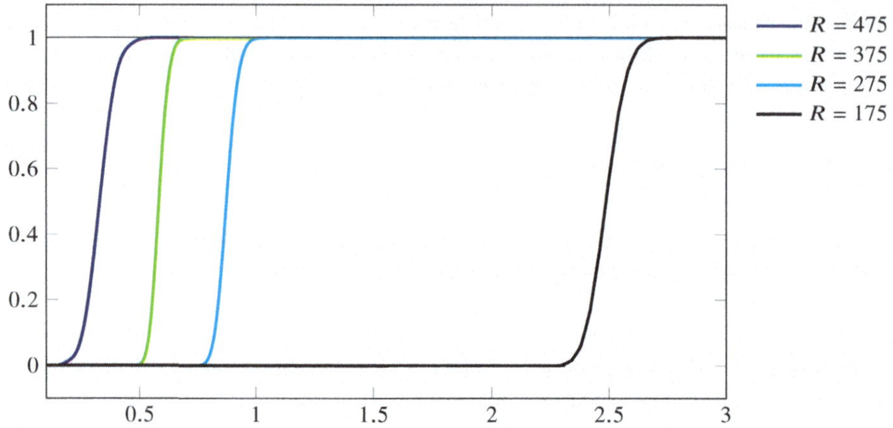

Fig. 3: Simulated graphs of the function $\lambda \mapsto \theta(\lambda, R, \gamma)$ for various values of R (in meters) for the Voronoi environment with parameter $\gamma = 20\,\mathrm{km}^{-1}$

4 Large deviations in high-density networks with interference and capacity constraints

In this section we present two examples of how to employ the theory of large deviations to better understand ad hoc networks with many users in terms of its bottleneck behavior. Typically, a first probabilistic analysis of a given network is concerned with its expected performance or with its normal behavior in terms of a law of large numbers. But from an operator's perspective, it appears equally important to understand the system in its rare but disruptive configurations. Corresponding questions are then for example of the form: How unlikely is a situation where a large percentage of users is disconnected? What does a typical user configuration look like in such an undesired event? These types of problems can be tackled with large deviations theory as introduced in Section 2.5.

In this section, we will focus on the frustration event of too many users being unable to send their messages to a single base station placed at the center of some bounded volume $W \subset \mathbb{R}^2$. The network is assumed to carry a relaying functionality, that is, messages do not have to be delivered to the base station directly but can also use one intermediate relaying step. In our two examples we investigate two related but different reasons for unsuccessful message transmission. In the first case in Section 4.1, message transmissions are constrained due to interference. We adopt a properly adjusted SIR setup as introduced in Section 2.4 and call a user frustrated if any possible message route from him is blocked due to too low SIR along the message trajectory. In particular we do not put a limit on the number of messages forwarded by any relay. In the second example, as presented in Section 4.2, we *do* limit the number of messages per relay and replace the interference constraint by a capacity constraint. To avoid confusion, in this setting, by capacity we simply mean the maximal number of messages a relay can handle at any point in time. In both cases, we position the users according to a Poisson point process X^λ in a centered cube $W \subset \mathbb{R}^2$, with intensity measure $\lambda\mu$, and let λ tend to infinity and thus look at a high-density scenario.

4.1 Interference constraints

In this section we present a selection of results from [13] where a large cloud of Poisson points $X^\lambda = (X_i)_{i\in I^\lambda}$ in a square domain W try to submit messages to the unique central base station. The system is assumed to be interference constrained in that message transmission can fail due to insufficiently high SIR. We consider the high-density situation as $\lambda \longrightarrow \infty$. For reasons that we explained in Section 2.4, we put $\gamma = \lambda^{-1}$ in (2.2). Using the normalized empirical measure $L_\lambda = \frac{1}{\lambda}\sum_{i\in I^\lambda}\delta_{X_i}$ we can rewrite the SIR as $\mathrm{SIR}(X_i, x, L_\lambda)$, where for any measure ν on W, we have

$$\mathrm{SIR}(X_i, x, \nu) = \frac{\ell(|X_i - x|)}{\nu[\ell(|\cdot - x|)]},$$

with $\nu[f]$ a short-hand notation for the integral of a function f with respect to ν. Since any bounded background noise N would vanish in the high-density limit, there is no difference in considering the SIR instead of the SINR. For the same reason, we also added without loss the contributions from X_i and x in the interference term in the denominator.

Now we want to allow each message from some X_i to the base station at the origin o to make one direct step or at most one relaying step into some relay X_j. We have to make a choice to describe the *quality of service* of the two-hop trajectory $X_i \longrightarrow X_j \longrightarrow o$. In the following model, we decide to require that each of the two steps has to satisfy the interference condition that the SIR is not smaller than a given threshold $\tau \in (0, \infty)$. See Section 5 for another choice. Let us write this in terms of the minimum SIR in a trajectory $x \longrightarrow y \longrightarrow o$

$$\mathrm{D}(x, y, o, \nu) = \min\{\mathrm{SIR}(x, y, \nu), \mathrm{SIR}(y, o, \nu)\}.$$

The maximum over the two trajectories $x \longrightarrow o$ and $x \longrightarrow y \longrightarrow o$ for some y is then given by

$$\mathrm{R}(x, o, \nu) = \max\Big\{\mathrm{SIR}(x, o, \nu), \max_{y\in X^\lambda} \mathrm{D}(x, y, o, \nu)\Big\}.$$

We will render the transmission of a message from X_i to o successful if $\mathrm{R}(X_i, o, L_\lambda) \geq \tau$ for a given threshold $\tau \in (0, \infty)$, i.e., if either the SIR of the direct or both SIRs of the two hops of some indirect two-hop link from X_i to o are larger than τ. Note that the source of the interference is, for any considered hop, the totality of all the users present in W. That is, we assume that every user is sending out his/her message to the origin at a joint time instant. In Figure 4 we present a realization of the network indicating direct and indirect uplinks.

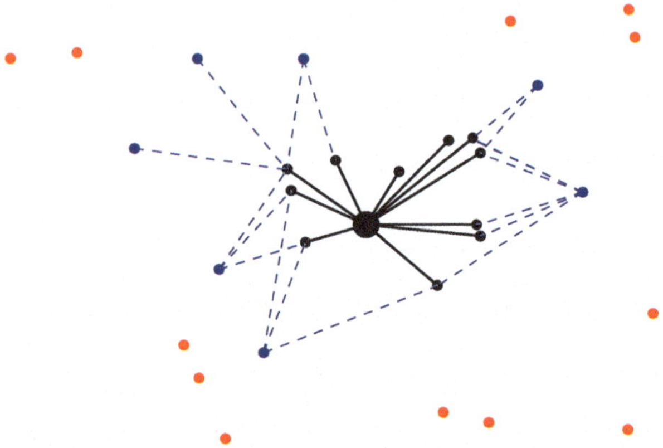

Fig. 4: Realization of user configuration with black users directly linked to the origin, blue users indirectly connected and red users disconnected

Let us note that in [13] a much more general setting is considered, where the users X_i are randomly independently moving in time and perform Lipschitz trajectories on a finite time interval. This makes temporal considerations possible, e.g., the quality of service over part of the given time. Furthermore, other choices of measuring the quality of a two-hop trajectory are discussed and compared, as well as the (subtle, but decisive) difference between an uplink (as considered above) and a downlink scenario. In [22] the results of [13] are further extended to include random fading in the SIR.

We are interested in the large-deviation behavior of the empirical measure of frustrated users, that is, of those ones whose message do not reach o. Their empirical measure may be written in terms of the function

$$\varphi_{\nu,\tau}(x) = \mathbb{1}\{\mathrm{R}(x,o,\nu) < \tau\}, \qquad x \in W,$$

as the measure with density $\varphi_{L_\lambda,\tau}$ with respect to L_λ, that is,

$$\mathfrak{M}_{L_\lambda}(\mathrm{d}x) = \varphi_{L_\lambda,\tau}(x)\, L_\lambda(\mathrm{d}x) = \frac{1}{\lambda}\sum_{i\in I_\lambda} \delta_{X_i}(\mathrm{d}x)\mathbb{1}\{\mathrm{R}(X_i,o,L_\lambda) < \tau\}.$$

Note that L_λ appears here at two places: As the ground measure and as inducing interference; a general definition of the measure $\mathfrak{M}_\nu$ for arbitrary measures ν is obvious. Examples of events that we could handle now are of the form

$$A_\epsilon = \{\mathfrak{M}_{L_\lambda}(W) - \mathfrak{M}_\mu(W) > \epsilon\},$$

in words, the event that the proportion of disconnected users is by $\epsilon > 0$ higher than expected. For a similar event, [13, Theorem 1.1] implies that, for any sub-cube B of W, subject to some mild topological constraint,

$$\lim_{\lambda \uparrow \infty} \frac{1}{\lambda} \log \mathbb{P}(\mathfrak{M}_{L_\lambda}(B) > b) = -\inf\{H(\nu|\mu) \colon \mathfrak{M}_\nu(B) > b\}, \qquad b \in (0, \infty). \quad (4.1)$$

Here, the Kullback-Leibler divergence H is given by (2.4). This result is very much in line with the large-deviation principle that we briefly discussed in Section 2.5, and its proof would be rather easy if this LDP would exist in the literature, and the map $\nu \mapsto \mathfrak{M}_\nu$ would be continuous in the topology of the LDP. However, this is not true, and the proof had to go via a sequence of technical discretization steps. Likewise, it is by no means clear that the right-hand side of (4.1) is indeed negative, and provable criteria required extra assumptions: Indeed, if the proportion of disconnected users is too large, in the sense that $\mu[\varphi_{(1+\epsilon)\mu,T}] < b$ for some $\epsilon > 0$, then the probability of such an atypical event indeed decays exponentially as the right-hand side of (4.1) is negative. See also [15] for a case study.

As mentioned in Section 2.4, apart from the exponential decay, the minimizers ν of $\inf\{H(\nu|\mu) \colon \mathfrak{M}_\nu(B) > b\}$ describe the typical behavior of the system conditioned on the atypical event $\{\mathfrak{M}_{L_\lambda}(W) > b\}$, see [21, Section 5.3]. It can be shown that for this event, in a domain W given by a centered disk, in the direct uplink case where R = SIR, any minimizer must be rotationally invariant if μ = Leb, see [13, Proposition 7.1]. However, finding more explicit properties of the minimizers is a hard analytic task and requires additional research.

Note that characteristic variational formulas derived from large-deviations principles can in principle also be used to obtain illustrating simulations for the rare events under consideration, if one is able to handle the particular problems coming from the entropy term. This may be indeed enormously helpful, as standard simulation techniques require to produce a long sequence of independent copies of the entire particle cloud, until the desired rare event is realized, which is very costly. But also in cases in which this is the only feasible way, a large-deviation result as in (4.1) is helpful by giving a rough estimate of how long the sequence of simulations will be.

Interestingly, there is strong numerical evidence that the rotational invariance of the minimizers of the right-hand side of (4.1) might be broken if we allow for relaying. To see this, consider a rotationally invariant a priori measure μ that puts positive weight only close to the base station, at the boundary of the disc and in an inner annulus. Then, in order to produce the required disproportional amount of disconnected users, it can be entropically beneficial not only to increase users at the boundary (since they are more easily disconnected) or to increase users near the base station (since they create interference and therefore shield the base station), but also to decrease the users in the inner annulus (since they serve many boundary users as relays). As a consequence, the disconnected users cluster at some part of the boundary and violate rotation invariance, see Figure 5.

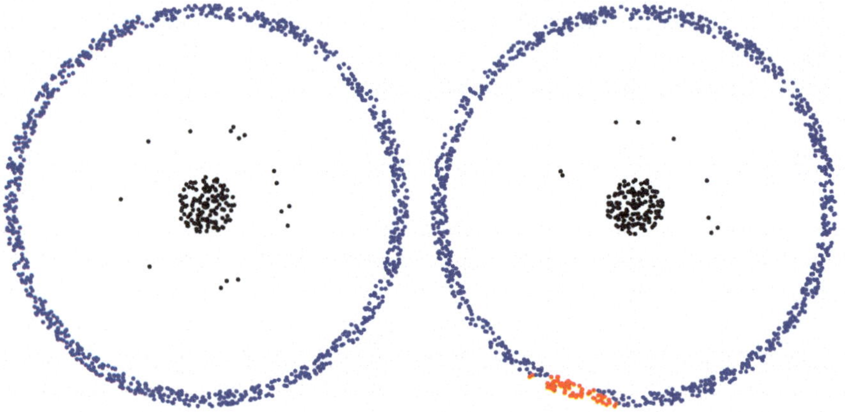

Fig. 5: Realizations of user configurations under a rotation-invariant a priori measure concentrated on three annuli. On the left, a typical (i.e., usual, unconstrained) realization with black users being directly and blue users being indirectly connected to the base station at the origin. On the right, a realization of the atypical event of very many disconnected users. Due to the lack of users in the inner annulus, disconnected users in red are clustered in a non-translation invariant region.

4.2 Capacity constraints

Now we report on some large-deviations results first presented in [11, 14]. We are in almost the same situation as in the previous subsection, where a large number of Poisson points tries to submit messages to a central base station. However, now every message makes exactly one relaying hop, and the relays are given by an independent deterministic process. The main difference is that now the bottleneck behavior comes from an overload at the relays, as we put a restriction for the maximal number of messages that any relay can handle at a given time. In particular, we introduce time, more precisely the individual times of transmissions. The rule that we impose is a very simple one, as it already renders the mathematical model quite complex: We consider a transmission via a relay as successful only if the chosen relay is not handling another message that arrived earlier. In the analysis of [11], additional later transmission attempts are possible, however in this presentation we only allow one attempt.

More precisely, we assume that every transmitter $X_i \in X^\lambda$ carries a stochastically independent alarm clock that randomly rings at two times $S_i \le T_i$, marking the start and end of its transmission attempt in a fixed finite time interval $[0, J]$. We tacitly extend X^λ to a Poisson point process on $V = W \times [0, J]^2$ with intensity measure $\lambda(\mu^{\rm space} \otimes \mu^{\rm times})(\mathrm{d}x, \mathrm{d}s, \mathrm{d}t)$. The normalized empirical measure $L_\lambda = \lambda^{-1} \sum_{i \in I_\lambda} \delta_{(X_i, S_i, T_i)}$ is now a measure on V that carries information not only on the location of transmitters, but also on their time intervals of transmission.

To avoid discontinuities in the model coming from the Poissonian placement of relays, see [13] and Section 4.1, we assume that the relays are given by a deterministic point process $Y^\lambda = (Y_i)_i$ with limiting density μ_{R}, i.e., we assume that $l_\lambda = \lambda^{-1} \sum_i \delta_{Y_i}$ converges to μ_{R} as $\lambda \to \infty$. The routing scheme is encoded in terms of a *preference kernel* $\kappa\colon X^\lambda \times Y^\lambda \to (0,\infty)$ by putting

$$\kappa(Y_j|X_i) = \frac{\kappa(X_i, Y_j)}{\lambda l_\lambda[\kappa(X_i, \cdot)]},$$

where we recall the notation $\nu[f]$ for the integral of a function f with respect to a measure ν. The measure $\kappa(\cdot|X_i)$ defines the probability that a transmitter X_i chooses the relay $Y_j \in Y^\lambda$. The function κ is assumed to satisfy some continuity conditions, but it can for example be used to adjust the range or direction of transmission depending on the location. An illustration of the evolution of the process is provided in Figure 6.

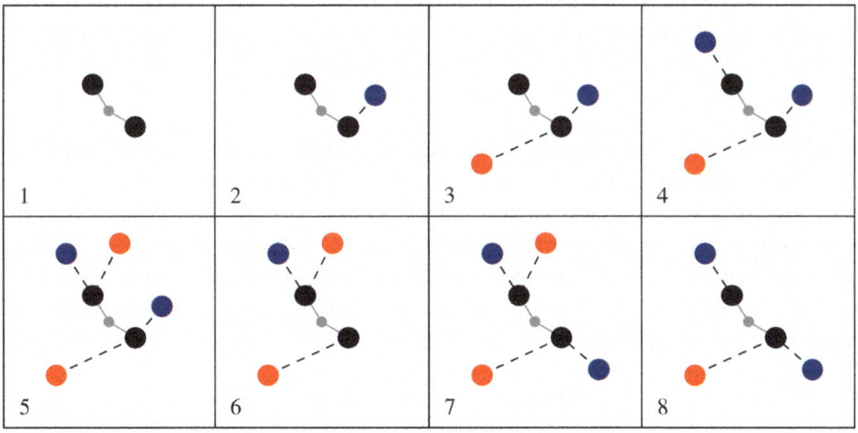

Fig. 6: A collection of five transmitters (blue and red) attempting to transmit data to two relays (black) in discrete time steps $1, \ldots, 8$. The relays are hardwired to a central base station (gray). Only the first transmitter (blue) can establish a connection. Later transmitters are unable to connect and become frustrated (red) unless the transmitters that initially established the connection have stopped their transmission.

A transmitter whose one and only relaying attempt hits an occupied relay and therefore cannot establish a connection is called frustrated. We are interested in the normalized empirical measure of frustrated marked transmitters,

$$\mathfrak{F}L_\lambda = \frac{1}{\lambda} \sum_{i \in I_\lambda} \delta_{(X_i, S_i, T_i)} \mathbb{1}\{Y(X_i) \text{ is occupied at time } S_i\},$$

where $Y(X_i) \in Y^\lambda$ denotes the relay chosen by the transmitter X_i based on the preference kernel κ. Note that $\mathfrak{F}_{L_\lambda}$ is a random counting measure on V.

In order to state our main results, we have to specify the type of questions we are able to answer for $\mathfrak{F}_{L_\lambda}$. This is captured in the choice of topology on the set of Borel measures $\mathcal{M}$ on V. In our case, this is the *τ-topology*, where convergence is tested on bounded and measurable functions, see [6, Section 6.2]. This topology is finer (stronger) then the weak topology, where we test only for continuous functions, and its associated Borel-σ-algebra is thus also richer than the one for the weak topology. More precisely, the τ-topology is the coarsest topology such that the evaluation maps $\nu \mapsto \nu(B)$ are continuous for all Borel measurable sets $B \subset V$. Thus, working with the τ-topology allows us to infer the large-deviations behavior of $\mathfrak{F}_{L_\lambda}$ evaluated on any Borel measurable event B.

The rate function I is again given by a minimization of the Kullback-Leibler divergence as defined in (2.4), but this time on a larger space of measures. Giving an explicit formula for the rate function requires some additional notation and concepts. First, we consider the process of transmitter requests to a relay at location y, which is a Poisson point process on $W^2 \times [0, J]^2$ with intensity measure $\lambda\mu(l_\lambda)$ where

$$\mu(l_\lambda)(\mathrm{d}y, \mathrm{d}x, \mathrm{d}s, \mathrm{d}t) = \kappa_{l_\lambda}(\mathrm{d}y|x)(\mu^{\mathrm{space}} \otimes \mu^{\mathrm{times}})(\mathrm{d}x, \mathrm{d}s, \mathrm{d}t)$$

with $\kappa_{l_\lambda}(\mathrm{d}y|x) = \kappa(y|x)l_\lambda(\mathrm{d}y)$. Suppose for a moment that there are several relays available at location y, then the probability for the transmission of a user x at time s to fail is given by the proportion of occupied relays at y at time s, since the preference kernel does not distinguish between different relays at y. This is a Markovian structure and it is indeed possible to characterize the random measure of frustrated transmitters as a function of the empirical measure of the Poisson point process with intensity measure

$$\mu(\mu_{\mathrm{R}}) = \mu(\mu_{\mathrm{R}}) \otimes U([0, 1])$$

on the extended state space $V' = W^2 \times [0, J]^2 \times [0, 1]$, see [11, Proposition 2.1]. Here the additional choice variable, which is drawn from the uniform distribution $U([0, 1])$ on $[0, 1]$, captures the random selection of the relay. Now, let $\mathcal{M}'$ denote the set of Borel measures on V' and consider a general measure $\nu' \in \mathcal{M}'$ which is absolutely continuous with respect to $\mu(\mu_{\mathrm{R}})$. Denote by ν'_y the associated measure of transmitters choosing a relay at y, i.e., $\nu'(\mathrm{d}y, \mathrm{d}x, \mathrm{d}s, \mathrm{d}t, \mathrm{d}u) = \nu'_y(\mathrm{d}x, \mathrm{d}s, \mathrm{d}t, \mathrm{d}u)\mu_{\mathrm{R}}(\mathrm{d}y)$. Then, the measure of frustrated transmitters can be accumulated over all relay locations y, i.e.,

$$\nu(\nu')(\mathrm{d}s, \mathrm{d}t, \mathrm{d}x) = \int_W \nu'(\mathrm{d}s, \mathrm{d}t, \mathrm{d}x, \mathrm{d}y, [1 - \beta_s(\nu'_y), 1]),$$

where the integration is performed w.r.t. $\mathrm{d}y$. For every y, the trajectory $\beta(\nu'_y)$ is a solution to the differential equation

$$\beta_t = \int_0^t \nu'_y(W, \mathrm{d}s, [t, J], [0, 1 - \beta_{s-}]),$$

constructed via a space-time discretization procedure and it describes the evolution of the occupied relays at location y. The rate function I is now given by the entropically optimal choice of a distribution ν' resulting in the target measure ν, more precisely,

$$\mathrm{I}(\nu) = \inf_{\nu' \in \mathcal{M}' : \nu(\nu') = \nu} H\big(\nu' | \mu(\mu_{\mathrm{R}})\big).$$

We can now state our main result: The family of random measures $\{\mathfrak{F}_{L_\lambda}\}_\lambda$ satisfies the LDP in the τ-topology with good rate function I.

As mentioned in Sections 2.4 and 4.1, for some atypical event A, which in this setting only has to be an element of the Borel-σ-algebra associated to the τ-topology, apart from the exponential decay, the minimizers ν of $\inf\{\mathrm{I}(\nu) : \nu \in A\}$ describe the typical behavior of the system conditioned on the event A. Such an atypical event could for example be that an unexpected large proportion of transmitters is frustrated in a certain region in W and/or in a certain time interval for transmission beginnings and/or transmission endings.

5 Random message routing

One of the most important problems for telecommunication networks is about how to send as many as possible of the messages successfully through the system in short time. In other words, one tries to optimize the throughput, the number of messages per time unit that the system can cope with. Observe that this question is not a connectivity question, since it is not about one single path, but about problems that come from the simultaneous delivery of many messages.

5.1 The model and its motivation

As we explained in Section 2.4, there is a serious upper bound on the success of a transmission put by interference constraints. Additionally each of the nodes may put an upper bound on that success in terms of its own capacity restrictions. One may try to keep the number of messages high by using several channels at the same time, but additional technical and mathematical problems arise, and we will not tackle this issue here. The capacity problem manifests itself also in congestion problems, i.e., restrictions of each of the network nodes concerning the number of messages that it can deal with at the same time.

The main task is then to define a strategy how to conduct each of the messages through the system via multihop trajectories to their goal in such a way that both, the interference and the congestion constraints are obeyed at any time. One way to resolve this is to use a decomposition into time instances, i.e., to decide for any single time instance which of the many messages is allowed to make a hop and which ones are not. Strategies like that are pretty complex and are not of the type that we want to discuss here.

Another aspect of the problem is the routing of the messages according to capacity and congestion, also considering interference, but not time, to keep the complexity of the problem simple. This becomes one of the prominent problems from traffic theory: Given a geometric network (the street system) and origins and goals of a number of messages, how should one choose all the trajectories optimally such that the total congestion (measured in terms of delay) becomes minimal, and/or the total interference stays small? Here one gives to any street segment (direct local connection between users) a weight that expresses its interference (capacity, time to travers it, ...), and the energy that is to be minimized is the total sum of the weights over all trajectories and over all hops. This is a deterministic optimization problem, and the solution is a theoretical optimum, which reflects the most important properties of the throughput properties of the system. *A priori*, the optimum depends on all the details of the network.

However, in [16] and [17], a different philosophy is followed: It is assumed there that all the messages are random and follow a joint law that depends in a soft way on the important properties (interference, congestion, occurrence) in a form of making a best compromise. That is, all the message trajectories are independent and uniformly picked among all possible trajectories that hop from user to user, and the probability gets an exponential weight that punishes high interference and high congestion. (Observe that the locations of the users are considered to be fixed, i.e., not random.) This is a modus of a common welfare, where no trajectory plays any particular role, but is handled with equal rights, and the joint family of trajectories tries to make the best out of it. The way how interference and congestion are punished reminds on the structure of the probability measure of a Gibbs measure, since it is of the form

$$\mathbb{P}_{\beta,\gamma}\big((s^i)_{i\in I}\big) = \frac{1}{Z_{\beta,\gamma}} \exp\Big\{ -\gamma\mathfrak{S}\big((s^i)_{i\in I}\big) - \beta\mathfrak{M}\big((s^i)_{i\in I}\big)\Big\}, \tag{5.1}$$

where s^i is a trajectory from user X_i to the base station placed at $o \in W$, and $\beta, \gamma \in [0, \infty)$ are strength parameters, and $Z_{\beta,\gamma}$ is the normalization that turns $\mathbb{P}_{\beta,\gamma}$ into a probability measure on all families of trajectories $X_i \to o$. The terms $\mathfrak{S}((s^i)_{i\in I})$ and $\mathfrak{M}((s^i)_{i\in I})$ describe the total interference of all the s^i respectively the total congestion of the family $s = (s^i)_{i\in I}$. Their precise choices are a bit of a matter of taste, but here is a quite natural one. Each trajectory $s^i = (k_i; s^i_0 = X_i, s^i_1, \dots, s^i_{k_i-1}, s^i_{k_i} = o)$ has its individual hop number $k_i \in \{1, \dots, k_{\max}\}$ and runs within X, i.e., from user to user. We assume that, at the given time instant that we consider, all the messages run simultaneously through the system and that each user sends out precisely one message. Then a natural way to quantify the total interference punishment is to put

$$\gamma\mathfrak{S}\big((s^i)_{i\in I}\big) = \sum_{i\in I}\sum_{l=1}^{k_i} \frac{1}{\mathrm{SIR}(s^i_{l-1}, s^i_l, X)}, \tag{5.2}$$

where we recall that the definition (2.2) of the SIR includes the parameter γ. This is the sum of the inverse SIR of any hop in the system with respect to the transmission of one message from any user in $X = (X_i)_{i\in I}$; it is large if the total interference is bad. In a model that takes some time development into account, one would assign all the hops to time steps and would consider only the interference that comes from the hops that are made at that time instance, but this model is much more complicated to study. In such a model, one would also have to make a decision at what time which hops are to be done.

The congestion term is a bit more complicated: For $i \in I$, let $m_i(s)$ be the number of hops into X_i of any of the trajectories s^j, $j \in I$, then we put

$$\mathfrak{M}\big((s^i)_{i\in I}\big) = \sum_{i\in I} m_i(s)(m_i(s) - 1), \tag{5.3}$$

which is the sum of pairs of hops that use the same user as a relay.

The model of random message routing that we set up in (5.1) can be used for studying the family of message trajectories that are controlled entirely by two of the most important properties, but are otherwise completely freely picked, may be even in very ridiculous ways with many detours. One expects that the desire to minimize interference will in the end force the trajectories to avoid detours, and to avoid too long and too many hops as well. It will be interesting to see how long the hops decide to be 'on their own', i.e., just following the probability mechanism and not some deterministic rule of any operator. Furthermore, the tendency to minimize congestion is expected to smoothen the profile of the trajectory flow. The main questions are about geometric properties of the main flow of the messages.

In this model, we choose to consider the situation where all the messages go from a user to one fixed base station, but one could certainly also study the situation where the messages intend to go to any other user. Also our assumption that each user sends out precisely one message can in principle be replaced by anything else.

5.2 A law of large numbers for the message flow in the high-density limit

The description of the message flow is rather difficult in general, since the measure $\mathbb{P}_{\beta,\gamma}$ depends on all the details of all the trajectories. Instead, we strive to derive the general picture if the number of users is large, with simplifying asymptotic formulas. Two settings are natural in this respect: The high-density limit and the thermodynamic limit. In [16] and [17], the former is considered, and we will report on that here. Hence, let the underlying user process $X = (X_i)_{i\in I_\lambda}$ be dependent on some parameter $\lambda \to \infty$, such that its normalized empirical measure $L_\lambda = \frac{1}{\lambda}\sum_{i\in I_\lambda} \delta_{X_i}$ approaches some measure μ on the communication area W. (One way to achieve this is to assume that X is a Poisson point process in W with intensity measure $\lambda\mu$.) Note that the two exponential terms in the definition of the Gibbs measure $\mathbb{P}_{\gamma,\beta}$ are divergent in

λ, and also all probability terms will turn out to be exponential in λ, hence we will be concerned with a large-deviations statement. The vehicle through which we are studying the family of trajectories $s = (s^i)_{i\in I_\lambda}$ is their normalized empirical measure of the k-step trajectories,

$$R_{k,\lambda}(s) = \frac{1}{\lambda} \sum_{i\in I_\lambda :\, k_i=k} \delta_{s^i}, \qquad k \in \{1, \dots, k_{\max}\}. \tag{5.4}$$

Note that $R_{k,\lambda}(s)$ is a measure on the set W^k, whose first marginal $\pi_0 R_{k,\lambda}$ is equal to L_λ, where we write π_l for taking the l-th marginal measure. Now we have to rewrite all the three terms (entropy and the two energy terms) as a function of $(R_{k,\lambda}(s))_{k=1}^{k_{\max}}$. From now on, we will drop the congestion term and restrict to the interference term, in order to simplify the presentation.

If the strength parameter γ in the definition (2.2) of the SIR is fixed, then, in the limit $\lambda \to \infty$, the interference term $\mathfrak{S}(s)$ is easily seen to be of order λ^2, since I_λ is of order λ. Hence, we will replace the strength parameter γ by γ/λ, whose interpretation we discussed at the end of Section 2.4. Then the interference term can be written as a function of the empirical measures as

$$\mathfrak{S}(s) = \gamma \sum_{k=1}^{k_{\max}} \int_W R_{\lambda,k}(s)(\mathrm{d}x_0, \dots, \mathrm{d}x_{k-1}) \sum_{l=1}^{k} \frac{\int_W \ell(|y - x_l|) L_\lambda(\mathrm{d}y)}{\ell(|x_{l-1} - x_l|)}, \qquad x_k = o. \tag{5.5}$$

In the limit $\lambda \to \infty$, we will write ν_k for a generic measure on W^k that plays the role of $R_{k,\lambda}$, and the interference term can easily be approximated in terms of $\Sigma = (\nu_k)_{k=1}^{k_{\max}}$ with the function

$$\mathrm{S}(\Sigma) = \sum_{k=1}^{k_{\max}} \int_W \nu_k(\mathrm{d}x_0, \dots, \mathrm{d}x_{k-1}) \sum_{l=1}^{k} g(x_{l-1}, x_l), \qquad x_k = o, \tag{5.6}$$

where $g(x, y) = \int_W \mu(\mathrm{d}z)\ell(|z - y|)/\ell(|x - y|)$. Furthermore, the following entropy term describes the negative exponential rate of the counting complexity

$$\mathrm{J}(\Sigma) = \sum_{k=1}^{k_{\max}} \int_{W^k} \mathrm{d}\nu_k \log \frac{\mathrm{d}\nu_k}{\mathrm{d}\mu^{\otimes k}} + \log \mu(W) \sum_{k=1}^{k_{\max}} (k-1)\nu_k(W) \in [0, \infty]. \tag{5.7}$$

Now we can formulate the first result about the high-density limit: The empirical measures $(R_{\lambda,k}(S))_{k=1}^{k_{\max}}$ satisfy a law of large numbers under $\mathbb{P}^\lambda_{\gamma,0}$. More precisely, they converge weakly to the unique minimizer of the variational formula

$$\inf_{\Sigma=(\nu_k)_{k=1}^{k_{\max}} :\, \sum_{k=1}^{k_{\max}} \pi_0 \nu_k = \mu} \Big(\mathrm{J}(\Sigma) + \gamma \mathrm{S}(\Sigma) \Big). \tag{5.8}$$

For $k_{\max} > 1$, the minimizer is given as

$$\nu_k(\mathrm{d}x_0,\dots,\mathrm{d}x_{k-1}) = \mu(\mathrm{d}x_0)A(x_0)\prod_{l=1}^{k-1}\frac{\mu(\mathrm{d}x_l)}{\mu(W)}\mathrm{e}^{-\gamma\sum_{l=1}^{k} g(x_{l-1},x_l)}, \quad x_k = o, \tag{5.9}$$

where the normalizing function A is defined as

$$\frac{1}{A(x_0)} = \sum_{k=1}^{k_{\max}}\int_{W^{k-1}}\prod_{l=1}^{k-1}\frac{\mu(\mathrm{d}x_l)}{\mu(W)}\mathrm{e}^{-\gamma\sum_{l=1}^{k} g(x_{l-1},x_l)}, \quad x_0 \in W. \tag{5.10}$$

This law of large numbers is based on a large-deviations principle for the empirical measures under the counting measure, which is derived by tedious, but elementary combinatorics, combined with a discretization procedure. If the congestion term is also present (i.e., if we do not put $\beta = 0$), then an analogous variational formula and law of large numbers is proved, but the characterization of the minimizers is substantially more involved; actually its uniqueness is unknown.

5.3 Analytic properties of the message trajectory flow

Hence, in the high-density limit, the situation has become much simpler. The typical message flow is described in terms of a law of large numbers for its normalized empirical measures, and the (deterministic) limit is characterized as the minimizer of a characteristic variational formula. Luckily, its description in terms of the Euler-Lagrange equations can be derived in a standard way and turns out to be quite explicit. It is amenable to a further study, which we are going to do now. We restrict to presenting simulations and to comment on what can be proved without too great difficulties. In [17], we analyze the typical message trajectory distribution $(\nu_k)_{k=1}^{k_{\max}}$ in two regimes:

5.3.1 Large distances, many hops

That is, W is large, and $k_{\max} = \infty$, and the starting site x_0 is far from o. Here we answer the following questions:

- How many hops will be taken?
- How large are the hops on an average? Different lengths at the beginning and the end of the trajectory?
- Does the long trajectory approach a straight line?

Here we have interesting and partially surprising results, which show that, in this limit, the typical length of a hop diverges to ∞ logarithmically in the number of hops k as $k \to \infty$, and the trajectories approximate a straight line from the starting site x_0 to the origin o. We also give qualitative (even exponential) estimates for deviations of the trajectory from the straight line. This effect seems to come from the fact that *a priori* every hop has the same positive probability, before the interference is punished, and this gives a large effect, since W is large. The proof methods rely on elementary analytic tools like convexity and the Laplace method.

5.3.2 High interference punishment

That is, $\gamma \to \infty$. Here we answer the following questions:

- Do the trajectories approach a straight line?
- How costly are deviations from the straight line?

Our answers are in the affirmative under appropriate, quite mild assumptions on the path-loss functions ℓ, and we feel that there is much more room for positive results for much more general choices of ℓ. Again, the proof methods are elementary analytic tools.

5.4 Illustrations

Let us give simulations to illustrate our findings about the influence of strong interference punishment in the special case $d = 1, W = [-5,5], \ell(r) = \min\{1, r^{-4}\}, \mu = \mathrm{Leb}|_W$, and $k_{\max} = 2$. First we show the distribution of those starting points from which all the messages jump to the base station at the origin in just one single hop for some values of γ. In Figure 7, note the strong effect of the interference penalization

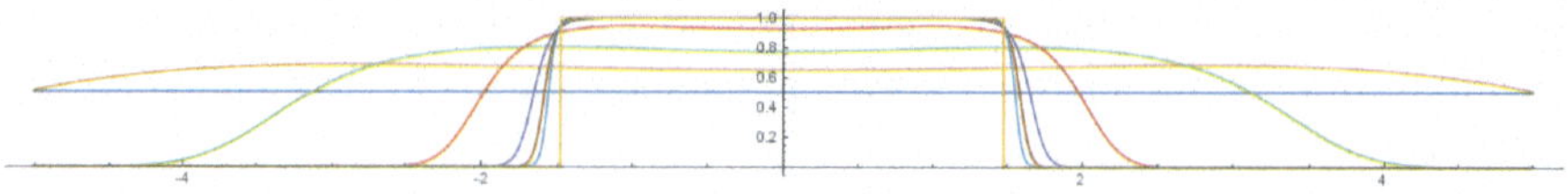

Fig. 7: The graph of the map $x_0 \mapsto \nu_1(\mathrm{d}x_0)/\mu(\mathrm{d}x_0)$. Blue line: $\gamma = 1$, orange line: $\gamma = 1.5$, green line: $\gamma = \infty$

already for small values of γ. There is a sharp transition from one hop (distance ≤ 1.5) to two hops (distance ≥ 1.45). Actually, our theoretical results show that such sharp transitions indeed occur in the limit $\gamma \to \infty$ in broad generality.

Now we show the joint distribution of the initial site and the intermediate hop site for two-steps trajectories: In Figure 8, the distribution of x_1 appears to be a bit noisy for $|x_0| \in (1.45, 1.5)$, i.e., the decision where to place the intermediate hop is not clearly decided, but truly random. One can observe a concentration on $|x_1| \approx \frac{1}{2}|x_0|$ for $|x_0| \in (1.5, 2.5)$ and on $|x_1| \approx c|x_0|$ for $|x_0| \in (2.5, 5]$ with some $c \in (0, 1)$.

Acknowledgements The authors like to thank their coauthors, in particular Christian Hirsch and András Tóbiás, for their contributions to the subject. BJ thanks Alexander Wapenhans for providing Figure 2.

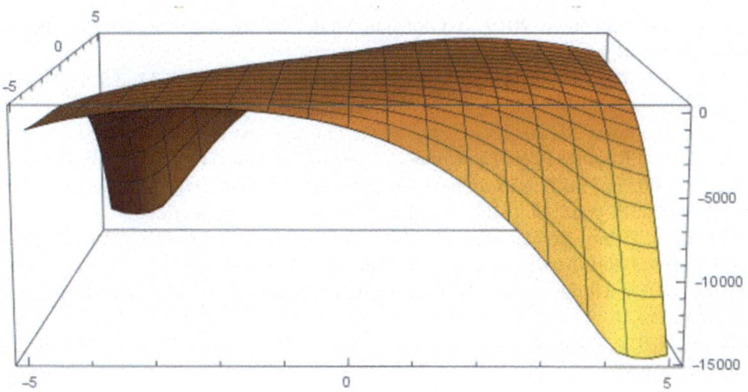

Fig. 8: Graph of the map $(x_0, x_1) \mapsto \log \nu_2(\mathrm{d}x_0, \mathrm{d}x_1)/\mu(\mathrm{d}x_0)\mu(\mathrm{d}x_1)$ for $\gamma = 1$

References

1. Balister, P., Bollobás, B., and Walters, M. 2005. Continuum Percolation with Steps in the Square or the Disc *Random Structures Algorithms* 26:392–403
2. Baccelli, F. and Błaszczyszyn, B. 2009. *Stochastic Geometry and Wireless Networks: Volume II: Applications* Now Publishers Inc.
3. Baccelli, F. and Błaszczyszyn, B. 2009. *Stochastic Geometry and Wireless Networks: Volume I: Theory* Now Publishers Inc.
4. Błaszczyszyn, B. and Yogeshwaran, D. 2013. Clustering and Percolation of Point Processes *Electron. J. Probab.* 18:1–20
5. Cali, E., Gafur, N.N., Jahnel, B., En-Najjary, T. and Patterson, R. 2018. Percolation for D2D Networks on Street Systems *Proceedings of WiOpt/SpaSWiN*
6. Dembo, A. and Zeitouni, O. 1998. *Large Deviations Techniques and Applications*, 2nd edition, Springer, Berlin
7. Den Hollander, F. 2008. *Large Deviations*, American Mathematical Society, Fields Institute monographs
8. Franceschetti, M. and Meester, R. 2008. *Random Networks for Communication: From Statistical Physics to Information Systems* Cambridge University Press
9. Gloaguen, C., Fleischer, F., Schmidt, H., and Schmidt, V. 2000. Fitting of Stochastic Telecommunication Network Models via Distance Measures and Monte–Carlo Tests *Telecommunication Systems* 31:4
10. Gupta, P. and Kumar, P.R. 2000. The Capacity of Wireless Networks *IEEE Trans. Inform. Theory* 46:2
11. Hirsch, C. and Jahnel, B. 2019. Large deviations for the Capacity in Dynamic Spatial Relay Networks *Markov Process. Related Fields* 25:33–73
12. Hirsch, C., Jahnel, B. and Cali, E. 2018. Continuum Percolation for Cox Point Processes *Stoch. Process. Their Appl.*
13. Hirsch, C., Jahnel, B., Keeler, P. and Patterson, R. 2018. Large Deviations in Relay-augmented Wireless Networks *Queueing Systems* 88:3–4
14. Hirsch, C., Jahnel, B. and Patterson, R. 2018. Space-time Large Deviations in Capacity-constrained Relay Networks *Lat. Am. J. Probab.* 15:1–29
15. Keeler, P., Jahnel, B., Maye, O., Aschenbach, D. and Brzozowski, M. 2018. Disruptive Events in High-density Cellular Networks *Proceedings of WiOpt/SpaSWiN*

16. König, W. and Tóbiás, A. 2019. A Gibbsian Model for Message Routing in Highly Dense Multihop Networks *Lat. Am. J. Probab.*
17. König, W. and Tóbiás, A. 2018. Routeing Properties in a Gibbsian Model for Highly Dense Multihop Networks *arXiv:1801.04985*
18. Last, G. and Penrose, M. 2017. *Lectures on the Poisson Process* Cambridge University Press
19. Meester, R. and Roy, R. 1996. *Continuum Percolation* Cambridge University Press
20. Okabe, A. 2000. *Spatial Tessellations: Concepts and Applications of Voronoi Diagrams* Wiley Series in Probability and Statistics: Applied Probability and Statistics
21. Rassoul-Agha, F. and Seppäläinen, T. 2015. *A Course on Large Deviation Theory with an Introduction to Gibbs Measures* Graduate Studies in Mathematics
22. Tóbiás, A. 2016. Highly Dense Mobile Communication Networks with Random Fadings *arXiv:1606.06473*
23. Tóbiás, A. 2018. Signal to Interference Ratio Percolation for Cox Point Processes *arXiv:1808.09857*

Mathematical Modeling of Semiconductors: From Quantum Mechanics to Devices

Markus Kantner, Alexander Mielke, Markus Mittnenzweig, and Nella Rotundo

Abstract We discuss recent progress in the mathematical modeling of semiconductor devices. The central result of this paper is a combined quantum-classical model that self-consistently couples van Roosbroeck's drift-diffusion system for classical charge transport with a Lindblad-type quantum master equation. The coupling is shown to obey fundamental principles of non-equilibrium thermodynamics. The appealing thermodynamic properties are shown to arise from the underlying mathematical structure of a damped Hamitlonian system, which is an isothermal version of so-called GENERIC systems. The evolution is governed by a Hamiltonian part and a gradient part involving a Poisson operator and an Onsager operator as geoemtric structures, respectively. Both parts are driven by the conjugate forces given in terms of the derivatives of a suitable free energy.

1 Introduction

The development of semiconductor devices has been strongly supported by mathematical modeling and numerical simulations over the last decades. Mathematical models provide insights into the internal physical mechanisms in semiconductor devices, can help to optimize particular designs and decrease the development costs

Markus Kantner
Weierstrass Institute for Applied Analysis and Stochastics (WIAS), Mohrenstr. 39, 10117 Berlin, Germany, e-mail: kantner@wias-berlin.de

Alexander Mielke
WIAS and Humboldt University of Berlin, Department of Mathematics, Rudower Chaussee 25, 12489 Berlin, Germany, e-mail: e-mail: mielke@wias-berlin.de

Markus Mittnenzweig
WIAS, e-mail: mittnenzweig@wias-berlin.de

Nella Rotundo
WIAS, e-mail: rotundo@wias-berlin.de

M. Hintermüller, J. F. Rodrigues (eds.), *Topics in Applied Analysis and Optimisation*, CIM Series in Mathematical Sciences, https://doi.org/10.1007/978-3-030-33116-0_11

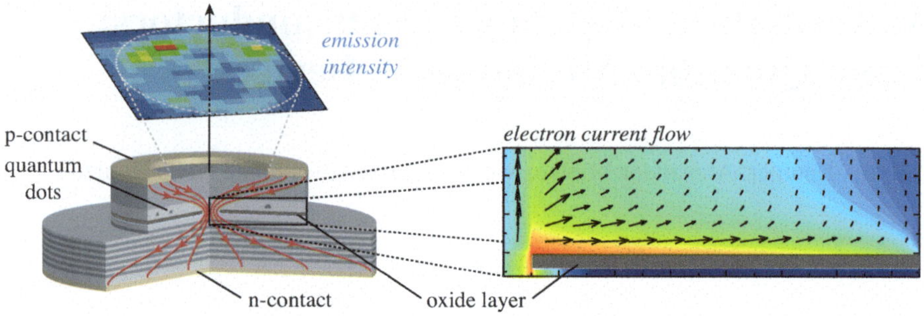

Fig. 1: Lateral current spreading in an oxide–confined single–photon source leading to unwanted optical activity of parasitic quantum dots in the outer parts of the structure [13].

by reducing the demand for the expensive processing of a large number of prototypes. For instance, the progress in performance and miniaturization of silicon transistors following Moore's law over the last decades was inconceivable without modern TCAD (technology computer-aided design) simulation tools. The on-going reduction of the characteristic length scales of semiconductor devices as well as the integration of semiconductor nanostructures such as quantum dots [2], requires an extension of the classical semiconductor device equations towards the inclusion of quantum mechanical models.

Many modern opto-electronic devices such as, e.g., quantum light sources and nanolasers [6], employ semiconductor quantum dots as an optically active element embedded in photonic micro-resonators. The transport of charge carriers in such devices can be described by semi-classical drift-diffusion-reaction models, such as the van Roosbroeck system [38]. For example, in [13] the current injection into an electrically driven single-photon emitting diode has been investigated in order to understand the experimentally observed malfunction of the design, see Fig. 1. The device features an oxide aperture that is intended to confine the injection current into a narrow region above the aperture, where a single quantum dot shall be electrically excited. The experimentally observed electroluminescence, however, indicates a counterintuitive rapid lateral current spreading. On the basis of the van Roosbroeck system, the phenomenon was reproduced in numerical simulations and eventually understood as an inherent feature of the design under the typical operation conditions of the device. Finally, based on mathematical modeling, a revised design with superior current confinement was suggested [13]. While this example convincingly substantiates the importance of carrier transport modeling, many other important properties of the single-photon source can not be described by the van Roosbroeck system. In particular, the quantum optical features of the radiation generated by the quantum dot, namely the correlation statistics of the emitted photons that allow to quantify non-classical phenomena like "photon anti-bunching" [37], are not accessible by the van Roosbroeck system. This requires a microscopic modeling framework, that describes the evolution of open quantum systems.

A broad class of problems in semiconductor quantum optics can be described by quantum master equations [4]. These are evolution equations for the quantum mechanical density matrix $\boldsymbol{\rho} \in \mathbb{C}^{n\times n}$, that is an Hermitian operator which describes the state of a quantum system. Unlike the Schrödinger equation, which models the Hamiltonian evolution of closed quantum systems, quantum master equations allow for the consideration of dissipative dynamics that arise due to the coupling of the quantum system with its macroscopic environment. The simplest class of quantum master equation, which guarantees the preservation of trace $\operatorname{tr}\boldsymbol{\rho} = 1$, self-adjointness $\boldsymbol{\rho} = \boldsymbol{\rho}^*$, and positivity $\boldsymbol{\rho} \geq 0$ of the density matrix, is the Lindblad master equation [18, 19], see (4.9). While providing access to the microscopic dynamics and the quantum optical figures of merit of open quantum systems, the Lindblad equation (1.1d) complements the classical modeling approaches to semiconductor devices based on the van Roosbroeck system for the electrostatic potential ϕ and the charge carrier densities n and p for electrons and holes, respectively.

Our interest lies in a mathematically systematic and thermodynamically correct derivation of coupled systems of the form

$$0 = \operatorname{div}(\varepsilon\nabla\phi) + e_0\big(C + p - n + \rho_{\mathrm{qd}}\operatorname{tr}(\mathsf{Z}\boldsymbol{\rho})\big), \tag{1.1a}$$

$$\dot n = \operatorname{div}\big(\mathbb{M}_n(\nabla n - n\nabla\phi)\big) - R(n,p) + R_n^{\text{quant-class}}(n,p,\boldsymbol{\rho}), \tag{1.1b}$$

$$\dot p = \operatorname{div}\big(\mathbb{M}_p(\nabla p + p\nabla\phi)\big) - R(n,p) + R_p^{\text{quant-class}}(n,p,\boldsymbol{\rho}), \tag{1.1c}$$

$$\dot{\boldsymbol{\rho}} = \jmath[\boldsymbol{\rho}, \mathsf{H} + e_0\phi\mathsf{Z}] + \mathfrak{D}_0\boldsymbol{\rho} + \mathfrak{D}(n,p)\boldsymbol{\rho}. \tag{1.1d}$$

In [14], a hybrid quantum-classical modeling approach was introduced that self-consistently combines these two approaches and allows for a comprehensive description of quantum dot-based semiconductor devices for quantum optical applications. Here we want to show that this model is a special case of a general class of models that have the form of damped Hamiltonian system, in the sense explained now.

From a mathematical point of view the thermodynamic consistency of complex physical systems like (1.1) can be encoded in the GENERIC framework. GENERIC is an acronym for *General Equations for Non-Equilibrium Reversible Irreversible Coupling* and provides a thermodynamically consistent way of coupling reversible Hamiltonian dynamics with irreversible dissipative dynamics, see [9,35] and Section 3.1. In Section 3.2 we introduce the concept of *damped Hamiltonian systems* as a simplified, isothermal version of GENERIC systems. They are defined by a quadruple $(\boldsymbol{Q}, \mathcal{F}, \mathbb{J}, \mathbb{K})$ where $\boldsymbol{Q}$ is the state space and $\mathcal{F}(q)$ is the free energy functional on it. Moreover, the state space carries two geometric structures, namely the Poisson structure $\mathbb{J}$ that generates the Hamiltonian evolution and the Onsager operator $\mathbb{K}$ driving the dissipative dynamics. The time evolution of the damped Hamiltonian system $(\boldsymbol{Q}, \mathcal{F}, \mathbb{J}, \mathbb{K})$ is given via

$$\dot q = \big(\mathbb{J}(q) - \mathbb{K}(q)\big)\mathrm{D}\mathcal{F}(q). \tag{1.2}$$

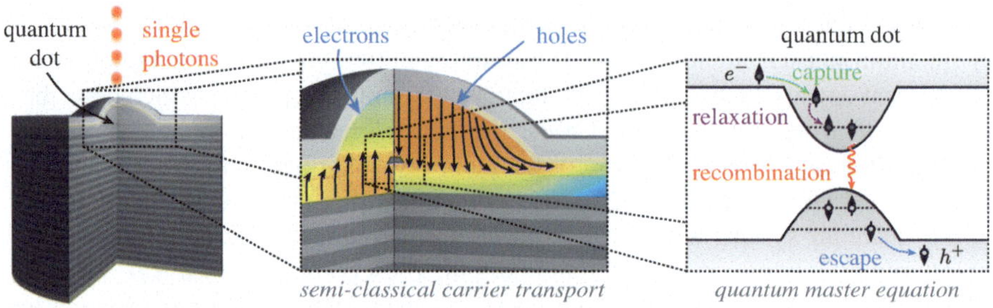

Fig. 2: The hybrid quantum-classical modeling approach for quantum light sources combines semi-classical carrier transport theory with microscopic models for the quantum dot-photon system [12].

The Poisson operator $\mathbb{J}(q)$ is skew-symmetric and satisfies the Jacobi identity, while the Onsager operator $\mathbb{K}(q)$ is symmetric and positive semidefinite, which encodes the second law of thermodynamics.

The aim of the paper is to realize (1.1) in the form (1.2) with the state variable by $q = (n, p, \boldsymbol{\rho})$. For this, we first introduce in details the van Roosbroeck system (1.1), but without (1.1d), in Section 2. Next, we shortly summarize the abstract, thermodynamical modeling via the GENERIC framework in Section 3.1 and via so-called *damped Hamiltonian systems* in Section 3.2. A special emphasis to the additive structure of dissipative processes and to admissible couplings are given in Sections 3.3 and 3.4, respectively.

In Section 4 we then apply the abstract theory to (1.1), first to certain subparts and finally to the full coupled system. Based on [1] it was shown in [23] that the van Roosbroeck system for (ϕ, n, p) can be written as a gradient system (i.e. with $\mathbb{J} \equiv 0$), see Section 4.2. Extensions to more general reaction systems or to more general carrier statistics are discussed next. The most recent building block of the theory was provided in [29], where it was shown that all Markovian quantum master equation in Lindblad form that satisfy a suitable detailed-balance condition can be written as a damped Hamiltonian, see Section 4.5. In Section 4.6, we present the analysis from [14] which allows to show that the free energy is a Liapunov function, without referring to Onsager operator. Finally, Section 4.7 contains the nontrivial coupling between quantum and classical system via Onsager operators.

The ideas in this paper provide a series of mathematical concepts that are useful in modeling complex physical system, where different components interact in nontrivial ways. Here, we apply these concepts to models for semiconductor physics, but we believe that they are also relevant and helpful many other application areas.

2 The van Roosbroeck system

The van Roosbroeck system is a system of drift-diffusion-reaction equations that describe the transport of charge carriers in semiconductor devices in their self-consistently generated electrostatic field. The model equations are posed on a domain $\Omega \in \mathbb{R}^d$ with $d = 1, 2$ or 3, time $t \in [0, T]$ and read

$$0 = \operatorname{div}(\varepsilon \nabla \phi) + e_0\big(C + p - n\big), \tag{2.1a}$$

$$\dot{n} = -\operatorname{div} \boldsymbol{J}_{\mathrm{e}} - R(n, p), \tag{2.1b}$$

$$\dot{p} = -\operatorname{div} \boldsymbol{J}_{\mathrm{h}} - R(n, p). \tag{2.1c}$$

The continuity equations (2.1b)–(2.1c) model the transport and recombination dynamics of the electron density n and hole density p, where $\boldsymbol{J}_{\mathrm{e}}$ and $\boldsymbol{J}_{\mathrm{h}}$ are the respective carrier flux densities. The reaction rate R describes the recombination and generation kinetics of electron-hole pairs, which can be created or annihilated in several radiative and non-radiative processes [20, 40]. The electrostatic interaction between the negatively charged electrons and the positively charge holes (which are missing electrons), is mediated by Poisson's equation (2.1a) for the electrostatic potential ϕ. Here, e_0 is the elementary charge, ε is the dielectric permittivity of the semiconductor material and $C : \Omega \to \mathbb{R}$ is the built-in doping profile.

2.1 Carrier flux densities and chemical potentials

The van Roosbroeck system needs to be supplemented by state equations for the carrier densities and the carrier flux densities. The drift-diffusion flux densities read

$$\boldsymbol{J}_n = \mathbb{M}_n n \nabla \phi - \mathbb{D}_n \nabla n, \qquad \boldsymbol{J}_p = -\mathbb{M}_p p \nabla \phi - \mathbb{D}_p \nabla p, \tag{2.2}$$

where the gradient of the electrostatic potential generates the drift transport of the charge carriers within the electric field $\boldsymbol{E} = -\nabla \phi$. As the charges of electrons and holes have different signs, they drift into opposite directions. The electric conductivity is determined by the carrier mobilities $\mathbb{M}_{\mathrm{e}}$ and $\mathbb{M}_{\mathrm{h}}$ that are material-dependent parameters. A second process leading to the transport of charge carriers is diffusion, which is driven by the gradients of the carrier densities. In the simplest case the diffusion constant matrices $\mathbb{D}_{\mathrm{e}}$ and $\mathbb{D}_{\mathrm{h}}$ are linked to the mobility matrices via the Einstein relation

$$\mathbb{D}_{\mathrm{e}} = \frac{k_{\mathrm{B}}\theta}{e_0}\mathbb{M}_{\mathrm{e}}, \qquad \mathbb{D}_{\mathrm{h}} = \frac{k_{\mathrm{B}}\theta}{e_0}\mathbb{M}_{\mathrm{h}}, \tag{2.3}$$

which is a manifestation of the fluctuation-dissipation theorem. Here, k_{B} is the Boltzmann constant and θ is the temperature.

The carrier densities are linked to the electrostatic potential ϕ and the electrochemical potentials μ_{e} and μ_{h}, often denoted also as *quasi-Fermi energies*, by the state equations

$$n = N_e \mathfrak{F}\left(\frac{\mu_e - (E_e - e_0\phi)}{k_B\theta}\right), \qquad p = N_h \mathfrak{F}\left(\frac{\mu_h - (E_h + e_0\phi)}{k_B\theta}\right). \tag{2.4}$$

The equations for the carrier densities (2.4) are based on the assumption of a quasi-equilibrium distribution of the electrons and holes in the material. The function $\mathfrak{F}$ contains information on the statistical distribution function underlying the nature of the particles (Fermi–Dirac statistics for fermions, Bose–Einstein statistics for bosons or Maxwell–Boltzmann statistics for classical particles), the energy band structure and the density of states provided by the material. The effective density of states N_e, N_h and the band-edge energy levels E_e, E_h are material specific parameters. Throughout this work we focus on the most simple case of non-degenerate semiconductors in which the Maxwell–Boltzmann statistics is considered, this means that $\mathfrak{F}(x) = \exp(x)$ is an exponential function. At high carrier densities or at cryogenic temperatures, degeneration effects due to the Pauli exclusion principle come into play. In this case $\mathfrak{F}$ is typically given by a Fermi–Dirac integral in conventional semiconductor crystals [40], or the Gauss–Fermi integral in disordered organic materials [21]. As a consequence, the Einstein relation (2.3) must be generalized in degenerate semiconductors to account for the density-dependent nonlinear diffusion [15,21].

Using the Einstein relation (2.3) and considering the Maxwell–Boltzmann statistics in the state equations (2.4), the flux densities can be written as gradients of the chemical potentials

$$\boldsymbol{J}_e = -\frac{1}{e_0}\mathbb{M}_e n \nabla \mu_e, \qquad \boldsymbol{J}_h = -\frac{1}{e_0}\mathbb{M}_h p \nabla \mu_h. \tag{2.5}$$

This reflects a basic principle of linear irreversible thermodynamics, where the gradients of the chemical potentials are the thermodynamic forces driving the carrier flux [10,31].

In the thermodynamic equilibrium, the chemical potentials approach a common global constant $\mu_e^{eq} \equiv \mu_h^{eq} \equiv 0$ such that the net current flow vanishes. Microscopically, this feature emerges from the principle of detailed balance principle in the thermodynamic equilibrium. The corresponding equilibrium carrier densities of the non-degenerate semiconductor read

$$n_{eq} = N_e \exp\left(-\frac{E_e - e_0\phi_{eq}}{k_B\theta}\right), \qquad p_{eq} = N_h \exp\left(-\frac{E_h + e_0\phi_{eq}}{k_B\theta}\right), \tag{2.6}$$

where ϕ_{eq} solves (2.1a) with equilibrium boundary conditions (see Section 2.2). Finally, the reaction rate R takes the form

$$R(n,p) = r(n,p)\left(np - n_{eq}p_{eq}\right), \tag{2.7}$$

with $r(n,p) = \sum_{m=1}^{m_*} r_m(n,p) \geq 0$, where $m = 1,\ldots,m_*$ labels various recombination processes (e.g, Shockley–Read–Hall recombination, direct band-to-band recombination, Auger recombination etc., see [36, 40]). For the bi-polar van Roosbroeck system with non-degenerate carrier statistics one often introduces the intrinsic carrier density n_{intr} according to $n_{\text{intr}}^2 = n_{\text{eq}} p_{\text{eq}}$.

2.2 Boundary conditions

The van Roosbroeck system (2.1) is supplemented with initial conditions at time $t = 0$

$$\phi(x,0) = \phi^I(x), \qquad \mu_{\text{e}}(x,0) = \mu_{\text{e}}^I(x), \qquad \mu_{\text{h}}(x,0) = \mu_{\text{h}}^I(x) \quad \text{for} \quad x \in \Omega,$$

where ϕ^I, μ_{e}^I and μ_{h}^I are the initial distributions.

Regarding the boundary conditions modeling electrical contacts or semiconductor-insulator interfaces, we assume a decomposition of the domain boundary as

$$\partial\Omega = \Gamma_D \cup \Gamma_N,$$

with Dirichlet boundary conditions imposed on Γ_D and Neumann conditions on Γ_N. Ideal Ohmic contacts are modeled as Dirichlet boundary conditions. For a device featuring several Ohmic contacts $\Gamma_D = \bigcup_\alpha \Gamma_{D,\alpha}$, on imposes

$$\phi(x,t) = \phi_0(x) + U_\alpha(t), \qquad \mu_{\text{e}}(x,t) = -e_0 U_\alpha(t), \qquad \mu_{\text{h}}(x,t) = e_0 U_\alpha(t) \tag{2.8}$$

for all $x \in \Gamma_{D,\alpha}$. Here, $U_\alpha(t)$ is the (possibly time-dependent) voltage applied to the α-th contact and $\phi_0(x)$ is the built-in electrostatic potential that enforces local charge neutrality on $\Gamma_{D,\alpha}$, i.e., it holds

$$0 = C + N_{\text{h}} \mathfrak{F}\left(\frac{-E_{\text{h}} - e_0\phi_0}{k_B\theta}\right) - N_{\text{e}} \mathfrak{F}\left(\frac{-E_{\text{e}} + e_0\phi_0}{k_B\theta}\right)$$

everywhere on Γ_D. In the case of Maxwell–Boltzmann statistics, the built-in potential can be explicitly obtained as

$$\phi_0 = -\frac{E_{\text{h}} - E_{\text{e}}}{2e_0} + \frac{k_B\theta}{2e_0} \log\left(\frac{N_{\text{h}}}{N_{\text{e}}}\right) + \frac{k_B\theta}{e_0} \operatorname{arsinh}\left(\frac{C}{2n_{\text{intr}}}\right).$$

For degenerate semiconductors ϕ_0 must be obtained numerically. See [36, 39, 40] for other boundary conditions modeling electrical contacts, e.g., Gate contacts or Schottky contacts. On the boundary segments Γ_N one typically imposes no-flux boundary conditions

$$\boldsymbol{\nu} \cdot \nabla\phi = 0, \qquad \boldsymbol{\nu} \cdot \boldsymbol{J}_{\text{e}} = 0, \qquad \boldsymbol{\nu} \cdot \boldsymbol{J}_{\text{h}} = 0, \tag{2.9}$$

which are homogeneous Neumann boundary conditions that guarantee that the computational domain is self contained. Here, $\boldsymbol{\nu}$ is the outward-oriented normal vector on the boundary segment. The boundary conditions (2.9) are no physical boundaries and must be chosen carefully in order to restrict the computational domain to a reasonably small region [39].

3 Mathematical modeling based on thermodynamical principles

The classical approach to thermodynamical modeling starts from balance laws (e.g., for mass of different species, linear momentum, charges, energy, etc.) and then adds suitable constitutive relations to connect the state variables and the fluxes. In a second step, the constitutive laws are restricted to satisfy the second law of thermodynamics, i.e., the entropy is non-decreasing in non-isothermal systems. Correspondingly for systems at constant temperature the total energy is not conserved, but a suitable free energy is decreasing.

Here we will use a different approach that starts from energy and entropy functionals and uses their derivatives with respect to the state variables (also called thermodynamical conjugate forces) as driving forces. We first discuss the more general modeling framework GENERIC and then restrict to the isothermal version, which we refer to as *damped Hamiltonian systems*.

3.1 The GENERIC framework

The framework of GENERIC was introduced by Morrison [30] under the name *metriplectic systems*, see [3, Sec. 15.4] for an outline of these early developments. In [9, 35] Öttinger and Grmela introduced the name GENERIC to emphasize the thermodynamical modeling aspects that were relevant for their applications in fluid mechanics. More mathematical formulations are given in [22, 25] with applications to thermoplasticity and optoelectronics, respectively.

A GENERIC system is a quintuple $(\boldsymbol{Q}, \mathcal{E}, \mathcal{S}, \widehat{\mathbb{J}}, \widehat{\mathbb{K}})$, where the smooth functionals $\mathcal{E}$ and $\mathcal{S}$ on the state space $\boldsymbol{Q}$ denote the total energy and the total entropy, respectively. Moreover, $\boldsymbol{Q}$ carries two geometric structures, namely a Poisson structure $\widehat{\mathbb{J}}$ and a dissipative Onsager structure $\widehat{\mathbb{K}}$, i.e., for each $q \in \boldsymbol{Q}$ the operators $\widehat{\mathbb{J}}(q)$ and $\widehat{\mathbb{K}}(q)$ map the cotangent space $\mathrm{T}^*_q\boldsymbol{Q}$ into the tangent space $\mathrm{T}_q\boldsymbol{Q}$. The evolution of the system is given as the sum of the Hamiltonian part $\widehat{\mathbb{J}}(q)\mathrm{D}\mathcal{E}(q)$ and the gradient-flow $\widehat{\mathbb{K}}(q)\mathrm{D}\mathcal{S}(q)$, namely

$$\dot{q} = \widehat{\mathbb{J}}(q)\mathrm{D}\mathcal{E}(q) + \widehat{\mathbb{K}}(q)\mathrm{D}\mathcal{S}(q). \tag{3.1}$$

The basic conditions on the geometric structures $\widehat{\mathbb{J}}$ and $\widehat{\mathbb{K}}$ are the symmetries

$$\widehat{\mathbb{J}}(q) = -\widehat{\mathbb{J}}^*(q) \qquad \text{and} \qquad \widehat{\mathbb{K}}(q) = \widehat{\mathbb{K}}^*(q) \tag{3.2}$$

and the structural properties

$$\begin{aligned}&\widehat{\mathbb{J}} \text{ satisfies Jacobi's identity,}\\&\widehat{\mathbb{K}}(q) \text{ is positive semi-definite, i.e., } \langle \xi, \widehat{\mathbb{K}}(q)\xi\rangle \geq 0.\end{aligned} \tag{3.3}$$

Here, Jacobi's identity for $\widehat{\mathbb{J}}$ holds, if for all $\eta_j \in \mathrm{T}^*_q\boldsymbol{Q}$ we have

$$\langle \eta_1, \mathrm{D}\widehat{\mathbb{J}}(q)[\widehat{\mathbb{J}}(q)\eta_2]\eta_3\rangle + \langle \eta_2, \mathrm{D}\widehat{\mathbb{J}}(q)[\widehat{\mathbb{J}}(q)\eta_3]\eta_1\rangle + \langle \eta_3, \mathrm{D}\widehat{\mathbb{J}}(q)[\widehat{\mathbb{J}}(q)\eta_1]\eta_2\rangle = 0. \tag{3.4}$$

The central conditions connecting $\widehat{\mathbb{J}}$, $\widehat{\mathbb{K}}$, $\mathcal{E}$, and $\mathcal{S}$ ask that the energy functional does not contribute to dissipative mechanisms and that the entropy functional does not contribute to reversible dynamics, which is encoded in the following *non-interaction conditions*:

$$\textbf{(NIC)} \qquad \forall\, q \in \boldsymbol{Q}: \quad \widehat{\mathbb{J}}(q)\mathrm{D}\mathcal{S}(q) = 0 \quad \text{and} \quad \widehat{\mathbb{K}}(q)\mathrm{D}\mathcal{E}(q) = 0. \tag{3.5}$$

In summary, the quintuple $(\boldsymbol{Q}, \mathcal{E}, \mathcal{S}, \widehat{\mathbb{J}}, \widehat{\mathbb{K}})$ is called a GENERIC system, if the conditions (3.2)–(3.5) hold.

Of course, the structure of GENERIC is geometric in the sense that it is invariant under coordinate transformations, see [22]. The first observation is that (3.3) and (3.5) imply energy conservation and entropy increase:

$$\frac{\mathrm{d}}{\mathrm{d}t}\mathcal{E}(q(t)) = \langle \mathrm{D}\mathcal{E}(q), \dot q\rangle = \langle \mathrm{D}\mathcal{E}(q), \widehat{\mathbb{J}}(q)\mathrm{D}\mathcal{E}(q) + \widehat{\mathbb{K}}(q)\mathrm{D}\mathcal{S}(q)\rangle = 0 + 0 = 0, \tag{3.6}$$

$$\frac{\mathrm{d}}{\mathrm{d}t}\mathcal{S}(q(t)) = \langle \mathrm{D}\mathcal{S}(q), \dot q\rangle = \langle \mathrm{D}\mathcal{S}(q), \widehat{\mathbb{J}}(q)\mathrm{D}\mathcal{E}(q) + \widehat{\mathbb{K}}(q)\mathrm{D}\mathcal{S}(q)\rangle = 0 + \langle \mathrm{D}\mathcal{S}, \widehat{\mathbb{K}}\,\mathrm{D}\mathcal{S}\rangle \geq 0. \tag{3.7}$$

Of course, to guarantee energy conservation and positivity of the entropy production one needs much less than the two conditions (3.3) and (3.5).

However, the *maximum entropy principle* really relies on (3.5). It states that a maximizer q_* of $\mathcal{S}$ subject to the constraint $\mathcal{E}(q) = E_0$ is an equilibrium of (3.1), the so-called thermodynamic equilibrium for the given energy. Indeed, if q_* maximizes $\mathcal{S}$ under the constraint $\mathcal{E}(q) = E_0$, then we obtain a Lagrange multiplier $\lambda_* \in \mathbb{R}$ such that $\mathrm{D}\mathcal{S}(q_*) = \lambda_*\mathrm{D}\mathcal{E}(q_*)$. Since $\mathrm{D}\mathcal{S}(q) \neq 0$ for all q (e.g., by $\partial_\theta S > 0$), we have $\lambda_* \neq 0$ and conclude $\widehat{\mathbb{J}}(q_*)\mathrm{D}\mathcal{E}(q_*) = \frac{1}{\lambda_*}\widehat{\mathbb{J}}(q_*)\mathrm{D}\mathcal{S}(q_*) = 0$ and $\widehat{\mathbb{K}}(q_*)\mathrm{D}\mathcal{S}(q_*) = \lambda_*\widehat{\mathbb{K}}(q_*)\mathrm{D}\mathcal{E}(q_*) = 0$, where we have used the non-interaction condition (3.5). Vice versa, every steady state q_* of (3.1) satisfies

$$\widehat{\mathbb{J}}(q_*)\mathrm{D}\mathcal{E}(q_*) = 0 \qquad \text{and} \qquad \widehat{\mathbb{K}}(q_*)\mathrm{D}\mathcal{S}(q_*) = 0. \tag{3.8}$$

Thus, in a steady state there cannot be any balancing between reversible and irreversible forces, both have to vanish independently.

The Onsager operator $\widehat{\mathbb{K}}(q)$ defines the linear kinetic relation between the entropic driving force $\eta = \mathrm{D}\mathcal{S}(q)$ and the dissipative flux $\dot q^{\mathrm{diss}} = \widehat{\mathbb{K}}(q)\eta$. In many applications one needs to consider nonlinear kinetic relations $\dot q^{\mathrm{diss}} \leftrightarrow \eta$. For this we use so-called dual dissipation potentials $\mathcal{R}^*(q,\cdot) : \mathrm{T}^*_q\boldsymbol{Q} \to [0,\infty[$, which are lower semicontinuous, convex, and satisfy $\mathcal{R}^*(q,0) = 0$. Then, the kinetic relation is given in the form

$$\dot{q}^{\mathrm{diss}} = \mathrm{D}_\eta \mathcal{R}^*(q,\eta),$$

see [8, 9, 17, 22]. The linear Onsager case is included via quadratic dual dissipation potentials $\mathcal{R}^*(q,\eta) = \frac{1}{2}\langle \eta, \widehat{\mathbb{K}}(q)\eta\rangle$. The generalized version of the second non-interaction condition in (3.5) then reads $\mathcal{R}^*(q,\lambda\mathrm{D}\mathcal{E}(q)) = 0$ for all $\lambda \in \mathbb{R}$.

In many situations the evolution equations satisfy additional conservation laws, such as mass or charge balance. If $C : \boldsymbol{Q} \to \mathbb{R}$ is a such a conserved quantity, then the GENERIC system should also satisfy both $\langle \mathrm{D}C(q), \widehat{\mathbb{J}}(q)\mathrm{D}\mathcal{E}(q)\rangle = 0$ and $\langle \mathrm{D}C(q), \widehat{\mathbb{K}}(q)\mathrm{D}\mathcal{S}(q)\rangle = 0$ for all $q \in \boldsymbol{Q}$.

3.2 Damped Hamiltonian systems

If temperature effects are negligible, GENERIC systems can be simplified by assuming constant temperature θ and using the free energy $\mathcal{F}(q) := \mathcal{E}(q) - \theta\mathcal{S}(q)$. We call a quadruple $(\boldsymbol{Q}, \mathcal{F}, \mathbb{J}, \mathbb{K})$ a *damped Hamiltonian system*, if $\mathcal{F}$ is a sufficiently smooth functional on $\boldsymbol{Q}$ and if $\mathbb{J}$ and $\mathbb{K}$ satisfy (3.2) and (3.3), but no non-interaction condition is needed. If we have a linear Onsager operator $\mathbb{K}$ or a more general dual dissipation potential $\widehat{\mathcal{R}}^*$, the associated evolution equations read

$$\dot{q} = \big(\mathbb{J}(q) - \mathbb{K}(q)\big)\mathrm{D}\mathcal{F}(q) \quad \text{or} \quad \dot{q} = \mathbb{J}(q)\mathrm{D}\mathcal{F}(q) + \mathrm{D}_\eta\mathcal{R}^*(q, -\mathrm{D}\mathcal{F}(q)), \tag{3.9}$$

respectively. Of course, we still may have conserved quantities $C : \mathbf{Q} \to \mathbb{R}$; then we always ask to have $\langle \mathrm{D}C(q), \mathbb{J}(q)\mathrm{D}\mathcal{F}(q)\rangle = 0 = \langle \mathrm{D}C(q), \mathbb{K}(q)\mathrm{D}\mathcal{F}(q)\rangle$ for all $q \in \boldsymbol{Q}$.

Every damped Hamiltonian system can be augmented to become a GENERIC system with constant temperature $\theta > 0$ as follows. We introduce a scalar entropy variable $s \in \mathbb{R}$ and set

$$y = (q,s), \quad \boldsymbol{Y} = \mathbf{Q}\times\mathbb{R}, \quad \mathcal{E}(q,s) = \mathcal{F}(q) + \theta s, \quad \mathcal{S}(q,s) = s,$$

$$\widehat{\mathbb{J}}(q,s) = \begin{pmatrix} \mathbb{J}(q) & 0 \\ 0 & 0 \end{pmatrix}, \qquad \widehat{\mathbb{K}}(q,s) = \begin{pmatrix} \mathbb{K}(q) & -\mathbb{K}(q)\mathrm{D}\mathcal{F}(q) \\ -\langle \square, \mathbb{K}(q)\mathrm{D}\mathcal{F}(q)\rangle & \frac{1}{\theta}\langle \mathrm{D}\mathcal{F}(q), \mathbb{K}(q)\mathrm{D}\mathcal{F}(q)\rangle \end{pmatrix}.$$

Then $(\boldsymbol{Y}, \mathcal{E}, \mathcal{S}, \widehat{\mathbb{J}}, \widehat{\mathbb{K}})$ is a GENERIC system generating the evolution (3.9) for q and the entropy balance $\dot{s} = \frac{1}{\theta}\langle \mathrm{D}\mathcal{F}(q), \widehat{\mathbb{K}}(q)\mathrm{D}\mathcal{F}(q)\rangle$, which immediately gives $\dot{s} \geq 0$.

3.3 Additive structure of dissipative contributions

As observed in [24, Sec. 2.2] the representation of the dissipative parts of the dynamics in terms of Onsager operators or dual dissipation potentials has the major advantage that there is often an additive structure. Indeed, given $\mathcal{E}$ the set of operators $\mathbb{K}$ satisfying (3.2), (3.3), and (3.5) is a convex cone, i.e., if $\mathbb{K}_1$ and $\mathbb{K}_2$ satisfy the conditions, then $\alpha_1\mathbb{K}_1 + \alpha_2\mathbb{K}_2$ does so as well for all $\alpha_1, \alpha_2 \geq 0$. A similar statement is not true for the Poisson structures $\mathbb{J}$, because Jacobi's identity (3.4) is nonlinear.

The additive structure will be useful in modeling complex semiconductor devices, since we are able to consider different dissipative processes as independent building blocks each giving rise to one $\mathbb{K}_j$ and then simply add these operators, or similarly for dual dissipation potentials. Thus, we will use the form

$$\mathbb{K} = \mathbb{K}_{\text{diffusion}} + \mathbb{K}_{\text{reaction}} + \mathbb{K}_{\text{quant-class}} + \mathbb{K}_{\text{heat.cond.}} + \mathbb{K}_{\text{bulk-interface}}.$$

In the present work we will ignore temperature effects and bulk-interface interactions and refer to [22–24] for the modeling of heat transfer via $\mathbb{K}_{\text{heat.cond.}}$ and to [7, 17] for bulk-interface interactions encoded in $\mathbb{K}_{\text{bulk-interface}}$.

3.4 Dissipative coupling between different components

The construction of couplings between different components of a system, such as classical charge carriers and the state of a quantum system, can be done efficiently in terms of Onsager operators, thus building the Onsager symmetry into the system automatically. Assume that $\mathbf{Q}$ is given in the form $q = (q_1, q_2) \in \mathbf{Q}_1 \times \mathbf{Q}_2 = \mathbf{Q}$ and denote by $\eta_1 \in \mathbf{Q}_1^*$ and $\eta_2 \in \mathbf{Q}_2^*$ the corresponding conjugate thermodynamical driving forces.

As we want to couple η_1 and η_2 we introduce a third linear space $\mathbf{X}$ (which may also be $\mathbf{Q}_1^*$ or $\mathbf{Q}_2^*$) and linear mappings $\mathsf{A}_j(q) : \mathbf{Q}_j^* \to \mathbf{X}^*$ and an Onsager operator $\mathbb{K}_{\mathbf{X}}(q) : \mathbf{X}^* \to \mathbf{X}$ to define a dual dissipation potential

$$\mathcal{R}^*_{\text{coupl}}(q; \eta_1, \eta_2) := \frac{1}{2}\big\langle \mathsf{A}_1(q)\eta_1 + \mathsf{A}_2(q)\eta_2, \mathbb{K}_{\mathbf{X}}(q)(\mathsf{A}_1(q)\eta_1 + \mathsf{A}_2(q)\eta_2)\big\rangle_{\mathbf{X}}.$$

For a GENERIC system one additionally has to ask for the non-interaction condition $\mathcal{R}^*(q, \lambda \mathrm{D}\mathcal{E}(q)) = 0$.

The dual dissipation potential $\mathcal{R}^*_{\text{coupl}}$ defines the Onsager operator $\mathbb{K}_{\text{coupl}}(q) : \mathbf{Q}^* \to \mathbf{Q}$, which takes the following block structure with respect to to the decomposition $\mathbf{Q} = \mathbf{Q}_1 \times \mathbf{Q}_2$:

$$\mathbb{K}_{\text{coupl}}(q) = \begin{pmatrix} \mathsf{A}_1^*(q)\mathbb{K}_{\mathbf{X}}(q)\mathsf{A}_1(q) & \mathsf{A}_1^*(q)\mathbb{K}_{\mathbf{X}}(q)\mathsf{A}_2(q) \\ \mathsf{A}_2^*(q)\mathbb{K}_{\mathbf{X}}(q)\mathsf{A}_1(q) & \mathsf{A}_2^*(q)\mathbb{K}_{\mathbf{X}}(q)\mathsf{A}_2(q) \end{pmatrix}.$$

In addition to quantum-classical coupling this method can also be used to couple the interaction between bulk effects and interfacial effects, where A_j may be a trace operator, see [7].

4 Semiconductor modeling via damped Hamiltonian systems

We now show how the above concepts of damped Hamiltonian systems can be used to construct thermodynamically consistent models for semiconductor devices including arbitrarily many charge carriers $\mathbf{c} = (c_1, \ldots, c_{i_*})$, where the number of species is denoted by i_*, as well as a quantum system described by a finite-dimensional density matrix $\boldsymbol{\rho}$.

First, we recover the van Roosbroeck system for the carrier densities $\mathbf{c} = (n, p)$ considered in Section 2 to highlight the simplicity of the gradient structure constructed in [23]. Then, we show how this structure generalizes to arbitrarily many species and to general statistics. Next, we review recent results on the gradient structure for the dissipative part of open quantum systems described by the Lindblad master equation [29]. Finally, we show how the coupling strategy developed in Section 3.4 can be adapted to model the interaction of macroscopic thermodynamical systems and open quantum systems.

4.1 The state variables and free energy

Throughout we consider a domain $\Omega \subset \mathbb{R}^d$ in which all the charge carriers move and interact. The carrier densities are denoted by $c_i(t, x)$ with $t \in [0, T]$, $x \in \Omega$ and $i \in \{1, \ldots, i_*\}$. Moreover, we assume that there is a population of identical quantum dots each of which is described by a density matrix

$$\rho \in \mathfrak{R}_n := \left\{ \rho \in \mathbb{C}^{n \times n} \,\middle|\, \rho = \rho^*,\ \rho \geq 0,\ \operatorname{tr} \rho = 1 \right\}.$$

We assume that the distance between the quantum dots is sufficiently big, such that they do not interact directly. However, we assume that there are still enough quantum dots such that we can model the states by a continuum description via a function $\rho(t, x) \in \mathfrak{R}_n$.

In total our state space $\mathbf{Q}$ for $q = (\rho, \mathbf{c})$ will be given as

$$\boldsymbol{Q} = \boldsymbol{Q}_{\text{quant}} \times \boldsymbol{Q}_{\text{carr}} \quad \text{with } \boldsymbol{Q}_{\text{quant}} := \mathrm{L}^1(\Omega; \mathfrak{R}_n) \text{ and } \boldsymbol{Q}_{\text{carr}} := \mathrm{L}^1(\Omega; [0, \infty[^{i_*}).$$

Of course, from the modeling perspective there are many other options to "localize" the charge carriers of the quantum system, e.g., by constructing models with several dimensions. In the case of wetting layers, certain charge carrier species may live only on a submanifold, see e.g. [7]. A single quantum dot embedded into a bulk material can be considered be means of a weight function as it was done in [14]. Such situations can also be modeled by the approach presented here, but for notational simplicity we stick with the setup as given above.

On the state space $\boldsymbol{Q} = \boldsymbol{Q}_{\text{quant}} \times \boldsymbol{Q}_{\text{carr}}$ we consider the free energy functional

$$\mathcal{F}(\rho,\mathbf{c}) = \int_{\Omega} \Big(\frac{\varepsilon}{2}|\nabla\Phi_{\rho,\mathbf{c}}|^2 + \rho_{\mathrm{qd}}\operatorname{tr}\big(\rho\mathsf{H} + E_\beta\rho\log\rho\big) + F_{\mathrm{carr}}(\mathbf{c})\Big)\,\mathrm{d}x, \tag{4.1}$$

where $E_\beta = k_{\mathrm{B}}\theta = \beta^{-1}$ is the inverse thermal energy, ρ_{qd} is the volume density of quantum dots, H is the Hamiltonian of the quantum system and $F_{\mathrm{carr}}(\mathbf{c})$ is the free energy density of the classical carrier system. The electrostatic potential $\Phi_{\rho,\mathbf{c}} = \phi_{\rho,\mathbf{c}} - \phi_{\mathrm{eq}}$ is defined in terms of the the Poisson problem

$$-\operatorname{div}(\varepsilon\nabla\phi_{\rho,\mathbf{c}}) = e_0\left(C + \mathbf{z}\cdot\mathbf{c} + \rho_{\mathrm{qd}}\operatorname{tr}(\mathsf{Z}\rho)\right) \text{ on } \Omega, \quad \nu\cdot\nabla\phi_{\rho,\mathbf{c}} = 0 \text{ on } \partial\Omega. \tag{4.2}$$

Here $\mathbf{z} = (z_1,\ldots,z_{i_*}) \in \mathbb{Z}^{i_*}$ is the vector of charge numbers associated with $\mathbf{c}$, while $\mathsf{Z} \in \mathbb{C}^{n\times n}_{\mathrm{herm}}$ is the charge number operator for the quantum system. The vector $\mathbf{c}_{\mathrm{eq}}$ of equilibrium carrier densities and the equilibrium density matrix ρ_{eq} define the equilibrium electrostatic potential $\phi_{\mathrm{eq}} := \phi_{\rho_{\mathrm{eq}},\mathbf{c}_{\mathrm{eq}}}$, such that under equilibrium conditions we have $\Phi_{\rho,\mathbf{c}} \equiv 0$. For the sake of simplicity, we have restricted ourselves to a closed system with homogeneous Neumann boundary conditions (2.9).

4.2 The van Roosbroeck system as gradient system

The free energy density associated with the van Roosbroeck system (2.1) for the density of electrons and holes $\mathbf{c} = (n,p)$ consists of the sum of the relative entropies for the two species and the electrostatic energy

$$\mathcal{F}_{\mathrm{vR}}(n,p) = \int_{\Omega} \Big(\frac{\varepsilon}{2}|\nabla\Phi_{n,p}|^2 + E_\beta\lambda_{\mathrm{B}}(n/n_{\mathrm{eq}})n_{\mathrm{eq}} + E_\beta\lambda_{\mathrm{B}}(p/p_{\mathrm{eq}})p_{\mathrm{eq}}\Big)\,\mathrm{d}x$$

with $\lambda_{\mathrm{B}}(z) := z\log z - z + 1$. The electrostatic potential $\Phi_{n,p}$ solves the Poisson problem (4.2) without the quantum mechanical part. Here we restrict our considerations to non-degenerate carrier ensembles (Maxwell–Boltzmann statistics) with $F_{\mathrm{carr}}(\mathbf{c}) = F_{\mathrm{B}}(n,p)$. For the structure of the free energy functional in the case of Fermi–Dirac statistics we refer to [1].

When taking the variational derivatives of $\mathcal{F}_{\mathrm{vR}}$ with respect to n and p we have to take into account the linear dependence of $\Phi_{n,p}$ on n and p given in terms of (4.2). Then we find

$$\mu = \begin{pmatrix}\mu_{\mathrm{e}}\\ \mu_{\mathrm{h}}\end{pmatrix} = \mathrm{D}\mathcal{F}_{\mathrm{vR}}(n,p) = \begin{pmatrix}E_\beta\log(n/n_{\mathrm{eq}}) - e_0\Phi_{n,p}\\ E_\beta\log(p/p_{\mathrm{eq}}) + e_0\Phi_{n,p}\end{pmatrix}, \tag{4.3}$$

where μ_{e} and μ_{h} are the electro-chemical potentials that are thermodynamically conjugate to n and p, respectively (see Section 2.1). The different signs in front of $e_0\Phi_{n,p}$ reflect the charges of electrons and holes and arise because of the different signs in front of n and p in (4.2).

The gradient structure for the van Roosbroeck system developed in [23] is completed by the Onsager operator

$$\mathbb{K}_{\mathrm{vR}}(n,p)\begin{pmatrix}\mu_{\mathrm{e}}\\ \mu_{\mathrm{h}}\end{pmatrix} = -\begin{pmatrix}\operatorname{div}(\frac{1}{e_0}\mathbb{M}_{\mathrm{e}}n\nabla\mu_{\mathrm{e}})\\ \operatorname{div}(\frac{1}{e_0}\mathbb{M}_{\mathrm{h}}p\nabla\mu_{\mathrm{h}})\end{pmatrix} + \frac{r(n,p)}{k_{\mathrm{B}}\theta}\Lambda(np, n_{\mathrm{eq}}p_{\mathrm{eq}})\begin{pmatrix}1 & 1\\ 1 & 1\end{pmatrix}\begin{pmatrix}\mu_{\mathrm{e}}\\ \mu_{\mathrm{h}}\end{pmatrix}, \quad (4.4)$$

where $\Lambda(a,b) = \int_0^1 a^s b^{1-s}\,\mathrm{d}s = (a-b)/\log(a/b)$ is the logarithmic mean of a and b. Note that $\mathbb{K}_{\mathrm{vR}}$ is the sum of a transport part, which gives rise to drift and diffusion, and a reaction part.

Using the identities $c\nabla(\log c) = \nabla c$ and $\Lambda(a,b)(\log a - \log b) = a - b$, it is easy to see that the equations of motion generated by (4.3) and (4.4) read

$$\begin{pmatrix}\dot{n}\\ \dot{p}\end{pmatrix} = -\mathbb{K}_{\mathrm{vR}}(n,p)\mathrm{D}\mathcal{F}_{\mathrm{vR}}(n,p) = -\begin{pmatrix}\operatorname{div}\boldsymbol{J}_{\mathrm{e}} + R(n,p)\\ \operatorname{div}\boldsymbol{J}_{\mathrm{h}} + R(n,p)\end{pmatrix},$$

which are the continuity equations (2.1b)–(2.1c). Here we have used (2.5) and (2.7). Hence, the van Roosbroeck system has indeed a gradient flow structure that is generated by a gradient system $(\boldsymbol{Q}_{\mathrm{vR}}, \mathcal{F}_{\mathrm{vR}}, \mathbb{K}_{\mathrm{vR}})$ with $\boldsymbol{Q}_{\mathrm{vR}} = \{\,(n,p) \in \mathrm{L}^1(\Omega)^2 \mid p,n \geq 0 \text{ a.e. }\}$. The total charge $C(n,p) = \int_\Omega (n-p)\,\mathrm{d}x$ is a conserved quantity.

In the following subsection we will show how this can be generalized to an arbitrary number of charge carrier densities c_i, $i = 1,\ldots,i_*$. Based on the fundamental work [1] it is shown in [23, Sec. 4.2] how temperature effects can be taken into account by using the physical entropy as a driving functional.

4.3 Reactions between and transport of charge carriers

The charge carriers $\mathbf{c} = (n,p)$ can react in various ways, in particular they can by annihilated in recombination processes. The generation of electron-hole pairs is written as $\emptyset \rightharpoonup X_n + X_p$ and the recombination of electron-hole pairs reads $X_n + X_p \rightharpoonup \emptyset$. In the limit of small carrier densities, the reaction rate equation is of so-called *mass-action type* and reads

$$\dot{n}|_{\mathrm{react}} = -\left(k^{\mathrm{fw}}np - k^{\mathrm{bw}}\right), \quad \dot{p}|_{\mathrm{react}} = -\left(k^{\mathrm{fw}}np - k^{\mathrm{bw}}\right).$$

The forward and backward coefficients k^{fw} and k^{bw} may depend on the variables of the system giving $r(n,p) = k^{\mathrm{fw}}$ and $n_{\mathrm{eq}}p_{\mathrm{eq}} = k^{\mathrm{bw}}/k^{\mathrm{fw}}$ in (2.7). Moreover, they involve several material-dependent parameters. An important feature is the conservation of charge, namely $(\dot{p}-\dot{n})|_{\mathrm{react}} \equiv 0$.

More general, we consider m_* reactions of mass-action type for the charge carrier species $X_1, X_2, \ldots, X_{i_*}$. They are defined in terms of the forward and backward stoichiometric coefficients α_i^m and β_i^m, respectively, via

$$\alpha_1^m X_1 + \cdots + \alpha_{i_*}^m X_{i_*} \rightleftharpoons \beta_1^m X_1 + \cdots + \beta_{i_*}^m X_{i_*}.$$

With $\mathbf{c}^\gamma := \prod_{i=1}^{i_*} c_i^{\gamma_i}$ this leads to the reaction-rate equation

$$\dot{\mathbf{c}}|_{\text{react}} = -\mathbf{R}(\mathbf{c}) := -\sum_{m=1}^{m_*} \left(k_m^{\text{fw}} \mathbf{c}^{\alpha^m} - k_m^{\text{bw}} \mathbf{c}^{\beta^m}\right)\left(\alpha^m - \beta^m\right), \tag{4.5}$$

where the stoichiometric vectors α^m and β^m lie in $\mathbb{N}_0^{i_*}$ and satisfy the condition of electro-neutrality $(\alpha^m - \beta^m) \cdot \mathbf{z} \equiv 0$.

The condition of detailed balance means that there exists a positive density vector $\mathbf{c}_{\text{eq}} \in]0, \infty[^{i_*}$ such that all reactions are in equilibrium, i.e., $\overline{k}_m = k_m^{\text{fw}} \mathbf{c}_{\text{eq}}^{\alpha^m} = k_m^{\text{bw}} \mathbf{c}_{\text{eq}}^{\beta^m}$. It was observed in [23,45] that (4.5) with its polynomial right-hand side is generated by the gradient system $(]0, \infty[^{i_*}, F_{\text{B}}, \mathbb{K}_{\text{react}})$ if the m_* reactions satisfy this detailed balance condition: Setting $F_{\text{carr}}(\mathbf{c}) = F_{\text{B}}(\mathbf{c}) := k_{\text{B}}\theta \sum_{i=1}^{i_*} \lambda_{\text{B}}(c_i/c_i^{\text{eq}})c_i^{\text{eq}}$ with $\lambda_{\text{B}}(z) = z \log z - z + 1$, we obtain

$$\begin{aligned} \dot{\mathbf{c}}|_{\text{react}} &= -\mathbf{R}(\mathbf{c}) = -\mathbb{K}_{\text{react}}(\mathbf{c}) \mathrm{D}F_{\text{B}}(\mathbf{c}) \\ \text{and} \quad \mathbb{K}_{\text{react}}(\mathbf{c}) &= \frac{1}{k_{\text{B}}\theta} \sum_{m=1}^{m_*} \overline{k}_m \Lambda\left(\widehat{\mathbf{c}}^{\alpha^m}, \widehat{\mathbf{c}}^{\beta^m}\right) \left(\alpha^m - \beta^m\right) \otimes \left(\alpha^m - \beta^m\right) \geq 0. \end{aligned} \tag{4.6}$$

Here $\widehat{\mathbf{c}} := (c_i/c_i^{\text{eq}})_{i=1,\ldots,i_*}$ is the vector of relative densities, and $\Lambda(a, b)$ is the logarithmic mean. We emphasize that $\mathbb{K}_{\text{react}}$ is a sum over the individual contribution of each of the m_* reactions. Of course, the coefficients $\overline{k}_m$ may depend on the whole state $\mathbf{c}$ without destroying the gradient structure.

Because of charge neutrality it doesn't matter whether we use the chemical potentials $\eta = \mathrm{D}F_{\text{B}}(\mathbf{c})$ or the electro-chemical potentials $\mu = \mathrm{D}\mathcal{F}(\mathbf{c}) = \eta + e_0 \Phi_{\mathbf{c}} \mathbf{z}$ as driving forces. Moreover, the total charge $C(\mathbf{c}) := \int_\Omega \mathbf{z} \cdot \mathbf{c}(x) \, \mathrm{d}x$ is a conserved quantity.

Remark 1 In [27] a gradient structure for reaction-rate equations was derived via a large-deviation principle for the underlying chemical master equation. It leads to the same free energy F_{B} but to a non-quadratic dual dissipation potential, namely

$$\begin{aligned} \mathcal{R}^*(\mathbf{c}, \mu) &= \sum_{m=1}^{m_*} \frac{\overline{k}_m}{k_{\text{B}}\theta} \left(\widehat{\mathbf{c}}^{\alpha^m} \widehat{\mathbf{c}}^{\beta^m}\right)^{1/2} C_*\left((\alpha^m - \beta^m) \cdot \mu\right) \\ \text{with} \quad C_*(\xi) &= 4 \cosh\left(\frac{\xi}{2}\right) - 4. \end{aligned} \tag{4.7}$$

See also [8, Eqn. (69)] for the occurrence of the cosh potential in chemical reactions.

4.4 Gradient structure for general carrier statistics

We emphasize that the structure of (4.6) looks very special and is chosen in order to produce a simple polynomial right-hand side $\mathbf{R}$ in the case of Maxwell–Boltzmann statistics. However, the gradient structure is still valid for more general statistics. Indeed we may consider a free energy density $F_{\text{carr}}(\mathbf{c})$ for the charge carriers that involves more general statistical distribution functions (e.g., Fermi–Dirac statistics) and then choose $\mathbb{K}_{\text{react}}(\mathbf{c}) = \sum_{m=1}^{m_*} \kappa_m(\mathbf{c})(\alpha^m - \beta^m)\otimes(\alpha^m - \beta^m)$ with suitable scalar coefficients $\kappa_m(\mathbf{c})$. In this way one keeps the gradient structure and hence the thermodynamic principles even without the mass-action type kinetics.

The transport of charge carriers occurs by diffusion as well as by drift in the electric field $\boldsymbol{E} = -\nabla\Phi_{\text{c}}$. The thermodynamical driving force is the electro-chemical potential

$$\mu = \mathrm{D}\mathcal{F}(\mathbf{c}) = \eta + e_0\Phi_{\text{c}}\mathbf{z} \quad \text{with} \quad \eta = \mathrm{D}F_{\text{carr}}(\mathbf{c}),$$

that can be split into the chemical potentials $\eta = \mathrm{D}F_{\text{carr}}(\mathbf{c})$ and the electrostatic forces involving the charge numbers z_i of the respective carrier species. The associated fluxes $\boldsymbol{J}_i \in \mathbb{R}^d$ can be combined into a matrix $\mathsf{J}_{\text{c}} \in \mathbb{R}^{i_* \times d}$. Within the framework of linear irreversible thermodynamics [31], the classical ansatz is

$$\mathsf{J}_{\text{c}} = -\mathsf{M}(\mathbf{c})\nabla\mu = -\mathsf{M}(\mathbf{c})\big(\nabla\eta + \mathbf{z}\otimes\nabla\Phi_{\text{c}}\big), \tag{4.8}$$

where the conductivity tensor M is symmetric and positive semi-definite mapping $\mathbb{R}^{i_* \times d}$ into itself. In the isotropic case one may choose $\mathsf{M}(\mathbf{c}) \in \mathbb{R}^{i_* \times i_*}$, e.g., $\mathsf{M}(\mathbf{c}) = \operatorname{diag}(m_i c_i)_{i=1,\dots,i_*}$.

The Onsager operator $\mathbb{K}_{\text{transp}}$ for transport of charge carriers is now given in the form

$$\mathbb{K}_{\text{transp}}(\mathbf{c})\mu = -\operatorname{div}\big(\mathsf{M}(\mathbf{c})\nabla\mu\big).$$

Thus, for general situations the reactions and transport for the carrier density vector $\mathbf{c}$ can be written as a gradient system in the form

$$\begin{aligned}\dot{\mathbf{c}} &= -\big(\mathbb{K}_{\text{transp}}(\mathbf{c}) + \mathbb{K}_{\text{react}}(\mathbf{c})\big)\mathrm{D}\mathcal{F}(\mathbf{c}) \\ &= \operatorname{div}\big(\mathsf{M}(\mathbf{c})(\nabla \mathrm{D}F_{\text{carr}}(\mathbf{c}) + \mathbf{z}\otimes\nabla\Phi_{\text{c}})\big) - \mathbb{K}_{\text{react}}(q)\mathrm{D}F_{\text{carr}}(\mathbf{c}),\end{aligned}$$

where we used electro-neutrality of the reactions, i.e., $\mathbb{K}_{\text{react}}(\mathbf{c})\mathbf{z} \equiv 0$. In the case of nontrivial carrier statistics $F_{\text{carr}} \not\equiv F_{\text{B}}$ the Hessian of the carrier's free energy density $\nabla \mathrm{D}F_{\text{carr}}(\mathbf{c}) = \mathrm{D}^2 F_{\text{carr}}(\mathbf{c})\nabla\mathbf{c} \in \mathbb{R}^{i_* \times d}$ gives rise to nonlinear diffusion [15,21].

4.5 Dissipative quantum mechanics

Here we show how dissipative quantum systems subject to the Lindblad master equation can be written as a damped Hamiltonian system with respect to the von Neumann entropy, if the dissipative part satisfies a suitable detailed balance condition.

A Markovian quantum master equation (see, e.g., [4,19,41,43]) in Lindblad form reads

$$\dot{\rho} = \jmath[\rho, \mathsf{H}] + \mathfrak{D}\rho, \quad \text{where} \quad \mathfrak{D}\rho = \sum_{k=1}^{k_*} \alpha_k \mathbb{G}_Q \rho \tag{4.9}$$

with

$$\mathbb{G}_Q A := \frac{1}{2}\left([Q, AQ^*] + [QA, Q^*]\right). \tag{4.10}$$

The coupling operators $Q \in \mathbb{C}^{n\times n}$ and the transition rates $\alpha_k \in [0,\infty[$ are arbitrary and $\mathfrak{D}$ is called the dissipation superoperator. We use the notation $\jmath = i/\hbar$.

With the derivative $\mathrm{D}\mathcal{F}_{\text{quant}}(\rho) = \mathsf{H} + E_\beta \log\rho$ (up to a constant) of the free energy functional $\mathcal{F}_{\text{quant}}(\rho) = \operatorname{tr}\left(\mathsf{H}\rho + E_\beta\rho\log\rho\right)$, the Hamiltonian part has the desired form

$$\jmath[\rho,\mathsf{H}] = \jmath\left[\rho, \mathrm{D}\mathcal{F}_{\text{quant}}(\rho)\right] = \mathbb{J}_{\text{quant}}(\rho)\mathrm{D}\mathcal{F}_{\text{quant}}(\rho) \quad \text{with } \mathbb{J}_{\text{quant}}(\rho)A := \jmath[\rho, A].$$

The Jacobi identity (3.4) for $\mathbb{J}_{\text{quant}}$ follows from the elementary Jacobi identity for the commutator.

Thus it remains to write $\mathfrak{D}\rho$ in the form $-\mathbb{K}(\rho)\mathrm{D}\mathcal{F}_{\text{quant}}(\rho)$. This problem was first solved in [5, 29], again relying on a suitable non-commutative version of a detailed balance condition. In Section 4.7 we will show how the coefficients α_k may be chosen to depend on $\mathbf{c}$, if there are corresponding back-coupling terms in the reaction equation. For this we exploit the coupling technique introduced in Section 3.4.

The main structure is a non-commutative version of the chain-rule identities $c\nabla\log c = \nabla c$ for diffusion (cf. Section 4.2) and $\Lambda(c_i^\alpha, c_j^\beta)(\alpha\log c_i - \beta\log c_j) = c_i^\alpha - c_j^\beta$ for reactions of mass-action type (cf. Section 4.3). As highlighted in [33,34], the proper generalization is obtained via the Kubo–Mori multiplication operator

$$\mathfrak{C}_\rho : A \mapsto \int_0^1 \rho^s A \rho^{1-s}\,\mathrm{d}s.$$

Kubo's miracle identity (see [16, 44]) then states that

$$\forall Q \in \mathbb{C}^{n\times n}\ \forall \rho \in \mathfrak{R}_n : \quad \mathfrak{C}_\rho[Q,\log\rho] = [Q,\rho].$$

The above formula is not sufficient to treat terms of the form $[Q, \log\rho + \beta\mathsf{H}]$. However, when restricting to a suitable subclass of operators Q, a generalized miracle identity can be obtained. For this we choose eigenpairs (ω, Q) of the commutator operator $A \mapsto [A,\mathsf{H}]$, viz.

$$[Q,\mathsf{H}] = \hbar\omega Q. \tag{4.11}$$

Using two eigenstates ψ_1 and ψ_2 with $\mathsf{H}\psi_j = h_j\psi_j$ with energy levels $h_j \in \mathbb{R}$, we see that $Q = \psi_1 \otimes \psi_2$ satisfies (4.11) with $\hbar\omega = h_2 - h_1$. The following generalization of the miracle identity relies on doubling the dimension by considering

$$\mathbb{B}_Q := \begin{pmatrix} 0 & Q^* \\ Q & 0 \end{pmatrix} \in \mathbb{C}^{2n\times 2n}_{\text{Herm}} \quad \text{and} \quad \mathsf{R}(\rho,\gamma) := \begin{pmatrix} \mathrm{e}^{\gamma/2}\rho & 0 \\ 0 & \mathrm{e}^{-\gamma/2}\rho \end{pmatrix} \in \mathbb{C}^{2n\times 2n}_{\text{Herm}}.$$

Using (ω, Q) satisfying (4.11) and the doubling operator $\mathbb{X} : \mathbb{C}^{n\times n} \to \mathbb{C}^{2n\times 2n};\ A \mapsto \left(\begin{smallmatrix} A & 0 \\ 0 & A \end{smallmatrix}\right)$ we obtain the *generalized miracle identity*

$$\mathbb{C}_{\mathsf{R}(\rho,\beta\hbar\omega)}\big[\mathbb{B}_Q,\mathbb{X}(\mathsf{H}+E_\beta\log\rho)\big] = \big[\mathbb{B}_Q,\mathsf{R}(\rho,\beta\hbar\omega)\big]$$
$$= \begin{pmatrix} 0 & \mathrm{e}^{-\beta\hbar\omega/2}Q^*\rho - \mathrm{e}^{\beta\hbar\omega/2}\rho Q^* \\ \mathrm{e}^{\beta\hbar\omega/2}Q\rho - \mathrm{e}^{-\beta\hbar\omega/2}\rho Q & 0 \end{pmatrix}, \tag{4.12}$$

where again the right-hand side is linear in ρ. As shown in [29], see also [5] for related results, this identity follows simply from the miracle identity applied to $\mathsf{R}(\rho,\gamma)$ and $\mathbb{B}_Q$ and the fact that (4.11) implies the commutator result

$$\big[\mathbb{B}_Q,\log\mathsf{R}(\rho,\beta\hbar\omega)\big] = \left[\mathbb{B}_Q,\begin{pmatrix}\frac{1}{2}\beta\hbar\omega I+\log\rho & 0\\ 0 & -\frac{1}{2}\beta\hbar\omega I+\log\rho\end{pmatrix}\right]$$
$$= \beta\big[\mathbb{B}_Q,\mathbb{X}(\mathsf{H}+E_\beta\log\rho)\big].$$

This construction allows us to define dual dissipation potentials, where we can take a sum over a set (ω_k,Q_k) of eigenpairs satisfying (4.11), namely

$$\mathcal{R}^*_{\mathrm{quant}}(\rho,\sigma) = \beta\sum_{k=1}^{k_*}\frac{\alpha_k}{2}\,\mathrm{tr}\,\Big(\big([\mathbb{B}_{Q_k},\mathbb{X}\sigma]\big)^*\mathbb{C}_{\mathsf{R}(\rho,\beta\hbar\omega_k)}\big[\mathbb{B}_{Q_k},\mathbb{X}\sigma\big]\Big),$$

where $\alpha_k \geq 0$. Because the adjoint $\mathbb{X}^*$ satisfies $\mathbb{X}^*\left(\begin{smallmatrix}A & B\\ C & D\end{smallmatrix}\right) = A + D$ we obtain the associated Onsager operator, which depends highly nontrivial on ρ:

$$\mathbb{K}_{\mathrm{quant}}(\rho)\,\sigma = -\beta\sum_{k=1}^{k_*}\alpha_k\mathbb{X}^*\Big[\mathbb{B}_{Q_k},\mathbb{C}_{\mathsf{R}(\rho,\beta\hbar\omega_k)}\big[\mathbb{B}_{Q_k},\mathbb{X}\sigma\big]\Big]. \tag{4.13}$$

Exploiting the generalized miracle identity (4.12) we find the desired Lindblad form

$$\mathbb{K}_{\mathrm{quant}}(\rho)\,(\mathsf{H}+E_\beta\log\rho) = -\sum_{k=1}^{k_*}\alpha_k\big(\mathrm{e}^{\beta\hbar\omega_k/2}\mathbb{G}_{Q_k}\rho + \mathrm{e}^{-\beta\hbar\omega_k/2}\mathbb{G}_{Q_k^*}\rho\big),$$

that features special choices for the weights of $\mathbb{G}_{Q_k}$ and $\mathbb{G}_{Q_k^*}$ to guarantee the detailed balance condition.

4.6 The van Roosbroeck system coupled to a quantum system

Before we show how quantum mechanics can be coupled to the van Roosbroeck system via the formalism of damped Hamiltonian systems, we follow the approach in [14] and show that for simple couplings one can prove that the free energy $\mathcal{F}$ is indeed a Liapunov function. In Section 4.7 this will follow automatically, but from very elaborate construction, while the arguments in this subsection are more intuitive and direct.

For quantum master equations of type (4.9) without the detailed balance condition it is much more difficult to show that the relative entropy is a Liapunov function, see [18]. In [41, 42] explicit expressions for the dissipation were derived for systems satisfying a detailed balance conditions. Exactly these formulas stimulated the gradient structure developed in the previous subsection.

To highlight the idea of consistent coupling of the van Roosbroeck system and a quantum system developed in [14], we now look at one coupling mechanism, namely the capture and escape of a free electron into a quantum dot, which can be written as a forward-backward reaction

$$X_e + \psi_1 \rightleftharpoons \psi_2,$$

where $\psi_j \in \mathbb{C}^n$ denote normalized eigenstates with $\mathsf{H}\psi_j = h_j\psi_j$ of the Hamiltonian H. Here ψ_1 might denote a ground state (empty quantum dot) while ψ_2 denotes an excited state (electron captured by the quantum dot), see [14] for details. The eigenstates ψ_j should also satisfy $\mathsf{Z}\psi_j = (1 - j)\psi_j$ for $j = 1$ and 2. Hence, H and Z share the same eigenbasis such that

$$[\mathsf{H}, \mathsf{Z}] = 0. \tag{4.14}$$

Physically, the condition (4.14) implies that the Hamiltonian evolution leaves the charge of the quantum system invariant. Therefore, the exchange of charges is necessarily a dissipative process that couples the open quantum system to the macroscopic system in its environment.

Using the transfer operator $Q = \psi_1 \otimes \psi_2 \in \mathbb{C}^{n\times n}$ and setting $\hbar\omega = h_2 - h_1$ we obtain the commutator relations

$$[Q, \mathsf{H}] = \hbar\omega Q, \qquad [Q^*, \mathsf{H}] = -\hbar\omega Q^*, \qquad [Q, \mathsf{Z}] = -Q, \qquad [Q^*, \mathsf{Z}] = +Q^*.$$

The electron-exchange flux between the macroscopic system and the quantum dots is modeled by an additional reaction term on the right hand side of (2.1b)

$$R_n^{\text{quant-class}}(\rho, n) = -\rho_{\text{qd}}\,\mathrm{tr}\left(\mathsf{Z}\mathfrak{D}_{\text{cp}}(n)\rho\right) \tag{4.15}$$

where the coupling operator $\mathfrak{D}_{\text{cp}}$ is given via

$$\mathfrak{D}_{\text{cp}}(n)\rho = \kappa(n)\left(\mathrm{e}^{(\beta\hbar\omega+\log(n/n_{\text{eq}}))/2}\,\mathbb{G}_Q\rho + \mathrm{e}^{-(\beta\hbar\omega+\log(n/n_{\text{eq}}))/2}\,\mathbb{G}_{Q^*}\rho\right) \tag{4.16}$$

with $\kappa \geq 0$ non-negative. The "total" dissipation superoperator in (4.10) is the sum $\mathfrak{D}(n)\rho = \mathfrak{D}_0\rho + \mathfrak{D}_{\text{cp}}(n)\rho$. The processes described by $\mathfrak{D}_0$ do not couple the open quantum system and the macroscopic system as they are assumed to not exchange carriers between both subsystems, i.e., it holds

$$\mathrm{tr}\left(\mathsf{Z}\mathfrak{D}_0\rho\right) \equiv 0. \tag{4.17}$$

With this construction, the coupled system reads

$$0 = \operatorname{div}(\varepsilon\nabla\phi_{\rho,n,p}) + e_0\big(C + p - n + \rho_{\mathrm{qd}}\operatorname{tr}(\mathsf{Z}\rho)\big), \tag{4.18a}$$
$$\dot{n} = -\operatorname{div}\boldsymbol{J}_{\mathrm{e}} - R(n,p) + \rho_{\mathrm{qd}}\operatorname{tr}\big(\mathsf{Z}\mathfrak{D}_{\mathrm{cp}}(n)\rho\big), \tag{4.18b}$$
$$\dot{p} = -\operatorname{div}\boldsymbol{J}_{\mathrm{h}} - R(n,p), \tag{4.18c}$$
$$\dot{\rho} = \jmath[\rho, \mathsf{H} + e_0\Phi_{\rho,n,p}\mathsf{Z}] + \mathfrak{D}_0\rho + \mathfrak{D}_{\mathrm{cp}}(n)\rho. \tag{4.18d}$$

As before, the electrostatic potential $\Phi_{\rho,n,p} = \phi_{\rho,n,p} - \phi_{\mathrm{eq}}$ is defined via (4.2). The system (4.18) conserves the total charge, as it implies continuity equation

$$\dot{\varrho} + \operatorname{div}\boldsymbol{J}_{\varrho} = 0 \tag{4.19}$$

for the total charge density $\varrho = e_0\big(C+p-n+\rho_{\mathrm{qd}}\operatorname{tr}(\mathsf{Z}\rho)\big)$, where $\boldsymbol{J}_{\varrho} = e_0(\boldsymbol{J}_{\mathrm{h}} - \boldsymbol{J}_{\mathrm{e}})$ is the electrical charge current density. For this (4.14), (4.15), (4.17) and cyclic permutations under the trace have been used. We conclude that the total charge $C(\rho,n,p) = \int_\Omega \varrho\,\mathrm{d}x$ is a conserved quantity.

The system (4.18) has a steady state solution $(\rho,n,p) = (\rho_{\mathrm{eq}}, n_{\mathrm{eq}}, p_{\mathrm{eq}})$ where $[\rho_{\mathrm{eq}}, \mathsf{H}] \equiv 0$ and $\mathfrak{D}_{\mathrm{cp}}(n_{\mathrm{eq}})\,\rho_{\mathrm{eq}} \equiv 0$. Moreover, the steady state solution is assumed to satisfy

$$Q\rho_{\mathrm{eq}} = \mathrm{e}^{-\beta\hbar\omega}\rho_{\mathrm{eq}}Q.$$

Then, the charge exchange (4.15) between the classical and the quantum system vanishes in equilibrium, i.e.,

$$R_n^{\text{quant-class}}\big(\rho_{\mathrm{eq}}, n_{\mathrm{eq}}\big) = -\rho_{\mathrm{qd}}\kappa\,(n_{\mathrm{eq}})\left(\mathrm{e}^{\beta\hbar\omega/2}\operatorname{tr}(Q\rho_{\mathrm{eq}}Q^*) - \mathrm{e}^{-\beta\hbar\omega/2}\operatorname{tr}(Q^*\rho_{\mathrm{eq}}Q)\right) = 0,$$

which is a manifestation of the detailed balance condition.

We consider the free energy functional $\mathcal{F}(\rho,n,p)$ as given in (4.1) with the free energy density $F_{\mathrm{carr}}(\mathbf{c}) = F_{\mathrm{B}}(n,p)$ for the macroscopic carriers. The thermodynamic driving forces are obtained by the Gâteaux derivatives, namely

$$\begin{pmatrix}\sigma\\ \mu_n\\ \mu_p\end{pmatrix} = \begin{pmatrix}\mathrm{D}_\rho\mathcal{F}\\ \mathrm{D}_n\mathcal{F}\\ \mathrm{D}_p\mathcal{F}\end{pmatrix} = \begin{pmatrix}E_\beta\log\rho + \mathsf{H} + e_0\Phi_{\rho,n,p}\mathsf{Z}\\ E_\beta\log(n/n_{\mathrm{eq}}) - e_0\Phi_{\rho,n,p}\\ E_\beta\log(p/p_{\mathrm{eq}}) + e_0\Phi_{\rho,n,p}\end{pmatrix}.$$

Following [14] we consider solutions $t \mapsto (\rho(t,\cdot), n(t,\cdot), p(t,\cdot))$ of (4.18) and discuss the energy-dissipation relation. A direct computation gives

$$\begin{aligned}\frac{\mathrm{d}}{\mathrm{d}t}\mathcal{F}(\rho,n,p) &= \langle \mathrm{D}\mathcal{F}(\rho,n,p), (\dot{\rho},\dot{n},\dot{p})^\top\rangle\\ &= -\mathcal{D}_{\mathrm{transp}}(\rho,n,p) - \mathcal{D}_{\mathrm{react}}(\rho,n,p) - \mathcal{D}_{\text{quant-class}}(\rho,n,p) - \mathcal{D}_{\mathrm{quant}}(\rho,n,p),\end{aligned}$$

where we exploit the additive structure of the different dissipative mechanisms. Using (2.2), (2.3), (2.7), (4.14) and (4.17), the individual terms are given as follows:

$$
\begin{aligned}
\mathcal{D}_{\text{transp}} &= \int_\Omega \Big(\frac{1}{e_0} \mathbb{M}_{\text{e}} n \big| \nabla \big(E_\beta \log \big(n/n_{\text{eq}} \big) - e_0 \Phi_{\rho,n,p} \big) \big|^2 \\
&\qquad + \frac{1}{e_0} \mathbb{M}_{\text{h}} p \big| \nabla \big(E_\beta \log \big(p/p_{\text{eq}} \big) + e_0 \Phi_{\rho,n,p} \big) \big|^2 \Big) \,\mathrm{d}x, \\
\mathcal{D}_{\text{react}} &= E_\beta \int_\Omega r(n,p) \frac{\big(np - n_{\text{eq}} p_{\text{eq}}\big)^2}{\Lambda \big(np, n_{\text{eq}} p_{\text{eq}}\big)} \,\mathrm{d}x, \\
\mathcal{D}_{\text{quant}} &= \int_\Omega \rho_{\text{qd}} \,\mathrm{tr}\, \big(\big[-\mathsf{H} - E_\beta \log \rho \big] \mathfrak{D}_0 \rho \big) \,\mathrm{d}x, \\
\mathcal{D}_{\text{quant-class}} &= \int_\Omega \rho_{\text{qd}} \,\mathrm{tr}\, \Big(\big[- \big(\mathsf{H} + E_\beta \log \big(n/n_{\text{eq}} \big) \mathsf{Z} \big) - \log \rho \big] \mathfrak{D}_{\text{cp}}(q) \rho \Big) \,\mathrm{d}x.
\end{aligned}
$$

Here, $\mathcal{D}_{\text{transp}}$ and $\mathcal{D}_{\text{react}}$, which describe the dissipation due to transport via drift and diffusion and due to reactions (recombination) in the van Roosbroeck system, are trivially non-negative. The non-negativity of $\mathcal{D}_{\text{quant}}$ follows from

$$
\mathrm{tr}\,\Big(\big(\log \rho_0 - \log \rho \big) \mathfrak{D}_0 \rho \Big) \geq 0 \quad \text{for all } \rho \in \mathfrak{R}_n, \tag{4.20}
$$

whenever $\mathfrak{D}_0 \rho_0 \equiv 0$, see [41, Thm. 3]. For the charge conserving processes this is easily achieved for $\log \rho_0 = -\beta \mathsf{H}$. The non-negativity of $\mathcal{D}_{\text{quant-class}}$ follows analogously by a generalization of (4.20) to

$$
\mathrm{tr}\,\Big(\big(\log \widehat{\rho}_n - \log \rho \big) \mathfrak{D}_{\text{cp}}(n) \rho \Big) \geq 0 \quad \text{for all } \rho \in \mathfrak{R}_n,
$$

if $\mathfrak{D}_{\text{cp}}(n) \widehat{\rho}_n \equiv 0$. Indeed, by (4.16) this is satisfied for $\log \widehat{\rho}_n = -\beta \mathsf{H} - \log \big(n/n_{\text{eq}} \big) \mathsf{Z}$. The proof is based on the relation $Q \widehat{\rho}_n = \mathrm{e}^{-(\beta \hbar \omega + \log(n/n_{\text{eq}}))} \widehat{\rho}_n Q$. In conclusion, we recover the result of [14] and state consistency of the model system (4.18) with the second law of thermodynamics, as the free energy $\mathcal{F}$ is a Liapunov function, namely $\frac{\mathrm{d}}{\mathrm{d}t} \mathcal{F}(\rho, n, p) \leq 0$.

4.7 Quantum-classical coupling via Onsager operators

We can combine the ideas developed in Section 4.3 and Section 4.6 by looking at chemical reactions involving quantum states as well. For a single reaction we have

$$
\widetilde{\alpha}_1 X_1 + \cdots + \widetilde{\alpha}_{i_*} X_{i_*} + \psi_j \quad \rightleftharpoons \quad \widetilde{\beta}_1 X_1 + \cdots + \widetilde{\beta}_{i_*} X_{i_*} + \psi_k \tag{4.21}
$$

where $\mathsf{H} \psi_j = h_j \psi_j$, $\mathsf{Z} \psi_j = \zeta_j \psi_j$ and $[\mathsf{H}, \mathsf{Z}] = 0$. As before, the theory is based on coupling operators $Q_{j,k} = \psi_j \otimes \psi_k$ between the system's eigenstates that satisfy

$$
[Q_{j,k}, \mathsf{H}] = \hbar \omega_{j,k} Q_{j,k} \quad \text{and } [Q_{j,k}, \mathsf{Z}] = \ell_{j,k} Q_{j,k}, \tag{4.22}
$$

with $\hbar\omega_{j,k} = h_k - h_j \neq 0$ and $\ell_{j,k} = \zeta_k - \zeta_j \in \mathbb{Z}$. We focus on a single reaction in (4.21) with (j,k) such that $\omega_{j,k} \equiv \omega_0$ and $\ell_{j,k} \equiv \ell_0$ such that charge conservation means

$$\mathbf{z} \cdot (\widetilde{\boldsymbol{\beta}} - \widetilde{\boldsymbol{\alpha}}) + \ell_0 = 0. \tag{4.23}$$

For example, the capture of an electron from the macroscopic system with $\mathbf{c} = (n,p)$ to the quantum system considered in Section 4.6 is described by $n + \psi_1 \rightleftharpoons \psi_2$, where, as in Section 4.6, ψ_1 models the empty quantum dot and ψ_2 is a state occupied by a single electron. The charge neutrality condition (4.23) is satisfied since $\mathbf{z} = (-1,+1)^\top$, $\widetilde{\boldsymbol{\alpha}} = (1,0)^\top$, $\widetilde{\boldsymbol{\beta}} = (0,0)^\top$ and $\ell_0 = -1$.

Following the general coupling strategy outlined in Section 3.4, we construct the coupling via a dual dissipation potential by using a suitable linear combination of the electro-chemical potential $\boldsymbol{\mu} = \mathrm{D}F_{\mathrm{carr}}(\mathbf{c}) + e_0\Phi_{\rho,\mathbf{c}}\mathbf{z}$ and the driving force $\sigma = E_\beta \log\rho + \mathsf{H} + e_0\Phi_{\rho,\mathbf{c}}\mathsf{Z}$. In particular, $\boldsymbol{\mu}$ is mapped into a multiple of the Hamiltonian H or the charge operator Z, such that it is possible to exploit the commutator relations (4.22) with the transition operator $Q_0 = Q_{j,k} = \psi_j \otimes \psi_k$. The dual dissipation potential reads

$$\mathcal{R}^*_{\text{quant-class}}(\rho,\mathbf{c};\sigma,\boldsymbol{\mu}) := \beta\frac{\kappa(\mathbf{c})}{2}\,\mathrm{tr}\Big(\big([\mathbb{B}_{Q_0},\mathbb{X}(\sigma + \mathsf{A}\boldsymbol{\mu})]\big)^*\mathbb{C}_{\mathsf{R}(\rho,\gamma(\mathbf{c}))}\big[\mathbb{B}_{Q_0},\mathbb{X}(\sigma+\mathsf{A}\boldsymbol{\mu})\big]\Big),$$
$$\text{with } \mathsf{A}(\mathbf{c})\boldsymbol{\mu} := (\widetilde{\boldsymbol{\beta}}-\widetilde{\boldsymbol{\alpha}})\cdot\boldsymbol{\mu}\,\big(a(\mathbf{c})\mathsf{H} + b(\mathbf{c})\mathsf{Z}\big) \in \mathbb{C}^{n\times n}_{\mathrm{Herm}},$$

where the coupling strength $\kappa(\mathbf{c}) \geq 0$ will remain free, while the scalars $a(\mathbf{c})$, $b(\mathbf{c})$, and $\gamma(\mathbf{c})$ need to be chosen suitably as functions of $\mathbf{c}$ to be able to exploit electro-neutrality and the generalized miracle identity (4.12) for all $\mathbf{c}$. For this, we simply observe that replacing H by $b(\mathbf{c})\mathsf{H}$ does not change Q_0, but replaces ω_0 by $b(\mathbf{c})\omega_0$. Thus, with $(\sigma,\boldsymbol{\mu}) = (E_\beta\log\rho + \mathsf{H} + e_0\Phi_{\rho,\mathbf{c}}\mathsf{Z}, \mathrm{D}F_{\mathrm{carr}}(\mathbf{c}) + e_0\Phi_{\rho,\mathbf{c}}\mathbf{z})$ we obtain the commutator relation

$$\begin{aligned}
\big[Q_0,\sigma{+}\mathsf{A}(\mathbf{c})\boldsymbol{\mu}\big] &= E_\beta[Q_0,\log\rho] + E_\beta\gamma(\mathbf{c},\Phi_{\rho,\mathbf{c}})Q_0 \quad \text{with}\\
\gamma(\mathbf{c},\Phi_{\rho,\mathbf{c}}) &= \beta\Big(\hbar\omega_0 + (\widetilde{\boldsymbol{\beta}}{-}\widetilde{\boldsymbol{\alpha}}){\cdot}(\mathrm{D}F_{\mathrm{carr}}(\mathbf{c})\big(\hbar\omega_0 a(\mathbf{c}) + \ell_0 b(\mathbf{c})\big)\\
&\qquad + e_0\Phi_{\rho,\mathbf{c}}\big(\ell_0 + (\widetilde{\boldsymbol{\beta}}{-}\widetilde{\boldsymbol{\alpha}}){\cdot}\mathbf{z}\big(\hbar\omega_0 a(\mathbf{c}) + \ell_0 b(\mathbf{c})\big)\big)\Big).
\end{aligned}$$

Thus, to have electro-neutrality (4.23) the factor multiplying $e_0\Phi_{\rho,\mathbf{c}}$ has to vanish, which by (4.23) imposes the condition $\hbar\omega_0 a(\mathbf{c}) + \ell_0 b(\mathbf{c}) \equiv 1$. (For the case treated in Section 4.6 a natural choice is $a(n,p) \equiv 0$ and $b(n,p) = 1/\ell_0 = -1$.) Moreover, we find that γ does not depend on $\Phi_{\rho,\mathbf{c}}$ and takes the simple form

$$\gamma(\mathbf{c}) = \beta\hbar\omega_0 + \beta\,(\widetilde{\boldsymbol{\beta}}{-}\widetilde{\boldsymbol{\alpha}}){\cdot}\mathrm{D}F_{\mathrm{carr}}(\mathbf{c}). \tag{4.24}$$

The associated Onsager operator now has a block structure

$$\mathbb{K}_{\text{quant-class}}(\rho,\mathbf{c}) = \kappa(\mathbf{c})\begin{pmatrix} \mathbb{K}^{\mathrm{cq}}_{\mathrm{quant}}(\rho,\mathbf{c}) & \mathbb{K}^{\mathrm{cq}}_{\mathrm{quant}}(\rho,\mathbf{c})\mathsf{A}(\mathbf{c}) \\ \mathsf{A}^*(\mathbf{c})\mathbb{K}^{\mathrm{cq}}_{\mathrm{quant}}(\rho,\mathbf{c}) & \mathsf{A}^*(\mathbf{c})\mathbb{K}^{\mathrm{cq}}_{\mathrm{quant}}(\rho,\mathbf{c})\mathsf{A}(\mathbf{c})\end{pmatrix}$$

that clearly shows the symmetric coupling. Here $\mathbb{K}^{\text{cq}}_{\text{quant}}$ is constructed as in (4.13), namely

$$\mathbb{K}_{\text{quant}}(\rho,\mathbf{c})\,\sigma = -\beta\,\kappa(\mathbf{c})\mathbb{X}^*\Big[\mathbb{B}_{Q_0},\mathfrak{C}_{\mathsf{R}(\rho,\gamma(\mathbf{c}))}\big[\mathbb{B}_{Q_0},\mathbb{X}\sigma\big]\Big], \tag{4.25}$$

where now an explicit dependence on the macroscopic densities c_i occurs.

It is interesting to note that this construction yields a simple final result for the coupling terms in the equations if we use Maxwell–Boltzmann statistics for $\mathbf{c}$, because the arising terms for the macroscopic system are very similar to the expressions for the quantum system that are based on the von Neumann entropy. Using coupling strength function $\kappa(\mathbf{c}) = \kappa_0\big(\widehat{\mathbf{c}}^{\widetilde{\alpha}}\widehat{\mathbf{c}}^{\widetilde{\beta}}\big)^{1/2}$ (which also occurs in [27], see (4.7)), we obtain

$$\begin{aligned}\begin{pmatrix}\mathfrak{D}_{\text{quant-class}}(\mathbf{c})\rho\\ -\mathbf{R}_{\text{quant-class}}(\rho,\mathbf{c})\end{pmatrix} &= -\mathbb{K}_{\text{qu-cl}}(\rho,\mathbf{c})\begin{pmatrix}E_\beta\log\rho+\mathsf{H}+e_0\Phi_{\rho,\mathbf{c}}\mathsf{Z}\\ \mathrm{D}F_{\mathrm{B}}(\mathbf{c})+e_0\Phi_{\rho,\mathbf{c}}\mathbf{z}\end{pmatrix}\\ &= \kappa_0\begin{pmatrix}e^{\beta\hbar\omega_0/2}\widehat{\mathbf{c}}^{\widetilde{\alpha}}\mathbb{G}_{Q_0}\rho+\mathrm{e}^{-\beta\hbar\omega_0/2}\widehat{\mathbf{c}}^{\widetilde{\beta}}\mathbb{G}_{Q_0^*}\rho\\ \big(e^{\beta\hbar\omega_0/2}\widehat{\mathbf{c}}^{\widetilde{\alpha}}\,\mathrm{tr}(Q_0\rho Q_0^*)-\mathrm{e}^{-\beta\hbar\omega_0/2}\widehat{\mathbf{c}}^{\widetilde{\beta}}\,\mathrm{tr}(Q_0^*\rho Q_0)\big)\big(\widetilde{\alpha}-\widetilde{\beta}\big)\end{pmatrix},\end{aligned} \tag{4.26}$$

where $\widehat{\mathbf{c}} = (c_i/c_i^{\text{eq}})_{i=1,\ldots,i_*}$ is the vector of relative densities. It is surprising that all the terms are polynomial in $\mathbf{c}$ and linear in ρ. We refer to (4.15) and (4.18) in Section 4.6 for the special case with $\widetilde{\alpha} = (1,0)^\top$ and $\widetilde{\beta} = (0,0)^\top$.

The surprising fact is that by the generalized miracle identity (4.12) all the complicated nonlinearity cancels each other, and at the end polynomial vector fields remain. In particular the equation for ρ is in Lindblad form, where only the prefactors of the generators $\mathbb{G}_{Q_0}$ and $\mathbb{G}_{Q_0^*}$ depend on the macroscopic charge carrier densities. Similarly, the reactions of the charge carrier densities c_i obey the mass-action law, see (4.6). We refer to [12, 26] for more details concerning the coupling of charge carriers and quantum systems.

4.8 Further dissipative coupling strategies

Further dissipative processes can be modeled by similar approaches. For instance, in [29, Sec. 5.5] a dissipative Maxwell–Bloch system is considered, where a dissipative coupling between the electromagnetic radiation field and a quantum mechanical multi-level system is considered. A much simpler and more direct coupling of recombination and light generation is discussed in [28]. In [11] a different model for dissipative Maxwell equations is formulated within the framework of GENERIC.

Many applications involve the interaction of bulk and interface effects. We refer to [7, 24, 32]. In particular, the capture and escape of species from the bulk to the interface and back can be understood as a reaction in the sense of Section 4.3.

Acknowledgements The research was partially supported by the ERC grant AdG 267802 (AnaMultiScale) and the German Research Foundation (DFG) via project B4 within the collaborative research center SFB 787 *Semiconductor Nanophotonics*. The authors are grateful for helpful and stimulating discussions with Uwe Bandelow and Thomas Koprucki.

References

1. Albinus, G., Gajewski, H., Hünlich, R.: Thermodynamic design of energy models of semiconductor devices. Nonlinearity **15**(2), 367–383 (2002).
2. Bimberg, D., Grundmann, M., Ledentsov, N.N.: Quantum Dot Heterostructures. John Wiley & Sons, Chichester (1999)
3. Bloch, A.M., Morrison, P.J., Ratiu, T.S.: Gradient flows in the normal and Kähler metrics and triple bracket generated metriplectic systems. In: A. Johann, H.P. Kruse, F. Rupp, S. Schmitz (eds.) Recent Trends in Dynamical Systems, chap. 15, pp. 371–415. Springer, Basel, Heidelberg (2013).
4. Breuer, H.P., Petruccione, F.: The Theory of Open Quantum Systems. Oxford University Press, Oxford (2002).
5. Carlen, E.A., Maas, J.: Gradient flow and entropy inequalities for quantum Markov semigroups with detailed balance. J. Funct. Analysis **273**(5), 1810–1869 (2017).
6. Chow, W.W., Jahnke, F.: On the physics of semiconductor quantum dots for applications in lasers and quantum optics. Prog. Quantum Electron. **37**(3), 109–184 (2013).
7. Glitzky, A., Mielke, A.: A gradient structure for systems coupling reaction-diffusion effects in bulk and interfaces. Z. angew. Math. Phys. (ZAMP) **64**, 29–52 (2013).
8. Grmela, M.: Multiscale equilibrium and nonequilibrium thermodynamics in chemical engineering. Adv. Chem. Eng. **39**, 75–128 (2010).
9. Grmela, M., Öttinger, H.C.: Dynamics and thermodynamics of complex fluids. I. Development of a general formalism. Phys. Rev. E **56**(6), 6620–6632 (1997).
10. de Groot, S.R., Mazur, P.: Non-equilibrium thermodynamics. North-Holland Publishing, Amsterdam (1962)
11. Jelić, A., Hütter, M., Öttinger, H.C.: Dissipative electromagnetism from a nonequilibrium thermodynamics perspective. Phys. Rev. E **74**, 041126 (2006).
12. Kantner, M.: Modeling and simulation of electrically driven quantum dot based single-photon sources: From classical device physics to open quantum systems. Ph.D. thesis, Technical University Berlin, Berlin (2018). doi: 10.14279/depositonce-7516
13. Kantner, M., Bandelow, U., Koprucki, T., Schulze, J.H., Strittmatter, A., Wünsche, H.J.: Efficient current injection into single quantum dots through oxide-confined pn-diodes. IEEE Trans. Electron Devices **63**, 2036–2042 (2016).
14. Kantner, M., Mittnenzweig, M., Koprucki, T.: Hybrid quantum-classical modeling of quantum dot devices. Phys. Rev. B **96**, 205301/1–17 (2017).
15. Koprucki, T., Gärtner, K.: Discretization scheme for drift-diffusion equations with strong diffusion enhancement. Opt. Quant. Electron. **45**(7), 791–796 (2013).
16. Kubo, R.: Some aspects of the statistical-mechanical theory of irreversible processes. In: W.E. Brittin, L.G. Dunham (eds.) Lectures in Theoretical Physics. Interscience Publishers, New York (1959)
17. Liero, M., Mielke, A., Peletier, M.A., Renger, D.R.M.: On microscopic origins of generalized gradient structures. Discr. Cont. Dynam. Systems Ser. S **10**(1), 1–35 (2017).
18. Lindblad, G.: Completely positive maps and entropy inequalities. Comm. Math. Phys. **40**, 147–151 (1975).
19. Lindblad, G.: On the generators of quantum dynamical semigroups. Comm. Math. Phys. **48**(2), 119–130 (1976)
20. Markowich, P.A.: The Stationary Semiconductor Device Equations. Springer, Wien, New York (1986).

21. van Mensfoort, S.L.M., Coehoorn, R.: Effect of Gaussian disorder on the voltage dependence of the current density in sandwich-type devices based on organic semiconductors. Phys. Rev. B **78**(8), 085207 (2008).
22. Mielke, A.: Formulation of thermoelastic dissipative material behavior using GENERIC. Contin. Mech. Thermodyn. **23**(3), 233–256 (2011).
23. Mielke, A.: A gradient structure for reaction-diffusion systems and for energy-drift-diffusion systems. Nonlinearity **24**, 1329–1346 (2011).
24. Mielke, A.: Thermomechanical modeling of energy-reaction-diffusion systems, including bulk-interface interactions. Discr. Cont. Dynam. Systems Ser. S **6**(2), 479–499 (2013).
25. Mielke, A.: On thermodynamical couplings of quantum mechanics and macroscopic systems. In: P. Exner, W. König, H. Neidhardt (eds.) Mathematical Results in Quantum Mechanics, pp. 331–348. World Scientific, Singapore (2015). Proceedings of the QMath12 Conference
26. Mielke, A., Mittnenzweig, M., Rotundo, N.: On a thermodynamically consistent coupling of quantum systems to reaction-rate equation. In preparation (2017)
27. Mielke, A., Patterson, R.I.A., Peletier, M.A., Renger, D.R.M.: Non-equilibrium thermodynamical principles for chemical reactions with mass-action kinetics. SIAM J. Appl. Math. **77**(4), 1562–1585 (2017).
28. Mielke, A., Peschka, D., Rotundo, N., Thomas, M.: On some extension of energy-drift-diffusion models: Gradient structures for optoelectronic models of semiconductors. In: P. Quintela, P. Barral, D. Gómez, F.J. Pena, J. Rodríguez, P. Salgado, M. Vázquez-Mendéz (eds.) Progress in Industrial Mathematics at ECMI 2016, Mathematics in Industry Vol. 26, pp. 291–298. Springer (2017).
29. Mittnenzweig, M., Mielke, A.: An entropic gradient structure for Lindblad equations and couplings of quantum systems to macroscopic models. J. Stat. Phys. **167**(2), 205–233 (2017).
30. Morrison, P.J.: A paradigm for joined Hamiltonian and dissipative systems. Phys. D **18**(1-3), 410–419 (1986)
31. Onsager, L.: Reciprocal relations in irreversible processes, I. Phys. Rev. **37**, 405–426 (1931).
32. Öttinger, H.C.: Nonequilibrium thermodynamics for open systems. Phys. Rev. E **73**(3), 036126 (2006).
33. Öttinger, H.C.: The nonlinear thermodynamic quantum master equation. Phys. Rev. A **82**, 052119 (2010)
34. Öttinger, H.C.: The geometry and thermodynamics of dissipative quantum systems. Europhys. Lett. **94**, 10006 (2011).
35. Öttinger, H.C., Grmela, M.: Dynamics and thermodynamics of complex fluids. II. Illustrations of a general formalism. Phys. Rev. E **56**(6), 6633–6655 (1997)
36. Palankovski, V., Quay, R.: Analysis and Simulation of Heterostructure Devices. Computational Microelectronics. Springer Science & Business Media, Vienna (2004).
37. Paul, H.: Photonen: Eine Einführung in die Quantenoptik. Vieweg+Teubner Verlag, Stuttgart, Leipzig (1995).
38. van Roosbroeck, W.: Theory of the flow of electrons and holes in germanium and other semiconductors. Bell Syst. Tech. J. **29**(4), 560–607 (1950).
39. Schröder, D.: Modelling of interface carrier transport for device simulation. Series in Computational Microelectronics. Springer, Vienna (1994).
40. Selberherr, S.: Analysis and simulation of semiconductor devices. Springer, Wien, New York (1984).
41. Spohn, H.: Entropy production for quantum dynamical semigroups. J. Math. Phys. **19**(5), 1227–1230 (1978).
42. Spohn, H., Lebowitz, J.L.: Irreversible thermodynamics for quantum systems weakly coupled to thermal reservoirs. Adv. Chemical Physics XXXVIII pp. 109–142 (1978).
43. Weiss, U.: Quantum Dissipative Systems, 2nd edn. World Scientific, Singapore (1999).
44. Wilcox, R.M.: Exponential operators and parameter differentiation in quantum physics. J. Math. Phys. **8**(4), 962–982 (1967).
45. Yong, W.A.: An interesting class of partial differential equations. J. Math. Phys. **49**, 033503, 21 (2008).

Gradient Structures for Flows of Concentrated Suspensions

Dirk Peschka, Marita Thomas, Tobias Ahnert, Andreas Münch, Barbara Wagner

Abstract In this work we investigate a two-phase model for concentrated suspensions. We construct a PDE formulation using a gradient flow structure featuring dissipative coupling between fluid and solid phase as well as different driving forces. Our construction is based on the concept of flow maps that also allows it to account for flows in moving domains with free boundaries. The major difference compared to similar existing approaches is the incorporation of a non-smooth two-homogeneous term to the dissipation potential, which creates a normal pressure even for pure shear flows.

1 Introduction

Suspension flows of solid particles in a viscous liquid are omnipresent in nature and are involved in many technological processes, *e.g.*, in the food, pharmaceutical,

Dirk Peschka
Weierstrass Institute Berlin, Mohrenstr. 39, 10117 Berlin, Germany, e-mail: peschka@wias-berlin.de

Marita Thomas
Weierstrass Institute Berlin, Mohrenstr. 39, 10117 Berlin, Germany, e-mail: thomas@wias-berlin.de

Tobias Ahnert
Deloitte Consulting, Hohenzollerndamm 150-151, 14199 Berlin, Germany, e-mail: tahnert@deloitte.de

Andreas Münch
Mathematical Institute, University of Oxford, Andrew Wiles Building, Oxford OX2 6GG, UK e-mail: muench@maths.ox.ac.uk

Barbara Wagner
Weierstrass Institute Berlin, Mohrenstr. 39, 10117 Berlin, Germany, e-mail: Barbara.Wagner@wias-berlin.de

M. Hintermüller, J. F. Rodrigues (eds.), *Topics in Applied Analysis and Optimisation*,
CIM Series in Mathematical Sciences, https://doi.org/10.1007/978-3-030-33116-0_12

printing or oil industries. The fraction of volume occupied by solid particles $0 \le \phi_s \le 1$ relative to the combined solid and liquid content, as shown in Figure 1, strongly affects the suspension flow. For very small volume fraction ϕ_s the suspension is called *dilute*, and mutual interaction between particles is neglegible. For increasing volume fraction of the particles the suspension enters a number of flow regimes and rheological behaviours, from shear thinning, to discontinuous shear thickening until it enters the shear *jamming* transition, when a critical volume fraction ϕ_{crit} is reached. Suspensions in this state are called *dense* or *concentrated*. The actual value of ϕ_{crit} depends sensitively on the particle shape, surface and other material properties.

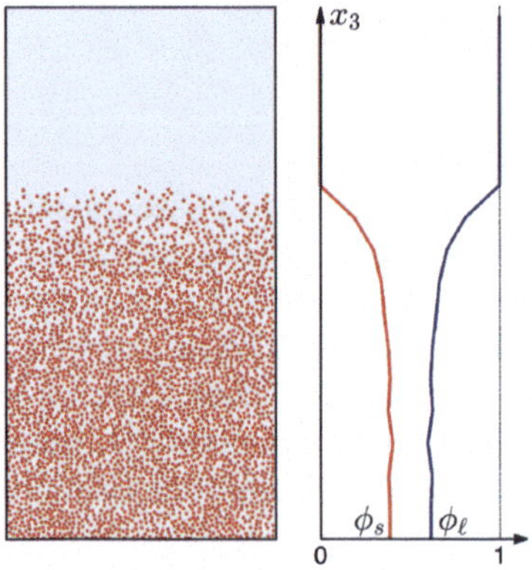

Fig. 1: Discrete solid particle distribution and corresponding volume fractions ϕ_s, ϕ_ℓ. Left: characterstic functions of particles $P : \Omega \to \{0, 1\}$ and Right: volume fractions $\phi_s : \Omega \to [0, 1] \equiv \langle P \rangle$ defined by a suitable average.

Predictive models therefore need to link the interaction of solid particles with the liquid and with other particles on the micro scale with the large-scale description of the dynamics of the liquid and solid phases on the continuum scale. In Figure 2 the numerical simulation of the sedimentation of two-dimensional particles in a viscous liquid is shown. The sedimentation of a particle is certainly influenced by the presence of other particles that create mututal long-ranged interactions due to the fluid flow. On the continuum scale, such a two-phase model works with averaged flow quantities such as averaged velocity $\mathbf{u}$, or effective viscosity μ_{eff} which relates the deviatoric[1] stress τ and the shear rate $\mathbb{D}\mathbf{u} = \frac{1}{2}(\nabla\mathbf{u} + \nabla\mathbf{u}^\top)$ via $\tau = 2\mu_{\mathrm{eff}}\,\mathrm{dev}\,\mathbb{D}\mathbf{u}$. For dilute suspensions of Newtonian liquids with viscosity μ and spherical particles Einstein [12] derived the effective viscosity law

$$\frac{\mu_{\mathrm{eff}}}{\mu} = 1 + \frac{5}{2}\phi_s. \tag{1.1}$$

[1] The deviator of a tensor/matrix in dimension d is defined as $\mathrm{dev}\,A = A - d^{-1}\,\mathrm{tr}(A)\mathbb{I}$ with $\mathbb{I}$ the $d \times d$ unit matrix. It is $\mathrm{tr}\,\mathrm{dev}\,A \equiv 0$ by construction. Subsequently we use σ to denote the total (Cauchy) stress, τ for the deviatoric stress, and p for the normal stress/pressure.

However, for many problems suspensions are not dilute but exhibit complex phenomena such as the formation of aggregates, creation of dense sedimentation layers, and shear-induced phase separation into highly concentrated and dilute regions. In fact, for any suspension where the liquid phase evaporates, Einstein's result (1.1) or its extensions [4] will eventually fail.

For many decades a great number of experimental and theoretical studies have been devoted to obtain expressions for an effective viscosity for the regime of concentrated suspensions, such as the Krieger-Dougherty law [21]. It has been observed experimentally that, as the suspension attains a solid-like state, it undergoes a jamming transition and develops further distinct phases [8, 16, 19, 27, 32]. These studies focussed on examining the role of friction and other properties of the particles interacting with each other and the liquid, reflecting how these microscopic properties control large-scale networked patterns. The dramatic increase in research devoted to this topic is rooted in the ground-breaking experimental study by Cassar et al. [7], where it was found that a dense suspension on an inclined plane sheared at a rate $2|\mathbb{D}\mathbf{u}|$ under a confining pressure p_c can be characterized by a single dimensionless control parameter, the viscous number

$$I_v = \frac{2\mu|\mathbb{D}\mathbf{u}|}{p_c}. \tag{1.2}$$

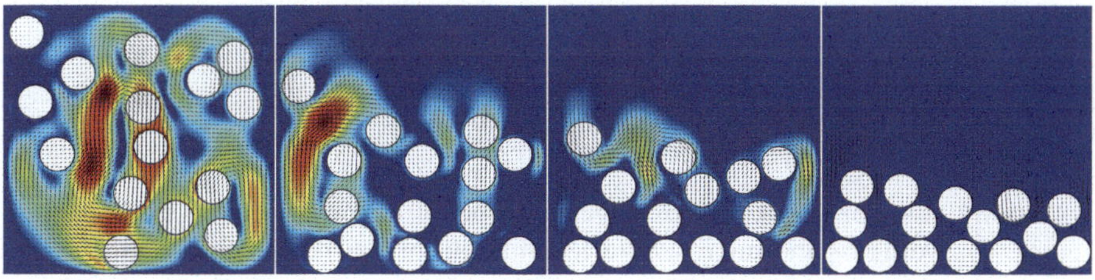

Fig. 2: Particulate flow with gravity in $\Omega \subset \mathbb{R}^2$ showing the sedimentation in a suspension with time advancing from left to right. The particle indictor function $P : \Omega \to \{0, 1\}$ is shown using white discs, the shading indicates the magnitude of the velocity field, shown using vectors.

This result was taken up by Boyer et al. [5], where a new constitutive friction law combining the rheology for non-Brownian suspensions and granular flows has been proposed, and for the first time offers to quantitatively capture the jamming transition. In Ahnert et al. [1], this new constitutive friction law was incorporated in the derivation of a new two-phase model for non-homogeneous shear flows and studied for simple shear flows such as plane Couette and Poiseuille flow. One key feature of these suspension models is the appearance of the normal or contact pressures p_c, with their role for particle migration discussed by Morris & Boulay [24].

Beyond simple effective models, predictive models on the length scale of these applications need to combine the interactions of the liquid and solid particles among each other on the microscale with a description of the dynamics of the liquid and solid phase on the continuum scale. This requires to incorporate phenomena such as transport of volume and mass with the balances of momenta and forces. The solid and liquid phase, i.e., their volume fractions ϕ_s and $\phi_\ell = 1 - \phi_s$, are transported by individual velocities $\mathbf{u}_s$ and $\mathbf{u}_\ell$. The velocities themselves obey the momentum balances of solid and liquid phase and are dissipative due to the presence of viscosity. Similar phenomena are known in the literature for classical mixture models.

To obtain further insight into the mathematical structure of this model we discuss in this article two-phase flow models from an energetic point of view and obtain that the general mathematical structure behind is of gradient-flow type. Hence, the evolution of the model system is characterized in terms of an energy functional and a dissipation potential. In particular, we will use the property that the model for the different regimes, from dilute to highly concentrated states, have a common general mathematical structure of variational type. In the long run, this will allow it to pursue the limit passage using variational convergence methods, and thus to carry out the transition from a dilute to a concentrated suspension as a rigorous scaling limit.

The focus of this work is to construct a class of thermodynamically and mechanically consistent models that support normal pressures using the framework of variational modelling. We present a method to construct suspension models with free boundaries and provide the underlying construction for gravity driven and surface-tension driven flows. Examples of such flows are given in applications such as in Murisic et al. [26].

2 Model for a concentrated suspension

We briefly summarise the dense suspension model that was derived in Ahnert et al. [1], by averaging the microscopic formulation of the flow with a liquid and a particulate solid phase along the lines of Drew [9] and Drew & Passman [10], in combination with a constitutive law for the solid phase stress-strain rate relation based on the results of the experiments by Boyer et al. [5] and a Kozeny-Carman relation for the interphase drag, see for example Brennen [6]. We assumed that the suspension consists of monodisperse, spherical, non-Brownian particles. It is also assumed that the mass densities of the solid ρ_s and liquid phase ρ_ℓ are constant. The equations are stated in non-dimensional variables as explained in detail in [1]; here we only give a brief summary of the scalings and the resulting equations. We use a velocity scale U, a length scale L, a time scale L/U and a viscous scale $\mu_\ell U/\rho_\ell$ for the pressure and stress field, where μ_ℓ is the liquid phase viscosity. The variables ϕ_s, $\mathbf{u}_s$, τ_s and p_s denote the volume fraction, velocity, deviatoric stress and normal stress for the solid phase, respectively, and analogously ϕ_ℓ, $\mathbf{u}_\ell$, τ_ℓ and p_ℓ for the liquid phase; t is the time. (The index ℓ is ommitted from the liquid pressure to be consistent with notation for the Lagrange multiplier in subsequent sections.) The bars $|\cdot|$ represent the componentwise Euclidean norm of a vector or tensor. Without inertia, the mass conservation and momentum balance equations for the two phases are

$$\partial_t \phi_\ell + \nabla \cdot (\phi_\ell \mathbf{u}_\ell) = 0, \tag{2.1a}$$
$$\partial_t \phi_s + \nabla \cdot (\phi_s \mathbf{u}_s) = 0, \tag{2.1b}$$
$$-\nabla \cdot \sigma_\ell + M_d + \phi_\ell \nabla \pi = 0, \tag{2.1c}$$
$$-\nabla \cdot \sigma_s - M_d + \phi_s \nabla \pi = 0, \tag{2.1d}$$

where the total stresses in liquid and solid phase are

$$\sigma_\ell(\mathbf{u}_\ell) = -p_\ell(\mathbf{u}_\ell)\mathbb{I} + \tau_\ell(\mathbf{u}_\ell), \tag{2.1e}$$
$$\sigma_s(\mathbf{u}_s) = -\big(p_c(\mathbf{u}_s) + p_s(\mathbf{u}_s)\big)\mathbb{I} + \tau_s(\mathbf{u}_s). \tag{2.1f}$$

The Lagrange multiplier π takes care of the constraint $\mathrm{div}_x(\phi_s \mathbf{u}_s + \phi_\ell \mathbf{u}_\ell) = 0$, which results from the condition $\phi_s + \phi_\ell = 0$ upon differentiation with respect to time using the transport equations. The drag M_d is given by the non-dimensional form of the Kozeny-Carman relation

$$M_d = \mathrm{Da}\, \frac{\phi_s^2}{\phi_\ell}(\mathbf{u}_\ell - \mathbf{u}_s). \tag{2.2}$$

The Darcy number which appears here is $\mathrm{Da} = L^2/K_p^2$, where K_p is proportional to the square of the particle diameter, so that Da is typically large. Next we specify the constitutive equations for the rheology of the liquid and the solid phase. For the liquid phase in three space dimensions, i.e., for $d = 3$, we have

$$p_\ell = -\frac{2}{3}\phi_\ell\, \mathrm{div}_x(\mathbf{u}_\ell), \qquad \tau_\ell = 2\phi_\ell\, \mathrm{dev}\, \mathbb{D}\mathbf{u}_\ell, \tag{2.3}$$

with $\mathbb{D}\mathbf{u}_\ell = \left(\nabla \mathbf{u}_\ell + \nabla \mathbf{u}_\ell^T\right)/2$ the shear rate. For the solid phase, if $|\mathbb{D}\mathbf{u}_s| > 0$, then

$$p_s = -\frac{2}{3}\phi_s \eta_s(\phi_s)\, \mathrm{div}_x\, \mathbf{u}_s, \qquad \tau_s = 2\phi_s \eta_s(\phi_s)\, \mathrm{dev}\, \mathbb{D}\mathbf{u}_s, \tag{2.4a}$$

with $\mathrm{dev}\, A = A - \frac{1}{3}\,\mathrm{tr} A$ the deviator of a matrix $A \in \mathbb{R}^{3\times 3}$; additionally there also acts a contact pressure given by

$$p_c = 2\phi_s \eta_n(\phi_s)|\mathbb{D}\mathbf{u}_s|. \tag{2.4b}$$

For $i = s, \ell$ note that $p_i = 0$ for divergence-free flows $\mathrm{div}_x\, \mathbf{u}_i = 0$, whereas p_c only vanishes when $\mathbb{D}\mathbf{u}_s$ does so. The constitutive material laws in the above definitions are

$$\eta_s(\phi_s) = 1 + \frac{5}{2}\frac{\phi_{\text{crit}}}{\phi_{\text{crit}} - \phi_s} + \mu_c(\phi_s)\frac{\phi_s}{(\phi_{\text{crit}} - \phi_s)^2}, \tag{2.4c}$$

$$\mu_c(\phi_s) = \mu_1 + \frac{\mu_2 - \mu_1}{1 + I_0\phi_s^2(\phi_{\text{crit}} - \phi_s)^{-2}}, \tag{2.4d}$$

$$\eta_n(\phi_s) = \left(\frac{\phi_s}{\phi_{\text{crit}} - \phi_s}\right)^2, \tag{2.4e}$$

with the non-dimensional parameters, $\mu_2 \geq \mu_1, I_0$, and the maximum volume fraction ϕ_{crit} for a random close packing. Instead, if $\mathbb{D}\mathbf{u}_s = 0$, we require

$$\phi_s = \phi_{\text{crit}}, \tag{2.4f}$$

and

$$|\sigma_s| \leq \mu_1 p_c. \tag{2.4g}$$

A typical value for the maximum random packing fraction is $\phi_{\text{crit}} = 0.63$. The values suggested in Boyer et al. [5] for the other parameters are $\mu_2 = 0.7$, and $I_0 = 0.005$, but these lead to a problem with ill-posedness even for plane Poiseuille flow [1].

The constitutive law (2.4) has the following implications: Given a fixed, positive finite contact pressure p_c, if the shear rate $\mathbb{D}\mathbf{u}_s$ tends to zero, then $\eta_n = p_c/(2\phi_s|\mathbb{D}\mathbf{u}_s|) \to \infty$ and thus $(\phi_{\text{crit}} - \phi_s) \to 0$. Since η_s has the same singular dependence on $\phi_{\text{crit}} - \phi_s$, it tends to infinity at the same rate and, therefore, $|\sigma_s|$ tends to a finite positive value, $\mu_1 p_c/\phi_s$, which gives rise to the yield stress in (2.4g). Across a yield surface, we require that ϕ_s, $\mathbf{u}_\ell$, $\mathbf{u}_s$, $|\mathbb{D}\mathbf{u}_s|$ and the projection of $-p_\ell\mathbb{I} + \tau_\ell$ and $-(p_s + p_c)\mathbb{I} + \tau_s$ onto the surface normal are continuous. While the suspension model above is stated for simplicity without any additional external forces, the later gradient flow construction will contain the full model with forces arising due to certain bulk or surface energies.

3 Gradient flow for two-phase flows of concentrated suspensions

Beyond flows of purely viscous liquids, the discussion of the proper mechanical statement of models for multi-phase flows has been studied extensively in the past, *e.g.*, [9, 10, 18, 20]. A major challenge from the modelling point of view is the construction of models that are mathematically, thermodynamically and mechanically meaningful. We here construct a class of models using a *variational approach* based on the energy and dissipation functionals related to the processes. In this way, we will deduce one possible model to describe flows of two-phase mixtures with free, evolving boundaries and provide the underlying construction for gravity-driven and surface-tension driven flows.

First variational descriptions of fluid flows are due to Helmholtz [15] and Rayleigh [31]. A general framework for the thermodynamic description of fluids has been layed out by Öttinger & Grmela [14, 28]. For the special construction of Euler flows using Poisson structures has been reviewed, for instance, by Morrison [25]. Peletier [29] gave a well-structured overview of systems which can be casted as gradient flows. For an extensive overview of different models for complex fluids and flow maps we refer to the recent review by Giga et al. [13]. It has to be stressed that the afore mentioned contributions consider the flow in fixed domains with fixed boundaries. In fact, our approach can be seen as a generalization of the one presented in [13] for single-phase fluid flow in a fixed domain to the problem of two-phase flows on evolving domains.

In the following we focus on the formal description of free boundary multi-phase flows on moving domains in terms of generalized gradient flows. This concept has been discussed e.g. by Mielke [22] in an abstract framework and formally applied to models arising in many different applications. Following [22], such a description is based the specification of a triple $(\mathbf{V}, \mathcal{R}, \mathcal{E})$ consisting of the (Banach) space of velocities, a dissipation potential $\mathcal{R} : \mathbf{Q} \times \mathbf{V} \to [0, \infty]$, and an energy functional $\mathcal{E} : \mathbf{Q} \to \mathbb{R}$ defined on the state space $\mathbf{Q}$. Elements of the state space are denoted by $q \in \mathbf{Q}$ and their corresponding velocities by $\dot{q} \in \mathbf{V}$. For all states $q \in \mathbf{Q}$ it is required that $\mathcal{R}(q; \cdot) : \mathbf{V} \to [0, \infty]$ is convex and that $\mathcal{R}(q; \dot{q} = 0) = 0$. With $\mathbf{V}^*$ we denote the dual space of $\mathbf{V}$ and define for fixed $q \in \mathbf{Q}$ the dual dissipation functional $\mathcal{R}^*(q, \cdot) : \mathbf{V}^* \to [0, \infty]$ as the convex conjugate of $\mathcal{R}(q; \cdot)$, i.e., for all $v^* \in \mathbf{V}^*$ it is $\mathcal{R}^*(q, v^*) := \sup_{v \in \mathbf{V}} \left(\langle v^*, v \rangle_{\mathbf{V}} - \mathcal{R}(q, v)\right)$. As in [22] we speak here of a generalized gradient flow as we neither require $\mathcal{R}$ to be quadratic nor classically differentiable. In this generalized setting it can be shown, cf. e.g. [22, 23], by exploiting the convexity of the functionals $\mathcal{R}(q, \cdot)$ and $\mathcal{R}^*(q, \cdot)$ that a solution $q : [0, T] \to \mathbf{Q}$ of $(\mathbf{V}, \mathcal{R}, \mathcal{E})$ is characterized by the following three equivalent problem formulations:

$$
\begin{aligned}
& \dot{q}(t) \in \partial \mathcal{R}^*(q(t), -\mathrm{D}_q \mathcal{E}(q(t))) && \text{in } \mathbf{V}, && (3.1a)\\
\Leftrightarrow \quad & -\mathrm{D}_q \mathcal{E}(q(t)) \in \partial \mathcal{R}(q(t), \dot{q}(t)) && \text{in } \mathbf{V}^*, && (3.1b)\\
\Leftrightarrow \quad & \langle -\mathrm{D}_q \mathcal{E}(q(t)), \dot{q}(t) \rangle_{\mathbf{V}} = \mathcal{R}(q(t), \dot{q}(t)) + \mathcal{R}^*(q(t), -\mathrm{D}_q \mathcal{E}(q(t))), && && (3.1c)
\end{aligned}
$$

where $\partial(\cdot)$ denotes the subdifferential of a convex functional with respect to $\dot{q}$ and $\mathrm{D}\mathcal{E}(q)$ the Fréchet-derivative of $\mathcal{E}$.

Since the Young-Fenchel inequality for convex functionals and their conjugate always ensures $\langle -\mathrm{D}_q \mathcal{E}(q), \dot{q} \rangle_{\mathbf{V}} \leq \mathcal{R}(q, \dot{q}) + \mathcal{R}^*(q, -\mathrm{D}_q \mathcal{E}(q))$ one can infer from (3.1c) that the time-derivative $\dot{q}$ of a solution q of (3.1) also satisfies

$$
\dot{q} \in \operatorname{argmin}_{\dot{q} \in \mathbf{V}} \left(\langle \mathrm{D}_q \mathcal{E}(q), \dot{q} \rangle_{\mathbf{V}} + \mathcal{R}(q, \dot{q}) \right), \tag{3.1d}
$$

since $\mathcal{R}^*(q(t), -\mathrm{D}_q \mathcal{E}(q(t)))$ is independent of $\dot{q}$.

Indeed, the setting of generalized gradient flows based on convex potentials with the formulation (3.1) provides a generalization of classical gradient flows characterized by quadratic potentials. For a given self-adjoint linear operator $G(q) : \mathbf{V} \to \mathbf{V}^*$ and quadratic functionals $\mathcal{R}(q, \dot{q}) = \frac{1}{2} \langle G(q)\dot{q}, \dot{q} \rangle$ a solution $q(t)$ of the gradient flow is given by a curve $q : [0, T] \to \mathbf{Q}$ satisfying (3.1b), which reads in this smooth, quadratic context as

$$\dot{q}(t) = -\nabla_{\mathcal{R}}\mathcal{E}\big(q(t)\big). \tag{3.1e}$$

where the gradient $v = \nabla_{\mathcal{R}}\mathcal{E}(q) \in \mathbf{V}$ of $\mathcal{E}$ with respect to the metric induced by $\mathcal{R}$ is defined by $\langle G(q)^* v, \tilde{\dot{q}}\rangle = \langle G(q)^* \mathrm{D}_q\mathcal{E}(q), \tilde{\dot{q}}\rangle$ for all $\tilde{\dot{q}} \in \mathbf{V}$ and $G(q) = G(q)^*$.

Formulation (3.1) provides the abstract framework that we are going to use in order to deduce two-phase suspension models on moving domains. More precisely, in this section we will show that, under suitable smoothness assumptions on the functions involved, flow models for suspensions as discussed in the previous Section 2 indeed arise as generalized gradient flows $(\mathbf{V}, \mathcal{R}, \mathcal{E})$ in the form (3.1b). Given a suitable triple $(\mathbf{V}, \mathcal{R}, \mathcal{E})$ we will rigorously derive a weak formulation of the corresponding PDE system (3.1b). At this point our presentation will stay on a formal level, as we will not address the existence and regularity of solutions for the resulting problem. Under further smoothness assumptions we will then formally deduce a pointwise formulation of the associated Cauchy problem and compare our resulting system with the one presented in Section 2. Indeed, we shall see that a dissipation potential suited to produce a critical pressure of 1-homogeneous nature is not of the standard smooth, quadratic nature.

3.1 Notation and states

We consider the motion of a liquid continuous phase (index ℓ) mixed with a solid dispersed phase of non-Brownian particles (index s) phase occupying at each time $t \in [0,T]$ a bounded set $\Omega(t) \subset \mathbb{R}^d$ where $d \in \mathbb{N}$. At the initial time $t = 0$ this subdomain is denoted by $\bar{\Omega} = \Omega(0)$. For each point in space $x \in \Omega(t)$ the state of the suspension is characterized by volume fractions $0 \leq \phi_s(t,x), \phi_\ell(t,x) \leq 1$ such that we have $\phi_s(t,x) + \phi_\ell(t,x) = 1$ pointwise. In the following we define the structures needed to model the evolution of ϕ_i and $\Omega(t)$ using a gradient flow structure. One key idea in this construction is the consistent use of *flow maps* as elements of an abstract state space.

Definition 1 (Evolution of shapes with flow maps) Let $\chi(t,\cdot) : \bar{\Omega} \to \Omega(t)$ a family of diffeomorphisms that map from $\bar{\Omega} \subset \mathbb{R}^d$ to $\Omega(t) \subset \mathbb{R}^d$ using

$$\Omega(t) = \chi(t,\bar{\Omega}) \equiv \{x \in \mathbb{R}^d : \exists X \in \bar{\Omega} \ s.t. \ x = \chi(t,X)\}. \tag{3.2}$$

The small letter x will always denote coordinates in $\Omega(t)$, whereas the capital letter X denotes coordinates in the reference configuration $X \in \bar{\Omega}$. We define the associated velocity $\mathbf{u}(t,\cdot) : \Omega(t) \to \mathbb{R}^d$ with

$$\mathbf{u}(t,x) = \big(\partial_t\chi\big)(t,\chi^{-1}(t,x)). \tag{3.3}$$

We call $\chi(t,\cdot)$ the *flow map* associated to the motion of $\Omega(t)$ and $\mathbf{u}(t,\cdot)$ the corresponding velocity vector field. Initial data are chosen such that $\chi(t = 0, X) = X$ and $\bar{\Omega} = \Omega(0)$. With the notation $F_\chi = \nabla_X\chi$ we indicate the gradient of the transformation and assume for its Jacobian determinant that $\det F_\chi > 0$. On the other hand, for

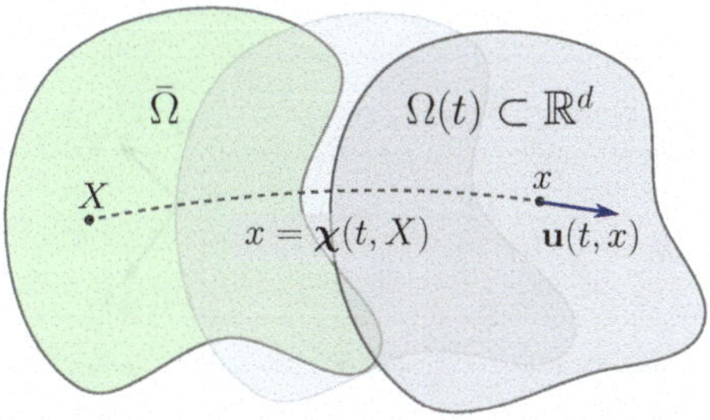

Fig. 3: Flow map $\chi(t,\cdot) : \bar{\Omega} \to \Omega(t)$ mapping a point $X \in \bar{\Omega} \subset \mathbb{R}^d$ from the reference domain $\bar{\Omega}$ (green shaded) to a point in the mapped configuration $\Omega(t)$ (gray shaded). When considering the trajectory $x(t) = \chi(t, X)$ (dashed line) for any given X, then $\mathbf{u} = \dot{x}(t)$ is the associated velocity (arrow) and $\mathbf{u}(t, x)$ the corresponding flow field.

given flow field $\mathbf{u}$ we have an associated ODE-Cauchy problem:

$$\partial_t \chi(t, X) = \mathbf{u}(t, \chi(t, X)) \qquad \text{for all } X \in \bar{\Omega} \text{ and } t \in [0, T], \tag{3.4a}$$

$$\chi(0, X) = X \qquad \text{for all } X \in \bar{\Omega}. \tag{3.4b}$$

Note that (3.4) is the *kinematic condition* for the domain motion.

In the presence of two phases, each phase is characterized by its own flow map $\chi_i : [0,T] \times \bar{\Omega} \to \Omega(t) \subset \mathbb{R}^d$ with $i \in \{s, \ell\}$ for solid and liquid phase. Correspondingly we use $\mathbf{u}_i$ and F_i to indicate the corresponding velocities and Jacobians. With this notation we further require the flow maps to satisfy the following assumptions: Multiple flow maps are defined on the same domain

$$\Omega(t) = \chi_s(t, \bar{\Omega}) = \chi_\ell(t, \bar{\Omega}), \tag{3.5}$$

which of course does not imply that the flow maps are equal. Furthermore we have the following assumptions on $\mathbf{u}_i$ and χ_i for $i \in \{s, \ell\}$ for all $t \in [0, T]$:

- $\Omega(t)$ and $\bar{\Omega}$ bounded and sufficiently smooth, (3.6a)
- $\chi_i(t,\cdot) : \bar{\Omega} \to \Omega(t)$ is a smooth diffeomorphism, (3.6b)
- $\det F_i(t,\cdot) > 0$, where $F_i(t, X) = \nabla_X \chi_i(t, X)$, (3.6c)
- $\mathbf{u}_s(t,\cdot) = \mathbf{u}_\ell(t,\cdot)$ on $\partial\Omega(t)$. (3.6d)

Note that for the gradient structure equality of tangential velocities on the boundary would suffice to ensure (3.5), but we require the slightly stronger condition (3.6d).

At each time $t \in [0, T]$ and each $x \in \Omega(t)$, the fraction of volume occupied by liquid and solid phase is characterized by the two phase indicators $\phi_i(t, x), i \in \{s, \ell\}$. Since ϕ_i represent volume fractions, they must satisfy

$$\phi_s(t, x), \phi_\ell(t, x) \in [0, 1] \quad \text{for all } t \in [0, T] \text{ and all } x \in \Omega(t) \tag{3.7a}$$

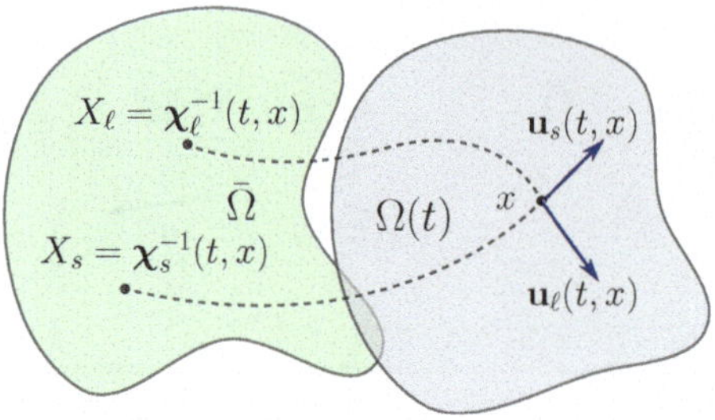

Fig. 4: Flow maps $\chi_i(t,\cdot)$ for solid $i = s$ phase and liquid $i = \ell$ phase mapping a point $X \in \bar{\Omega} \subset \mathbb{R}^d$ from the reference domain $\bar{\Omega}$ (green shaded) to a point in the mapped configuration $\Omega(t)$ (gray shaded). When considering the trajectories $x_i(t) = \chi_i(t,X)$ (dashed lines) for any given X and $i \in \{s,\ell\}$, then $\mathbf{u}_i = \dot{x}_i(t)$ are the associated velocities (arrow) and $\mathbf{u}_i(t,x)$ the corresponding flow field. At time t the trajectories meet at the same point x, when they started at $X_i = \chi_i^{-1}(t,x)$.

and fill the volume such that

$$\phi_s(t,x) + \phi_\ell(t,x) = 1 \quad \text{for all } t \in [0,T] \text{ and all } x \in \Omega(t) \tag{3.7b}$$

The evolution of the densities is defined via a local conservation law for the two volume fractions. We assume that the given initial volume fractions $\phi_i(t=0,X) = \bar{\phi}_i(X) \in [0,1]$ and that the flow maps $\chi_i(t,\cdot) : \bar{\Omega} \to \Omega(t)$ as well as their velocities $\mathbf{u}_i(t,\cdot) : \Omega(t) \to \mathbb{R}^d$ are sufficiently smooth for $i \in \{s,\ell\}$. For arbitrary $\bar{\omega} \subset \bar{\Omega}$ let $\omega_i(t) = \chi_i(t,\bar{\omega}) \subset \Omega(t)$. In the absence of reaction or diffusion processes we require the volume fraction $\phi_i(t,\cdot) : \Omega(t) \to \mathbb{R}$ to satisfy the *integral form of volume conservation* stated as

$$\int_{\omega(t)} \phi_i(t,x)\,\mathrm{d}x = \int_{\bar{\omega}} \bar{\phi}_i(X)\,\mathrm{d}X. \tag{3.8a}$$

Differentiating (3.8a) in time and using the Reynolds transport theorem, given the smoothness of all quantities involved, shows the equivalent *differential form of volume conservation*: For given $t \in [0,T]$ and any $x \in \Omega(t)$ the density $\phi_i(t,x)$ satisfies the (Cauchy problem for the) transport equation

$$\begin{aligned} &\partial_t \phi_i(t,x) + \mathrm{div}_x\big(\phi_i(t,x)\mathbf{u}_i(t,x)\big) = 0 \quad \text{in } \Omega(t), \\ &\phi_i(0,X) = \bar{\phi}_i(X) \quad \text{in } \bar{\Omega}, \end{aligned} \tag{3.8b}$$

with given, sufficiently smooth initial data $\bar{\phi}_i$, which also have to satisfy the volume constraints, i.e., we claim that

- $0 \le \bar{\phi}_i \le 1 \quad \text{for all } X \in \bar{\Omega},$ (3.9a)
- $\bar{\phi}_s + \bar{\phi}_\ell = 1 \quad \text{for all } X \in \bar{\Omega}.$ (3.9b)

The following lemma summarizes a few immediate consequences of the preceeding definitions, constraints, and assumptions. Moreover, Statement 4. below justifies why we can subsequently work with the divergence contraint (3.10) for the average velocity, cf. the definition of the dissipation potential (3.17), in order to equivalently guarantee the volume constraint (3.7b) for the phase indicators.

Lemma 3.1 *Let $i \in \{s, \ell\}$ and let all the quantities $\chi_i, \phi_i, \mathbf{u}_i, \bar{\phi}_i$ be sufficiently smooth.*

1. *Assume the densities ϕ_i fill the volume* (3.7b). *Then at each time $t \in [0, T]$ the average velocity defined as $\mathbf{u} = \phi_s \mathbf{u}_s + \phi_\ell \mathbf{u}_\ell$ satisfies the following divergence constraint*

$$\operatorname{div}_x \mathbf{u}(t, x) = 0, \quad \textit{for all } x \in \Omega(t). \tag{3.10}$$

2. *Let the sufficiently smooth flow map also satisfy the positivity assumption* (3.6c), *i.e.,* $\det F_i > 0$. *Then, the transport problem* (3.8) *for the volume fraction ϕ_i is equivalent to the following explicit representation for any given $t \in [0, T]$:*

$$\phi_i\big(t, \chi_i(t, X)\big) = \big(\det F_i(t, X)\big)^{-1} \bar{\phi}_i(X), \qquad \textit{for each } X \in \bar{\Omega}. \tag{3.11}$$

3. *Let the transport problem* (3.8) *as well as the volume constraint* (3.7b) *be satisfied. Then the two phase volumes are conserved, i.e.,*

$$V_i(t) = \int_{\Omega(t)} \phi_i(t, x)\,\mathrm{d}x = V_i(0), \qquad \textit{and} \qquad |\Omega(t)| = V_s + V_\ell = |\bar{\Omega}|. \tag{3.12}$$

4. *Let the transport problem* (3.8) *be satisfied. Then, the following equivalence holds true for the volume constraint on the phase indicator:*

$$\left\{ \begin{array}{l} (3.9\mathrm{b}) \textit{ for } \bar{\phi}_s, \bar{\phi}_\ell \textit{ at initial time} \\ \textit{\& divergence constraint } (3.10) \end{array} \right\} \Leftrightarrow \big\{ (3.7\mathrm{b}) \textit{ for } \phi_s, \phi_\ell \textit{ at any } t \in [0, T] \big\} \tag{3.13}$$

5. *Assume that $\bar{\phi}_i$ satisfies the convex constraint* (3.9a) *at initial time, that the transport relation* (3.8) *as well as volume constraint* (3.9b) *and divergence constraint* (3.10) *hold true. Then the convex constraint* (3.7a) *holds true also for $\phi_i(t, \cdot)$ in $\Omega(t)$ for any $t \in [0, T]$.*

Proof **To 1.:** Since we assumed $\phi_s(t, x) + \phi_\ell(t, x) = 1$ for any $x \in \Omega(t)$ we readily conclude $\partial_t(\phi_s + \phi_\ell) = 0 = \operatorname{div}_x(\phi_s \mathbf{u}_s + \phi_\ell \mathbf{u}_\ell)$.
To 2.: Using change of variables $X = \chi_i^{-1}(t, x)$ and volume conservation (3.8) we find for all $t \in [0, T]$, arbitrary $\bar{\omega} \subset \bar{\Omega}$ and $\omega(t) = \chi_i(t, \bar{\omega})$ that

$$\int_{\bar{\omega}} \bar{\phi}_i(X)\,\mathrm{d}X = \int_{\omega(t)} \phi_i(t, x)\,\mathrm{d}x = \int_{\bar{\omega}} \phi_i(t, \chi_i(t, X)) \det F_i(t, X)\,\mathrm{d}X\,.$$

The assertion follows due to the smoothness of ϕ_i and the positivity of $\det F_i$.
To 3.: This is a direct consequence of 1. and $\phi_s + \phi_\ell = 1$.

To 4.: Clearly, condition (3.7b) includes (3.9b) initial time. The divergence constraint again follows from (3.7b) and (3.8) along the lines of Item 1.. Hence, '⇐' in (3.13). To find also '⇒' we argue as follows: Transport problem (3.8) together with (3.10) implies $\partial_t(\phi_s + \phi_\ell) = 0$ in $[0,T] \times \Omega(t)$. Hence $\phi_s(t,x) + \phi_\ell(t,x) = c(x) = \phi_s(0,X) + \phi_\ell(0,X) = 1$, which is (3.7b) at any $t \in [0,T]$.

To 5.: The above argument implies $\phi_i \le 1$, if it is possible to show that $\phi_i \ge 0$. Indeed, the latter follows from (3.8) thanks to its equivalence to the representation (3.11). By the positivity of the determinant (3.6c) and the constraint (3.9a) satisfied by the initial data we may thus conclude that $\phi_i \ge 0$. □

Here we point out the crucial observation that the evolution of ϕ_i is not independent but rather defined using the flow maps χ_i. However, when considering functional depending on ϕ_i we need to be able to compute its variations. For this we recall the simple identity for change of variables for volume integrals.

Theorem 3.1 (Change of variables in volumes) *Let* $\chi : (t,\cdot)\bar{\Omega} \to \Omega$ *a flow map from* $\bar{\Omega} \subset \mathbb{R}^d$ *to* $\Omega(t) \subset \mathbb{R}^d$ *and let* $\phi(t,\chi(t,X)) = (\det F_\chi)^{-1}\bar{\phi}(X)$ *and* $f(x,\phi)$ *given.*

$$\int_{\chi(t,\bar{\Omega})} f(x,\phi)\,\mathrm{d}x = \int_{\bar{\Omega}} f\big(\chi(t,X),(\det F_\chi(t,X))^{-1}\bar{\phi}(X)\big)\,\det F_\chi(t,X)\,\mathrm{d}X.$$

For instance using $f(x,\phi_i) = \phi_i$ and $\chi = \chi_i$ shows that conservation of volume holds by construction since

$$\int_{\chi_i(t,\bar{\Omega})} \phi_i(t,x)\,\mathrm{d}x = \int_{\bar{\Omega}} \bar{\phi}_i(X)\,\mathrm{d}X = V_i.$$

3.2 The triple $(\mathbf{V},\mathcal{R},\mathcal{E})$ for flows of concentrated suspensions

In view of the discussion in Section 3.1 we denote in the following the vector of states by $q := (\chi_s,\chi_\ell) \in \mathbf{Q}$ and its associated vector of velocities by $\dot{q} := (\mathbf{u}_s,\mathbf{u}_\ell) \in \mathbf{V}$. Hereby, we will use the spaces

$$\mathbf{X} := \{\chi \in H^1(\bar{\Omega};\mathbb{R}^d),\ \chi = \mathrm{id}_{\bar{\Omega}} \text{ on } \partial\bar{\Omega}\backslash\bar{\Gamma}\}, \tag{3.14a}$$

$$\mathbf{Q} := \{(\chi_s,\chi_\ell) \in \mathbf{X}\times\mathbf{X},\ \chi_s = \chi_\ell \text{ on } \bar{\Gamma}\}, \tag{3.14b}$$

as the state space for the flow maps defined on the reference configuration $\bar{\Omega}$ and

$$\mathbf{V} := \left\{(\tilde{\mathbf{u}}_s,\tilde{\mathbf{u}}_\ell) \in H^1(\Omega(t);\mathbb{R}^d\times\mathbb{R}^d),\ \begin{array}{ll}\tilde{\mathbf{u}}_s = \tilde{\mathbf{u}}_\ell & \text{on } \Gamma(t)\\ \tilde{\mathbf{u}}_s = \tilde{\mathbf{u}}_\ell = 0 & \text{on } \partial\Omega(t)\backslash\Gamma(t)\end{array}\right\}, \tag{3.15}$$

as the function space for the velocities defined on the current configuration $\Omega(t)$ for all $t \in [0,T]$. Note that above we introduced a part of the boundary $\partial\bar{\Omega} \setminus \bar{\Gamma}$, on which the shape of the domain is fixed corresponding to a no-slip boundary condition. Moreover we stress that the upcoming definitions of functionals will always implicitely depend on a vector of given data $(\bar{\phi}_s, \bar{\phi}_\ell, \bar{\Omega})$, which consists of the reference configuration $\bar{\Omega} \subset \mathbb{R}^d$, and of the reference densities $\bar{\phi}_s, \bar{\phi}_\ell$ of solid and fluid phase.

Further using the notation from Definition 1 we consider an energy functional $\mathcal{E} : \mathbf{Q} \to [0, \infty]$ where

$$\mathcal{E}(q) := \mathcal{E}_{\text{bulk}}(q) + \mathcal{E}_{\text{surf}}(q) \quad \text{with}$$

$$\mathcal{E}_{\text{bulk}}(q) := \begin{cases} \int_{\Omega(t)} E(x, \phi_s)\,\mathrm{d}x & \text{if } \phi_i(t,x) = \frac{\bar{\phi}_i(t,X)}{\det F_i(t,X)}, \\ \infty & \text{otherwise,} \end{cases} \tag{3.16a}$$

$$\text{where } E(x, \phi_s) := g x_d(\phi_s \rho_s + (1 - \phi_s)\rho_\ell), \quad \text{and} \tag{3.16b}$$

$$\mathcal{E}_{\text{surf}}(q) := \begin{cases} \int_{\Gamma(t)} \vartheta\,\mathrm{d}\mathcal{H}^{d-1} & \text{if } \phi_i(t,x) = \frac{\bar{\phi}_i(t,X)}{\det F_i(t,X)}, \\ \infty & \text{otherwise.} \end{cases} \tag{3.16c}$$

In (3.16b) the constant g denotes the gravity constant, x_d is the dth component of the space variable $x \in \Omega(t) \subset \mathbb{R}^d$ in the current configuration $\Omega(t)$, and ρ_s, ρ_ℓ denote the mass densities of the solid and the fluid phase, respectively. Moreover, in (3.16c), the parameter ϑ denotes the surface tension and $\mathcal{H}^{d-1}$ is the $(d-1)$-dimensional Hausdorff measure.

In addition, we also introduce the dissipation potential $\mathcal{R} : \mathbf{Q} \times \mathbf{V} \to [0, \infty]$ as

$$\mathcal{R}(q; \dot{q}) := \int_{\Omega(t)} R(\phi_s, \phi_\ell; \tilde{\mathbf{u}}_s, \tilde{\mathbf{u}}_\ell, \tilde{e}_s, \tilde{e}_\ell)\,\mathrm{d}x + \mathcal{I}_{\mathbf{K}(q)}(\tilde{\mathbf{u}}_s, \tilde{\mathbf{u}}_\ell), \tag{3.17a}$$

where we used the indicator functional $\mathcal{I}_{\mathbf{K}(q)}$ and the constraint set $\mathbf{K}(q)$ defined as

$$\mathcal{I}_{\mathbf{K}(q)}(\tilde{\mathbf{u}}_s, \tilde{\mathbf{u}}_\ell) := \begin{cases} 0 & \text{if } (\tilde{\mathbf{u}}_s, \tilde{\mathbf{u}}_\ell) \in \mathbf{K}(q), \\ \infty & \text{otherwise,} \end{cases} \tag{3.17b}$$

$$\mathbf{K}(q) := \{(\tilde{\mathbf{u}}_s, \tilde{\mathbf{u}}_\ell) \in \mathbf{V},\ \mathrm{div}_x(\phi_s \tilde{\mathbf{u}}_s + \phi_l \tilde{\mathbf{u}}_\ell) = 0 \text{ a.e. in } \Omega(t)\}. \tag{3.17c}$$

The constraint set $\mathbf{K}(q) \subset \mathbf{V}$ depends on $q = (\chi_s, \chi_\ell) \in \mathbf{Q}$ through ϕ_s, ϕ_ℓ by (3.11). Indeed, with given, fixed ϕ_s, ϕ_ℓ it can be checked that $\mathbf{K}(q)$ is a closed linear subspace of $\mathbf{V}$.

Moreover, in (3.17a) there are the following contributions to the density R

$$R(\phi_s, \phi_\ell; \tilde{\mathbf{u}}_s, \tilde{\mathbf{u}}_\ell, \tilde{e}_s, \tilde{e}_\ell) := R_\ell(\phi_\ell; \tilde{e}_\ell) + R_s(\phi_s; \tilde{e}_s) + R_{s\ell}(\phi_s; \tilde{\mathbf{u}}_s, \tilde{\mathbf{u}}_\ell), \tag{3.17d}$$

$$R_\ell(\phi_\ell; \tilde{e}_\ell) := \frac{\tilde{\mu}_{\ell 1}(\phi_\ell)}{2} |\,\mathrm{dev}\, \tilde{e}_\ell|^2 + \frac{\tilde{\mu}_{\ell 2}(\phi_\ell)}{2} |\,\mathrm{tr} \tilde{e}_\ell|^2, \tag{3.17e}$$

$$R_{s\ell}(\phi_s; \tilde{\mathbf{u}}_s, \tilde{\mathbf{u}}_\ell) := \frac{\tilde{\mu}_{s\ell}(\phi_s)}{2} |\tilde{\mathbf{u}}_s - \tilde{\mathbf{u}}_\ell|^2, \tag{3.17f}$$

$$R_s(\phi_s; \tilde{e}_s) := \frac{\tilde{\mu}_s(\phi_s)}{2} \Big[\alpha |\,\mathrm{dev}\, \tilde{e}_s|^2 + \beta_+ (\,\mathrm{tr} \tilde{e}_s)_+^2 + \beta_- (\,\mathrm{tr} \tilde{e}_s)_-^2 + \gamma |\tilde{e}_s| (\,\mathrm{tr} \tilde{e}_s)_- \Big], \tag{3.17g}$$

where $e_i = e(\mathbf{u}_i)$, $\tilde{e}_i = e(\tilde{\mathbf{u}}_i)$, $e(\mathbf{u}) := \frac{1}{2}(\nabla \mathbf{u} + \nabla \mathbf{u}^\top)$ is the symmetric strain tensor, $\mathrm{tr} e_i := \sum_{k=1}^d e_{i,kk}$ is the trace of the matrix $e_i = (e_{i,kl})_{k,l=1}^d \in \mathbb{R}^{d \times d}$, and with the notation $\mathrm{dev}\, e_i := e_i - \frac{1}{d}\,\mathrm{tr} e_i \mathbb{I}$ we indicate its deviator. The functions $(\cdot)_\pm$ in (3.17g) denote the positive, resp. negative part, i.e.,

$$(a)_\pm := \max\{\pm a, 0\} \quad \text{for } a \in \mathbb{R}.$$

Observe that the contribution of the liquid (3.17e) and the coupled part (3.17f) are both quadratic, hence convex for strictly positive coefficient functions. Instead, the dissipation potential of the solid phase features, in addition to the quadratic terms, also the mixed term $|\operatorname{dev}\tilde{e}_s|(\operatorname{tr}\tilde{e}_s)_-$. Hence, convexity of the solid dissipation potential can only be ensured under additional assumptions on α, β_-, and γ.

We now specify conditions on the coefficients in (3.17), for which coercivity and convexity of $\mathcal{R}$ can be ensured. Under these conditions we give a characterization of its subdifferential.

Proposition 3.1 (Properties of $\mathcal{R}$) *Let $\mathcal{R}$ be given by* (3.17) *with the velocity space* $\mathbf{V}$ *as in* (3.15) *and let the states $q = (\chi_s, \chi_\ell)$ be given in accordance with* (3.6) *and* (3.11). *For given $\mathbf{u}_s \in H^1(\Omega(t);\mathbb{R}^d)$, resp. $\mathbf{u}_\ell \in H^1(\Omega(t);\mathbb{R}^d)$ with $\mathbf{u}_s = \mathbf{u}_\ell = 0$ on $\partial\Omega(t)\backslash\Gamma(t)$, set*

$$\mathbf{V}_{\mathbf{u}_s} := \{\tilde{\mathbf{u}}_\ell \in H^1(\Omega(t);\mathbb{R}^d),\ (\mathbf{u}_s, \tilde{\mathbf{u}}_\ell) \in \mathbf{K}(q)\}\,, \tag{3.18a}$$

$$\mathbf{V}_{\mathbf{u}_\ell} := \{\tilde{\mathbf{u}}_s \in H^1(\Omega(t);\mathbb{R}^d),\ (\tilde{\mathbf{u}}_s, \mathbf{u}_\ell) \in \mathbf{K}(q)\}\,. \tag{3.18b}$$

1. *Assume that $\tilde{\mu}_s, \tilde{\mu}_\ell, \tilde{\mu}_{s\ell} \in L^\infty(\mathbb{R})$ and that there is a constant $\tilde{\mu}_* > 0$ such that $\tilde{\mu}_s, \tilde{\mu}_{\ell 1}, \tilde{\mu}_{\ell 2}, \tilde{\mu}_{s\ell} > \tilde{\mu}_*$ a.e. in $\mathbb{R}$. Further assume that $\alpha, \beta_+, \beta_- > 0$, and $\gamma \geq 0$. Then, for given $q \in \mathbf{Q}$ the functional $\mathcal{R}(q;\cdot)$ is lower semicontinuous and coercive on $\mathbf{V}$, i.e., for all $(\mathbf{u}_s, \mathbf{u}_\ell) \in \mathbf{V}$ it is*

$$\mathcal{R}(q;\dot{q}) \geq \tfrac{1}{2}\tilde{\mu}_*\alpha_* C_{\mathrm{PF}}\big(\|\mathbf{u}_s\|^2_{H^1(\Omega(t);\mathbb{R}^d)} + \|\mathbf{u}_\ell\|^2_{H^1(\Omega(t);\mathbb{R}^d)}\big), \tag{3.19}$$

   where $\alpha_* := \min\{\alpha, \beta_-, \beta_+\}$ and C_{PF} is the Poincaré-Friedrichs constant for $\mathbf{V}$.
2. *Let the assumptions of Item 1. hold true. Then the functional $\mathcal{R}_s(q;\cdot) := \int_{\Omega(t)} R_s(q; e(\cdot))\,\mathrm{d}x$ with R_s from* (3.17g) *is strictly convex on $\mathbf{V}_{\mathbf{u}_\ell}$, cf.* (3.18b), *if $\frac{\gamma^2}{(1-\delta)\beta_-} \leq 4\min\{\alpha, \beta_+, \delta\beta_-\}$ for a constant $\delta \in (0,1)$.*
3. *Let the assumptions of Item 1. hold. Then $\mathcal{R}(q;\cdot)$ is strictly convex if $\frac{\gamma^2}{(1-\delta)\beta_-} \leq 4\min\{\alpha, \beta_+, \delta\beta_-\}$ for a constant $\delta \in (0,1)$.*
4. *Let the assumptions of Item 3. hold true. The subdifferential of $\mathcal{R}(q;\cdot)$ for an element $(\mathbf{u}_s, \mathbf{u}_\ell) \in \mathbf{V}$ is given by the elements $(\xi_s, \xi_\ell) + (\zeta_s, \zeta_\ell) \in \mathbf{V}^*$ such that for all $(\tilde{\mathbf{u}}_s, \tilde{\mathbf{u}}_\ell) \in \mathbf{V}$ it is*

$$\mathcal{R}(q;\dot{\tilde{q}}) - \mathcal{R}(q;\dot{q}) \geq \langle\!\langle(\xi_s + \zeta_s, \xi_\ell + \zeta_\ell), (\tilde{\mathbf{u}}_s, \tilde{\mathbf{u}}_\ell)\rangle\!\rangle_{\mathbf{V}},$$

   with $(\zeta_s, \zeta_\ell) \in \partial I_{\mathbf{K}(q)}(\mathbf{u}_s, \mathbf{u}_\ell)$ characterized for any $(\mathbf{u}_s, \mathbf{u}_\ell) \in \mathbf{K}(q)$ by elements $\pi \in L^2(\Omega(t))$ in the following way

$$0 = \langle\!\langle(\zeta_s, \zeta_\ell), (\tilde{\mathbf{u}}_s, \tilde{\mathbf{u}}_\ell)\rangle\!\rangle_{\mathbf{V}} = \int_{\Omega(t)} \pi\,\mathrm{div}_x(\phi_s\tilde{\mathbf{u}}_s + \phi_l\tilde{\mathbf{u}}_\ell)\,\mathrm{d}x \quad \textit{for all } (\tilde{\mathbf{u}}_s, \tilde{\mathbf{u}}_\ell) \in \mathbf{K}(q)\,. \tag{3.20a}$$

Moreover, the elements $(\xi_s, \xi_\ell) \in \mathbf{V}^*$ *are given by*

$$\langle \xi_\ell, \tilde{\mathbf{u}}_\ell \rangle_{H^1(\Omega(t);\mathbb{R}^d)} := \int_{\Omega(t)} \Big[\tilde{\mu}_{\ell 1}(\phi_s) \operatorname{dev} e_\ell : \operatorname{dev} \tilde{e}_\ell + \tilde{\mu}_{\ell 2}(\phi_s) \operatorname{tr} e_\ell \operatorname{tr} \tilde{e}_\ell \\ - \tilde{\mu}_{s\ell}(\phi_s)(\mathbf{u}_s - \mathbf{u}_\ell) \cdot \tilde{\mathbf{u}}_\ell \Big] \, \mathrm{d}x, \tag{3.20b}$$

$$\langle \xi_s, \tilde{\mathbf{u}}_s \rangle_{H^1(\Omega(t);\mathbb{R}^d)} := \int_{\Omega(t)} \Big[\tilde{\mu}_s(\phi_s)\alpha \operatorname{dev} e_s : \operatorname{dev} \tilde{e}_s + \tilde{\mu}_{s\ell}(\phi_s)(\mathbf{u}_s - \mathbf{u}_\ell) \cdot \tilde{\mathbf{u}}_s \\ + \hat{\beta}_+ (\operatorname{tr} e_s)_+ \operatorname{tr} \tilde{e}_s + \hat{\beta}_- (\operatorname{tr} e_s)_- \operatorname{tr} \tilde{e}_s \\ + \hat{\mu}_1 |e_s| \operatorname{tr} \tilde{e}_s + \hat{\mu}_2(e_s) : \tilde{e}_s (\operatorname{tr} e_s)_- \Big] \, \mathrm{d}x, \tag{3.20c}$$

$$\begin{aligned} &\textit{with } \hat{\beta}_+ \in L^\infty(\Omega(t)), && \hat{\beta}_+ = \beta_+ \tilde{\mu}_s(\phi_s) H(\operatorname{tr} e_s), \\ &\textit{and } \hat{\beta}_- \in L^\infty(\Omega(t)), && \hat{\beta}_- = -\beta_- \tilde{\mu}_s(\phi_s) H(-\operatorname{tr} e_s), \\ &\textit{and } \hat{\mu}_1 \in L^\infty(\Omega(t)), && \hat{\mu}_1 = -\tfrac{\tilde{\mu}_s(\phi_s)\gamma}{2} H(-\operatorname{tr} e_s), \\ &\textit{and } \hat{\mu}_2 \in L^\infty(\Omega(t);\mathbb{R}^{d\times d}), && \hat{\mu}_2(e_s) = \begin{cases} \frac{e_s}{|e_s|} \frac{\tilde{\mu}_s(\phi_s)\gamma}{2} \textit{ if } |e_s| > 0, \\ \hat{e} \in \mathbb{R}^{d\times d}_{\mathrm{sym}} \textit{ with } |\hat{e}| \le \frac{\tilde{\mu}_s(\phi_s)\gamma}{2} \textit{ if } |e_s| = 0, \end{cases} \end{aligned}$$

at a.e. point $x \in \Omega(t)$*, and where* H *denotes the Heaviside function*

$$H(a) \in \begin{cases} \{0\} & \textit{if } a < 0, \\ [0,1] & \textit{if } a = 0, \\ \{1\} & \textit{if } a > 0. \end{cases} \tag{3.21}$$

Proof **To 1.:** Observe that the functional $\mathcal{R}(q;\cdot)$ is continuous in $\mathbf{K}(q)$ due to the closedness of this subspace in $\mathbf{V}$. In particular this is immediate for all the quadratic contributions of the functional; the continuity of the product term can be seen by the following calculation

$$\begin{aligned} &\left| \int_{\Omega(t)} \tfrac{\tilde{\mu}_s(\phi_s)}{2} \big(|e_s| (\operatorname{tr} e_s)_- - |\tilde{e}_s| (\operatorname{tr} \tilde{e}_s)_- \big) \, \mathrm{d}x \right| \\ &\le \tilde{\mu}^* \int_{\Omega(t)} \big| |e_s| - |\tilde{e}_s| \big| \big| (\operatorname{tr} e_s)_- - (\operatorname{tr} \tilde{e}_s)_- \big| \, \mathrm{d}x \\ &\le \tilde{\mu}^* \| e_s - \tilde{e}_s \|_{L^2(\Omega(t);\mathbb{R}^{d\times d})} \| (\operatorname{tr} e_s)_- - (\operatorname{tr} \tilde{e}_s)_- \|_{L^2(\Omega(t))} \end{aligned}$$

by Hölder's inequality. This proves continuity in $\mathbf{K}(q)$. Lower semicontinuity in $\mathbf{V}$ then follows by the fact that $\mathcal{R}(q; \mathbf{u}_s, \mathbf{u}_\ell) = \infty$ for any $(\mathbf{u}_s, \mathbf{u}_\ell) \in \mathbf{V} \backslash (\mathbf{V}_{\mathbf{u}_\ell} \times \mathbf{V}_{\mathbf{u}_s})$.

Coercivity estimate (3.19) directly follows from all quadatic terms thanks to the positive bounds from below for the coefficient functions and by Poincaré-Friedrich's inequality in $\mathbf{V}$.

To 2.: Let $(\mathbf{u}_s, \mathbf{u}_\ell), (\tilde{\mathbf{u}}_s, \tilde{\mathbf{u}}_\ell) \in \mathbf{V}$ and $\lambda \in (0,1)$. In what follows, we abbreviate $e = e_s$ and $\tilde{e} = \tilde{e}_s$. First of all, we observe that the positive and the negative part $(\cdot)^\pm$ are convex functions so that $(\lambda \operatorname{tr} e + (1-\lambda)\operatorname{tr}\tilde{e})^\pm \le \lambda(\operatorname{tr} e)^\pm + (1-\lambda)(\operatorname{tr}\tilde{e})^\pm$. Since $|\cdot|^2$ is monotone, we find

$$\beta_\pm |(\operatorname{tr}(\lambda e + (1-\lambda)\tilde{e}))^\pm|^2 \le \beta_\pm |\lambda(\operatorname{tr} e)^\pm + (1-\lambda)(\operatorname{tr}\tilde{e})^\pm|^2 . \tag{3.22}$$

Furthermore, dev and tr are linear operators. Hence, with placeholders $a \in \{\operatorname{dev} e, \operatorname{tr} e, (\operatorname{tr} e)_\pm\}$ and $\tilde{a} \in \{\operatorname{dev}\tilde{e}, \operatorname{tr}\tilde{e}, (\operatorname{tr}\tilde{e})_\pm\}$ the uniform convexity of $|\cdot|^2$ can be checked:

$$|\lambda a + (1-\lambda)\tilde{a}|^2 = \lambda |a|^2 + (1-\lambda)|\tilde{a}|^2 - \lambda(1-\lambda)(a-\tilde{a})^2. \tag{3.23}$$

This also proves the convexity of R_ℓ. Moreover, the product term contained in R_s can be estimated by monotonicity of $|\cdot|$ and $(\cdot)_-$, and Young's inequality as follows:

$$\begin{aligned}
&|\lambda e + (1-\lambda)\tilde{e}|(\operatorname{tr}(\lambda e + (1-\lambda)\tilde{e}))_- \\
&\le (\lambda|e| + (1-\lambda)|\tilde{e}|)(\lambda(\operatorname{tr} e)_- + (1-\lambda)(\operatorname{tr}\tilde{e})_-) \\
&= \lambda|e|(\operatorname{tr} e)_- + (1-\lambda)|\tilde{e}|(\operatorname{tr}\tilde{e})_- \\
&\quad -\lambda(1-\lambda)\big(|e| - |\tilde{e}|\big)\big((\operatorname{tr} e)_- - (\operatorname{tr}\tilde{e})_-\big) \\
&\le \lambda|e|(\operatorname{tr} e)_- + (1-\lambda)|\tilde{e}|(\operatorname{tr}\tilde{e})_- \\
&\quad +\lambda(1-\lambda)\sqrt{\varepsilon}\big||e| - |\tilde{e}|\big|\big|(\operatorname{tr} e)_- - (\operatorname{tr}\tilde{e})_-\big|(\sqrt{\varepsilon})^{-1} \\
&\le \lambda|e|(\operatorname{tr} e)_- + (1-\lambda)|\tilde{e}|(\operatorname{tr}\tilde{e})_- \\
&\quad +\lambda(1-\lambda)\big(\tfrac{\varepsilon}{2}\big(|e| - |\tilde{e}|\big)^2 + \tfrac{1}{2\varepsilon}\big((\operatorname{tr} e)_- - (\operatorname{tr}\tilde{e})_-\big)^2\big),
\end{aligned}$$

where the positive terms in the very last line of this estimate have to be absorbed by the corresponding negative term obtained in (3.23). For this, it can be checked that

$$\begin{aligned}
&-\alpha(|\operatorname{dev} e| - |\operatorname{dev}\tilde{e}|)^2 - \beta_+((\operatorname{tr} e)_+ - (\operatorname{tr}\tilde{e})_+)^2 - (\delta + 1 - \delta)\beta_-((\operatorname{tr} e)_- - (\operatorname{tr}\tilde{e})_-)^2 \\
&\le -m_\delta(|e| - |\tilde{e}|)^2 - (1-\delta)\beta_-((\operatorname{tr} e)_- - (\operatorname{tr}\tilde{e})_-)^2
\end{aligned}$$

with $m_\delta := \min\{\alpha, \beta_+, \delta\beta_-\}$ for a constant $\delta \in (0,1)$. Thus, combining this estimate with the previous ones, we obtain

$$\begin{aligned}
R_s(\phi_s; \lambda e + (1-\lambda)\tilde{e}) &\le \lambda R_s(\phi_s; \lambda e) + (1-\lambda)R_s(\phi_s; \tilde{e}) \\
&\quad -\lambda(1-\lambda)\tfrac{\tilde{\mu}(\phi_s)}{2}\big((m_\delta - \tfrac{\gamma\varepsilon}{2})(|\operatorname{dev} e| - |\operatorname{dev}\tilde{e}|)^2 \\
&\qquad +((1-\delta)\beta_- - \tfrac{\gamma}{2\varepsilon})((\operatorname{tr} e)_- - (\operatorname{tr}\tilde{e})_-)^2 + \beta_+((\operatorname{tr} e)_+ - (\operatorname{tr}\tilde{e})_+)^2\big),
\end{aligned}$$

and we have to make sure that both $m_\delta - \frac{\gamma\varepsilon}{2} \ge 0$ and $(1-\delta)\beta_- - \frac{\gamma}{2\varepsilon} \ge 0$. This implies the constraint $\frac{\gamma}{2(1-\delta)\beta_-} \le \varepsilon \le \frac{2m_\delta}{\gamma}$, which finally gives $\frac{\gamma^2}{(1-\delta)\beta_-} \le 4m_\delta$ for strict convexity.

To 3.: Thanks to the previously proved statement of Item 2, the convexity properties of the full functional $\mathcal{R}(q;\cdot)$ now follow by the uniform convexity of the quadratic fluid and solid-fluid contributions.

To 4.: From Item 1. we recall that $\mathcal{R}(q;\cdot)$ is convex. The Moreau-Rockafellar Theorem for convex functionals, cf. e.g. [17, p. 200, Thm. 1], provides a sum rule for the subdifferential of convex functionals, i.e.: *If* $F_1,\ldots,F_k : U \to (-\infty,\infty]$ *are proper, convex functionals, all but possibly one of them continuous in a point* $\bar{v} \in \mathrm{dom}F_1 \cap \ldots \cap \mathrm{dom}F_k$, *then*

$$\partial F_1(v) + \ldots + \partial F_k(v) = \partial(F_1(v) + \ldots + F_k(v)) \quad \textit{for all } v \in U. \tag{3.24}$$

We observe that all the contributions to $\int_{\Omega(t)} R\,\mathrm{d}x$ are continuous on all of $\mathbf{V}$ and a possible discontinuity for some $(\mathbf{u}_s,\mathbf{u}_\ell) \in \mathrm{dom}\mathcal{R}$ arises by the constraint term $\mathcal{I}_{\mathbf{K}(q)}$. Hence, the prerequisites of the Moreau-Rockafellar Theorem are met and (3.24) applies to determine the contributions of its subdifferential.

In order to find the characterization (3.20a) of $(\zeta_s,\zeta_\ell) \in \partial\mathcal{I}_{\mathbf{K}(q)}(\mathbf{u}_s,\mathbf{u}_\ell)$ we note that for any $(\tilde{\mathbf{u}}_s,\tilde{\mathbf{u}}_\ell) \in \mathbf{V}$ it is

$$(\tilde{\mathbf{u}}_s,\tilde{\mathbf{u}}_\ell) \in \mathbf{K}(q) \quad \Leftrightarrow \quad \text{for all } \eta \in L^2(\Omega(t)) : \int_{\Omega(t)} \eta\,\mathrm{div}_x(\phi_s\tilde{\mathbf{u}}_s + \phi_\ell\tilde{\mathbf{u}}_\ell)\,\mathrm{d}x = 0\,. \tag{3.25}$$

This equivalently states that the annihilator $\mathbf{K}(q)^\perp$ of the linear subspace $\mathbf{K}(q)$ is given by

$$\mathbf{K}(q)^\perp = \Big\{\Big(\int_{\Omega(t)} \eta\,\mathrm{div}_x(\phi_s \bullet + \phi_\ell \bullet)\,\mathrm{d}x\Big) : \mathbf{K}(q) \to 0,\ \eta \in L^2(\Omega(t))\Big\}\,.$$

On the other hand, by the definition of the subdifferential of $\mathcal{I}_{\mathbf{K}(q)}$ for any $(\mathbf{u}_s,\mathbf{u}_\ell) \in \mathbf{K}(q)$ we have that $(\zeta_s,\zeta_\ell) \in \partial\mathcal{I}_{\mathbf{K}(q)}(\mathbf{u}_s,\mathbf{u}_\ell)$ is a support function, i.e.,

$$(\zeta_s,\zeta_\ell) \in \partial\mathcal{I}_{\mathbf{K}(q)}(\mathbf{u}_s,\mathbf{u}_\ell) \quad \Leftrightarrow \quad \text{for all } (\tilde{\mathbf{u}}_s,\tilde{\mathbf{u}}_\ell) \in \mathbf{K}(q) : \\ \langle(\zeta_s,\zeta_\ell),(\mathbf{u}_s,\mathbf{u}_\ell) - (\tilde{\mathbf{u}}_s,\tilde{\mathbf{u}}_\ell)\rangle_{\mathbf{V}} \ge 0\,.$$

With the specific choices $(\tilde{\mathbf{u}}_s,\tilde{\mathbf{u}}_\ell) = (0,0) \in \mathbf{K}(q)$ and $(\tilde{\mathbf{u}}_s,\tilde{\mathbf{u}}_\ell) = -2(\mathbf{u}_s,\mathbf{u}_\ell) \in \mathbf{K}(q)$ we find that $\langle(\zeta_s,\zeta_\ell),(\mathbf{u}_s,\mathbf{u}_\ell)\rangle_{\mathbf{V}} = -\langle(\zeta_s,\zeta_\ell),(\mathbf{u}_s,\mathbf{u}_\ell)\rangle_{\mathbf{V}} \ge 0$ and hence $(\zeta_s,\zeta_\ell) = 0$ on $\mathbf{K}(q)$. This means that $(\zeta_s,\zeta_\ell) \in \mathbf{K}(q)^\perp$ and hence (3.20a) is deduced.

It remains to determine the other contributions to the subdifferential given by (3.20b) & (3.20c). For this, we further make use of the chain rule for the subdifferential of convex functionals, cf. e.g. [17, p. 201, Thm. 2]: *Assume that* $A : U \to W$ *is linear,* $\mathcal{F} : W \to (-\infty,\infty]$ *is convex and there is* $u \in U$ *such that* $\mathcal{F}$ *is continuous in* Au. *Then,*

$$\partial(\mathcal{F} \circ A)(u) = (A^*\partial\mathcal{F})(Au), \tag{3.26}$$

where $A^* : W^* \to U^*$ *is the adjoint of* A, *defined by* $\langle A^*u^*, v\rangle = \langle u^*, Av\rangle$.

Since the dissipation potential of the liquid (3.17e) and the coupling term (3.17f) are Fréchet-differentiable, we directly find (3.20b) and the second summand of

(3.20c). To deduce the remaining terms of (3.20c) we shall apply the above theorem to the dissipation potential of the solid $\mathcal{R}_s(q;\cdot)$. To this aim we set $U = \mathbf{V}_{\mathbf{u}_\ell}$ with given $\mathbf{u}_\ell$, $A : \mathbf{V}_{\mathbf{u}_\ell} \to W = L^2(\Omega(t);\mathbb{R}^{d\times d}) \times L^2(\Omega(t)) \times L^2(\Omega(t);\mathbb{R}^{d\times d})$, $Av := (\operatorname{dev} e(v), \operatorname{tr} e(v), e(v))^\top$ and set $\widetilde{\mathcal{R}}_s(a_1,a_2,a_3) := \int_{\Omega(t)} \frac{\tilde{\mu}(\phi_s)}{2}\big(\alpha|a_1|^2 + \beta_+|(a_2)_+|^2 + \beta_-|(a_2)_-|^2 + \gamma|a_3|(a_2)_-\big)\,\mathrm{d}x$. Thus, $\mathcal{R}_s(q;v) = (\widetilde{\mathcal{R}}_s \circ A)(v)$ for all $v \in \mathbf{V}_{\mathbf{u}_\ell}$. Thanks to the previously proved continuity and convexity properties of $\mathcal{R}(q;\cdot)$ we see that also $\widetilde{\mathcal{R}}_s$ is convex and continuous on W. Hence chain rule (3.26) is applicable. To ultimately conclude (3.20c), we note that for all $(a_1,a_2,a_3),(\tilde{a}_1,\tilde{a}_2,\tilde{a}_3) \in W$ it is

$$\begin{aligned}
&\langle \partial\widetilde{\mathcal{R}}_s(a_1,a_2,a_3),(\tilde{a}_1,\tilde{a}_2,a_3)\rangle_W \\
&\quad= \langle \partial_{a_1}\widetilde{\mathcal{R}}_s(a_1,a_2,a_3),\tilde{a}_1)\rangle_{L^2(\Omega(t))} + \langle \partial_{a_2}\widetilde{\mathcal{R}}_s(a_1,a_2,a_3),\tilde{a}_2)\rangle_{L^2(\Omega(t))} \quad \text{with} \\
&\langle \partial_{a_1}\widetilde{\mathcal{R}}_s(a_1,a_2,a_3),\tilde{a}_1)\rangle_{L^2(\Omega(t))} = \int_{\Omega(t)} \tilde{\mu}_s(\phi_s)\alpha a_1 : \tilde{a}_1\,\mathrm{d}x\,, \\
&\langle \partial_{a_2}\widetilde{\mathcal{R}}_s(a_1,a_2,a_3),\tilde{a}_2)\rangle_{L^2(\Omega(t))} = \int_{\Omega(t)} \big(\hat{\beta}_+(a_2)_+\tilde{a}_2 + \hat{\beta}_-(a_2)_-\tilde{a}_2 + \hat{\mu}_1|a_3|\tilde{a}_2\big)\,\mathrm{d}x\,, \\
&\langle \partial_{a_3}\widetilde{\mathcal{R}}_s(a_1,a_2,a_3),\tilde{a}_3)\rangle_{L^2(\Omega(t))} = \int_{\Omega(t)} (a_2)_-\hat{\mu}_2(a_3) : \tilde{a}_3\,\mathrm{d}x\,,
\end{aligned}$$

with the coefficient functions $\hat{\beta}_\pm, \hat{\mu}_1, \hat{\mu}_2$ as stated in (3.20c). □

In order to state (3.1b) for the system of concentrated suspensions it remains to calculate the derivative of the energy functional.

Proposition 3.2 (Functional derivative of $\mathcal{E}$) *Let the energy functional $\mathcal{E}(q)$ be given as in (3.16) and consider the family of flow maps $q(h)$ defined by*

$$\chi_i(h,X) = X + h\,\mathbf{u}_i(X), \tag{3.27}$$

for any arbitrary $\mathbf{u}_i \in H^1(\bar{\Omega};\mathbb{R}^d)$ representing an element $\dot{q} = (\mathbf{u}_s,\mathbf{u}_\ell) \in \mathbf{V}$. Then the variation of $\mathcal{E}$ in an arbitrary direction $\dot{q} \in \mathbf{V}$ is given by

$$\langle \mathrm{D}_q\mathcal{E}(q),\dot{q}\rangle = \lim_{h\to 0}\frac{1}{h}\left[\mathcal{E}\big(q(h)\big) - \mathcal{E}\big(q(0)\big)\right]. \tag{3.28}$$

1. *The functional derivative of $\mathcal{E}_{\mathrm{bulk}}$ from (3.16a) reads*

$$\langle \mathrm{D}_q\mathcal{E}_{\mathrm{bulk}}(q),\dot{q}\rangle = \int_\Omega (\nabla_x E)\cdot\mathbf{u}_s + (E - \phi_s\partial_{\phi_s}E)(\nabla\cdot\mathbf{u}_s)\,\mathrm{d}x \tag{3.29}$$

where $E = E(x,\phi_s)$.

2. *The functional derivative of $\mathcal{E}_{\mathrm{surf}}$ from (3.16c) reads*

$$\langle \mathrm{D}_q\mathcal{E}_{\mathrm{surf}}(q),\dot{q}\rangle = \int_\Gamma \Big(\mathbf{u}\cdot\nabla_x\vartheta + \vartheta\,\mathrm{div}_\Gamma(\mathbf{u})\Big)\,\mathrm{d}\mathcal{H}^{d-1} \tag{3.30}$$

with surface energy $\vartheta = \vartheta(x)$ and $\mathbf{u} = \phi_s\mathbf{u}_s + \phi_\ell\mathbf{u}_\ell$.

Proof **To 1.:** First we use change of variables to express the integral

$$\begin{aligned}\mathcal{E}_{\text{bulk}}\big(q(h)\big) &= \int_{\Omega(h)} E(x, \phi_s(h,x))\,\mathrm{d}x \\ &= \int_{\bar{\Omega}} E\left(\chi_s(h,X), \frac{\bar{\phi}_s(X)}{\det F_s(h,X)}\right) \det F_s(h,X)\,\mathrm{d}X.\end{aligned}$$

This allow us to use (3.28) for a fixed domain. Then the differentiation of the integrand gives the expression

$$\begin{aligned}\langle \mathrm{D}_q\mathcal{E}_{\text{bulk}}(q), \dot{q}\rangle &= \int_{\bar{\Omega}} \lim_{h\to 0} \tfrac{1}{h}\left[E\left(\chi_s(h,X), \tfrac{\bar{\phi}_s(X)}{\det F_s(h,X)}\right)\det F_s(h,X) - E\left(X, \bar{\phi}_s(X)\right)\right]\mathrm{d}X \\ &= \int_{\bar{\Omega}} (\nabla_x E)\cdot \mathbf{u}_s + \left[(\partial_{\phi_s} E)\left(-\frac{\bar{\phi}_s}{\det F_s}\right) + E\right](\partial_h \det F_s)_{h=0}\,\mathrm{d}X \\ &= \int_{\bar{\Omega}} (\nabla_x E)\cdot \mathbf{u}_s + (E - \phi_s \partial_{\phi_s} E)\nabla\cdot \mathbf{u}_s\,\mathrm{d}x\end{aligned}$$

where at $h = 0$ we have $x \equiv X$. We used a simple version of Jacobi's formula $\partial_h \det F_s = (\det F_s)\,\mathrm{tr}(F_s^{-1}\partial_h F_s)$, $\det F_s = 1$, and $\mathrm{tr}\partial_h F_s = \nabla\cdot\mathbf{u}_s$ for $h = 0$. The result remains valid for arbitrary q if the final integral is expressed in x-coordinates.

To 2.: We use again change of variables to express the integrals

$$\mathcal{E}_{\text{surf}}(q(h)) = \int_{\Gamma(h)} \vartheta(x)\,\mathrm{d}\mathcal{H}^{d-1} = \int_{\bar{\Gamma}} \vartheta(\chi_i(t,X))\|\mathrm{Cof}(F_i(t,X))\cdot\mathbf{n}\|\,\mathrm{d}\mathcal{H}^{d-1},$$

where $\mathrm{Cof}F_i = (\det F_i)(F_i^{-1})^\top$ is the cofactor of the Jacobian. The differentiation of this term is slightly more technical and can be found, for instance, in [30]. The resulting expression is

$$\langle \mathcal{E}_{\text{surf}}, \dot{q}\rangle = \int_\Gamma (\mathbf{u}\cdot\nabla_x\vartheta + \vartheta\,\mathrm{div}_\Gamma\,\mathbf{u})\,\mathrm{d}\mathcal{H}^{d-1},$$

where we used the surface divergence $\mathrm{div}_\Gamma\,\mathbf{u}$. Observe that via the *divergence theorem on manifolds* one can rewrite

$$\int_\Gamma \mathrm{div}_\Gamma\,\mathbf{u}\,\mathrm{d}s = -\int_\Gamma \mathbf{H}\cdot\mathbf{u}\,\mathrm{d}s, \tag{3.31}$$

where $\mathbf{H} = H\mathbf{n} \equiv -\mathbf{n}(\nabla_\Gamma\cdot\mathbf{n})$ is the mean curvature vector and H the scalar mean curvature (with respect to $\mathbf{n}$). Also note that $\mathbf{u}$ can be replaced with $\mathbf{u}_s$ or $\mathbf{u}_\ell$ since by (3.6d) they all agree on Γ for a given variation $\dot{q}$. □

3.3 PDE system obtained by the gradient flow formulation

In this section we combine the results of Propositions 3.1 & 3.2 in order to obtain formulation of the force balance (3.1b) for the problem.

Weak formulation of the problem.

Force balance (3.1b)

$$-\mathrm{D}\mathcal{E}(q) \in \partial\mathcal{R}(\dot{q}) \quad \text{in } \mathbf{V}^*$$

is now directly obtained from the results of Propositions 3.1 & 3.2. For shorter notation we observe that the gradient terms arising in the differential of the bulk dissipation, cf. (3.20), define viscous stresses of solid and liquid phase. We here gather them in terms of the stress tensors σ_s, σ_ℓ given by

$$\sigma_\ell = \tilde{\mu}_{\ell 1}\,\mathrm{dev}(e(\mathbf{u}_\ell)) + \tilde{\mu}_{\ell 2}\,\mathrm{tr}(e(\mathbf{u}_\ell))\mathbb{I}, \tag{3.32a}$$

$$\sigma_s = \tilde{\mu}_s\alpha\,\mathrm{dev}(e(\mathbf{u}_s)) + \left[\hat{\beta}_+(\,\mathrm{tr}e(\mathbf{u}_s))_+ + \hat{\beta}_-(\,\mathrm{tr}e(\mathbf{u}_s))_- + \hat{\mu}_1|e(\mathbf{u}_s)|\right]\mathbb{I} + \sigma_s^*, \tag{3.32b}$$

$$\sigma_s^* = \hat{\mu}_2(e(\mathbf{u}_s))(\,\mathrm{tr}e(\mathbf{u}_s))_-, \tag{3.32c}$$

where the coefficient functions $\hat{\beta}_\pm, \hat{\mu}_1, \hat{\mu}_2$ are defined in (3.20c). In this way, the weak formulation induced by (3.1b) reads as follows:

$$\begin{aligned}
&\langle(\xi_s + \zeta_s, \xi_\ell + \zeta_\ell), (\tilde{\mathbf{u}}_s, \tilde{\mathbf{u}}_\ell)\rangle_{\mathbf{V}} \\
&= \int_{\Omega(t)} \sigma_\ell : e(\tilde{\mathbf{u}}_\ell) + \sigma_s : e(\tilde{\mathbf{u}}_s) + \tilde{\mu}_{s\ell}(\mathbf{u}_\ell - \mathbf{u}_s)\cdot(\tilde{\mathbf{u}}_\ell - \tilde{\mathbf{u}}_s) + \pi\,\mathrm{div}_x(\phi_s\tilde{\mathbf{u}}_s + \phi_\ell\tilde{\mathbf{u}}_\ell)\,\mathrm{d}x \\
&= -\int_{\Omega(t)} (\nabla_x E)\cdot\tilde{\mathbf{u}}_s - \pi_s\,\mathrm{div}_x\,\tilde{\mathbf{u}}_s\,\mathrm{d}x - \int_{\Gamma(t)} \vartheta\,\mathrm{div}_\Gamma\,\tilde{\mathbf{u}}\,\mathrm{d}\mathcal{H}^{d-1} = -\langle \mathrm{D}\mathcal{E}(q), (\tilde{\mathbf{u}}_s, \tilde{\mathbf{u}}_\ell)\rangle_{\mathbf{V}},
\end{aligned} \tag{3.33}$$

for all $(\tilde{\mathbf{u}}_s, \tilde{\mathbf{u}}_\ell) \in \mathbf{V}$, with $\pi_s := -(E - \phi_s\partial_{\phi_s}E)$ as an effective pressure of the solid phase, $\tilde{\mathbf{u}} = \phi_s\tilde{\mathbf{u}}_s + \phi_\ell\tilde{\mathbf{u}}_\ell$, and with $(\xi_s + \zeta_s, \xi_\ell + \zeta_\ell) \in \partial\mathcal{R}(q, \dot{q})$.

Pointwise formulation of the problem.

Suppose now that all the functions involved in (3.33) are sufficiently smooth, so that we can integrate by parts in (3.33) in order to move the gradients from the test functions to the stress and pressure terms. This leads to the classical, pointwise formulation of the problem, again involving the stresses σ_ℓ, σ_s from (3.32).

In order to reconstruct the pointwise PDE formulation we first rewrite the derivative of $\mathcal{E}_{\mathrm{bulk}}$ as

$$\langle \mathrm{D}_q \mathcal{E}_{\mathrm{bulk}}(q), \dot{\tilde{q}} \rangle = \int_\Omega (\nabla_x E) \cdot \tilde{\mathbf{u}}_s + (E - \phi_s \partial_{\phi_s} E)(\nabla \cdot \tilde{\mathbf{u}}_s)\,\mathrm{d}x$$
$$= \int_\Omega (\nabla_x E + \nabla p^*) \cdot \tilde{\mathbf{u}}_s\,\mathrm{d}x + \int_{\partial\Omega} (-p^*)\tilde{\mathbf{u}}_s \cdot \mathbf{n}\,\mathrm{d}\mathcal{H}^{d-1}, \tag{3.34a}$$

where the effective pressure is defined

$$p^*(x, \phi_s) = \phi_s \partial_{\phi_s} E(x, \phi_s) - E(x, \phi_s). \tag{3.34b}$$

The derivative of $\mathcal{E}_{\mathrm{surf}}$ we already characterized in (3.31) using the mean curvature.
In particular, for all $t \in [0, T]$, a.e. in $\Omega(t)$ the following PDE-system has to be satisfied:

$$-\operatorname{div}_x \sigma_s + \tilde{\mu}_{s\ell}(\mathbf{u}_s - \mathbf{u}_\ell) = -\phi_s \nabla(\pi + \pi_s), \tag{3.35a}$$
$$-\operatorname{div}_x \sigma_\ell - \tilde{\mu}_{s\ell}(\mathbf{u}_s - \mathbf{u}_\ell) = -\phi_\ell \nabla\pi, \tag{3.35b}$$
$$\operatorname{div}_x \left(\phi_s \mathbf{u}_s + \phi_\ell \mathbf{u}_\ell\right) = 0, \tag{3.35c}$$

together with the following boundary conditions:

$$(\sigma_s + \sigma_\ell)\,\mathbf{n} = (d-1)\vartheta\kappa + \pi + \pi_s \quad \text{on } \Gamma(t), \tag{3.35d}$$
$$\mathbf{u}_\ell = \mathbf{u}_s \quad \text{on } \Gamma(t), \tag{3.35e}$$
$$\mathbf{u}_\ell = \mathbf{u}_s = 0 \quad \text{on } \partial\Omega(t)\backslash\Gamma(t). \tag{3.35f}$$

Comparison of models.

Even though the model in (3.35) already appears very similar to the one in (2.1), we perform a short discussion on the terms in the stress and the pressures. Firstly, the gradient flow model (3.35) offers a systematic way to include forces due to bulk energies $\mathcal{E}_{\mathrm{bulk}}$ and surface energies and $\mathcal{E}_{\mathrm{surf}}$, leading to the coupling term p^*, p_s and the corresponding boundary terms in (3.35d). The easiest to identify are the pressure terms in (3.35), if we decompose the contribution $\sigma_s - \sigma_s^*$ to the solid stress into a volumetric and a deviatoric part

$$\sigma_s - \sigma_s^* := -p_s \mathbb{I} + 2\mu_s^* \operatorname{dev} \mathbb{D}\mathbf{u}_s \tag{3.36a}$$

with

$$\mu_s^* = \frac{1}{2}\tilde{\mu}_s \alpha, \tag{3.36b}$$

$$p_s = -\beta_+(\operatorname{div}_x \mathbf{u}_s)_+ - \beta_-(\operatorname{div}_x \mathbf{u}_s)_- - \hat{\mu}_1 |\mathbb{D}\mathbf{u}_s| \overset{\operatorname{div}_x \mathbf{u}_s = 0}{=} \frac{\gamma}{2}\tilde{\mu}_s(\phi_s)|\mathbb{D}\mathbf{u}_s| H(-\operatorname{div}_x \mathbf{u}_s), \tag{3.36c}$$

where we used the material law $\hat{\mu}_1 = -\frac{\gamma}{2}\tilde{\mu}_s H(-\operatorname{div}_x \mathbf{u}_s)$ with the Heaviside function H as defined in (3.21). Note that H is mutli-valued in $\operatorname{div}_x \mathbf{u}_s = 0$ with $H(0) \in [0, 1]$.

In order to compare this with stresses p_c, τ_s in (2.1) we have to identify

$$-p_c\mathbb{I} + \phi_s\tau_s \stackrel{!}{=} -p_s\mathbb{I} + 2\mu_s^* \operatorname{dev}\mathbb{D}(\mathbf{u}_s). \tag{3.37}$$

with p_s, μ_s^* from (3.36). This shows how the normal pressure p_c emerges from p_s in (3.36c) and also gives rise to a novel coupling term σ_s^* from (3.32c), i.e.,

$$\sigma_s^* = \begin{cases} \frac{e_s}{|e_s|}\frac{\tilde{\mu}_s(\phi_s)\gamma}{4}(\operatorname{tr} e_s)_- & \text{if } |e_s| > 0, \\ 0 \in \mathbb{R}^{d\times d}_{\text{sym}} & \text{if } |e_s| = 0. \end{cases} \tag{3.38}$$

The comparison for liquid stresses is entirely similar.

4 Conclusion

This paper focusses on a two-phase model that was derived in [1] using the general averaging approach introduced in [10, 11]. The key ingredient is a stress-strain relation that features a normal pressure p_c which is proportional to the solid shear rate $|\mathbb{D}(\mathbf{u}_s)|$ and becomes singular as the solid volume fraction approaches a critical value $\phi_s \to \phi_{\text{crit}}$. In stationary shear flow situations with prescribed normal pressure p_c this produces a yield threshold due to zones where $\phi_s = \phi_{\text{crit}}$. This law extends a rheological relation inferred by [5] from scaling arguments and experimental measurements of constant shear flow to the general case where the average liquid and solid phase flow fields can be different. Unfortunately, previous investigations also showed that even in these simple flow situations the equations are not well-posed suggesting that some physics is missing.

In this paper, we reformulate the model within a variational framework based on the concepts of gradient flows and energy dissipation. This allows us to infer useful properties about the model and, as a long-term goal, access the rich analytical machinery that has been developed for models formulated within this framework. For example, we can deduce a general form of the normal pressure which includes the relation formulated in [1]. In fact, we observe that the model creates a novel contribution σ_s^* to the solid shear stress σ_s. The dissipation potential is only ensured to be convex for certain parameter ranges $(\alpha, \beta_\pm, \gamma)$, thus offering an analytical reason for the loss of well-posedness that may provide clues which kind of additional physics is required. Thus, we provide an alternative route for discussing this phenomenon which has also been observed for granular flow models based on the similar $\mu(I)$ rheology [2,3]. Moreover, the variational framework provides appropriate boundary conditions for free interfaces of the suspension.

References

1. Ahnert, T., Münch, A., Wagner, B.: Models for the two-phase flow of concentrated suspensions. European Journal of Applied Mathematics pp. 1–33 (2018).
2. Barker, T., Schaeffer, D.G., Bohorquez, P., Gray, J.M.N.T.: Well-posed and ill-posed behaviour of the μ(I)-rheology for granular flow. Journal of Fluid Mechanics **779**, 794–818 (2015).
3. Barker, T., Schaeffer, D.G., Shearer, M., Gray, J.M.N.T.: Well-posed continuum equations for granular flow with compressibility and μ(I)-rheology. Proc. R. Soc. A **473**, 20160846 (2017).
4. Batchelor, G.K., Green, J.T.: The determination of the bulk stress in a suspension of spherical particles to order c^2. J. Fluid Mech. **56**(03), 401–427 (1972).
5. Boyer, F., Guazzelli, É., Pouliquen, O.: Unifying suspension and granular rheology. Phys. Rev. Lett. **107**(18), 188301 (2011).
6. Brennen, C.: Fundamentals of Multiphase Flow. Cambridge University Press (2005)
7. Cassar, C., Nicolas, M., Pouliquen, O.: Submarine granular flows down inclined planes. Phys. Fluids **17**(10), 103301 (2005).
8. DeGiuli, E., Düring, G., Lerner, E., Wyart, M.: Unified theory of inertial granular flows and non-Brownian suspensions. Physical Review E **91**(6) (2015).
9. Drew, D.: Mathematical modeling of two-phase flow. Annu. Rev. Fluid Mech. **15**(1), 261–291 (1983).
10. Drew, D., Passman, S.: Theory of Multicomponent Fluids, *Applied Mathematical Sciences*, vol. 135. Springer (1999).
11. Drew, D., Segel, L.: Averaged equations for two-phase media. Stud. Appl. Math. **50**(2), 205–231 (1971)
12. Einstein, A.: Eine neue Bestimmung der Moleküldimensionen. Ann. Phys. (Berlin) **324**(2), 289–306 (1905).
13. Giga, M.H., Kirshtein, A., Liu, C.: Variational modeling and complex fluids. Handbook of Mathematical Analysis in Mechanics of Viscous Fluids pp. 73–113 (2018)
14. Grmela, M., Öttinger, H.: Dynamics and thermodynamics of complex fluids. i. development of a general formalism. Physical Review E **56**(6), 6620 (1997)
15. Helmholtz, H.v.: Zur Theorie der stationären Ströme in reibenden Flüssigkeiten. Wiss. Abh. **1**, 223–230 (1868)
16. Hermes, M., Guy, B., Poon, W., Poy, G., Cates, M., Wyart, M.: Unsteady flow and particle migration in dense, non-Brownian suspensions. Journal of Rheology **60**(5), 905–916 (2016).
17. Ioffe, A.D., Tihomirov, V.M.: Theory of extremal problems, *Studies in Mathematics and its Applications*, vol. 6. North-Holland Publishing Co., Amsterdam (1979). Translated from the Russian by Karol Makowski
18. Ishii, M., Hibiki, T.: Thermo-Fluid Dynamics of Two-Phase Flow. Springer (2011).
19. James, N., Han, E., Jureller, J., Jaeger, H.: Interparticle hydrogen bonding can elicit shear jamming in dense suspensions. arXiv:1707.09401 [cond-mat] (2017).
20. Joseph, D., Renardy, Y.: Fundamentals of two-fluid dynamics: Part i: Mathematical theory and applications, vol. 3. Springer Science & Business Media (2013)
21. Krieger, I., Dougherty, T.: A mechanism for non-newtonian flow in suspensions of rigid spheres. Transactions of the Society of Rheology **3**(1), 137–152 (1959).
22. Mielke, A.: On evolutionary Gamma convergence for gradient systems. In: A.Z. A. Muntean J.D.M. Rademacher (ed.) Macroscopic and Large Scale Phenomena: Coarse Graining, Mean Field Limits and Ergodicity, *Lecture Notes in Applied Mathematics and Mechanics*, vol. 3, pp. 187–249. Springer International Publishing, Heidelberg et al. (2016)
23. Mielke, A., Roubíček, T.: Rate-independent Systems: Theory and Application, *Applied Mathematical Sciences*, vol. 193. Springer (2015)
24. Morris, J., Boulay, F.: Curvilinear flows of noncolloidal suspensions: The role of normal stresses. J. Rheol. **43**, 1213–1237 (1999).
25. Morrison, P.J.: Hamiltonian description of the ideal fluid. Reviews of Modern Physics **70**, 467–521 (1998).

26. Murisic, N., Pausader, B., Peschka, D., Bertozzi, A.L.: Dynamics of particle settling and resuspension in viscous liquid films. J. Fluid Mech. **717**, 203–231 (2013).
27. Oh, S., Song, Y., Garagash, D., Lecampion, B., Desroches, J.: Pressure-Driven Suspension Flow near Jamming. Physical Review Letters **114**(8) (2015).
28. Öttinger, H., Grmela, M.: Dynamics and thermodynamics of complex fluids. ii. illustrations of a general formalism. Physical Review E **56**(6), 6633 (1997)
29. Peletier, M.: Variational modelling: Energies, gradient flows, and large deviations. arXiv:1402.1990 (2014)
30. Sokolowski, J., Zolesio, J.P.: Introduction to shape optimization. In: Introduction to Shape Optimization, pp. 5–12. Springer (1992)
31. Strutt, J.: Some general theorems relating to vibrations. Proceedings of the London Mathematical Society **1**(1), 357–368 (1871)
32. Wyart, M., Cates, M.E.: Discontinuous shear thickening without inertia in dense non-brownian suspensions. Phys. Rev. Lett. **112**, 098302 (2014).

Variational and Quasi-Variational Inequalities with Gradient Type Constraints

José Francisco Rodrigues* and Lisa Santos

Abstract This survey on stationary and evolutionary problems with gradient constraints is based on developments of monotonicity and compactness methods applied to large classes of scalar and vectorial solutions to variational and quasi-variational inequalities. Motivated by models for critical state problems and applications to free boundary problems in Mechanics and in Physics, in this work several known properties are collected and presented and a few novel results and examples are found.

Key words: Variational inequalities, Quasi-variational inequalities, Gradient constraints, Variational methods in Mechanic and in Physics.

MSC 35R35, 35J92, 35J87, 35J88, 35K92, 35K86, 35K87, 35L86, 47J20, 47J35, 47N50, 49J40, 74G25, 74H20, 76D99, 78M30, 80M30, 82D55, 82D99.

1 Introduction

The mathematical analysis of the unilateral problems were initiated in 1964 simultaneously by Fichera, to solve the Signorini problem in elastostatics [35], and by Stampacchia [86], as an extension of the Lax-Milgram lemma with application to the obstacle problem for elliptic equations of second order. The evolution version,

---

José Francisco Rodrigues
CMAFcIO and Departamento de Matemática, Faculdade de Ciências, Universidade de Lisboa, P-1749-016 Lisboa, Portugal, e-mail: jfrodrigues@ciencias.ulisboa.pt

Lisa Santos
CMAT and Departamento de Matemática, Escola de Ciências, Universidade do Minho, Campus de Gualtar, 4710-057 Braga, Portugal, e-mail: lisa@math.uminho.pt

* JFR acknowledges the hospitality of the Weierstrass Institute in Berlin (WIAS) during a visit when part of this work was developed.

M. Hintermüller, J. F. Rodrigues (eds.), *Topics in Applied Analysis and Optimisation*, CIM Series in Mathematical Sciences, https://doi.org/10.1007/978-3-030-33116-0_13

coining the expression variational inequalities and introducing weak solutions, was first treated in the pioneer paper of 1966 of Lions and Stampacchia [61], immediately followed by many others, including the extension to pseudo-monotone operators by Brézis in 1968 [17] (see also [58], [9], [52] or [85]). The importance of the new concept was soon confirmed by its versatility of their numerical approximations and in the first applications to optimal control of distributed systems in 1966-1968 by Lions and co-workers [57] and to solve many problems involving inequalities in Mechanics and Physics, by Duvaut and Lions in their book of 1972 [31], as well as several free boundary problems which can be formulated as obstacle type problems (see the books [9], [52], [36] or [74]).

Quasi-variational inequalities are a natural extension of the variational inequalities when the convex sets where the solutions are to be found depend on the solutions themselves. They were introduced by Bensoussan and Lions in 1973 to solve impulse control problems [16] and were developed, in particular, for certain free boundary problems, as the dam problem by Baiocchi in 1974 (see, for instance [9] and its references), as implicit unilateral problems of obstacle type, stationary or evolutionary [62], in which the constraints are only on the solutions.

While variational inequalities with gradient constraints appeared already to formulate the elastic-plastic torsion problem with an arbitrary cross section in the works of Lanchon, Duvaut and Ting around 1967 (see [31] or [74], for references), the first physical problem with gradient constraints formulated with quasi-variational inequalities of evolution type were proposed for the sandpile growth in 1986 by Prighozhin, in [69] (see also [70]). However, only ten years later the first mathematical results appeared, first for variational inequalities, see [71] and the independent work [5], together with a similar one for the magnetisation of type-II superconductors [72]. This last model has motivated a first existence result for the elliptic quasi-variational inequality in [56], which included other applications in elasto-plasticity and in electrostatics, and was extended to the parabolic framework for the p-Laplacian with an implicit gradient constraint in [77]. This result was later extended to quasi-variational solutions for first order quasilinear equations in [78], always in the scalar cases, and extended recently to a more general framework in [66]. The quasi-variational approach to the sand pile and the superconductors problems, with extensions to the simulation of lakes and rivers, have been successfully developed also with numerical approximations (see [73], [10], [11], [13], [14], for instance).

Although the literature on elliptic variational inequalities with gradient constraints is large and rich, including the issue of the regularity of the solution and their relations with the obstacle problem, it is out of the scope of this work to make its survey. Recent developments on stationary quasi-variational inequalities can be found in [47], [64], [50], [40], [6], [34], [55], [4] and the survey [53].

With respect to evolutionary quasi-variational problems with gradient constraint, on one hand, Kenmochi and co-workers, in [49], [38], [51], [53] and [54], have obtained interesting results by using variational evolution inclusions in Hilbert spaces with sub-differentials with a non-local dependence on parameters, and on the other hand, Hintermüller and Rautenberg in [41], using the pseudo-monotonicity and the $\mathscr{C}^0$-semigroup approach of Brézis-Lions, in [42], using contractive iteration arguments that yield uniqueness results and numerical approximations in interesting but special situations, and in [43], by time semi-discretisation of a monotone in time problem, have developed interesting numerical schemes that show the potential of the quasi-variational method. Other recent results on evolutionary quasi-variational inequalities can be also found in [51] and [54], both in more abstract frameworks and oriented to unilateral type problems and, therefore, with limited interest to constraints on the derivatives of the solutions.

This work is divided into two parts on stationary and evolutionary problems, respectively. The first one, after introducing the general framework of partial differential operators of p-Laplacian type and the respective functional spaces, exposes a brief introduction to the well-posedness of elliptic variational inequalities, with precise estimates and the use of the Mosco convergence of convex sets. Next section surveys old and recent results on the Lagrange multiplier problem associated with the gradient constraint, as well as its relation with the double obstacle problem and the complementarity problem. The existence of solutions to stationary quasi-variational inequalities is presented in the two following sections, one by using a compactness argument and the Leray-Schauder principle, extending [56], and the other one, for a class of Lipschitz nonlocal nonlinearity, by the Banach fixed point applied to the contractive property of the variational solution map in the case of smallness of data, following an idea of [40]. The first part is completed with three physical problems: a nonlinear Maxwell quasi-variational inequality motivated by a superconductivity model; a thermo-elastic system for a locking material in equilibrium and an ionisation problem in electrostatics. The last two problems, although variants of examples of [56], are new.

The second part treats evolutionary problems, of parabolic, hyperbolic and degenerate type. The first section treats weak and strong solutions of variational inequalities with time dependent convex sets, following [66] and giving explicit estimates on the continuous dependence results. The next two sections are, respectively, dedicated to the scalar problems with gradient constraint, relating the original works [83] and [84] to the more recent inequality for the transport equation of [79] for the variational case, and to the scalar quasi-variational strong solutions presenting a synthesis of [77] with [78] and an extension to the linear first order problem as a new corollary. The following section, based on [66], briefly describes the regularisation penalisation method to obtain the existence of weak solutions by compactness and monotonicity. The next section also develops the method of [42] in two concrete functional settings with nonlocal Lipschitz nonlinearities to obtain, under certain explicit conditions, novel results on the existence and uniqueness of strong (and weak) solutions of evolutionary quasi-variational inequalities. Finally, the last section presents also three physical problems with old and new observations, as applications of the previous results, namely on the dynamics of the sandpile of granular material, where conditions for the finite time stabilisation are described, on an evolutionary superconductivity model, in which the threshold is temperature dependent, and a variant of the Stokes flow for a thick fluid, for which it is possible to explicit conditions for the existence and uniqueness of a strong quasi-variational solution.

2 Stationary problems

2.1 A general p-framework

Let Ω be a bounded open subset of $\mathbb{R}^d$, with a Lipschitz boundary, $d \geq 2$. We represent a real vector function by a bold symbol $\boldsymbol{u} = (u_1, \ldots, u_m)$ and we denote the partial derivative of u_i with respect to x_j by $\partial_{x_j} u_i$. Given real numbers a, b, we set $a \vee b = \max\{a, b\}$.

For $1 < p < \infty$, let L be a linear differential operator of order one in the form

$$\mathrm{L} : \boldsymbol{V}_p \to L^p(\Omega)^\ell \text{ such that } (\mathrm{L}\boldsymbol{u})_i = \sum_{j=1}^{d} \sum_{k=1}^{m} \alpha_{ijk} \partial_{x_j} u_k, \tag{2.1}$$

where $\alpha_{ijk} \in L^\infty(\Omega)$, $i = 1, \ldots, \ell$, $j = 1, \ldots, d$, $k = 1, \ldots, m$, with $\ell, m \in \mathbb{N}$, and

$$\boldsymbol{V}_p = \left\{ \boldsymbol{u} \in L^p(\Omega)^m : \mathrm{L}\boldsymbol{u} \in L^p(\Omega)^\ell \right\}$$

is endowed with the graph norm.

We consider a Banach subspace $\mathbb{X}_p$ verifying

$$\mathscr{D}(\Omega)^m \subset \mathbb{X}_p \subset W^{1,p}(\Omega)^m \subset \boldsymbol{V}_p \tag{2.2}$$

where

$$\|\boldsymbol{w}\|_{\mathbb{X}_p} = \|\mathrm{L}\boldsymbol{w}\|_{L^p(\Omega)^\ell} \tag{2.3}$$

is a norm in $\mathbb{X}_p$ equivalent to the one induced from $\boldsymbol{V}_p$. In order that (2.3) holds, we suppose there exists $c_p > 0$ such that

$$\|\boldsymbol{w}\|_{L^p(\Omega)^m} \leq c_p \|\mathrm{L}\boldsymbol{w}\|_{L^p(\Omega)^\ell} \qquad \forall \boldsymbol{w} \in \boldsymbol{V}_p. \tag{2.4}$$

To fix ideas, here the framework (2.1) for the operator L can be regarded as any one of the following cases:

***Example* 1**

$\mathrm{L}u = \nabla u$ (gradient of u), $m = 1$, $\ell = d$;

$\mathrm{L}\boldsymbol{u} = \nabla\times\boldsymbol{u}$ (curl of $\boldsymbol{u}$), $m = \ell = d = 3$;

$\mathrm{L}\boldsymbol{u} = D\boldsymbol{u} = \frac{1}{2}(\nabla\boldsymbol{u} + \nabla\boldsymbol{u}^T)$ (symmetrised gradient of $\boldsymbol{u}$), $m = d$ and $\ell = d^2$. □

When $\mathrm{L}u = \nabla u$, we consider

$$\mathbb{X}_p = W_0^{1,p}(\Omega) \quad \text{and} \quad \|u\|_{\mathbb{X}_p} = \|\nabla u\|_{L^p(\Omega)^d}$$

is equivalent to the $\boldsymbol{V}_p = W^{1,p}(\Omega)$ norm, by Poincaré inequality.

In the case $\mathrm{L}\boldsymbol{u} = \nabla \times \boldsymbol{u}$, for a simply connected domain Ω, the vector space $\mathbb{X}_p$ may be

$$\mathbb{X}_p = \left\{ \boldsymbol{w} \in L^p(\Omega)^3 : \nabla \times \boldsymbol{w} \in L^p(\Omega)^3,\ \nabla \cdot \boldsymbol{w} = 0,\ \boldsymbol{w} \cdot \boldsymbol{n}_{|\partial\Omega} = 0 \right\}, \tag{2.5}$$

or

$$\mathbb{X}_p = \{w \in L^p(\Omega)^3 : \nabla \times w \in L^p(\Omega)^3, \nabla \cdot w = 0, w \times n_{|\partial\Omega} = \mathbf{0}\}, \tag{2.6}$$

corresponding to different boundary conditions, where $\nabla \cdot w$ means the divergence of w. Both spaces are closed subspaces of $W^{1,p}(\Omega)^3$ and a Poincaré type inequality is satisfied in $\mathbb{X}_p$ (for details see [2]).

When $\mathrm{L}u = Du$, we may have

$$\mathbb{X}_p = W_0^{1,p}(\Omega)^d \quad \text{or} \quad \mathbb{X}_p = W_{0,\sigma}^{1,p}(\Omega)^d = \{w \in W_0^{1,p}(\Omega)^d : \nabla \cdot w = 0\}$$

and $\|Dw\|_{L^p(\Omega)^{d^2}}$ is equivalent to the norm induced from $W^{1,p}(\Omega)^d$ by Poincaré and Korn's inequalities.

Given $\nu > 0$, we introduce

$$L_\nu^\infty(\Omega) = \{w \in L^\infty(\Omega) : w \geq \nu\}. \tag{2.7}$$

For $G : \mathbb{X}_p \to L_\nu^\infty(\Omega)$, we define the nonempty closed convex set

$$\mathbb{K}_{G[u]} = \{w \in \mathbb{X}_p : |\mathrm{L}w| \leq G[u]\}, \tag{2.8}$$

where $|\cdot|$ is the Euclidean norm in $\mathbb{R}^\ell$ and we denote, for $w \in V_p$,

$$Ł_p u = |\mathrm{L}w|^{p-2}\mathrm{L}w. \tag{2.9}$$

We may associate with $Ł_p$ a strongly monotone operator, and there exist positive constants d_p such that for all $w_1, w_2 \in V_p$

$$\begin{aligned} &\int_\Omega (Ł_p w_1 - Ł_p w_2) \cdot \mathrm{L}(w_1 - w_2) \\ &\qquad \geq \begin{cases} d_p \displaystyle\int_\Omega |\mathrm{L}(w_1 - w_2)|^p & \text{if } p \geq 2, \\ d_p \displaystyle\int_\Omega (|\mathrm{L}w_1| + |\mathrm{L}w_2|)^{p-2} |\mathrm{L}(w_1 - w_2)|^2 & \text{if } 1 \leq p < 2. \end{cases} \end{aligned} \tag{2.10}$$

For $1 < p < \infty$ and $f \in L^1(\Omega)^m$, we shall consider the quasi-variational inequality

$$u \in \mathbb{K}_{G[u]} : \quad \int_\Omega Ł_p u \cdot \mathrm{L}(w - u) \geq \int_\Omega f \cdot (w - u) \quad \forall w \in \mathbb{K}_{G[u]}. \tag{2.11}$$

2.2 Well-posedness of the variational inequality

For $g \in L_\nu^\infty(\Omega)$, it is well-know that the variational inequality, which is obtained by taking $G[u] \equiv g$ in (2.8) and in (2.11),

$$\boldsymbol{u} \in \mathbb{K}_g : \quad \int_\Omega Ł_p \boldsymbol{u} \cdot \mathrm{L}(\boldsymbol{w} - \boldsymbol{u}) \geq \int_\Omega \boldsymbol{f} \cdot (\boldsymbol{w} - \boldsymbol{u}) \quad \forall \boldsymbol{w} \in \mathbb{K}_g, \tag{2.12}$$

has a unique solution (see, for instance, [58] or [52]). The solution is, in fact, Hölder continuous on $\overline{\Omega}$ by recalling the (compact) Sobolev imbeddings

$$W^{1,p}(\Omega) \hookrightarrow \begin{cases} L^q(\Omega) \text{ for } 1 \leq q < \frac{dp}{d-p} \text{ if } p < d, \\ L^r(\Omega) \text{ for } 1 \leq r < \infty \text{ if } p = d, \\ \mathscr{C}^{0,\alpha}(\overline{\Omega}) \text{ for } 0 \leq \alpha < 1 - \frac{d}{p} \text{ if } p > d. \end{cases} \tag{2.13}$$

Indeed, in the three examples above we have, for any $p > d$ and $0 \leq \alpha < 1 - \frac{d}{p}$,

$$\mathbb{K}_g \subset W^{1,p}(\Omega)^m \subset \mathscr{C}^{0,\alpha}(\overline{\Omega})^m. \tag{2.14}$$

We note that, even if $\mathrm{L}\boldsymbol{u}$ is bounded in Ω, in general, this does not imply that the solution $\boldsymbol{u}$ of (2.12) is Lipschitz continuous. However, this holds, for instance, not only in the scalar case L= ∇, but, more generally if in (2.1) $m = 1$ and $\alpha_{ij} = \eta_i \delta_{ij}$ with $\eta_i \in L^\infty_\nu(\Omega)$, $i = 1, \ldots, d$ and δ_{ij} the Kronecker symbol.

We present now two continuous dependence results on the data. In particular, when (2.14) holds, any solution to (2.12) or (2.11) is a priori continuously bounded and therefore we could take not only $\boldsymbol{f} \in L^1(\Omega)^m$ but also $\boldsymbol{f}$ in the space of Radon measures.

Theorem 2.1 *Under the framework* (2.1), (2.2) *and* (2.3) *let* $\boldsymbol{f}_1$ *and* $\boldsymbol{f}_2$ *belong to* $L^1(\Omega)^m$ *and* $g \in L^\infty_\nu(\Omega)$. *Denote by* $\boldsymbol{u}_i$, $i = 1, 2$, *the solutions of the variational inequality* (2.12) *with data* $(\boldsymbol{f}_i, g)$. *Then*

$$\|\boldsymbol{u}_1 - \boldsymbol{u}_2\|_{\mathbb{X}_p} \leq C \|\boldsymbol{f}_1 - \boldsymbol{f}_2\|_{L^1(\Omega)^m}^{\frac{1}{p \vee 2}}, \tag{2.15}$$

being C *a positive constant depending on* p, Ω *and* $\|g\|_{L^\infty(\Omega)}$.

Proof We use $\boldsymbol{u}_2$ as test function in the variational inequality (2.12) for $\boldsymbol{u}_1$ and reciprocally, obtaining, after summation,

$$\int_\Omega (Ł_p \boldsymbol{u}_1 - Ł_p \boldsymbol{u}_2) \cdot \mathrm{L}(\boldsymbol{u}_1 - \boldsymbol{u}_2) \leq \int_\Omega (\boldsymbol{f}_1 - \boldsymbol{f}_2) \cdot (\boldsymbol{u}_1 - \boldsymbol{u}_2).$$

For $p \geq 2$, using (2.10), since $\boldsymbol{u}_i \in L^\infty(\Omega)^m$, we have

$$\|\boldsymbol{u}_1 - \boldsymbol{u}_2\|_{\mathbb{X}_p} \leq C \|\boldsymbol{f}_1 - \boldsymbol{f}_2\|_{L^1(\Omega)^m}^{\frac{1}{p}}.$$

If $1 \leq p < 2$, using (2.10) and $|\mathrm{L}\boldsymbol{u}_i| \leq M$, where $M = \|g\|_{L^\infty(\Omega)}$, we have first

$$d_p (2M)^{p-2} \int_\Omega |\mathrm{L}(\boldsymbol{u}_1 - \boldsymbol{u}_2)|^2 \leq \int_\Omega (\boldsymbol{f}_1 - \boldsymbol{f}_2) \cdot (\boldsymbol{u}_1 - \boldsymbol{u}_2)$$

and then, with $\omega_p = |\Omega|^{\frac{2-p}{2p}}$,

$$\|\boldsymbol{u}_1 - \boldsymbol{u}_2\|_{\mathbb{X}_p} \leq \omega_p \|\boldsymbol{u}_1 - \boldsymbol{u}_2\|_{\mathbb{X}_2} \leq C\|\boldsymbol{f}_1 - \boldsymbol{f}_2\|_{L^1(\Omega)^m}^{\frac{1}{2}},$$

concluding the proof. □

Remark 1 Since $|\mathrm{L}\boldsymbol{u}_i| \leq M$ we can always extend (2.15) for any $r > d$, obtaining for some positive constants $C_\alpha > 0, C_r > 0$ and $\alpha = 1 - \frac{d}{r} > 0$,

$$\|\boldsymbol{u}_1 - \boldsymbol{u}_2\|_{\mathscr{C}^\alpha(\overline{\Omega})^m} \leq C_\alpha \|\boldsymbol{u}_1 - \boldsymbol{u}_2\|_{\mathbb{X}_r} \leq C_r \|\boldsymbol{f}_1 - \boldsymbol{f}_2\|_{L^1(\Omega)^m}^{\frac{1}{r}}.$$

Indeed, it is sufficient to use the Sobolev imbedding and to observe that, for $r > p$,

$$\int_\Omega |\mathrm{L}(\boldsymbol{u}_1 - \boldsymbol{u}_2)|^r \leq (2M)^{r-p} \int_\Omega |\mathrm{L}(\boldsymbol{u}_1 - \boldsymbol{u}_2)|^p.$$

Theorem 2.2 *Under the framework* (2.1), (2.2) *and* (2.3) *let* $\boldsymbol{f} \in L^1(\Omega)^m$ *and* $g_1, g_2 \in L^\infty_\nu(\Omega)$. *Denote by* $\boldsymbol{u}_i$, $i = 1, 2$, *the solutions of the variational inequality* (2.12) *with data* $(\boldsymbol{f}, g_i)$. *Then*

$$\|\boldsymbol{u}_1 - \boldsymbol{u}_2\|_{\mathbb{X}_p} \leq C_\nu \|g_1 - g_2\|_{L^\infty(\Omega)}^{\frac{1}{p\vee 2}}. \tag{2.16}$$

Proof Calling $\beta = \|g_1 - g_2\|_{L^\infty(\Omega)}$, then for $i, j \in \{1, 2\}$, $i \neq j$, and

$$\boldsymbol{u}_{i_j} = \frac{\nu}{\nu + \beta}\boldsymbol{u}_i \in \mathbb{K}_{g_j},$$

$\boldsymbol{u}_{i_j}$ can be used as test function in the variational inequality (2.12) satisfied by $\boldsymbol{u}_j$, obtaining

$$\int_\Omega (\text{Ł}_p\boldsymbol{u}_1 - \text{Ł}_p\boldsymbol{u}_2) \cdot \mathrm{L}(\boldsymbol{u}_1 - \boldsymbol{u}_2) \leq \int_\Omega \text{Ł}_p\boldsymbol{u}_1 \cdot L(\boldsymbol{u}_{2_1} - \boldsymbol{u}_2)$$
$$+ \int_\Omega \text{Ł}_p\boldsymbol{u}_2 \cdot \mathrm{L}(\boldsymbol{u}_{1_2} - \boldsymbol{u}_1) + \int_\Omega f\big((\boldsymbol{u}_1 - \boldsymbol{u}_{1_2}) + (\boldsymbol{u}_2 - \boldsymbol{u}_{2_1})\big).$$

But

$$|\boldsymbol{u}_i - \boldsymbol{u}_{i_j}| + |\mathrm{L}(\boldsymbol{u}_i - \boldsymbol{u}_{i_j})| = \frac{\beta}{\nu + \beta}\big(|u_i| + \mathrm{L}\boldsymbol{u}_i|\big) \leq \frac{2M}{\nu}\beta,$$

where $M = \max_{i=1,2}\{\|g_i\|_{L^\infty(\Omega)}, \|\boldsymbol{u}_i\|_{L^\infty(\Omega)^m}\}$, since $\boldsymbol{u}_i \in \mathbb{K}_{g_i} \subset L^\infty(\Omega)^m$, and the conclusion follows. □

We can also consider a degenerate case, by letting $\delta \to 0$ in

$$\boldsymbol{u}^\delta \in \mathbb{K}_g : \quad \delta \int_\Omega \text{Ł}_p\boldsymbol{u}^\delta \cdot \mathrm{L}(\boldsymbol{w} - \boldsymbol{u}^\delta) \geq \int_\Omega \boldsymbol{f} \cdot (\boldsymbol{w} - \boldsymbol{u}^\delta) \quad \forall \boldsymbol{v} \in \mathbb{K}_g. \tag{2.17}$$

Indeed, since $\|\mathrm{L}\boldsymbol{u}^\delta\|_{L^\infty(\Omega)} \leq M$, where $M = \|g\|_{L^\infty(\Omega)}$, independently of $0 < \delta \leq 1$, we can extract a subsequence

$$\boldsymbol{u}^\delta \underset{\delta\to 0}{\rightharpoonup} \boldsymbol{u}^0 \quad \text{in } \mathbb{X}_p\text{-weak}$$

for some $\boldsymbol{u}^0 \in \mathbb{K}_g$. Then, we can pass to the limit in (2.17) and we may state:

Theorem 2.3 *Under the framework* (2.1), (2.2) *and* (2.3), *for any* $\boldsymbol{f} \in L^1(\Omega)^m$, *there exists at least a solution* $\boldsymbol{u}^0$ *to the problem*

$$\boldsymbol{u} \in \mathbb{K}_g : \quad 0 \geq \int_\Omega \boldsymbol{f} \cdot (\boldsymbol{w} - \boldsymbol{u}) \quad \forall \boldsymbol{w} \in \mathbb{K}_g. \tag{2.18}$$

In general, the strict positivity condition on the threshold $g = g(x)$, which is included in (2.7), is necessary in many interesting results, as the continuous dependence result (2.16), which can also be obtained in a weaker form by using the Mosco convergence and observing that, for $g_n \geq \nu > 0$,

$$\mathbb{K}_{g_n} \xrightarrow[n]{M} \mathbb{K}_g \quad \text{is implied by } g_n \underset{n}{\longrightarrow} g \text{ in } L^\infty(\Omega).$$

We recall that $\mathbb{K}_{g_n} \xrightarrow[n]{M} \mathbb{K}_g$ iff i) for any sequences $\mathbb{K}_{g_n} \ni w_n \underset{n}{\rightharpoonup} w$ in $\mathbb{X}_p$-weak, then $w \in \mathbb{K}_g$ and ii) for any $w \in \mathbb{K}_g$ there exists $w_n \in \mathbb{K}_{g_n}$ such that $w_n \underset{n}{\longrightarrow} w$ in $\mathbb{X}_p$.

However, the particular structure of the scalar case L=∇ in $\mathbb{X}_p = W_0^{1,p}(\Omega)$, i.e., with $\text{Ł}_p v = \nabla_p v = |\nabla v|^{p-2}\nabla v$ and a Mosco convergence result of [7] allows us to extend the continuous dependence of the solutions of the variational inequality with nonnegative continuous gradient constraints, as an interesting result of Mosco type (see [67]). Note that in the following result the g_n may vanish in some region, but the technique of proof in [7] requires a more regular boundary, restriction that would be interesting to remove.

Theorem 2.4 *Let* Ω *be an open domain with a* $\mathscr{C}^2$ *boundary,* L$= \nabla$, $f \in L^{p'}(\Omega)$ *and* $g_\infty, g_n \in \mathscr{C}(\overline{\Omega})$, *with* $g_n \geq 0$ *for* $n \in \mathbb{N}$ *and* $n = \infty$. *If* u_n *denotes the unique solution to*

$$u_n \in \mathbb{K}_{g_n} : \quad \int_\Omega \nabla_p u_n \cdot \nabla(w - u_n) \geq \int_\Omega f \cdot (w - u_n) \quad \forall\, w \in \mathbb{K}_{g_n} \tag{2.19}$$

then, as $n \to \infty$, $g_n \underset{n}{\longrightarrow} g_\infty$ *in* $\mathscr{C}(\overline{\Omega})$ *implies* $u_n \underset{n}{\longrightarrow} u_\infty$ *in* $W_0^{1,p}(\Omega)$.

Proof By Theorem 3.12 of [7], we have $\mathbb{K}_{g_n} \xrightarrow[n]{M} \mathbb{K}_{g_\infty}$. Since $|\nabla u_n| \leq g_n$ in Ω, we have $\|u_n\|_{W_0^{1,p}(\Omega)} \leq C|\Omega|^{\frac{1}{p}} \|g_n\|_{\mathscr{C}(\overline{\Omega})} \leq M$ independently of n and, therefore, we may take a subsequence $u_n \underset{n}{\rightharpoonup} u_*$ in $W_0^{1,p}(\Omega)$. Then $u_* \in \mathbb{K}_{g_\infty}$. For any $w_\infty \in \mathbb{K}_{g_\infty}$, take $w_n \in \mathbb{K}_{g_n}$ with $w_n \underset{n}{\longrightarrow} w_\infty$ in $W_0^{1,p}(\Omega)$ and, using Minty's Lemma and letting $n \to \infty$ in

$$\int_\Omega \nabla_p w_n \cdot \nabla(w_n - u_n) \geq \int_\Omega f(w_n - u_n),$$

we conclude that $u_* = u_\infty$ is the unique solution of (2.19) for $n = \infty$. The strong convergence follows easily, by choosing $v_n \underset{n}{\longrightarrow} u_\infty$ with $v_n \in \mathbb{K}_{g_n}$, from

$$\int_\Omega |\nabla(u_n - u_\infty)|^p \leq \int_\Omega f(u_n - v_n) + \int_\Omega \nabla_p u_n \cdot \nabla(v_n - u_\infty) - \int_\Omega \nabla_p u_\infty \cdot \nabla(u_n - u_\infty) \underset{n}{\longrightarrow} 0.$$

2.3 Lagrange multipliers

In the special case $p = 2$, $Ł_2 = \mathrm{L}$, consider the variational inequality ($\delta > 0$)

$$\boldsymbol{u}^\delta \in \mathbb{K}_g : \quad \delta \int_\Omega \mathrm{L}\boldsymbol{u}^\delta \cdot \mathrm{L}(\boldsymbol{w} - \boldsymbol{u}^\delta) \geq \int_\Omega \boldsymbol{f} \cdot (\boldsymbol{w} - \boldsymbol{u}^\delta) \quad \forall \boldsymbol{w} \in \mathbb{K}_g \tag{2.20}$$

and the related Lagrange multiplier problem, which is equivalent to the problem of finding $(\lambda^\delta, \boldsymbol{u}^\delta) \in \big(L^\infty(Q_T)^m\big)' \times \mathbb{X}_\infty$ such that

$$\langle \lambda^\delta \mathrm{L}\boldsymbol{u}^\delta, \mathrm{L}\boldsymbol{\varphi}\rangle_{(L^\infty(\Omega)^m)' \times L^\infty(\Omega)^m} = \int_\Omega \boldsymbol{f} \cdot \boldsymbol{\varphi} \quad \forall \boldsymbol{\varphi} \in \mathbb{X}_\infty, \tag{2.21a}$$

$$|\mathrm{L}\boldsymbol{u}^\delta| \leq g \text{ a.e. in } \Omega, \quad \lambda^\delta \geq \delta, \quad (\lambda^\delta - \delta)(|\mathrm{L}\boldsymbol{u}^\delta| - g) = 0 \quad \text{in } \big(L^\infty(\Omega)^m\big)', \tag{2.21b}$$

where we set $\mathbb{X}_\infty = \big\{\boldsymbol{\varphi} \in L^2(\Omega)^m : \mathrm{L}\boldsymbol{\varphi} \in L^\infty(\Omega)^\ell\big\}$ and define

$$\langle \lambda\boldsymbol{\alpha}, \boldsymbol{\beta}\rangle_{(L^\infty(\Omega)^m)' \times L^\infty(\Omega)^m} = \langle \lambda, \boldsymbol{\alpha} \cdot \boldsymbol{\beta}\rangle_{L^\infty(\Omega)' \times L^\infty(\Omega)} \quad \forall \lambda \in L^\infty(\Omega)' \ \forall \boldsymbol{\alpha}, \boldsymbol{\beta} \in L^\infty(\Omega)^m.$$

In fact, arguing as in [8, Theorem 1.3], which corresponds only to the particular scalar case $\mathrm{L} = \nabla$, we can prove the following theorem:

Theorem 2.5 *Suppose that Ω is a bounded open subset of $\mathbb{R}^d$ with Lipschitz boundary and the assumptions* (2.1) *and* (2.2) *are satisfied with $p = 2$. Given $\boldsymbol{f} \in L^2(\Omega)^m$ and $g \in L^\infty_\nu(\Omega)$,*

1. *if $\delta > 0$, problem* (2.21) *has a solution*

$$(\lambda^\delta, u^\delta) \in L^\infty(\Omega)' \times \mathbb{X}_\infty;$$

2. *at least for a subsequence $(\lambda^\delta, u^\delta)$ of solutions of problem* (2.21)*, we have*

$$\lambda^\delta \underset{\delta \to 0}{\rightharpoonup} \lambda^0 \quad \text{in} \quad L^\infty(\Omega)', \qquad u^\delta \underset{\delta \to 0}{\rightharpoonup} u^0 \quad \text{in} \quad \mathbb{X}_\infty.$$

In addition, $\boldsymbol{u}^\delta$ also solves (2.20) *for each $\delta \geq 0$ and (λ^0, u^0) solves problem* (2.21) *for $\delta = 0$.*

We observe that the last condition in (2.21) on the Lagrange multiplier λ^δ corresponds, in the case of integrable functions, to say that a.e. in Ω

$$\lambda^\delta \in \mathscr{K}_\delta(|\mathrm{L}\boldsymbol{u}^\delta| - g) \tag{2.22}$$

where, for $\delta \geq 0$, $\mathscr{K}^\delta$ is the family of maximal monotone graphs given by $\mathscr{K}_\delta(s) = \delta$ if $s < 0$ and $\mathscr{K}_\delta(s) = [\delta, \infty[$ if $s = 0$. In general, further properties for λ^δ are unknown except in the scalar case with $\mathrm{L} = \nabla$.

The model of the elastic-plastic torsion problem corresponds to the variational inequality with gradient constraint (2.20) with $\delta = 1$, $p = 2 = d$, L$= \nabla$, $g \equiv 1$ and $f = \beta$, a positive constant. In [18], Brézis proved the equivalence of this variational inequality with the Lagrange multiplier problem (2.21) with these data and assuming Ω simply connected, showing also that $\lambda \in L^\infty(\Omega)$ is unique and even continuous in the case of Ω convex. This result was extended to multiply connected domains by Gerhardt in [39]. Still for $g \equiv 1$, Chiadò Piat and Percival extended the result for more general operators in [26], being $f \in L^r(\Omega), r > d \geq 2$, proving that λ is a Radon measure but leaving open the uniqueness. Keeping $g \equiv 1$ but assuming $\delta = 0$, problem (2.21) is the Monge-Kantorovich mass transfer problem (see [33] for details) and the convergence $\delta \longrightarrow 0$ in the theorem above links this problem to the limit of Lagrange multipliers for elastic-plastic torsion problems with coercive constant $\delta > 0$. In [29], for the case $\delta = 0$, assuming Ω convex and $f \in L^q(\Omega)$, $2 \leq q \leq \infty$ with $\int_\Omega f = 0$, Pascale, Evans and Pratelli proved the existence of $\lambda^0 \in L^q(\Omega)$ solving (2.21). In [6], for Ω any bounded Lipschitz domain, it was proved the existence of solution $(\lambda, u) \in L^\infty(\Omega)' \times W^{1,\infty}_0(\Omega)$ of the problem (2.21), with $\delta = 1$, $f \in L^2(\Omega)$, $g \in W^{2,\infty}(\Omega)$ and in [8] this result was extended for $\delta \geq 0$, with $f \in L^\infty(\Omega)$ and g only in $L^\infty(\Omega)$, as it is stated in the theorem above, but for L$= \nabla$. Besides, when $g \in \mathscr{C}^2(\Omega)$ and $\Delta g^2 \leq 0$, in [8] it is also shown that $\lambda^\delta \in L^q(\Omega)$, for any $1 \leq q < \infty$ and $\delta \geq 0$.

Problem (2.21) is also related to the equilibrium of the table sandpiles problem (see [71], [24], [30]) and other problems in the Monge-Kantorovich theory (see [33], [1], [10], [44]).

In the degenerate case $\delta = 0$, problem (2.21) is also associated with the limit case $p \longrightarrow \infty$ of the $p-$Laplace equation and related problems to the infinity Laplacian (see, for instance [15] or [46] and their references), as well as in some variants of the optimal transport probem, like the obstacle Monge-Kantorovich equation (see [22], [37] and [45]).

There are other problems with gradient constraint that are related with the scalar variational inequality (2.20) with L$= \nabla$. To simplify, we assume $\delta = 1$.

When f is constant and $g \equiv 1$, it is well known that the variational inequality (2.20) is equivalent to the two obstacles variational inequality

$$u \in \mathbb{K}^{\overline{\varphi}}_{\underline{\varphi}}: \quad \int_\Omega \nabla u \cdot \nabla(w - u) \geq \int_\Omega f \cdot (w - u) \quad \forall\, w \in \mathbb{K}^{\overline{\varphi}}_{\underline{\varphi}}, \tag{2.23}$$

where

$$\mathbb{K}^{\overline{\varphi}}_{\underline{\varphi}} = \{v \in H^1_0(\Omega) : \underline{\varphi} \leq v \leq \overline{\varphi}\}, \tag{2.24}$$

with $\underline{\varphi}(x) = -d(x, \partial\Omega)$ and $\overline{\varphi}(x) = d(x, \partial\Omega)$, being d the usual distance if Ω is convex and the geodesic distance otherwise. This result was proved firstly by Brézis and Sibony in 1971 in [20], developed by Caffarelli and Friedman in [21] in the framework of elastic-plastic problems, and it was also extended in [87] for certain perturbations of convex functionals.

In [32], Evans proved the equivalence between (2.23) with the complementary problem (2.25) below, with $g = 1$. However, for non constant gradient constraint, the example below shows that the problem

$$\max\big\{ -\Delta u - f, |\nabla u| - g\big\} = 0 \tag{2.25}$$

for $f, g \in L^\infty(\Omega)$ is not always equivalent to (2.20), as well as the equivalence with the double obstacle variational inequality (2.24) defined with a general constraing g is not always true. We give the definition of the obstacles for g nonconstant: given $x,\ z \in \overline{\Omega}$, let

$$d_g(x,z) = \inf\Big\{ \int_0^\delta g(\xi(s))ds : \ \delta > 0,\ \xi : [0,\delta] \to \Omega,\ \xi \text{ smooth}, \\ \xi(0) = x,\ \xi(\delta) = z,\ |\xi'| \leq 1\Big\}. \tag{2.26}$$

This function is a pseudometric (see [59]) and the obstacles we consider are

$$\overline{\varphi}(x,t) = d_g(x,\partial\Omega) = \bigvee\{w(x) : w \in \mathbb{K}_g\} \tag{2.27}$$

and

$$\underline{\varphi}(x,t) = -d_g(x,\partial\Omega) = \bigwedge\{w(x) : w \in \mathbb{K}_g\}. \tag{2.28}$$

***Example* 2**
Let $f, g : (-1,1) \to \mathbb{R}$ be defined by $f(x) = 2$ and $g(x) = 3x^2$. Notice that $g(0) = 0$ and so $g \notin L^\infty_\nu(-1,1)$. However the solutions of the three problems under consideration exist. The two obstacles (with respect to this function g) are

$$\overline{\varphi}(x) = \begin{cases} x^3 + 1 & \text{if } x \in [-1,0[, \\ 1 - x^3 & \text{if } x \in [0,1], \end{cases} \quad \text{and} \quad \underline{\varphi}(x,t) = \begin{cases} -x^3 - 1 & \text{if } x \in [-1,0[, \\ x^3 - 1 & \text{if } x \in [0,1]. \end{cases}$$

The function

$$u(x) = \begin{cases} 1 - x^2 & \text{if } |x| \geq \frac{2}{3} \text{ and } |x| \leq 1, \\ \overline{\varphi}(x) - \frac{4}{27} & \text{otherwise} \end{cases}$$

is $\mathscr{C}^1$ and solves (2.20) with L$= \nabla$ and $\delta = 1$.

The function $z(x) = 1 - x^2$ belongs to $\mathbb{K}^{\overline{\varphi}}_{\underline{\varphi}}$ and, because $z'' = -2$, it solves (2.23).

Neither u nor z solve (2.25). In fact, as $-u''(x) = -6x$ in $(-\frac{2}{3}, \frac{2}{3})$, then $-u''(x) \not\leq 2$ a.e. and $|z'| \not\leq g$. □

Sufficient conditions to assure the equivalence among these problems will be given in Section 3 in the framework of evolution problems.

Nevertheless, the relations between the gradient constraint problem and the double obstacle problem are relevant to study the regularity of the solution, as in the recent works of [3] and [27], as well as for the regularity of the free boundary in the elastic-plastic torsion problem (see [36] or [74] and their references). Indeed, in this case, when $g = 1$ and $f = -\tau < 0$ are constants, it is well-known that the elastic and the plastic regions are, respectively, given by the subsets of $\Omega \subset \mathbb{R}^2$

$$\{|\nabla u| < 1\} = \{u > \underline{\varphi}\} = \{\lambda > 1\} \quad \text{and} \quad \{|\nabla u| = 1\} = \{u = \underline{\varphi}\} = \{\lambda = 1\}.$$

The free boundary is their common boundary in Ω and, by a result of Caffarelli and Rivière [23], consists locally of Jordan arcs with the same smoothness as the nearest portion of $\partial\Omega$. In particular, near reentrant corners of $\partial\Omega$, the free boundary is locally analytic. As a consequence, it was observed in [74, p.240] that those portions of the free boundary are stable for perturbations of data near the reentrant corners and near the connected components of $\partial\Omega$ of nonpositive mean curvature.

Also using the equivalence with the double obstacle problem, recently, Safdari has extended some properties on the regularity and the shape of the free boundary in the case L= ∇ with the pointwise gradient constraint $(\partial_{x_1} u)^q + (\partial_{x_2} u)^q \leq 1$, for $q > 1$ (see [82] and its references).

2.4 The quasi-variational solution via compactness

We start with an existence result for the quasi-variational inequality (2.11), following the ideas in [56].

Theorem 2.6 *Under the framework* (2.1), (2.2) *and* (2.3), *let* $\boldsymbol{f} \in L^{p'}(\Omega)^m$ *and* $p' = \frac{p}{p-1}$. *Then there exists at least one solution of the quasi-variational inequality* (2.11), *provided one of the following conditions is satisfied:*

1. *the functional* $G : \mathbb{X}_p \longrightarrow L^\infty_\nu(\Omega)$ *is completely continuous;*
2. *the functional* $G : \mathscr{C}(\overline{\Omega})^m \longrightarrow L^\infty_\nu(\Omega)$ *is continuous, when* $p > d$, *or it satisfies also the growth condition*

$$\|G[\boldsymbol{u}]\|_{L^r(\Omega)} \leq c_0 + c_1 \|\boldsymbol{u}\|^\alpha_{L^{\sigma p}(\Omega)^m}, \tag{2.29}$$

for some constants c_0, $c_1 \geq 0$, $\alpha \geq 0$, *with* $r > d$ *and* $\sigma \geq \frac{1}{p}$, *when* $p = d$, *or* $\frac{1}{p} \leq \sigma \leq \frac{d}{d-p}$, *when* $1 < p < d$.

Proof Let $\boldsymbol{u} = S(\boldsymbol{f}, g)$ be the unique solution of the variational inequality (2.12) with $g = G[\boldsymbol{\varphi}]$ for $\boldsymbol{\varphi}$ given in $\mathbb{X}_p$ or $\mathscr{C}(\overline{\Omega})^m$. Since $\mathbb{X}_p \subset W^{1,p}(\Omega)^m$, by Sobolev embeddings, and it is always possible to take $\boldsymbol{w} = \boldsymbol{0}$ in (2.12), we have

$$k_s \|\boldsymbol{u}\|_{L^s(\Omega)^m} \leq \|\boldsymbol{u}\|_{\mathbb{X}_p} \leq \left(c_p \|\boldsymbol{f}\|_{L^{p'}(\Omega)^m}\right)^{\frac{1}{p-1}} \equiv c_f, \tag{2.30}$$

independently of $g \in L^\infty_\nu(\Omega)$, with $s = \frac{dp}{d-p}$ if $p < d$, for any $s < \infty$ if $p = d$, or $s = \infty$ if $p > d$, for a Sobolev constant $k_s > 0$, being c_p the Poincaré constant. By Theorem 2.2, the solution map $S : L^\infty_\nu(\Omega) \ni g \mapsto \boldsymbol{u} \in \mathbb{X}_p$ is continuous.

Case 1. The map $T_p = S \circ G : \mathbb{X}_p \to \mathbb{X}_p$ is then also completely continuous and such that $T_p(D_{c_f}) \subset D_{c_f} = \{\varphi \in \mathbb{X}_p : \|\varphi\|_{\mathbb{X}_p} \leq c_f\}$. Then, by the Schauder fixed point theorem, there exists $\boldsymbol{u} = T_p(\boldsymbol{u})$, which solves (2.12).

Case 2. Set $T = S \circ G : \mathscr{C}(\overline{\Omega})^m \to \mathbb{X}_p$ and $\mathscr{S} = \{\boldsymbol{w} \in \mathscr{C}(\overline{\Omega})^m : \boldsymbol{w} = \lambda T\boldsymbol{w}, \lambda \in [0,1]\}$, which by (2.29) is bounded in $\mathscr{C}(\overline{\Omega})^m$. Indeed, if $\boldsymbol{w} \in \mathscr{S}$, $\boldsymbol{u} = T\boldsymbol{w}$ solves (2.12) with $g = G[\boldsymbol{w}]$ and we have, by the Sobolev inequality, (2.30) and $\boldsymbol{w} = \lambda\boldsymbol{u}$,

$$\begin{aligned}\|\boldsymbol{w}\|_{\mathscr{C}(\overline{\Omega})^m} &\leq C\lambda\||\mathrm{L}\boldsymbol{u}|\|_{L^r(\Omega)} \leq C\|g\|_{L^r(\Omega)} \leq C\big(c_0 + c_1\|\boldsymbol{w}\|^{\alpha}_{L^{\sigma p}(\Omega)^m}\big)\\ &\leq C\big(c_0 + c_1 k^{\alpha}_{\sigma p}\|\boldsymbol{u}\|^{\alpha}_{\mathbb{X}_p}\big) \leq C\big(c_0 + c_1 k^{\alpha}_{\sigma p} c^{\alpha}_f\big).\end{aligned}$$

Therefore T is a completely continuous mapping into some closed ball of $\mathscr{C}(\overline{\Omega})^m$ and it has a fixed point by the Leray-Schauder principle. □

Remark 2 The Sobolev's inequality also yields a version of Theorem 2.6 for $G : L^q(\Omega)^m \to L^\infty_\nu(\Omega)$ also merely continuous for any $q \geq 1$ when $p \geq d$ and $1 \leq q < \frac{dq}{d-p}$ when $1 < q < d$ (see [56]).

We present now examples of functionals G satisfying *1.* or *2.* of the above theorem.

***Example* 3**
Consider the functional $G : \mathbb{X}_p \to L^\infty_\nu(\Omega)$ defined as follows

$$G[\boldsymbol{u}](x) = F(x, \textstyle\int_\Omega \boldsymbol{K}(x,y)\cdot \mathrm{L}\boldsymbol{u}(y)dy),$$

where $F : \Omega \times \mathbb{R} \to \mathbb{R}$ is a bounded function in $x \in \Omega$ and continuous in $w \in \mathbb{R}$, uniformly in Ω, satisfying $0 < \nu \leq F$, and $\boldsymbol{K} \in \mathscr{C}(\overline{\Omega}; L^{p'}(\Omega)^\ell)$. This functional is completely continuous as a consequence of the fact that $\varphi : \mathbb{X}_p \to \mathscr{C}(\overline{\Omega})$ defined by

$$w(x) = \varphi(\boldsymbol{u})(x) = \int_\Omega \boldsymbol{K}(x,y)\cdot \mathrm{L}\boldsymbol{u}(y)dy, \quad \boldsymbol{u} \in \mathbb{X}_p, \quad x \in \overline{\Omega},$$

is also completely continuous. Indeed, if $\boldsymbol{u}_n \underset{n}{\rightharpoonup} \boldsymbol{u}$ in $\mathbb{X}_p$-weak, then $w_n \underset{n}{\longrightarrow} w$ in $\mathscr{C}(\overline{\Omega})$, because $L\boldsymbol{u}_n$, being bounded in $L^p(\Omega)^\ell$, implies w_n is uniformly bounded in $\mathscr{C}(\overline{\Omega})$, by

$$|w_n(x)| \leq \|\mathrm{L}\boldsymbol{u}_n\|_{L^p(\Omega)^\ell}\|\boldsymbol{K}(x)\|_{L^{p'}(\Omega)^\ell} \leq C\|\boldsymbol{K}\|_{\mathscr{C}(\overline{\Omega};L^{p'}(\Omega)^\ell)} \quad \forall\, x \in \overline{\Omega}$$

and equicontinuous in $\overline{\Omega}$ by

$$|w_n(x) - w_n(z)| \leq C\|\boldsymbol{K}(x,\cdot) - \boldsymbol{K}(z,\cdot)\|_{L^{p'}(\Omega)^\ell} \quad \forall\, x, z \in \overline{\Omega}.$$

***Example* 4**
Let $F : \Omega \times \mathbb{R}^m \to \mathbb{R}$ be a Carathéodory function $F = F(x, \boldsymbol{w})$, bounded in x for all $\boldsymbol{w} \in \mathbb{R}^m$ and continuous in $\boldsymbol{w}$ uniformly in $x \in \Omega$. If, for a.e. $x \in \Omega$ and all $\boldsymbol{w} \in \mathbb{R}^m$, F satisfies $0 < \nu \leq F(x, \boldsymbol{w})$, for $p > d$ and, for $p \leq d$ also

$$F(x, \boldsymbol{w}) \leq c_0 + c_1|\boldsymbol{w}|^\alpha,$$

for some constants $c_0, c_1 \geq 0$, $0 \leq \alpha \leq \frac{p}{d-p}$ if $1 < p < d$ or $\alpha \geq 0$ if $p = d$, then the Nemytskii operator

$$G[\boldsymbol{u}](x) = F(x, \boldsymbol{u}(x)), \quad \text{for } \boldsymbol{u} \in \mathscr{C}(\overline{\Omega})^m, \quad x \in \Omega,$$

yields a continuous functional $G : \mathscr{C}(\overline{\Omega})^m \to L^\infty_\nu(\Omega)$, which satisfies (2.29). □

***Example* 5.**
Suppose $p > d$. For fixed $g \in L^\infty_\nu(\Omega)$, defining

$$G[\boldsymbol{u}](x) = g(x) + \inf_{\substack{y \geq x \\ y \in \Omega}} |\boldsymbol{u}(y)|, \quad \boldsymbol{u} \in \mathscr{C}(\overline{\Omega})^m, \quad x \in \Omega,$$

where $y \geq x$ means $y_i \geq x_i$, $1 \leq i \leq d$ (see [60]), we have an example of case 2. of Theorem 2.6 above. □

2.5 The quasi-variational solution via contraction

In the special case of "small variations" of the convex sets, it is possible to apply the Banach fixed point theorem, obtaining also the uniqueness of the solution to the quasi-variational inequality for $1 < p \leq 2$. Here we simplify and develop the ideas of [40], by starting with a sharp version of the continuous dependence result of Theorem 2.1 for the variational inequality (2.12).

Proposition 2.1 *Under the framework of Theorem 2.1, let* $\boldsymbol{f}_1, \boldsymbol{f}_2 \in L^{p'}(\Omega)^m$, *with* $p' = \frac{p}{p-1} \geq 2$. *Then we have*

$$\|\boldsymbol{u}_1 - \boldsymbol{u}_2\|_{\mathbb{X}_p} \leq C_p \|\boldsymbol{f}_1 - \boldsymbol{f}_2\|_{L^{p'}(\Omega)^m}, \quad 1 < p \leq 2 \tag{2.31}$$

with

$$C_2 = c_2 \quad \text{and} \quad C_p = (2M)^{2-p} c_p \frac{\omega_p^2}{d_p}, \tag{2.32}$$

where c_p *and* d_p *are the constants, respectively, of* (2.4) *and* (2.10), $\omega_p = |\Omega|^{\frac{2-p}{2p}}$ *and* $M = \|g\|_{L^\infty(\Omega)}$.

Proof Using (2.10) and Hölder's and Poincaré's inequalities, from (2.12) for $\boldsymbol{u}_1$ with $\boldsymbol{w} = \boldsymbol{u}_2$ and for $\boldsymbol{u}_2$ with $\boldsymbol{w} = \boldsymbol{u}_1$, we easily obtain, first for $p = 2$,

$$\|\mathrm{L}(\boldsymbol{u}_1 - \boldsymbol{u}_2)\|^2_{L^2(\Omega)^m} \leq \|\boldsymbol{f}_1 - \boldsymbol{f}_2\|_{L^2(\Omega)^m} \|\boldsymbol{u}_1 - \boldsymbol{u}_2\|_{L^2(\Omega)^m}.$$

Hence (2.31) follows immediately for $p = 2$ since $d_2 = 1$.

Observing that, for $1 < p < 2$, Hölder inequality yields $\|\mathrm{L}\boldsymbol{w}\|_{L^p(\Omega)^\ell} \leq \omega_p \|\mathrm{L}\boldsymbol{w}\|_{L^2(\Omega)^\ell}$, using (2.10) and the Hölder inverse inequality, we get

$$(2M)^{p-2}d_p\,|\Omega|^{\frac{p-2}{p}}\,\|\mathrm{L}(\boldsymbol{u}_1-\boldsymbol{u}_2)\|^2_{L^p(\Omega)^\ell}\leq\int_\Omega|\mathrm{L}(\boldsymbol{u}_1-\boldsymbol{u}_2)|^2\big(|\mathrm{L}\boldsymbol{u}_1|+|\mathrm{L}\boldsymbol{u}_2|\big)^{p-2}$$
$$\leq c_p\|\boldsymbol{f}_1-\boldsymbol{f}_2\|_{L^{p'}(\Omega)^m}\|\mathrm{L}(\boldsymbol{u}_1-\boldsymbol{u}_2)\|_{L^p(\Omega)^\ell}.$$

and (2.31) follows easily by recalling that $\|\boldsymbol{w}\|_{\mathbb{X}_p}=\|\mathrm{L}\boldsymbol{w}\|_{L^p(\Omega)^\ell}$ for $\boldsymbol{w}\in\mathbb{X}_p$. □

We consider now a special case by separation of variables in the global constraint G. For $R>0$, denote

$$D_R=\{\boldsymbol{v}\in\mathbb{X}_p:\|\boldsymbol{v}\|_{\mathbb{X}_p}\leq R\}.$$

Theorem 2.7 *Let $1<p\leq 2$, $\boldsymbol{f}\in L^{p'}(\Omega)^m$ and*

$$G[\boldsymbol{u}](x)=\gamma(\boldsymbol{u})\varphi(x),\quad x\in\Omega, \tag{2.33}$$

where $\gamma:\mathbb{X}_p\to\mathbb{R}^+$ is a functional satisfying

i) $0<\eta(R)\leq\gamma\leq M(R)\quad\forall\,\boldsymbol{u}\in D_R$,
ii) $|\gamma(\boldsymbol{u}_1)-\gamma(\boldsymbol{u}_2)|\leq\Gamma(R)\|\boldsymbol{u}_1-\boldsymbol{u}_2\|_{\mathbb{X}_p}\quad\forall\,\boldsymbol{u}_1,\boldsymbol{u}_2\in D_R$,

for a sufficiently large $R\in\mathbb{R}^+$, being η, M and Γ monotone increasing positive functions of R, and $\varphi\in L^\infty_\nu(\Omega)$ is given. Then, the quasi-variational inequality (2.11) *has a unique solution, provided that*

$$\Gamma(R_f)p\,C_p\,\|\boldsymbol{f}\|_{L^{p'}(\Omega)^m}<\eta(R_f), \tag{2.34}$$

where $C_2=c_2$ and $C_p=\big(2M(R_f)\|\varphi\|_{L^\infty(\Omega)}\big)^{2-p}c_p\,\frac{\omega_p^2}{d_p}$ are given as in (2.32)*, with $R_f=\big(c_p\|\boldsymbol{f}\|_{L^{p'}(\Omega)}\big)^{\frac{1}{p-1}}$.*

Proof Let

$$\begin{aligned}S:D_R&\longrightarrow\mathbb{X}_p\\ \boldsymbol{v}&\mapsto\boldsymbol{u}=S(\boldsymbol{f},G[\boldsymbol{v}])\end{aligned}$$

where $\boldsymbol{u}$ is the unique solution of the variational inequality (2.12) with $g=G[\boldsymbol{v}]$.

By (2.30), any solution $\boldsymbol{u}$ to the variational inequality (2.12) is such that $\|\boldsymbol{u}\|_{\mathbb{X}_p}\leq R_f$ and therefore $S(D_{R_f})\subset D_{R_f}$.

Given $\boldsymbol{v}_i\in D_{R_f}$, $i=1,2$, let $\boldsymbol{u}_i=S(\boldsymbol{f},\gamma(\boldsymbol{v}_i)\varphi)$ and set $\mu=\frac{\gamma(\boldsymbol{v}_2)}{\gamma(\boldsymbol{v}_1)}$. We may assume $\mu>1$ without loss of generality. Setting $g=\gamma(\boldsymbol{v}_1)\varphi$, then $\mu g=\gamma(\boldsymbol{v}_2)\varphi$ and $S(\mu^{p-1}\boldsymbol{f},\mu g)=\mu S(\boldsymbol{f},g)$. Using (2.31) with $\boldsymbol{f}_1=\boldsymbol{f}$ and $\boldsymbol{f}_2=\mu^{p-1}\boldsymbol{f}$, we have

$$\begin{aligned}\|\boldsymbol{u}_1-\boldsymbol{u}_2\|_{\mathbb{X}_p}&\leq\|S(\boldsymbol{f},g)-S(\mu^{p-1}\boldsymbol{f},\mu g)\|_{\mathbb{X}_p}+\|S(\mu^{p-1}\boldsymbol{f},\mu g)-S(\boldsymbol{f},\mu g)\|_{\mathbb{X}_p}\\&\leq(\mu-1)\|\boldsymbol{u}_1\|_{\mathbb{X}_p}+(\mu^{p-1}-1)C_p\|\boldsymbol{f}\|_{L^{p'}(\Omega)^m}\\&\leq(\mu-1)p\,C_p\|\boldsymbol{f}\|_{L^{p'}(\Omega)^m},\end{aligned}\tag{2.35}$$

since $\mu^{p-1}-1\leq(p-1)(\mu-1)$, because $1<p\leq 2$, and $\|\boldsymbol{u}_1\|_{\mathbb{X}_p}\leq C_p\|\boldsymbol{f}\|_{L^{p'}(\Omega)^m}$ from the estimate (2.31) with $\boldsymbol{f}_1=\boldsymbol{f}$ and $\boldsymbol{f}_2=\boldsymbol{0}$, where C_p is given by (2.32) with $M=M(R_f)\|\varphi\|_{L^\infty(\Omega)}$.

Observing that, from the assumptions *i)* and *ii)*,

$$\mu - 1 = \frac{\gamma(\boldsymbol{v}_2) - \gamma(\boldsymbol{v}_1)}{\gamma(\boldsymbol{v}_1)} \leq \frac{\Gamma(R_f)}{\eta(R_f)}\|\boldsymbol{v}_1 - \boldsymbol{v}_2\|_{\mathbb{X}_p},$$

we get from (2.35)

$$\|S(\boldsymbol{v}_1) - S(\boldsymbol{v}_2)\|_{X_p} = \|\boldsymbol{u}_1 - \boldsymbol{u}_2\|_{X_p} \leq \frac{\Gamma(R_f)}{\eta(R_f)} p\, C_p \|\boldsymbol{f}\|_{L^{p'}(\Omega)^m} \|\boldsymbol{v}_1 - \boldsymbol{v}_2\|_{X_p}.$$

Therefore the application S is a contraction provided (2.34) holds and its fixed point $\boldsymbol{u} = S(\boldsymbol{f}, G[\boldsymbol{u}])$ solves uniquely (2.11). □

Remark 3 The assumptions *i)* and *ii)* are similar to the conditons in Appendix B of [40], where the contractiveness of the solution application S was obtained in an implicit form under the assumptions on the norm of $\boldsymbol{f}$ to be sufficiently small. Our expression (2.34) quantifies not only the size of the $L^{p'}$-norm of $\boldsymbol{f}$, but also the constants of the functional γ, the φ and the domain Ω, through its measure and the size of its Poincaré constant.

2.6 Applications

We present three examples of physical applications.

***Example* 6 (A nonlinear Maxwell quasi-variational inequality)** (see [64]) Consider a nonlinear electromagnetic field in equilibrium in a bounded simply connected domain Ω of $\mathbb{R}^3$. We consider the stationary Maxwell's equations

$$\boldsymbol{j} = \nabla \times \boldsymbol{h}, \quad \nabla \times \boldsymbol{e} = \boldsymbol{f} \quad \text{and} \quad \nabla \cdot \boldsymbol{h} = 0 \quad \text{in } \Omega,$$

where $\boldsymbol{j}$, $\boldsymbol{e}$ and $\boldsymbol{h}$ denote, respectively, the current density, the electric and the magnetic fields. For type-II superconductors we may assume constitutive laws of power type and an extension of the Bean critical-state model, in which the current density cannot exceed some given critical value $j \geq \nu > 0$. When j may vary with the absolute value $|\boldsymbol{h}|$ of the magnetic field (see Prigozhin, [72]) we obtain a quasi-variational inequality. Here we suppose

$$\boldsymbol{e} = \begin{cases} \delta|\nabla\times\boldsymbol{h}|^{p-2}\nabla\times\boldsymbol{h} & \text{if } |\nabla\times\boldsymbol{h}| < j(|\boldsymbol{h}|), \\ \left(\delta\, j^{p-2} + \lambda\right)\nabla\times\boldsymbol{h} & \text{if } |\nabla\times\boldsymbol{h}| = j(|\boldsymbol{h}|), \end{cases}$$

where $\delta \geq 0$ is a given constant and $\lambda \geq 0$ is an unknown Lagrange multiplier associated with the inequality constraint. The region $\{|\nabla \times \boldsymbol{h}| = j(|\boldsymbol{h}|)\}$ corresponds to the superconductivity region. We obtain the quasi-variational inequality (2.11) with $\mathbb{X}_p$ defined in (2.5) or (2.6), depending whether we are considering a domain with perfectly conductive or perfectly permeable walls.

The existence of solution is immediate by Theorem 2.6. 1., if we assume $j : \mathbb{X}_p \to \mathbb{R}^+$ continuous, with $j \geq \nu > 0$, for any $p > 3$ and, for $1 < p \leq 3$ if j also has the growth condition of F in Example 5. above. Therefore, setting L= $\nabla\times$, for any $\boldsymbol{f} \in L^{p'}(\Omega)^3$ and any $\delta \geq 0$, we have at least a solution to

$$\begin{cases} \boldsymbol{h} \in \mathbb{K}_{j(|\boldsymbol{h}|)} = \left\{\boldsymbol{w} \in \mathbb{X}_p : |\nabla \times \boldsymbol{w}| \leq j(|\boldsymbol{h}|) \text{ in } \Omega\right\}, \\ \delta \int_\Omega |\nabla \times \boldsymbol{h}|^{p-2} \nabla \times \boldsymbol{h} \cdot \nabla \times (\boldsymbol{w} - \boldsymbol{h}) \geq \int_\Omega \boldsymbol{f} \cdot (\boldsymbol{w} - \boldsymbol{h}) \quad \forall \boldsymbol{w} \in \mathbb{K}_{j(|\boldsymbol{h}|)}. \end{cases}$$

***Example* 7 (Thermo-elastic equilibrium of a locking material)**
Analogously to perfect plasticity, in 1957 Prager introduced the notion of an ideal locking material as a linear elastic solid for stresses below a certain threshold, which cannot be overpassed. When the threshold is attained,"there is locking in the sense that any further increase in stress will not cause any changes in strain" [68]. Duvaut and Lions, in 1972 [31], solved the general stationary problem in the framework of convex analysis. Here we consider a simplified situation for the displacement field $\boldsymbol{u} = \boldsymbol{u}(x)$ for $x \in \Omega \subset \mathbb{R}^d$, $d = 1, 2, 3$, which linearized strain tensor $D\boldsymbol{u}$ =L$\boldsymbol{u}$ is its symmetrized gradient. We shall consider $\mathbb{X}_2 = H^1_0(\Omega)^d$ with norm $\|D\boldsymbol{u}\|_{L^2(\Omega)^{d^2}}$ and, for an elastic solid with Lamé constants $\mu > 0$ and $\lambda \geq 0$, we consider the quasi-variational inequality

$$\begin{cases} \boldsymbol{u} \in \mathbb{K}_{b(\vartheta[\boldsymbol{u}])} = \left\{\boldsymbol{w} \in H^1_0(\Omega)^d : |D\boldsymbol{w}| \leq b(\vartheta[\boldsymbol{u}]) \text{ in } \Omega\right\} \\ \int_\Omega \Big(\mu D\boldsymbol{u} \cdot D(\boldsymbol{w} - \boldsymbol{u}) + \lambda(\nabla \cdot \boldsymbol{u})(\nabla \cdot (\boldsymbol{w} - \boldsymbol{u}))\Big) \\ \qquad\qquad \geq \int_\Omega \boldsymbol{f} \cdot (\boldsymbol{w} - \boldsymbol{u}) \quad \forall \boldsymbol{w} \in \mathbb{K}_{b(\vartheta[\boldsymbol{u}])}. \end{cases} \tag{2.36}$$

Here $b \in \mathscr{C}(\mathbb{R})$, such that $b(\vartheta) \geq \nu > 0$, is a continuous function of the temperature field $\vartheta = \vartheta[\boldsymbol{u}](x)$, supposed also in equilibrium under a thermal forcing depending on the deformation $D\boldsymbol{u}$. We suppose that $\vartheta[\boldsymbol{u}]$ solves

$$-\Delta\vartheta = h(x, D\boldsymbol{u}(x)) \text{ in } \Omega, \quad \vartheta = 0 \text{ on } \partial\Omega, \tag{2.37}$$

where $h : \Omega \times \mathbb{R}^{d^2} \to \mathbb{R}$ is a given Carathéodory function such that

$$|h(x, D)| \leq h_0(x) + C|D|^s, \qquad \text{for a.e. } x \in \Omega \text{ and } D \in \mathbb{R}^{d^2}, \tag{2.38}$$

for some function $h_0 \in L^r(\Omega)$, with $r > \frac{d}{2}$ and $0 < s < \frac{2}{r}$.

First, with $\boldsymbol{w} \equiv \boldsymbol{0}$ in (2.36), we observe that any solution to (2.36) satisfies the a priori bound $\|D\boldsymbol{u}\|_{L^2(\Omega)^{d^2}} \leq \frac{k}{\mu}\|\boldsymbol{f}\|_{L^2(\Omega)^{d^2}}$, where k is the constant of $\|\boldsymbol{u}\|_{L^2(\Omega)^d} \leq k\|D\boldsymbol{u}\|_{L^2(\Omega)^{d^2}}$ from Korn's inequality.

Therefore, for each $\boldsymbol{u} \in H^1_0(\Omega)^d$, the unique solution $\vartheta \in H^1_0(\Omega)$ to (2.37) is in the Hölder space $\mathscr{C}^\alpha(\overline{\Omega})$, for some $0 < \alpha < 1$, since $h = h(x, D\boldsymbol{u}(x)) \in L^{\frac{p}{2}}(\Omega)$ by (2.38), with the respective continuous dependence in $H^1_0(\Omega) \cap \mathscr{C}^\alpha(\overline{\Omega})$ for the strong topologies, by De Giorgi-Stamppachia estimates (see, for instance, [74, p. 170] and its references). By the a priori bound of $\boldsymbol{u}$ and the compactness of $\mathscr{C}^\alpha(\overline{\Omega}) \subset \mathscr{C}(\overline{\Omega})$, if we define $G : \mathbb{X}_2 \to \mathscr{C}(\overline{\Omega}) \cap L^\infty_\nu(\Omega)$ by $G[\boldsymbol{u}] = b(\vartheta[\boldsymbol{u}])$, we easily conclude that G is a completely continuous operator and we can apply Theorem 2.6 to conclude that, for any $\boldsymbol{f} \in L^2(\Omega)^d$, $b \in \mathscr{C}(\mathbb{R})$, $b \geq \nu > 0$ and any h satisfying (2.38), there exists at least one solution $(\boldsymbol{u}, \vartheta) \in H^1_0(\Omega)^d \times \big(H^1_0(\Omega) \cap \mathscr{C}^\alpha(\overline{\Omega})\big)$ to the coupled problem (2.36)-(2.37). □

***Example* 8 (An ionization problem in electrostatics)** (a new variant of [56])
Let Ω be a bounded Lipschitz domain of $\mathbb{R}^d$, $d = 2$ or 3, being $\partial\Omega = \Gamma_0 \cup \Gamma_1 \cup \Gamma_\#$, with $\overline{\Gamma_0} \cap \overline{\Gamma_\#} \neq \emptyset$, both sets with positive $d-1$ Lebesgue measure. Denote by $\boldsymbol{e}$ the electric field, which we assume to be given by a potential $\boldsymbol{e} = -\nabla u$. We impose a potential difference between Γ_0 and $\Gamma_\#$ and that Γ_1 is insulated. So

$$u = 0 \text{ on } \Gamma_0, \quad \boldsymbol{j} \cdot \boldsymbol{n} = 0 \text{ on } \Gamma_1 \quad \text{and} \quad u = u_\# \text{ on } \Gamma_\#, \tag{2.39}$$

with $\boldsymbol{n}$ being the outer unit normal vector to $\partial\Omega$. Here the trace $u_\#$ on $\Gamma_\#$ is an unknown constant to be found as part of the solution, by giving the total current τ across $\Gamma_\#$,

$$\tau = \int_{\Gamma_\#} \boldsymbol{j} \cdot \boldsymbol{n} \in \mathbb{R}. \tag{2.40}$$

We set L= ∇, $V_2 = H^1(\Omega)$ and, as in [75], we define

$$\mathbb{X}_2 = H^1_\# = \big\{w \in H^1(\Omega) : w = 0 \text{ on } \Gamma_0 \text{ and } w = w_\# = \text{constant on } \Gamma_\#\big\}, \tag{2.41}$$

where the Poincaré inequality (2.4) holds, as well as the trace property for $w_\# = w_{|_{\Gamma_\#}}$, for some $c_\# > 0$:

$$|w_\#| \leq c_\# \|\nabla w\|_{L^2(\Omega)^d} \quad \forall w \in \mathbb{X}_2.$$

We assume, as in [31, p.333] that

$$\boldsymbol{j} = \begin{cases} \sigma \boldsymbol{e} & \text{if } |\boldsymbol{e}| < \gamma, \\ (\sigma + \lambda)\boldsymbol{e} & \text{if } |\boldsymbol{e}| = \gamma, \end{cases} \tag{2.42}$$

where σ is a positive constant, $\lambda \geq 0$ is a Lagrange multiplier and γ a positive ionization threshold. However, this is only an approximation of the true ionization law. In [56], it was proposed to let γ vary locally with $|\boldsymbol{e}|^2$ in a neighbourhood of each point of the boundary, but here we shall consider instead that the ionization threshold depends on the difference of the potential on the opposite boundaries Γ_0 and $\Gamma_\#$, i.e.

$$\gamma = \gamma(u_\#) \text{ with } \gamma \in \mathscr{C}(\mathbb{R}) \text{ and } \gamma \geq \nu > 0. \tag{2.43}$$

Therefore we are led to search the electric potential u as the solution of the following quasi-variational inequality:

$$u \in \mathbb{K}_{\gamma(u_\#)} = \big\{w \in H^1_\#(\Omega) : |\nabla w| \leq \gamma(u_\#) \text{ in } \Omega\big\}, \tag{2.44}$$

$$\sigma \int_\Omega \nabla u \cdot \nabla (w-u) \geq \int_\Omega f(w-u) - \tau(w_\# - u_\#) \quad \forall w \in \mathbb{K}_{\gamma(u_\#)}, \tag{2.45}$$

by incorporating the ionization law (2.42) with the conservation law of the electric charge $\nabla \cdot \boldsymbol{j} = f$ in Ω and the boundary conditions (2.39) and (2.40) (see [75], for details).

From (2.45) with $w = 0$, we also have the a priori bound

$$\|\nabla u\|_{L^2(\Omega)^d} = \|u\|_{\mathbb{X}_2} \leq \tfrac{c_2}{\sigma}\|f\|_{L^2(\Omega)} + \tfrac{c_\#}{\sigma} \equiv R_\#. \tag{2.46}$$

Then, setting $G[u] = \gamma(u_\#)$ for $u \in \mathbb{X}_2 = H^1_\#$, by the continuity of the trace on $\Gamma_\#$ and the assumption (2.43), we easily conclude that $G : \mathbb{X}_2 \to [\nu, \gamma_\#]$ is a completely continuous operator, where $\gamma_\# = \max_{|r|\leq c_\# R_\#} \gamma(r)$, with $R_\#$ from (2.46). Consequently, by Theorem 2.6, there exists at least a solution to the ionization problem (2.44)-(2.45), for any $f \in L^2(\Omega)$ and any $\tau \in \mathbb{R}$.

From (2.45), if we denote by w_1 and w_2 the solutions of the variational inequality for (f_1, τ_1) and (f_2, τ_2) corresponding to the same convex $\mathbb{K}_g$ defined in (2.44), we easily obtain the following version of Proposition 2.1:

$$\|w_1 - w_2\|_{H^1_\#(\Omega)} \leq \tfrac{c_2}{\sigma}\|f_1 - f_2\|_{L^2(\Omega)} + \tfrac{c_\#}{\sigma}|\tau_1 - \tau_2|.$$

If, in addition, $\gamma \in \mathscr{C}^{0,1}(\mathbb{R})$ and we set $\gamma'_\# = \sup_{|r|\leq c_\# R_\#} |\gamma'(r)|$ we have

$$|\gamma(w_{1\#}) - \gamma(w_{2\#})| \leq \gamma'_\# |w_{1\#} - w_{2\#}| \leq \gamma'_\# c_\# \|w_1 - w_2\|_{H^1_\#(\Omega)}$$

and the argument of Theorem 2.7 yields that the solution u of (2.44)-(2.45) is unique provided that

$$2\gamma'_\# c_\# \big(\tfrac{c_2}{\sigma}\|f\|_{L^2(\Omega)} + \tfrac{c_\#}{\sigma}|\tau|\big) < \nu.$$

3 Evolutionary problems

3.1 The variational inequality

For $T > 0$ and $t \in (0, T)$, we set $Q_t = \Omega \times (0, t)$ and, for $\nu > 0$, we define

$$L^\infty_\nu(Q_T) = \{w \in L^\infty(Q_T) : w \geq \nu\}.$$

Given $g \in L^\infty_\nu(Q_T)$, for a.e. $t \in (0, T)$ we set

$$\boldsymbol{w} \in \mathbb{K}_g \text{ iff } \boldsymbol{w}(t) \in \mathbb{K}_{g(t)} = \big\{\boldsymbol{w} \in \mathbb{X}_p : |\mathrm{L}\boldsymbol{w}| \leq g(t)\big\}.$$

We define, for $1 < p < \infty$ and $p' = \frac{p}{p-1}$,

$$\mathscr{V}_p = L^p\big(0,T;\mathbb{X}_p\big), \quad \mathscr{V}_p' = L^{p'}\big(0,T;\mathbb{X}_p'\big), \quad \mathscr{Y}_p = \big\{w \in \mathscr{V}_p : \partial_t w \in \mathscr{V}_p'\big\}$$

and we assume that there exists an Hilbert space $\mathbb{H}$ such that

$$\mathbb{H} \subseteq L^2(\Omega)^m, \quad (\mathbb{X}_p, \mathbb{H}, \mathbb{X}_p') \text{ is a Gelfand triple}, \quad \mathbb{X}_p \hookrightarrow \mathbb{H} \text{ is compact.} \tag{3.1}$$

As a consequence, by the embedding results of Sobolev-Bochner spaces (see, for instance [81]), we have then

$$\mathscr{Y}_p \subset \mathscr{C}\big([0,T];\mathbb{H}\big) \subset \mathscr{H} \equiv L^p\big(0,T;\mathbb{H}\big)$$

and the embedding of $\mathscr{Y}_p \subset \mathscr{H}$ is also compact for $1 < p < \infty$.

For $\delta \geq 0$, given $\boldsymbol{f} : Q_T \to \mathbb{R}$ and $\boldsymbol{u}_0 : \Omega \to \mathbb{R}$, $\boldsymbol{u}_0 \in \mathbb{K}_{g(0)}$, we consider the weak formulation of the variational inequality, following [61],

$$\begin{cases} \boldsymbol{u}^\delta \in \mathbb{K}_g, \\ \displaystyle\int_0^T \langle \partial_t \boldsymbol{w}, \boldsymbol{w} - \boldsymbol{u}^\delta \rangle_p + \delta \int_{Q_T} \text{Ł}_p \boldsymbol{u}^\delta \cdot \text{L}(\boldsymbol{w} - \boldsymbol{u}^\delta) \geq \int_{Q_T} \boldsymbol{f} \cdot (\boldsymbol{w} - \boldsymbol{u}^\delta) \\ \qquad\qquad \displaystyle -\frac{1}{2}\int_\Omega |\boldsymbol{w}(0) - \boldsymbol{u}_0|^2, \quad \forall \boldsymbol{w} \in \mathbb{K}_g \cap \mathscr{Y}_p \end{cases} \tag{3.2}$$

and we observe that the solution $\boldsymbol{u}^\delta \in \mathscr{V}_p$ is not required to have the time derivative $\partial_t \boldsymbol{u}^\delta$ in the dual space $\mathscr{V}_p'$ and satisfies the initial condition in a very weak sense. In (3.2), $\langle \cdot, \cdot \rangle_p$ denotes the duality pairing between $\mathbb{X}_p'$ and $\mathbb{X}_p$, which reduces to the inner product in $L^2(\Omega)^m$ if both functions belong to this space.

When $\partial_t \boldsymbol{u}^\delta \in L^2\big(0,T;L^2(\Omega)^m\big)$ (or more generally when $\boldsymbol{u}^\delta \in \mathscr{Y}_p$), the strong formulation reads

$$\begin{cases} \boldsymbol{u}^\delta(t) \in \mathbb{K}_{g(t)}, \ t \in [0,T], \ \boldsymbol{u}(0) = \boldsymbol{u}_0, \\ \displaystyle\int_\Omega \partial_t \boldsymbol{u}^\delta(t) \cdot (\boldsymbol{w} - \boldsymbol{u}^\delta(t)) + \delta \int_\Omega \text{Ł}_p \boldsymbol{u}^\delta(t) \cdot \text{L}(\boldsymbol{w} - \boldsymbol{u}^\delta(t)) \\ \qquad\qquad \displaystyle\geq \int_\Omega \boldsymbol{f}(t) \cdot (\boldsymbol{w} - \boldsymbol{u}^\delta(t)), \quad \forall\, \boldsymbol{w} \in \mathbb{K}_{g(t)} \text{ for a.e. } t \in (0,T). \end{cases} \tag{3.3}$$

Integrating (3.3) in $t \in (0,T)$ with $\boldsymbol{w} \in \mathbb{K}_g \cap \mathscr{Y}_p \subset \mathscr{C}\big([0,T];L^2(\Omega)^m\big)$ and using

$$\begin{aligned} \int_0^t \langle \partial_t \boldsymbol{u}^\delta - \partial_t \boldsymbol{w}, \boldsymbol{w} - \boldsymbol{u}^\delta \rangle_p &= \frac{1}{2}\int_\Omega |\boldsymbol{w}(0) - \boldsymbol{u}_0|^2 - \frac{1}{2}\int_\Omega |\boldsymbol{w}(t) - \boldsymbol{u}^\delta(t)|^2 \\ &\leq \frac{1}{2}\int_\Omega |\boldsymbol{w}(0) - \boldsymbol{u}_0|^2 \end{aligned}$$

we immediately conclude that a strong solution is also a weak solution, i.e., it satisfies (3.2). Reciprocally, if $\boldsymbol{u}^\delta \in \mathbb{K}_g$ with $\partial_t \boldsymbol{u}^\delta \in L^2(Q_T)^m$ (or if $\boldsymbol{u}^\delta \in \mathscr{Y}_p$) is a weak solution with $\boldsymbol{u}^\delta(0) = \boldsymbol{u}_0$, replacing in (3.2) $\boldsymbol{w}$ by $\boldsymbol{u}^\delta + \theta(\boldsymbol{z} - \boldsymbol{u}^\delta)$ for $\theta \in (0,1]$ and $\boldsymbol{z} \in \mathbb{K}_g \cap \mathscr{Y}_p$, and letting $\theta \to 0$, we conclude that $\boldsymbol{u}^\delta$ also satisfies

$$\int_{Q_T} \partial_t \boldsymbol{u}^\delta \cdot (z - \boldsymbol{u}^\delta) + \delta \int_{Q_T} \mathrm{Ł}_p \boldsymbol{u}^\delta \cdot \mathrm{L}(z - \boldsymbol{u}^\delta) \geq \int_{Q_T} \boldsymbol{f} \cdot (z - \boldsymbol{u}^\delta)$$

and, by approximation, when $g \in \mathscr{C}\big([0,T]; L^\infty_\nu(\Omega)\big)$ (see [66, Lemma 5.2]), also for all $z \in \mathbb{K}_g$.

For any $w \in \mathbb{K}_{g(t)}$, for fixed $t \in (0,T)$ and arbitrary s, $0 < s < t < T - s$, we can use as test function in (3.3) $z \in \mathbb{K}_g$ such that $z(\tau) = 0$ if $\tau \notin (t-s, t+s)$ and $z(\tau) = \frac{\nu}{\nu+\varepsilon_s} w$ if $\tau \in (t-s, t+s)$, with $\varepsilon_s = \sup_{t-s<\tau<t+s} \|g(t) - g(\tau)\|_{L^\infty(\Omega)}$. Hence, dividing by $2s$ and letting $s \to 0$, we can conclude the equivalence between (3.3) and (3.2).

We have the following existence and uniqueness result whose proof, under more general assumptions for monotone operators, can be found in [66].

Theorem 3.8 *Suppose that* $\delta \geq 0$ *and* (2.1), (2.2), (2.3) *and* (3.1) *are satisfied. Assume that*

$$\boldsymbol{f} \in L^2(Q_T)^m, \quad g \in \mathscr{C}\big([0,T]; L^\infty_\nu(\Omega)\big), \quad \boldsymbol{u}_0 \in \mathbb{K}_{g(0)}. \tag{3.4}$$

Then, for any $\delta \geq 0$, *the variational inequality* (3.2) *has a unique weak solution*

$$\boldsymbol{u}^\delta \in \mathscr{V}_p \cap \mathscr{C}\big([0,T]; L^2(\Omega)^m\big).$$

If, in addition,

$$g \in W^{1,\infty}\big(0,T; L^\infty(\Omega)\big), \quad g \geq \nu > 0 \tag{3.5}$$

then the variational inequality (3.3) *has a unique strong solution*

$$\boldsymbol{u}^\delta \in \mathscr{V}_p \cap H^1\big(0,T; L^2(\Omega)^m\big).$$

Remark 4 For the scalar case L$= \nabla$, with $p = 2$, a previous result for strong solutions was obtained in [84] with $g \in \mathscr{C}(\overline{Q_T}) \cap W^{1,\infty}\big(0,T; L^\infty(\Omega)\big)$, $g \geq \nu > 0$ for the coercive case $\delta > 0$. More recently, a similar result was obtained with the time-dependent subdifferential operator techniques by Kenmochi in [49], also for $\delta > 0$ and for the scalar case L$=\nabla$, getting weak solutions for $1 < p < \infty$ with $g \in \mathscr{C}(\overline{Q_T})$ and strong solutions with $g \in \mathscr{C}(\overline{Q_T}) \cap H^1\big(0,T; \mathscr{C}(\overline{\Omega})\big)$.

The next theorem gives a quantitative result on the continuous dependence on the data, which essentially establishes the Lipschitz continuity of the solutions with respect to $\boldsymbol{f}$ and $\boldsymbol{u}_0$ and the Hölder continuity (up to $\frac{1}{2}$ only) with respect to the threshold g. This estimate in $\mathscr{V}_p$ was obtained first in [84] with L$= \nabla$ and $p = 2$ and developed later in several other works, including [65], [49] and [66]. Here we give an explicit dependence of the constants with respect to the data.

Theorem 3.9 *Suppose that* $\delta \geq 0$ *and* (2.1), (2.2), (2.3) *and* (3.1) *are satisfied. Let* $i = 1, 2$, *and suppose that* $\boldsymbol{f}_i \in L^2(\Omega)^m$, $g_i \in \mathscr{C}\big([0,T]; L^\infty_\nu(\Omega)\big)$ *and* $\boldsymbol{u}_{0i} \in \mathbb{K}_{g_i(0)}$. *If* $\boldsymbol{u}_i^\delta$ *are the solutions of the variational inequality* (3.2) *with data* $(\boldsymbol{f}_i, \boldsymbol{u}_{0i}, g_i)$ *then there exists a constant* B, *which depends only in a monotone increasing way on* T, $\|\boldsymbol{u}_{0i}\|^2_{L^2(\Omega)^m}$ *and* $\|\boldsymbol{f}_i\|^2_{L^2(\Omega)^m}$, *such that*

$$\|\boldsymbol{u}_1^\delta - \boldsymbol{u}_2^\delta\|^2_{L^\infty(0,T;L^2(\Omega)^m)} \leq (1+Te^T)\Big(\|\boldsymbol{f}_1-\boldsymbol{f}_2\|^2_{L^2(Q_T)^m} + \|\boldsymbol{u}_{01}-\boldsymbol{u}_{02}\|^2_{L^2(\Omega)^m} + \tfrac{B}{\nu}\|g_1-g_2\|_{L^\infty(Q_T)}\Big). \quad (3.6)$$

Besides, if $\delta > 0$,

$$\|\boldsymbol{u}_1^\delta - \boldsymbol{u}_2^\delta\|^{p\vee 2}_{\mathscr{V}_p} \leq \tfrac{a_p}{\delta}\Big(\|\boldsymbol{f}_1-\boldsymbol{f}_2\|^2_{L^2(Q_T)^m} + \|\boldsymbol{u}_{01}-\boldsymbol{u}_{02}\|^2_{L^2(\Omega)^m} + \tfrac{B}{\nu}\|g_1-g_2\|_{L^\infty(Q_T)}\Big), \quad (3.7)$$

where

$$a_p = \tfrac{(1+T+T^2e^T)}{2d_p}\big(c_g\,|Q_T|^{\frac{1}{p}}\big)^{(2-p)^+} \qquad 1<p<\infty, \quad (3.8)$$

being d_p *given by* (2.10) *and* $c_g = \|g_1\|_{L^\infty(Q_T)} + \|g_2\|_{L^\infty(Q_T)}$.

Proof We prove first the result for strong solutions, approximating the function g_i in $\mathscr{C}\big([0,T];L^\infty_\nu(\Omega)\big)$ by a sequence $\{g_i^n\}_n$ belonging to $W^{1,\infty}\big(0,T;L^\infty(\Omega)\big)$.

Given two strong solutions $\boldsymbol{u}_i^\delta$, $i=1,2$, setting $\beta = \|g_1-g_2\|_{L^\infty(Q_T)}$, denoting $\overline{\boldsymbol{u}} = \boldsymbol{u}_1^\delta - \boldsymbol{u}_2^\delta$, $\overline{\boldsymbol{u}}_0 = \overline{\boldsymbol{u}}_{01} - \overline{\boldsymbol{u}}_{02}$, $\overline{\boldsymbol{f}} = \boldsymbol{f}_1 - \boldsymbol{f}_2$, $\overline{g} = g_1 - g_2$ and $\alpha = \frac{\nu}{\nu+\beta}$ and using the test functions $\boldsymbol{w}_{i_j} = \frac{\nu \boldsymbol{u}_i^\delta}{\nu+\beta} \in \mathbb{K}_j$, for $i,j=1,2$, $i\neq j$, we obtain the inequality

$$\int_\Omega \partial_t\overline{\boldsymbol{u}}(t)\cdot\overline{\boldsymbol{u}}(t) + \delta\int_\Omega \big(\text{Ł}_p\boldsymbol{u}_1^\delta(t) - \text{Ł}_p\boldsymbol{u}_2^\delta(t)\big)\cdot \mathrm{L}\overline{\boldsymbol{u}}(t) \leq \int_\Omega \overline{\boldsymbol{f}}(t)\cdot\overline{\boldsymbol{u}}(t) + \Theta(t), \quad (3.9)$$

where

$$\Theta(t) = (\alpha-1)\int_\Omega \big(\partial_t(\boldsymbol{u}_1^\delta\cdot\boldsymbol{u}_2^\delta) + \delta\text{Ł}_p\boldsymbol{u}_1^\delta\cdot\mathrm{L}\boldsymbol{u}_2^\delta + \delta\text{Ł}_p\boldsymbol{u}_2^\delta\cdot\mathrm{L}\boldsymbol{u}_1^\delta - \boldsymbol{f}_1\cdot\boldsymbol{u}_2^\delta - \boldsymbol{f}_2\cdot\boldsymbol{u}_1^\delta\big)(t) \quad (3.10)$$

and, because $1-\alpha = \frac{\beta}{\beta+\nu} \leq \frac{1}{\nu}\|g_1-g_2\|_{L^\infty(Q_T)}$, then for any $t\in(0,T)$

$$\int_0^t \Theta \leq \tfrac{B}{2\nu}\|g_1-g_2\|_{L^\infty(Q_T)}, \quad (3.11)$$

where the constant B depends on $\|\boldsymbol{f}_i\|_{L^2(Q_T)^m}$ and $\|\boldsymbol{u}_{0i}\|_{L^2(\Omega)}$. From (3.9), we have

$$\int_\Omega |\overline{\boldsymbol{u}}(t)|^2 \leq \int_0^t\int_\Omega |\overline{\boldsymbol{u}}|^2 + \int_\Omega |\overline{\boldsymbol{u}}_0|^2 + \int_{Q_T}|\overline{\boldsymbol{f}}|^2 + 2\int_0^T \Theta,$$

proving (3.6) by applying the integral Gronwall inequality.

If $\delta>0$ and $p\geq 2$, using the monotonicity of Ł_p, then

$$\int_\Omega |\overline{\boldsymbol{u}}(t)|^2 + 2\delta\, d_p\int_{Q_t}|\mathrm{L}\overline{\boldsymbol{u}}|^p_{L^p(Q_t)^\ell} \leq \int_{Q_t}|\overline{\boldsymbol{f}}|^2 + \int_{Q_t}|\overline{\boldsymbol{u}}|^2 + \int_\Omega|\overline{\boldsymbol{u}}_0|^2 + 2\int_0^t\Theta$$

and, by the estimates (3.6) and (3.11), by integrating in t we easily obtain (3.7).

For $\delta > 0$ and $1 < p < 2$ set $c_g = \|g_1\|_{L^\infty(Q_T)} + \|g_2\|_{L^\infty(Q_T)}$. So, using the monotonicity (2.10) of $\mathrm{Ł}_p$ and the Hölder inverse inequality,

$$\int_\Omega |\overline{\boldsymbol{u}}(t)|^2 + 2\,\delta\, d_p\, c_g^{p-2} |Q_T|^{\frac{p-2}{p}} \Big(\int_{Q_T} |\mathrm{Ł}\overline{\boldsymbol{u}}(t)|^p \Big)^{\frac{2}{p}}$$

$$\leq \int_{Q_t} |\overline{\boldsymbol{f}}|^2 + \int_{Q_t} |\overline{\boldsymbol{u}}|^2 + \int_\Omega |\overline{\boldsymbol{u}}_0|^2 + \tfrac{B}{\nu} \|g_1 - g_2\|_{L^\infty(Q_T)}$$

and using the estimate (3.6) to control $\|\overline{\boldsymbol{u}}\|^2_{L^2(Q_T)^m}$ as above, we conclude the proof for strong solutions.

To prove the results for weak solutions, it is enough to recall that they can be approximated by strong solutions in $\mathscr{C}([0,T]; L^2(\Omega)^m) \cap \mathscr{V}_p$. □

Using the same proof for the case $\nabla\times$ of [65], which was a development of the scalar case with $p = 2$ of [84], we can prove the asymptotic behaviour of the strong solution of the variational inequality when $t \longrightarrow \infty$. Consider the stationary variational inequality (2.12) with data $\boldsymbol{f}_\infty$ and g_∞ and denoting its solution by $\boldsymbol{u}_\infty$, we have the following result.

Theorem 3.10 *Suppose that the assumptions* (2.1), (2.2), (2.3) *and* (3.1) *are satisfied and*

$$\boldsymbol{f} \in L^\infty(0,\infty; L^2(\Omega)^m), \qquad g \in W^{1,\infty}(0,\infty; L^\infty(\Omega)),\ g \geq \nu > 0,$$

$$\boldsymbol{f}_\infty \in L^2(\Omega)^m, \qquad g_\infty \in L^\infty_\nu(\Omega),$$

$$\int_{\frac{t}{2}}^{t} \xi^{p'}(\tau)d\tau \underset{t\to\infty}{\longrightarrow} 0, \quad \textit{if } p > 2 \quad \textit{and} \quad \int_t^{t+1} \xi^2(\tau)d\tau \underset{t\to\infty}{\longrightarrow} 0 \quad \textit{if } 1 < p \leq 2,$$

where

$$\xi(t) = \|\boldsymbol{f}(t) - \boldsymbol{f}_\infty\|_{L^2(\Omega)^m}. \tag{3.12}$$

Assume, in addition, that there exist D and γ positive such that

$$\|g(t) - g_\infty\|_{L^\infty(\Omega)} \leq \frac{D}{t^\gamma}, \qquad \text{where } \gamma > \begin{cases} \frac{3}{2} & \text{if } p > 2 \\ \frac{1}{2} & \text{if } 1 < p \leq 2. \end{cases} \tag{3.13}$$

Then, for $\delta > 0$ and $\boldsymbol{u}^\delta$ the solution of the variational inequality (3.3), *with $t \in [0,\infty)$,*

$$\|\boldsymbol{u}^\delta(t) - \boldsymbol{u}^\delta_\infty\|_{L^2(\Omega)^m} \underset{t\to\infty}{\longrightarrow} 0.$$

In the special case of (3.3) with $\delta = 0$ and $g(t) = g$ for al $t \geq T^*$, observing that we can apply a result of Brézis [19, Theorem 3.11] to extend the Theorem 3.4 of [30]), in which $g \equiv 1$, and obtain the following asymptotic behaviour of the solution $\boldsymbol{u}(t) \in \mathbb{K}_g$ with $\boldsymbol{u}(0) = \boldsymbol{u}_0$ of

$$\int_\Omega \partial_t \boldsymbol{u}(t) \cdot (\boldsymbol{v} - \boldsymbol{u}(t)) \geq \int_\Omega \boldsymbol{f}(t) \cdot (\boldsymbol{v} - \boldsymbol{u}(t)) \quad \forall \boldsymbol{v} \in \mathbb{K}_g, \tag{3.14}$$

which corresponds, in the scalar case, to the sandpile problem with space variable slope.

Theorem 3.11 *Suppose that the assumptions* (2.1), (2.2), (2.3) *and* (3.1) *are satisfied,* $f \in L^1_{loc}(0,\infty; L^2(\Omega)^m)$, $g \in L^\infty_\nu(\Omega)$, $\boldsymbol{u}_0 \in \mathbb{K}_g$ *and let* $\boldsymbol{u}$ *be the solution of the variational inequality* (3.14). *If there exists a function* $\boldsymbol{f}_\infty$ *such that* $\boldsymbol{f} - \boldsymbol{f}_\infty \in L^1\big(0,\infty; L^2(\Omega)^m\big)$ *then*

$$\boldsymbol{u}(t) \underset{t\to\infty}{\longrightarrow} \boldsymbol{u}_\infty \quad \text{in } L^2(\Omega)^m,$$

where $\boldsymbol{u}_\infty$ *solves the variational inequality* (2.18) *with* $\boldsymbol{f}_\infty$.

3.2 Equivalent formulations when L=∇

In this section, we summarize the main results of [84], assuming $\partial\Omega$ is of class $\mathscr{C}^2$, $p = 2$ and $\mathrm{L} = \nabla$ and considering the strong variational inequality (2.12) in this special case,

$$\begin{cases} u(t) \in \mathbb{K}_{g(t)},\ u(0) = u_0, \\ \displaystyle\int_\Omega \partial_t u(t)\cdot(v - u(t)) + \int_\Omega \nabla u(t)\cdot\nabla(v - u(t)) \geq \int_\Omega f(t)\cdot(v - u(t)), \\ \qquad\qquad \forall\, v \in \mathbb{K}_{g(t)} \text{ for a.e. } t \in (0,T), \end{cases} \tag{3.15}$$

where

$$\mathbb{K}_{g(t)} = \big\{v \in H^1_0(\Omega) : |\nabla v| \leq g(t)\big\}.$$

As in the stationary case, we can consider three related problems. The first one is the Lagrange multiplier problem

$$\begin{gathered} \int_{Q_T} \partial_t u\varphi + \langle \lambda\nabla u, \nabla\varphi\rangle_{(L^\infty(Q_T)' \times L^\infty(Q_T)} = \int_{Q_T} f\varphi, \quad \forall \varphi \in L^\infty\big(0,T; W^{1,\infty}_0(\Omega)\big), \\ \lambda \geq 1, \quad (\lambda - 1)(|\nabla u| - g) = 0 \quad \text{in } \big(L^\infty(Q_T)\big)', \\ u(0) = u_0, \text{ a.e. in } \Omega \quad |\nabla u| \leq g \text{ a.e. in } Q_T, \end{gathered} \tag{3.16}$$

which is equivalent to the variational inequality (3.15). This was first proved in [83] in the case $g \equiv 1$, where it was shown the existence of $\lambda \in L^\infty(Q_T)$ satisfying (3.16), in the case of a compatible and smooth nonhomogeneous boundary condition for u. When $u_{|\partial\Omega\times(0,T)}$ is independent of $x \in \partial\Omega$ then, by Theorem 3.11 of [83], λ is unique. In this framework, it was also shown in [83] that the solution $u \in L^p\big(0,T; W^{2,p}_{loc}(\Omega)\big) \cap \mathscr{C}^{1+\alpha,\alpha/2}(Q_T)$ for all $1 \leq p < \infty$ and $0 \leq \alpha < 1$.

Secondly, we define two obstacles as in (2.27) and (2.28) using the pseudometric $d_{g(t)}$ introduced in (2.26),

$$\overline{\varphi}(x,t) = d_{g(t)}(x,\partial\Omega) = \bigvee\{w(x) : w \in \mathbb{K}_{g(t)}\} \tag{3.17}$$

and

$$\underline{\varphi}(x,t) = d_{g(t)}(x, \partial\Omega) = \bigwedge\{w(x) : w \in \mathbb{K}_{g(t)}\}, \tag{3.18}$$

where the variable $\mathbb{K}^{\overline{\varphi}(t)}_{\underline{\varphi}(t)}$ is defined by (2.23) for each $t \in [0,T]$ and we consider the double obstacle variational inequality

$$\begin{cases} u(t) \in \mathbb{K}^{\overline{\varphi}(t)}_{\underline{\varphi}(t)}, \; u(0) = u_0, \\ \displaystyle\int_\Omega \partial_t u(t) \cdot (v - u(t)) + \int_\Omega \nabla u(t) \cdot \nabla(v - u(t)) \geq \int_\Omega f(t) \cdot (v - u(t)), \\ \qquad\qquad \forall\, v \in \mathbb{K}^{\overline{\varphi}(t)}_{\underline{\varphi}(t)} \quad \text{for a.e. } t \in (0,T). \end{cases} \tag{3.19}$$

The third and last problem is the following complementary problem

$$\begin{aligned} &(\partial_t u - \Delta u - f) \vee (|\nabla u| - g) = 0 \quad \text{in } Q_T, \\ &u(0) = u_0 \quad \text{in } \Omega, \quad u = 0 \quad \text{on } \partial\Omega \times (0,T). \end{aligned} \tag{3.20}$$

In [88], Zhu studied a more general problem in unbounded domains, for large times, with a zero condition at a fixed instant T, motivated by stochastic control.

These different formulations of gradient constraint problems are not always equivalent and were studied in [84], where sufficient conditions were given for the equivalence of each one with (3.15).

Assume that

$$\begin{aligned} &g \in W^{1,\infty}\big(0,T; L^\infty(\Omega)\big) \cap L^\infty\big(0,T; \mathscr{C}^2(\overline{\Omega})\big), \quad g \geq \nu > 0, \\ &|\nabla w_0| \leq g(0), \quad f \in L^\infty(Q_T). \end{aligned} \tag{3.21}$$

The first result holds with an additional assumption on the gradient constraint g, which is, of course, satisfied in the case of $g \equiv$constant> 0, by combining Theorem 3.9 of [84] and Theorem 3.11 of [83].

Theorem 3.12 *Under the assumptions* (3.21), *with* $f \in L^\infty(0,T)$ *and*

$$\partial_t(g^2) \geq 0, \qquad -\Delta(g^2) \geq 0, \tag{3.22}$$

problem (3.16) *has a solution* $(\lambda, u) \in L^\infty(Q_T) \times L^\infty\big(0,T; W^{1,\infty}_0(\Omega) \cap H^2_{loc}(\Omega)\big)$. *Besides, u is the unique solution of* (3.15) *and if g is constant then λ is unique.*

The equivalence with the double obstacle problem holds with a slightly weaker assumption on g.

Theorem 3.13 *Assuming* (3.21), *problem* (3.19) *has a unique solution. If* $f \in L^\infty(0,T)$ *and*

$$\partial_t(g^2) - \Delta(g^2) \geq 0, \tag{3.23}$$

then problem (3.19) *is equivalent to problem* (3.15).

Finally, the sufficient conditions for the equivalence of the complementary problem (3.20) and the gradient constraint scalar problem (3.15) require stronger assumptions on the data.

Theorem 3.14 *Suppose that* $f \in W^{1,\infty}\big(0,T;L^\infty(\Omega)\big)$, $w_0 \in H^1_0(\Omega)$, *and*

$$\Delta u_0 \in L^\infty(\Omega),\ -\Delta u_0 \leq f \text{ a.e. in } Q_T,$$
$$g \in W^{1,\infty}\big(0,T;L^\infty(\Omega)\big)\ g \geq \nu > 0 \text{ and } \partial_t(g^2) \leq 0.$$

Then problem (3.20) *has a unique solution. If, in addition,* $g = g(x)$ *and* $\Delta g^2 \leq 0$ *then this problem is equivalent to problem* (3.15).

The counterexample given at the end Section 2.3, concerning the non-equivalence among these problems, can be generalized easily for the evolutionary case, as we have stabilization in time to the stationary solution (see [84]).

3.3 The scalar quasi-variational inequality with gradient constraint

In [78], Rodrigues and Santos proved existence of solution for a quasi-variational inequality with gradient constraint for first order quasilinear equations ($\delta = 0$), extending the previous results for parabolic equations of [77].

For $\boldsymbol{\Phi} = \boldsymbol{\Phi}(x,t,u) : \overline{Q_T} \times \mathbb{R} \to \mathbb{R}^d$, $F = F(x,t,u) : \overline{Q_T} \times \mathbb{R} \to \mathbb{R}$ assume that

$$\boldsymbol{\Phi} \in W^{2,\infty}\big(Q_T \times (-R,R)\big)^d, \qquad F \in W^{1,\infty}\big(Q_T \times (-R,R)\big). \tag{3.24}$$

In addition, $\nabla \cdot \boldsymbol{\Phi}$ and F satisfy the growth condition in the variable u

$$\big|\big(\nabla \cdot \boldsymbol{\Phi}\big)(x,t,u) + F(x,t,u)\big| \leq c_1|u| + c_2, \tag{3.25}$$

uniformly in (x,t), for all $u \in \mathbb{R}$ and a.e. (x,t), being c_1 and c_2 positive constants. The gradient constraint $G = G(x,u) : \Omega \times \mathbb{R} \to \mathbb{R}$ is bounded in x and continuous in u and the initial condition $u_0 : \Omega \to \mathbb{R}$ are such that

$$G \in \mathscr{C}\big(\mathbb{R}; L^\infty_\nu(\Omega)\big), \qquad u_0 \in \mathbb{K}_{G(u_0)} \cap \mathscr{C}(\overline{\Omega}), \qquad \delta\Delta_p u_0 \in M(\Omega), \tag{3.26}$$

being

$$\mathbb{K}_{G(u(t))} = \big\{w \in H^1_0(\Omega) : |\nabla w| \leq G(u(t))\big\}$$

and $M(\Omega)$ denotes the space of bounded measures in Ω.

Theorem 3.15 *Assuming* (3.24), (3.25) *and* (3.26), *for each* $\delta \geq 0$ *and any* $1 < p < \infty$, *the quasi-variational inequality*

$$\begin{cases} u(t) \in \mathbb{K}_{G(u(t))} \textit{ for a.e. } t \in (0,T), u(0) = u_0, \\ \langle \partial_t u(t), w - u \rangle_{M(\Omega)\times\mathscr{C}(\overline{\Omega})} + \int_\Omega \big(\delta \nabla_p u(t) + \mathbf{\Phi}(u(t))\big) \cdot \nabla(w(t) - u(t)) \\ \qquad \geq \int_\Omega F(u(t))(w - u(t)) \qquad \forall w \in \mathbb{K}_{G(u(t))}, \textit{ for a.e. } t \in (0,T), \end{cases} \tag{3.27}$$

has a solution $u \in L^\infty\big(0,T; W_0^{1,\infty}(\Omega)\big) \cap \mathscr{C}(\overline{Q_T})$ *such that* $\partial_t u \in L^\infty\big(0,T; M(\Omega)\big)$.

Although this result was proved in [78] for $\delta = 0$ and only in the case $p = 2$ for $\delta > 0$, it can be proved for $p \neq 2$ exactly in the same way as in the previous framework of [77], which corresponds to (3.27) when $\mathbf{\Phi} \equiv \mathbf{0}$, with $G(x,u) = G(u)$ and $F(x,t,u) = f(x,t)$, with only $f \in L^\infty(Q_T)$ and $\partial_t f \in M(\Omega)$.

We may consider the corresponding stationary quasi-variational inequality for $u_\infty \in \mathbb{K}_{G[u_\infty]}$, such that

$$\int_\Omega \big(\delta \nabla_p u_\infty + \mathbf{\Phi}_\infty(u_\infty)\big) \cdot \nabla(w - u_\infty) \geq \int_\Omega F_\infty(u_\infty)(w - u_\infty) \qquad \forall w \in \mathbb{K}_{G[u_\infty]} \tag{3.28}$$

for given functions $F_\infty = F_\infty(x,u) : \overline{\Omega} \times \mathbb{R}$ and $\mathbf{\Phi}_\infty = \mathbf{\Phi}_\infty(x,u) : \Omega \times \mathbb{R} \to \mathbb{R}^d$, continuous in u and bounded in x for all $|u| \leq R$. In order to extend the asymptotic stabilization in time (for subsequences $t_n \to \infty$) obtained in [77] and [78], we shall assume that (3.24) holds for $T = \infty$,

$$\mathbf{\Phi}(t) = \mathbf{\Phi}_\infty, \qquad \text{and} \qquad \partial_u F \leq -\mu < 0 \ \text{ for all } t > 0, \tag{3.29}$$

and

$$0 < \nu \leq G(x,u) \leq N, \ \text{ for a.e. } x \in \Omega \quad \text{and all } u \in \mathbb{R}, \tag{3.30}$$

or there exists $M > 0$ such that, for all $R \geq M$

$$\nabla \cdot \mathbf{\Phi}(x,R) + F(x,t,R) \leq 0, \qquad \nabla \cdot \mathbf{\Phi}(x,-R) + F(x,t,-R) \geq 0. \tag{3.31}$$

Setting $\xi_R(t) = \displaystyle\int_\Omega \sup_{|u|\leq R} \big|\partial_t F(x,t,u)\big| dx$ and supposing that, for $R \geq R_0$ and some constant $C_R > 0$, we have

$$\sup_{0<t<\infty} \int_t^{t+1} \xi_R(\tau)d\tau \leq C_R, \qquad \textit{and} \qquad \int_t^{t+1} \xi_R(\tau)d\tau \underset{t\to\infty}{\longrightarrow} 0, \tag{3.32}$$

and

$$F(x,t,u) \underset{t\to\infty}{\longrightarrow} F_\infty(x,u) \ \text{ for all } |u| \leq R \ \text{ and a.e. } x \in \Omega, \tag{3.33}$$

we may prove the following result

Theorem 3.16 *For any* $\delta \geq 0$ *and* $1 < p < \infty$, *under the assumptions* (3.24)-(3.26) *and* (3.29)-(3.33), *problem* (3.28) *has a solution* $u_\infty \in \mathbb{K}_{G[u_\infty]}$ *which is the weak-$*$ limit in* $W_0^{1,\infty}(\Omega)$ *and strong limit in* $\mathscr{C}^\alpha(\overline{\Omega})$, $0 \leq \alpha < 1$, *for some* $t_n \to \infty$ *of a sequence* $\{u(t_n)\}_n$ *with* u *being a global solution of* (3.27).

We observe that the degenerate case $\delta = 0$ corresponds to a nonlinear conservation law for which we could also consider formally the Lagrange multiplier problem of finding $\lambda = \lambda(x,t)$ associated with the constraint $|\nabla u| \le G[u]$ and such that

$$\lambda \ge 0, \qquad \lambda(G[u] - |\nabla u|) = 0$$

and

$$\partial_t u - \nabla \cdot \Phi(u) - \nabla \cdot (\lambda u) = F(u), \text{ in } Q_T,$$

which is an open problem. However, in contrast with conservation laws without the gradient restriction, this problem has no spatial shock fronts nor boundary layers for the vanishing viscosity limit, since Dirichlet data may be prescribed for u on the whole boundary $\partial\Omega$.

It is clear that both results of Theorems 3.15 and 3.16 apply, in particular, to the linear transport equation

$$\partial_t u + \boldsymbol{b} \cdot \nabla u + cu = f$$

for a given vector field $\boldsymbol{b} \in W^{2,\infty}(Q_T)^d$ and given functions $c = c(x,t)$ and $f = f(x,t)$ in $W^{1,\infty}(Q_T)$, corresponding to set

$$\boldsymbol{\Phi}(u) = -\boldsymbol{b}u \quad \text{and} \quad F(u) = f - (c + \nabla \cdot \boldsymbol{b})u. \tag{3.34}$$

Nevertheless, in this case, if we analyse the a priori estimates for the approximating problem in the proof of Theorem 3.15 as in Section 3.1 of [78], the assumptions on the coefficients of the linear transport operator can be significantly weakened and it is possible to prove the following result

Corollary 3.1 *If* $\boldsymbol{\Phi}$ *and* F *are given by* (3.34) *with* $\boldsymbol{b} \in L^\infty(Q_T)^d$, $\nabla \cdot \boldsymbol{b}$, c, $f \in L^\infty(Q_T)$, $\partial_t \boldsymbol{b} \in L^r(Q_T)^d$, $r > 1$ *and* $\partial_t c$, $\partial_t f \in L^1(Q_T)$, *assuming* (3.26) *for* $\delta \ge 0$ *and* $1 < p < \infty$, *the quasi-variational inequality* (3.27) *with linear lower order terms has a strong solution* $\boldsymbol{u} \in L^\infty(0,T; W_0^{1,\infty}(\Omega)) \cap \mathscr{C}(\overline{Q}_T)$ *such that* $\partial_t u \in L^1(0,T; M(\Omega))$.

However, in this case, for the corresponding variational inequality, i.e. when $G \equiv g(x,t)$, in [79] it was shown that the problem is well-posed and has similar stability properties, as in Section 3.1, with coefficients only in L^2.

Suppose that, for some $l \in \mathbb{R}$,

$$\boldsymbol{b} \in L^2(Q_T)^d, \quad c \in L^2(Q_T) \text{ and } c - \frac{1}{2}\nabla \cdot \boldsymbol{b} \ge l \text{ in } Q_T, \tag{3.35}$$

and

$$f \in L^2(Q_T), \quad g \in W^{1,\infty}(0,T; L^\infty(\Omega)), \; g \ge \nu > 0 \text{ and } u_0 \in \mathbb{K}_{g(0)}. \tag{3.36}$$

Theorem 3.17 [79] *With the assumptions* (3.35) *and* (3.36), *there exists a unique strong solution*

$$w \in L^\infty(0,T; W_0^{1,\infty}(\Omega)) \cap \mathscr{C}(\overline{Q}_T), \qquad \partial_t w \in L^2(Q_T),$$

to the variational inequality

$$\begin{cases} w(t) \in \mathbb{K}_{g(t)},\ t \in (0,T), \quad w(0) = u_0, \\ \displaystyle\int_\Omega \big(\partial_t w(t) + \boldsymbol{b}(t) \cdot \nabla w(t) + c(t)w(t)\big)(v - w(t)) \\ \qquad \geq \displaystyle\int_\Omega f(t)(v - w(t)), \quad \forall v \in \mathbb{K}_{g(t)},\ \textit{for a.e. } t \in (0,T). \end{cases} \tag{3.37}$$

The corresponding stationary problem for

$$w_\infty \in \mathbb{K}_{g_\infty} : \quad \int_\Omega (\boldsymbol{b}_\infty \cdot \nabla u_\infty + c_\infty w_\infty)(v - w_\infty) \geq \int_\Omega f_\infty (v - w_\infty) \quad \forall v \in \mathbb{K}_{g_\infty} \tag{3.38}$$

can be solved uniquely for L^1 data

$$\boldsymbol{b}_\infty \in L^1(\Omega)^d, \quad c_\infty \in L^1(\Omega) \text{ and } c_\infty - \frac{1}{2}\nabla \cdot \boldsymbol{b}_\infty \geq \mu \text{ in } \Omega, \tag{3.39}$$

with

$$f_\infty \in L^1(\Omega), \quad g_\infty \in L^\infty(\Omega),\ g_\infty \geq \nu > 0. \tag{3.40}$$

and is the asymptotic limit of the solution of (3.37).

Theorem 3.18 [79] *Under the assumptions* (3.39) *and* (3.40), *if*

$$\int_t^{t+1} \int_\Omega \big(|f(\tau) - f_\infty| + |\boldsymbol{b}(\tau) - \boldsymbol{b}_\infty| + |c(\tau) - c_\infty|\big) dx d\tau \underset{t \to \infty}{\longrightarrow} 0$$

and there exists $\gamma > \frac{1}{2}$ *such that, for some constant* $C > 0$,

$$\|g(t) - g_\infty\|_{L^\infty(\Omega)} \leq \frac{C}{t^\gamma}, \quad t > 0,$$

then

$$w(t) \underset{t \to \infty}{\longrightarrow} w_\infty \quad n\ L^2(\Omega)$$

where w *and* w_∞ *are, respectively, the solutions of the variational inequality* (3.37) *and* (3.38).

3.4 The quasi-variational inequality via compactness and monotonicity

The results in Section 3.3 are for scalar functions and $L = \nabla$. As the arguments in the proof that $\partial_t u$ is a Radon measure do not apply to the vector cases, we consider the weak quasi-variational inequality for a given $\delta \geq 0$, for $\boldsymbol{u} = \boldsymbol{u}^\delta$,

$$\begin{cases} \boldsymbol{u} \in \mathbb{K}_{G[\boldsymbol{u}]}, \\ \displaystyle\int_0^T \langle \partial_t \boldsymbol{v}, \boldsymbol{v} - \boldsymbol{u} \rangle_p + \delta \int_{Q_T} \text{Ł}_p \boldsymbol{u} \cdot \mathrm{L}(\boldsymbol{v} - \boldsymbol{u}) \geq \int_{Q_T} \boldsymbol{f} \cdot (\boldsymbol{v} - \boldsymbol{u}) \\ \qquad\qquad - \dfrac{1}{2} \displaystyle\int_\Omega |\boldsymbol{v}(0) - \boldsymbol{u}_0|^2, \forall\, \boldsymbol{v} \in \mathscr{Y}_p \text{ such that } \boldsymbol{v} \in \mathbb{K}_{G[\boldsymbol{u}]}, \end{cases} \tag{3.41}$$

where $\langle \cdot, \cdot \rangle_p$ denotes the duality pairing between $\mathbb{X}_p' \times \mathbb{X}_p$.

Theorem 3.19 *Suppose that assumptions* (2.1), (2.2), (2.3), (3.1) *are satisfied and* $\boldsymbol{f} \in L^2(Q_T)^m$, $\boldsymbol{u}_0 \in \mathbb{K}_{G(\boldsymbol{u}_0)}$. *Assume, in addition that* $G : \mathscr{H} \to L^1(Q_T)$ *is a nonlinear continuous functional whose restriction to* $\mathscr{V}_p$ *is compact with values in* $\mathscr{C}\big([0,T]; L^\infty(\Omega)\big)$ *and* $G(\mathscr{H}) \subset L^\infty_\nu(Q_T)$ *for some* $\nu > 0$.

Then the quasi-variational inequality (3.41) *has a weak solution*

$$\boldsymbol{u} \in \mathscr{V}_p \cap L^\infty\big(0,T; L^2(\Omega)^m\big).$$

Proof We give a brief idea of the proof. The details can be found, in a more general setting, in [66].

Assuming first $\delta > 0$, we consider the following family of approximating problems, defined for fixed $\boldsymbol{\varphi} \in \mathscr{H}$, such that $\boldsymbol{u}_0 \in \mathbb{K}_{G[\boldsymbol{\varphi}(0)]}$: to find $\boldsymbol{u}_{\varepsilon,\varphi}$ such that $\boldsymbol{u}_{\varepsilon,\varphi}(0) = \boldsymbol{u}_0$ and

$$\langle \partial_t \boldsymbol{u}_{\varepsilon,\varphi}(t), \boldsymbol{\psi} \rangle_p + \int_\Omega \big(\delta + k_\varepsilon\big(|\mathrm{L}\boldsymbol{u}_{\varepsilon,\varphi}(t)| - G[\boldsymbol{\varphi}](t)\big)\big) \text{Ł}_p \boldsymbol{u}_{\varepsilon,\varphi}(t) \cdot \mathrm{L}\boldsymbol{\psi}$$
$$= \int_\Omega \boldsymbol{f}(t) \cdot \boldsymbol{\psi}, \qquad \forall \boldsymbol{\psi} \in \mathbb{X}_p, \quad \text{for a.e. } t \in (0,T), \tag{3.42}$$

where $k_\varepsilon : \mathbb{R} \to \mathbb{R}$ is an increasing continuous function such that

$$k_\varepsilon(s) = 0 \text{ if } s \leq 0, \qquad k_\varepsilon(s) = e^{\frac{s}{\varepsilon}} - 1 \text{ if } 0 \leq s \leq \tfrac{1}{\varepsilon}, \qquad k_\varepsilon(s) = e^{\frac{1}{\varepsilon^2}} - 1 \text{ if } s \geq \tfrac{1}{\varepsilon}.$$

This problem has a unique solution $\boldsymbol{u}_{\varepsilon,\varphi} \in \mathscr{V}_p$, with $\partial_t \boldsymbol{u}_{\varepsilon,\varphi} \in \mathscr{V}_p'$. Let $S : \mathscr{H} \to \mathscr{Y}_p$ be the mapping that assigns to each $\boldsymbol{\varphi} \in \mathscr{H}$ the unique solution $\boldsymbol{u}_{\varepsilon,\varphi}$ of problem (3.42). Considering the embedding $i : \mathscr{Y}_p \to \mathscr{H}$, then $i \circ S$ is continuous, compact and we have a priori estimates which assures that there exists a positive R, independent of ε, such that $i \circ S(\mathscr{H}) \subset D_R$, where $D_R = \big\{\boldsymbol{w} \in \mathscr{H} : \|\boldsymbol{w}\|_{\mathscr{H}} \leq R\big\}$. By Schauder's fixed point theorem, $i \circ S$ has a fixed point $\boldsymbol{u}_\varepsilon$, which solves problem (3.42) with $\boldsymbol{\varphi}$ replaced by $\boldsymbol{u}_\varepsilon$.

The sequence $\{\boldsymbol{u}_\varepsilon\}_\varepsilon$ satisfies a priori estimates which allow us to obtain the limit $\boldsymbol{u}$ for subsequences in $\mathscr{V}_p \cap L^\infty\big(0,T; L^2(\Omega)^m\big)$. Another main estimate

$$\|k_\varepsilon(|\mathrm{L}\boldsymbol{u}_\varepsilon| - G[\boldsymbol{u}_\varepsilon])\|_{L^1(Q_T)} \leq C,$$

with C a constant independent of ε, yields $\boldsymbol{u} \in K_{G[\boldsymbol{u}]}$.

Using $\boldsymbol{u}_\varepsilon - \boldsymbol{v}$ as test function in (3.42) corresponding to a fixed point $\boldsymbol{\varphi} = \boldsymbol{u}_\varepsilon$, with an arbitrary $\boldsymbol{v} \in \mathscr{V}_p \cap \mathbb{K}_{G[\boldsymbol{u}]}$, we obtain, after integration in $t \in (0,T)$ and setting

$k_\varepsilon = k_\varepsilon\big(|\mathrm{L}\boldsymbol{u}_\varepsilon| - G[\boldsymbol{u}_\varepsilon]\big)$:

$$\delta \int_{Q_T} Ł_p \boldsymbol{u}_\varepsilon \cdot \mathrm{L}(\boldsymbol{u}_\varepsilon - \boldsymbol{v}) \le \frac{1}{2}\int_\Omega |\boldsymbol{u}_0 - \boldsymbol{v}(0)|^2 + \int_0^T \langle \partial_t \boldsymbol{v}, \boldsymbol{v} - \boldsymbol{u}_\varepsilon\rangle_p + \int_{Q_T} k_\varepsilon Ł_p \boldsymbol{u}_\varepsilon \cdot \mathrm{L}(\boldsymbol{u}_\varepsilon - \boldsymbol{v}) - \int_{Q_T} \boldsymbol{f}\cdot(\boldsymbol{v} - \boldsymbol{u}_\varepsilon). \quad (3.43)$$

The passage to the limit $\boldsymbol{u}_\varepsilon \underset{\varepsilon\to 0}{\longrightarrow} \boldsymbol{u}$ in order to conclude that (3.41) holds for $\boldsymbol{u}$ is delicate and requires a new lemma, which proof can be found in [66]: given $\boldsymbol{w} \in \mathscr{V}_p$ such that $\boldsymbol{w} \in \mathbb{K}_{G[\boldsymbol{w}]}$ and $\boldsymbol{z} \in \mathbb{K}_{G[\boldsymbol{w}(0)]}$, we may construct a regularizing sequence $\{\boldsymbol{w}_n\}_n$ and a sequence of scalar functions $\{G_n\}_n$ satisfying i) $\boldsymbol{w}_n \in L^\infty(0,T;\mathbb{X}_p)$ and $\partial_t \boldsymbol{w}_n \in L^\infty(0,T;\mathbb{X}_p)$, ii) $\boldsymbol{w}_n \underset{n}{\longrightarrow} \boldsymbol{w}$ in $\mathscr{V}_p$ strongly, iii) $\overline{\lim}_n \int_0^T \langle \partial_t \boldsymbol{w}_n, \boldsymbol{w}_n - \boldsymbol{w}\rangle_p \le 0$ and iv) $|\mathrm{L}\boldsymbol{w}_n| \le G_n$, where $G_n \in \mathscr{C}\big([0,T];L^\infty(\Omega)\big)$ and $G_n \underset{n}{\longrightarrow} G[\boldsymbol{w}]$ in $\mathscr{C}\big([0,T];L^\infty(\Omega)\big)$.

If $\{\boldsymbol{u}_n\}_n$ is a regularizing sequence associated to $\boldsymbol{u}$ and $G[\boldsymbol{u}]$ then there exists a constant C independent of ε and n such that

$$\int_{Q_T} k_\varepsilon Ł_p \boldsymbol{u}_\varepsilon \cdot \mathrm{L}(\boldsymbol{u}_n - \boldsymbol{u}_\varepsilon) \le \int_{Q_T} k_\varepsilon |\mathrm{L}\boldsymbol{u}_\varepsilon|^{p-1}\big(|\mathrm{L}\boldsymbol{u}_n| - |\mathrm{L}\boldsymbol{u}_\varepsilon|\big) \le C\|G_n - G[\boldsymbol{u}_\varepsilon]\|_{L^\infty(Q_T)} \underset{n}{\longrightarrow} 0,$$

by the compactness of the operator G. For all $n \in \mathbb{N}$ we have, setting $\boldsymbol{v} = \boldsymbol{u}_\varepsilon$,

$$\int_{Q_T} \delta Ł_p \boldsymbol{u}_\varepsilon \cdot \mathrm{L}(\boldsymbol{u}_\varepsilon - \boldsymbol{u}_n) \le \int_0^T \langle \partial_t \boldsymbol{u}_n, \boldsymbol{u}_n - \boldsymbol{u}\rangle_p + \int_{Q_T} \delta Ł_p \boldsymbol{u}_\varepsilon \cdot \mathrm{L}(\boldsymbol{u}_n - \boldsymbol{u}) + \int_{Q_T} k_\varepsilon Ł_p \boldsymbol{u}_\varepsilon \cdot \mathrm{L}(\boldsymbol{u}_n - \boldsymbol{u}_\varepsilon) - \int_{Q_T} \boldsymbol{f}\cdot(\boldsymbol{u}_n - \boldsymbol{u}),$$

concluding that

$$\overline{\lim_{\varepsilon\to 0}} \int_{Q_T} \delta Ł_p \boldsymbol{u}_\varepsilon \cdot \mathrm{L}(\boldsymbol{u}_\varepsilon - \boldsymbol{u}) \le 0.$$

This operator is bounded, monotone and hemicontinuous and so it is pseudo-monotone and we get, using (3.43),

$$\int_{Q_T} \delta Ł_p \boldsymbol{u} \cdot \mathrm{L}(\boldsymbol{u} - \boldsymbol{v}) \le \underline{\lim_{\varepsilon\to 0}} \int_{Q_T} \delta Ł_p \boldsymbol{u}_\varepsilon \cdot \mathrm{L}(\boldsymbol{u}_\varepsilon - \boldsymbol{v}), \quad \forall \boldsymbol{v} \in \mathbb{K}_{G[\boldsymbol{u}]}$$

and the proof that $\boldsymbol{u}$ solves the quasi-variational inequality (3.41) is now easy, by using the well-known monotonicity methods (see [17] or [58]).

The proof for the case $\delta = 0$ is more delicate and requires taking the limit of diagonal subsequences of solutions $\{(\varepsilon,\delta)\}_{\varepsilon,\delta}$ of (3.42) as $\varepsilon \to 0$ and as $\delta \to 0$, in order to use the monotonicity methods to obtain a solution of (3.43) in the degenerate case. □

Remark 5 Two general examples for the compact operator $G : \mathscr{V}_p \to \mathscr{C}\big([0,T]; L^\infty_\nu(\Omega)\big)$ in the form $G[\boldsymbol{v}] = g(x,t,\zeta(\boldsymbol{v}(x,t)))$, with $g \in \mathscr{C}(\overline{Q}_T \times \mathbb{R}^m)$, $g \geq \nu > 0$, were given in [66], namely with

$$\zeta(\boldsymbol{v})(x,t) = \int_0^t \boldsymbol{v}(x,s)K(t,s)ds, \qquad (x,t) \in \overline{Q}_T,$$

with K, $\partial_t K \in L^\infty\big((0,T) \times (0,T)\big)$, or with $\zeta = \zeta(\boldsymbol{v})$ given by the unique solution of the Cauchy-Dirichlet problem of a quasilinear parabolic scalar equation $\partial_t \zeta - \nabla \cdot a\big(x,t,\nabla\zeta)\big) = \varphi_0 + \boldsymbol{\psi} \cdot \boldsymbol{v} + \boldsymbol{\eta} \cdot \mathrm{L}\boldsymbol{v} \in L^p(Q_T)$, which has solutions in the Hölder space $\mathscr{C}^\lambda(\overline{Q}_T)$, for some $0 < \lambda < 1$, provided that $\boldsymbol{v} \in \mathscr{V}_p$, $p > \frac{d+2}{d}$ and $\varphi_0 \in L^p(Q_T)$, $\boldsymbol{\psi} \in L^\infty(Q_T)^m$, $\boldsymbol{\eta} \in L^\infty(Q_T)^\ell$ are given.

Remark 6 Using the sub-differential analysis in Hilbert spaces, Kenmochi and co-workers have also obtained existence results in [49] and [51] for evolutionary quasi-variational inequalities with gradient constraints under different assumptions.

3.5 The quasi-variational solution via contraction

For the evolutionary quasi-variational inequalities and for nonlocal Lipschitz non-linearities we can apply the Banach fixed point theorem in two different functional settings obtaining weak and strong solutions under certain conditions.

Let E be $L^2(Q_T)^m$ or $\mathscr{V}_p$ and

$$D_R = \{\boldsymbol{v} \in E : \|\boldsymbol{v}\|_E \leq R\}.$$

For $\eta, M, \Gamma : \mathbb{R} \to \mathbb{R}^+$ increasing functions, let $\gamma : E \to \mathbb{R}^+$ be a functional satisfying

$$\begin{aligned} &0 < \eta(R_*) \leq \gamma(\boldsymbol{u}) \leq M(R_*) \quad \forall\, \boldsymbol{u} \in D_{R_*}, \\ &|\gamma(\boldsymbol{u}_1) - \gamma(\boldsymbol{u}_2)| \leq \Gamma(R_*)\|u_1 - u_2\|_E \quad \forall\, \boldsymbol{u}_1, \boldsymbol{u}_2 \in D_{R_*}, \end{aligned} \tag{3.44}$$

for a sufficiently large $R_* \in \mathbb{R}^+$.

Theorem 3.20 *For* $p > 1$ *and* $\delta \geq 0$, *suppose that the assumptions* (2.1), (2.2), (2.3) *and* (3.1) *are satisfied,* $\boldsymbol{f} \in L^2(Q_T)^m$,

$$G[\boldsymbol{u}](x,t) = \gamma(\boldsymbol{u})\varphi(x,t), \quad (x,t) \in Q_T,$$

where $E = L^2(Q_T)^m$ *and* γ *is a functional satisfying* (3.44), $\varphi \in \mathscr{C}\big([0,T]; L^\infty_\nu(\Omega)\big)$, $\boldsymbol{u}_0 \in \mathbb{K}_{G[\boldsymbol{u}_0]}$ *and*

$$R_* = \sqrt{T + T^2 e^T}\,\big(\|\boldsymbol{f}\|_{L^2(Q_T)^m} + \|\boldsymbol{u}_0\|_{L^2(\Omega)^m}\big). \tag{3.45}$$

If

$$2\,R_*\Gamma(R_*) < \eta(R_*)$$

then the quasi-variational inequality (3.41) *has a unique weak solution* $\boldsymbol{u} \in \mathscr{V}_p \cap \mathscr{C}\big([0,T]; L^2(\Omega)^m\big)$*, which is also a strong solution* $\boldsymbol{u} \in \mathscr{V}_p \cap H^1\big(0,T; L^2(\Omega)^m\big)$*, provided* $\varphi \in W^{1,\infty}(0,T; L^\infty(\Omega))$ *with* $\varphi \geq \nu > 0$.

Proof For any $R > 0$ let $S : D_R \to L^2(Q_T)^m$ be the mapping that, by Theorem 3.8, assigns to each $\boldsymbol{v} \in D_R$ the unique solution of the variational inequality (3.2) (respectively (3.3)) with data $\boldsymbol{f}$, $G[\boldsymbol{v}]$ and $\boldsymbol{u}_0$. Denoting $\boldsymbol{u} = S(\boldsymbol{v}) = S(\boldsymbol{f}, G[\boldsymbol{v}], \boldsymbol{u}_0)$, using the stability result (3.6) with $\boldsymbol{u}_1 = \boldsymbol{u}$ and $\boldsymbol{u}_2 = 0$ we have the estimate

$$\begin{aligned}\|\boldsymbol{u}\|_{L^2(Q_T)^m} &\leq \sqrt{T}\|\boldsymbol{u}\|_{L^\infty(0,T;L^2(\Omega))}\\ &\leq \sqrt{T + T^2 e^T}\big(\|\boldsymbol{f}\|_{L^2(Q_T)^m} + \|\boldsymbol{u}_0\|_{L^2(\Omega)^m}\big) = R_*, \qquad (3.46)\end{aligned}$$

being R_* fixed from now on. For this choice of R_* we have $S(D_{R_*}) \subseteq D_{R_*}$.

For $\boldsymbol{v}_i \in D_{R_*}$, $i = 1,2$ and $\boldsymbol{u}_i = S(\boldsymbol{f}, G[\boldsymbol{v}_i], \boldsymbol{u}_0)$, set $\mu = \frac{\gamma(\boldsymbol{v}_2)}{\gamma(\boldsymbol{v}_1)}$ which we may assume to be greater than 1. Denoting $g = G[\boldsymbol{v}_1] = \gamma(\boldsymbol{v}_1)\varphi$, then $\mu u_1 = S(\mu \boldsymbol{f}, \mu g, \mu \boldsymbol{u}_0)$, $\boldsymbol{u}_2 = S(\boldsymbol{f}, \mu g, \boldsymbol{u}_0)$ and, using (3.6), we have

$$\begin{aligned}\|S(\boldsymbol{v}_1) - S(\boldsymbol{v}_2)\|_{L^2(Q_T)^m} &\leq \|\boldsymbol{u}_1 - \mu\boldsymbol{u}_1\|_{L^2(Q_T)^m} + \|\mu\boldsymbol{u}_1 - \boldsymbol{u}_2\|_{L^2(Q_T)^m}\\ &\leq (\mu - 1)\|\boldsymbol{u}_1\|_{L^2(Q_T)^m} + (\mu - 1)R_* \leq 2(\mu - 1)R_*.\end{aligned}$$

But

$$\mu - 1 = \frac{\gamma(\boldsymbol{v}_2) - \gamma(\boldsymbol{v}_1)}{\gamma(\boldsymbol{v}_1)} \leq \frac{\Gamma(R_*)}{\eta(R_*)}\|\boldsymbol{v}_1 - \boldsymbol{v}_2\|_{L^2(Q_T)^m}$$

and consequently S is a contraction as long as

$$\frac{2\,R_*\Gamma(R_*)}{\eta(R_*)} < 1.$$

Remark 7 These results are new. In particular, the one with φ more regular gives the existence and uniqueness of the strong solution $\boldsymbol{u} \in \mathscr{V}_p \cap H^1\big(0,T; L^2(\Omega)^m\big)$ to the quasi-variational inequality (3.41) and therefore also satisfies $\boldsymbol{u}(t) \in \mathbb{K}_{G[\boldsymbol{u}(t)]}$ and (3.3) with $g = G[\boldsymbol{u}(t)]$,

$$\int_\Omega \partial_t \boldsymbol{u}(t)\cdot(\boldsymbol{w} - \boldsymbol{u}(t)) + \delta\int_\Omega \text{Ł}_p\boldsymbol{u}(t)\cdot \mathrm{L}(\boldsymbol{w} - \boldsymbol{u}(t)) \geq \int_\Omega \boldsymbol{f}(t)\cdot(\boldsymbol{w} - \boldsymbol{u}(t)),$$

for all $\boldsymbol{w} \in \mathbb{K}_{G[\boldsymbol{u}](t)}$, a.e. $t \in (0,T)$.

Theorem 3.21 *For* $1 < p \leq 2$ *and* $\delta > 0$*, suppose that the assumptions* (2.1), (2.2), (2.3) *and* (3.1) *are satisfied,* $\boldsymbol{f} \in L^2(Q_T)^m$*,*

$$G[\boldsymbol{u}](x,t) = \gamma(\boldsymbol{u})\varphi(x,t), \quad (x,t) \in Q_T,$$

where $E = \mathscr{V}_p$, γ *is a functional satisfying* (3.44), $\varphi \in \mathscr{C}\big([0,T]; L^\infty_\nu(\Omega)\big)$ *and* $\boldsymbol{u}_0 \in \mathbb{K}_{G[\boldsymbol{u}_0]}$. *Then, the quasi-variational inequality* (3.41) *has a unique weak solution* $\boldsymbol{u} \in \mathscr{V}_p \cap \mathscr{C}\big([0,T]; L^2(\Omega)^m\big)$, *provided that*

$$\rho\,\Gamma(R_p) < \eta(R_p), \tag{3.47}$$

where

$$\rho = 2R_p + (2-p)\big(2\,M(R_p)\,\|\varphi\|_{L^\infty(Q_T)}\big)^{2-p}|Q_T|^{\frac{2-p}{p}}(R_p)^{p-1}$$

and

$$R_p = \Big(\tfrac{1+T+T^2e^T}{2\delta}\big(\|\boldsymbol{f}\|^2_{L^2(Q_T)^m} + \|\boldsymbol{u}_0\|^2_{L^2(\Omega)^m}\big)\Big)^{\frac{1}{p}},$$

which is also a strong solution in $\mathscr{V}_p \cap H^1\big(0,T; L^2(\Omega)^m\big)$ *if, instead, we have* $\varphi \in W^{1,\infty}(0,T; L^\infty(\Omega))$ *with* $\varphi \geq \nu > 0$.

Proof For $R > 0$ let $S : D_R \to \mathscr{V}_p$ be defined by $\boldsymbol{u} = S(\boldsymbol{v}) = S(Ł_p, \boldsymbol{f}, g, \boldsymbol{u}_0)$, the unique strong solution of the variational inequality (3.3), with the operator $Ł_p$ and data $(\boldsymbol{f}, g, \boldsymbol{u}_0)$, where $g = G[\boldsymbol{v}]$. Taking $\boldsymbol{w} = \boldsymbol{0}$ in (3.3) and using the estimate (3.46) we have the a priori estimate

$$\begin{aligned} 2\,\delta\|\boldsymbol{u}\|^p_{\mathscr{V}_p} &\leq \big(\|\boldsymbol{f}\|^2_{L^2(Q_t)^m} + \|\boldsymbol{u}\|^2_{L^2(Q_T)^m} + \|\boldsymbol{u}_0\|^2_{L^2(\Omega)^m}\big) \\ &\leq (1+T+T^2e^T)\big(\|\boldsymbol{f}\|^2_{L^2(Q_t)} + \|\boldsymbol{u}_0\|^2_{L^2(\Omega)}\big) \end{aligned}$$

and therefore

$$\|\boldsymbol{u}\|_{\mathscr{V}_p} \leq \Big(\tfrac{1+T+T^2e^T}{2\delta}\big(\|\boldsymbol{f}\|^2_{L^2(Q_T)^m} + \|\boldsymbol{u}_0\|^2_{L^2(\Omega)^m}\big)\Big)^{\frac{1}{p}} = R_p. \tag{3.48}$$

Given $\boldsymbol{v}_i \in D_{R_p}, i = 1,2$, let $\boldsymbol{u}_i = S(Ł_p, \boldsymbol{f}, \gamma(\boldsymbol{v}_i)\varphi, \boldsymbol{u}_0)$ and set $\mu = \frac{\gamma(\boldsymbol{v}_2)}{\gamma(\boldsymbol{v}_1)}$, assuming $\mu > 1$.

Setting $g = G[\boldsymbol{v}_1] = \gamma(\boldsymbol{v}_1)\varphi$, observe that $\mu\boldsymbol{u}_1 = S(\mu^{2-p}Ł_p, \mu\boldsymbol{f}, \mu g, \mu\boldsymbol{u}_0) = \boldsymbol{z}_1$ and $\boldsymbol{z}_2 = S(Ł_p, \mu\boldsymbol{f}, \mu g, \mu\boldsymbol{u}_0)$, we get

$$\|\boldsymbol{u}_1 - \boldsymbol{u}_2\|_{\mathscr{V}_p} \leq \|\boldsymbol{u}_1 - \boldsymbol{z}_1\|_{\mathscr{V}_p} + \|\boldsymbol{z}_1 - \boldsymbol{z}_2\|_{\mathscr{V}_p} + \|\boldsymbol{z}_2 - \boldsymbol{u}_2\|_{\mathscr{V}_p}.$$

By (3.48) and the continuous dependence result (3.7),

$$\|\boldsymbol{u}_1 - \boldsymbol{z}_1\|_{\mathscr{V}_p} = (\mu-1)\|\boldsymbol{u}_1\|_{\mathscr{V}_p} \leq (\mu-1)R_p \quad\text{and}\quad \|\boldsymbol{z}_2 - \boldsymbol{u}_2\|_{\mathscr{V}_p} = (\mu-1)R_p. \tag{3.49}$$

Since $\boldsymbol{z}_1, \boldsymbol{z}_2 \in \mathbb{K}_{\mu g}$, we can use them as test functions in the variational inequality (3.3) satisfied by the other one. Then

$$\begin{aligned} \frac{1}{2}\int_\Omega |\boldsymbol{z}_1(t) - \boldsymbol{z}_2(t)|^2 + \int_{Q_T} |\mathrm{L}(\boldsymbol{z}_1 - \boldsymbol{z}_2)|^2\big(|\mathrm{L}\boldsymbol{z}_1| + |\mathrm{L}\boldsymbol{z}_2|\big)^{p-2} \\ \leq (\mu^{2-p} - 1)\int_{Q_T} \mathrm{L}_p\boldsymbol{z}_1 \cdot \mathrm{L}(\boldsymbol{z}_1 - \boldsymbol{z}_2) \end{aligned}$$

and, by the Hölder inverse inequality,

$$\frac{1}{2}\int_\Omega |z_1(t)-z_2(t)|^2 + \|\mathrm{L}(z_1-z_2)\|^2_{L^p(Q_T)^\ell}\Big(\int_{Q_T}(|\mathrm{L}z_1|+|\mathrm{L}z_1|)^p\Big)^{\frac{p-2}{p}}$$
$$\leq (\mu^{2-p}-1)\|\mathrm{L}z_1\|^{p-1}_{L^p(Q_T)^\ell}\|\mathrm{L}(z_1-z_2)\|_{L^p(Q_T)^\ell}.$$

But

$$\Big(\int_{Q_T}(|\mathrm{L}z_1|+|\mathrm{L}z_1|)^p\Big)^{\frac{p-2}{p}} \geq (2\,M(R_p)\,\|\varphi\|_{L^\infty(Q_T)})^{p-2}|Q_T|^{\frac{p-2}{p}}$$

and $\mu^{2-p}-1 \leq (2-p)(\mu-1)$, so

$$\|z_1-z_2\|_{\mathscr{V}_p} \leq (\mu-1)\,(2-p)\,\big(2\,M(R_p)\,\|\varphi\|_{L^\infty(Q_T)}\big)^{2-p}|Q_T|^{\frac{2-p}{p}}(R_p)^{p-1}. \quad (3.50)$$

From (3.49) and (3.50), we obtain

$$\|S(\boldsymbol{v}_1)-S(\boldsymbol{v}_2)\|_{\mathscr{V}_p} \leq (\mu-1)\Big(2R_p + (2-p)\big(2\,M(R_p)\,\|\varphi\|_{L^\infty(Q_T)}\big)^{2-p}|Q_T|^{\frac{2-p}{p}}(R_p)^{p-1}\Big). \quad (3.51)$$

Defining

$$\rho = 2R_p + (2-p)\,\big(2\,M(R_p)\,\|\varphi\|_{L^\infty(Q_T)}\big)^{2-p}|Q_T|^{\frac{2-p}{p}}(R_p)^{p-1}$$

we get, with $\Gamma = \Gamma(R_p)$ and $\eta = \eta(R_p)$,

$$\|S(\boldsymbol{v}_1)-S(\boldsymbol{v}_2)\|_{\mathscr{V}_p} \leq \frac{\rho\Gamma}{\eta}\|\boldsymbol{v}_1-\boldsymbol{v}_2\|_{\mathscr{V}_p}$$

and S is a contraction if $\rho\Gamma < \eta$, which fixed point $\boldsymbol{u} \in \mathscr{V}_p \cap H^1\big(0,T;L^2(\Omega)^m\big)$ is the strong solution of the quasi-variational inequality.

In the case of $\varphi \in \mathscr{C}\big([0,T];L^\infty_\nu(\Omega)\big)$, the solution map S of Theorem 3.8 only gives a weak solution $\boldsymbol{u} \in \mathscr{V}_p \cap \big([0,T];L^2(\Omega)^m\big)$, which is a contraction exactly in the same case as (3.48). The proof is the same, since the continuous dependence estimate (3.50) still holds for weak solutions of the variational inequality as in Theorem 3.9.□

Remark 8 These results apply to nonlocal dependences on the derivatives of $\boldsymbol{u}$ as well, since φ is Lipschitz continuous on $\mathscr{V}_p$. The part corresponding to weak solutions is new, while the one for strong solutions extends [42, Theorem 3.2]. This work considers strong solutions in the abstract framework of [58], which also include obstacle problems, it is aimed to numerical applications, but requires stronger restrictions on φ.

3.6 Applications

***Example* 9 (The dynamics of the sandpile)**
Among the continuum models for granular motion, the one proposed by Prigozhin (see [69], [70] and [71]) for the pile surface $u = u(x,t)$, $x \in \Omega \subset \mathbb{R}^2$, growing on a rigid support $u_0 = u_0(x)$, satisfying the repose angle α condition, i.e., the surface slope $|\nabla u|$ cannot exceed $k = \tan\alpha > 0$ nor the support slope $|\nabla u_0|$. This leads to the implicit gradient constraint

$$|\nabla u(x,t) \leq G_0[u](x,t) \equiv \begin{cases} k & \text{if } u(x,t) > u_0(x) \\ k \vee |\nabla u_0(x)| & \text{if } u(x,t) \leq u_0(x). \end{cases} \tag{3.52}$$

Following [73], the pile surface dynamics is related to the thickness $v = v(x,t)$ of a thin surface layer of rolling particles and may be described by

$$\partial_t u + v\big(1 - \tfrac{|\nabla u|^2}{k}\big) \quad \text{and} \quad \varepsilon\partial_t v - \eta\nabla\cdot(v\nabla u) = f - v\big(1 - \tfrac{|\nabla u|^2}{k}\big), \tag{3.53}$$

where $\varepsilon \sim 0$ is the ratio of the thickness of the rolling grain layer and the pile size, $\eta > 0$, is a ratio characterizing the competition between rolling and trapping of the granular material, and f the source intensity, which is positive for the growing pile, but may be zero or negative for taking erosion effects into account. Assuming $v(x,t) = \overline{v} > 0$, from (3.53) we obtain

$$\partial_t u - \delta\Delta u = f \quad \text{if} \quad |\nabla u| < G_0,$$

where $\delta = \eta\overline{v} > 0$ may account for a small rolling of sand and hence some surface diffusion below the critical slope, or no surface flow if $\delta = 0$.

Assuming an homogeneous boundary condition, which means the sand may fall out of $\partial\Omega$, and the initial condition below the critical slope, i.e., $|\nabla u_0| \leq k$, the pile surface $u = u^\delta(x,t)$, $\delta \geq 0$, is the unique solution of the scalar variational inequality (3.3) with L= ∇, $p = 2$ and $g(t) \equiv k$, provided we prescribe $f \in L^2(Q_T)$. We observe that, by comparison of $u_1 = u^\delta$ with $\delta > 0$ and $u_2 = u^0$ with $\delta = 0$, as in Theorem 3.9, we have the estimate

$$\int_\Omega |u^\delta - u^0|^2(t) \leq 2\delta \int_0^t \int_\Omega |\nabla u^\delta||\nabla(u^\delta - u^0)| \leq 4\delta k^2 |Q_t|, \quad 0 < t < T.$$

We can also immediately apply for $t \to \infty$ the asymptotic results of Theorem 3.10 for $\delta > 0$ and Theorem 3.11 for $\delta = 0$.

Moreover, if $\delta = 0$ in the case of the growing pile with $f(x,t) = f(x) \geq 0$ it was observed in [25] not only that, if $t > s > 0$

$$u_0(x) \leq u(x,s) \leq u(x,t) \leq u_\infty(x) = \lim_{t\to\infty} u(x,t) \leq kd(x), \ x \in \Omega, \tag{3.54}$$

where $d(x) = d(x, \partial\Omega)$ is the distance function to the boundary, but the limit stationary solution is given by

$$u_\infty(x) = u_0(x) \vee u_f(x), \quad x \in \Omega,$$

where $u_f(x) = \max_{y \in \operatorname{supp} f} \big(d(y) - |x-y|\big)^+$, $x \in \Omega$. This model has also a very interesting property of the finite time stabilization of the sandpile, provided that f is positive in a neighborhood of the ridge Σ of Ω, i.e. the set of points $x \in \Omega$ where d is not differentiable (see [25, Theorem 3.3]): there exists a time $T < \infty$ such that, for any $u_0 \in \mathbb{K}_k = \{v \in H^1_0(\Omega) : |\nabla v| \le k\}$,

$$u(x,t) = kd(x), \quad \forall t \ge T, \tag{3.55}$$

provided $\exists r > 0 : f(x) \ge r$ a.e. $x \in B_r(y)$, for all $y \in \Sigma$.

Similar results were obtained in [79] for the transported sandpile problem, for $u(t) \in \mathbb{K}_k$, such that

$$\int_\Omega \big(\partial_t u(t) + \boldsymbol{b} \cdot \nabla u(t) - f(t)\big)(v - u(t)) \ge 0, \quad \text{a.e. } t \in (0,T), \tag{3.56}$$

for all $v \in \mathbb{K}_k$, with $\boldsymbol{b} \in \mathbb{R}^2$, $\partial\Omega \in \mathscr{C}^2$, $f = f(t) \ge 0$ nondecreasing and $f \in L^\infty(0,\infty)$, which also satisfies (3.54). Moreover, it was also shown in [79] that $u(t)$ equivalently solves (3.56) for the double obstacle problem, i.e. with $\mathbb{K}^\wedge_\vee = \{v \in H^1_0(\Omega) : -kd(x) \le v(x) \le kd(x),\ x \in \Omega\}$ and, moreover, has also the finite time stabilization property (3.55) under the additional assumptions $\boldsymbol{b} \cdot \nabla u_0 \le f(t)$ in $\{x \in \Omega : -kd(x) < u_0(x)\}$ for $t > 0$ and $\liminf_{t \to \infty} f(t) > |\boldsymbol{b}| + 2k\|d\|_{L^\infty(\Omega)}$.

It should be noted that if we replace $\mathbb{K}_k$ by the solution dependent convex set $\mathbb{K}_{G_0[u]}$, with G_0 defined in (3.52), to solve the corresponding quasi-variational inequality (3.56), even with $\boldsymbol{b} \equiv \boldsymbol{0}$ or with an additional δ-diffusion term is an open problem since the operator G_0 is not continuous in u. Recently, in [12], Barrett and Prigozhin succeeded to construct, by numerical analysis methods, approximate solutions, including numerical examples, that converge to a quasi-variational solution of (3.56) without transport ($\boldsymbol{b} \equiv \boldsymbol{0}$), for fixed $\varepsilon > 0$, with the continuous operator $G_\varepsilon : \mathscr{C}(\overline{\Omega}) \to \mathscr{C}(\overline{\Omega})$ given by

$$G_\varepsilon[u](x,t) = \begin{cases} k & \text{if } u(x) \ge u_\varepsilon(x) + \varepsilon, \\ k_\varepsilon(x) + (k - k_\varepsilon(x))\dfrac{u(x) - u_\varepsilon(x)}{\varepsilon} & \text{if } u_\varepsilon(x) \le u(x) < u_\varepsilon(x) + \varepsilon, \\ k_\varepsilon(x) \equiv k \vee |\nabla u_\varepsilon(x)| & \text{if } u(x) < u_\varepsilon(x), \end{cases}$$

where $u_\varepsilon \in \mathscr{C}^1(\overline{\Omega}) \cap W^{1,\infty}_0(\Omega)$ is an approximation of the initial condition $u_0 \in W^{1,\infty}_0(\Omega)$. We observe that the existence of a quasi-variational solution of (3.56) with this G_ε is also guaranteed by Theorem 3.15 or Corollary 3.1. □

***Example* 10 (An evolutionary electromagnetic heating problem** [65]
We consider now an evolutionary case of the Example 6 for the magnetic field $\boldsymbol{h} = \boldsymbol{h}(x,t)$ of a superconductor, which threshold may depend of a temperature field $\vartheta = \vartheta(x,t)$, $(x,t) \in Q_T$, subjected to a magnetic heating. This leads to the quasi-variational weak formulation

$$\begin{cases} \boldsymbol{h} \in \mathbb{K}_{j(\vartheta(\boldsymbol{h}))} \subset \mathscr{V}_p, \\ \displaystyle\int_0^T \langle \partial_t \boldsymbol{w}, \boldsymbol{w} - \boldsymbol{h} \rangle_p + \delta \int_{Q_T} |\nabla \times \boldsymbol{h}|^{p-2} \nabla \times (\boldsymbol{w} - \boldsymbol{h}) \geq \int_{Q_T} \boldsymbol{f} \cdot (\boldsymbol{w} - \boldsymbol{h}) \\ \qquad\qquad \displaystyle -\frac{1}{2} \int_\Omega |\boldsymbol{w}(0) - \boldsymbol{h}_0|^2 \quad \forall \boldsymbol{w} \in \mathscr{Y}_p \text{ such that } \boldsymbol{w} \in \mathbb{K}_{j(\vartheta(\boldsymbol{h}))} \end{cases} \tag{3.57}$$

coupled with a Cauchy-Dirichlet problem for the heat equation

$$\begin{aligned} &\partial_t \vartheta - \Delta \vartheta = \eta + \boldsymbol{\zeta} \cdot \boldsymbol{h} + \boldsymbol{\xi} \cdot \nabla \times \boldsymbol{h} \quad \text{ in } Q_T, \\ &\vartheta = 0 \text{ on } \partial\Omega \times (0, T), \quad \vartheta(0) = \vartheta_0 \text{ in } \Omega. \end{aligned} \tag{3.58}$$

Here, for a.e. $t \in (0, T)$, the convex set depends on $\boldsymbol{h}$ trough ϑ and is given, for some $j = j(x, t, \vartheta) \in \mathscr{C}(\overline{Q}_T \times \mathbb{R})$, $j \geq \nu > 0$ by

$$\mathbb{K}_{j(\vartheta(t))} = \left\{ \boldsymbol{w} \in \mathbb{X}_p : |\nabla \times \boldsymbol{w}| \leq j(\theta(t)) \text{ in } \Omega \right\} \tag{3.59}$$

where $\mathbb{X}_p$ is given by (2.5) or (2.6).

If we give $\vartheta_0 \in H_0^1(\Omega) \cap \mathscr{C}^\alpha(\overline{\Omega})$, $\eta \in L^p(Q_T)$ and $\boldsymbol{\zeta}$, $\boldsymbol{\xi} \in L^\infty(Q_T)^3$, the solution map that, for $p \geq \frac{5}{2}$, associates to each $\boldsymbol{h} \in \mathscr{V}_p$, the unique solution $\vartheta \in L^2\big(0, T; H_0^1(\Omega)\big) \cap \mathscr{C}^\lambda(\overline{Q}_T)$, for some $0 < \lambda < 1$, is continuous and compact as a linear operator from $\mathscr{V}_p$ in $\mathscr{C}(\overline{Q}_T)$. Therefore, with $\boldsymbol{f} \in L^2(Q_T)^3$ and $\vartheta_0 \in \mathbb{K}_{j(\vartheta_0)}$, Theorem 3.19 guarantees the existence of a weak solution $(\boldsymbol{h}, \vartheta) \in \big(\mathscr{V}_p \cap L^\infty(0, T; L^2(\Omega)^3)\big) \times \big(L^2(0, T; H_0^1(\Omega)) \cap \mathscr{C}^\lambda(\overline{Q}_T)\big)$ to the coupled problem (3.57)-(3.58).

We observe that, if the threshold j is independent of ϑ, the problem becomes variational and admits not only weak but also strong solutions, by Theorem 3.8.

However, if we set a direct local dependence of the type $j = j(|\boldsymbol{h}|)$, as in Example 7, the problem is open in vectorial case.

Nevertheless, if the domain $\Omega = \omega \times (-R, R)$, with $\omega \subset \mathbb{R}^2$, $\partial\omega \in \mathscr{C}^{0,1}$ and the magnetic field has the form $\boldsymbol{h} = (0, 0, u(y, t))$, $y \in \omega$, $0 < t < T$, the critical-state superconductor model has a longitudinal geometry, where u satisfies the scalar quasi-variational inequality (3.27) with $\boldsymbol{\Phi} \equiv \boldsymbol{0}$ and Theorem 3.15 provides in this case the existence of a strong solution $u \in \mathscr{C}(\overline{\omega} \times [0, T]) \cap \mathbb{K}_{j(|\boldsymbol{h}|)} \cap W^{1,\infty}\big(0, T; M(\Omega)\big)$, with $j \in \mathscr{C}\big(\mathbb{R}; L_\nu^\infty(\omega)\big)$, for $\delta \geq 0$. The case $\delta > 0$ was first given in [77] and $\delta = 0$ in [78]. □

***Example* 11 (Stokes flow for a thick fluid)**
The case where $\boldsymbol{u} = \boldsymbol{u}(x, t)$ represents the velocity field of an incompressible fluid in a limit case of a shear-thickening viscosity has been considered in [28], [76] and [63] by using variational inequalities. Those works consider a constant or variable positive threshold on the symmetric part of the velocity field L= D. Here we consider the more general situation of a nonlocal dependence on the total energy of displacement

$$|D\boldsymbol{u}(x, t)| \leq G[\boldsymbol{u}(x, t)] = \varphi(x, t)\Big(\eta + \delta \int_{Q_T} |D\boldsymbol{u}|^2\Big), \; x \in \Omega \subset \mathbb{R}^d, \; t \in (0, T), \tag{3.60}$$

for given $\delta, \eta > 0$, $\varphi \in W^{1,\infty}\big(0,T; L^\infty(\Omega)\big)$, $\varphi \geq \nu > 0$.

We set $\mathbb{X}_2 = \big\{ \boldsymbol{w} \in H_0^1(\Omega)^d : \nabla \cdot \boldsymbol{w} = 0 \big\}$, $d = 2, 3$, which is an Hilbert space for $\|D\boldsymbol{w}\|_{L^2(\Omega)^{d^2}}$ compactly embedded in $\mathbb{H} = \big\{ \boldsymbol{w} \in L^2(\Omega)^d : \nabla \cdot \boldsymbol{w} = 0 \big\}$. Defining $\mathbb{K}_{G[\boldsymbol{u}](t)}$ for each $t \in (0,T)$ by (2.8) and giving $\boldsymbol{f} \in L^2(Q_T)^d$ and $\boldsymbol{u}_0 \in \mathbb{X}_2$ satisfying (3.60) at $t = 0$, i.e. $\boldsymbol{u}_0 \in \mathbb{K}_{G[\boldsymbol{u}_0]}$, in order to apply Theorem 3.21, we set

$$R_2 = \tfrac{1+T+T^2e^T}{2\delta}\big(\|\boldsymbol{f}\|_{L^2(Q_T)^d} + \|\boldsymbol{u}_0\|_{L^2(\Omega)^d}\big) = \tfrac{\rho}{2}.$$

The nonlocal functional satisfies (3.44) with $E = L^2\big(0,T;\mathbb{X}_2)\big) = \mathscr{V}_2$ and $T = \delta\rho$, since we have

$$\begin{aligned}|\gamma(\boldsymbol{u}_1) - \gamma(\boldsymbol{u}_2)| &= \delta\Big|\int_{Q_T} |D\boldsymbol{u}_1|^2 - |D\boldsymbol{u}_2|^2\Big| = \delta\Big|\int_{Q_T} (D\boldsymbol{u}_1 - D\boldsymbol{u}_2)\cdot(D\boldsymbol{u}_1 + D\boldsymbol{u}_2)\Big| \\ &\leq \delta\rho\Big(\int_{Q_T} |D\boldsymbol{u}_1 - D\boldsymbol{u}_2|^2\Big)^{\frac{1}{2}}, \qquad \text{for } \boldsymbol{u}_1, \boldsymbol{u}_2 \in D_{R_2}.\end{aligned}$$

Hence, by Theorem 3.21, if $\delta\rho^2 < \eta$, i.e. if

$$(1 + T + T^2e^T)^2\big(\|\boldsymbol{f}\|_{L^2(Q_T)^d} + \|\boldsymbol{u}_0\|_{L^2(\Omega)^d}\big) < \tfrac{\eta}{\delta},$$

there exists a unique strong solution $\boldsymbol{u} \in \mathscr{V}_2 \cap H^1\big(0,T; L^2(\Omega)^d\big) \cap \mathbb{K}_{G[\boldsymbol{u}]}$, with $\boldsymbol{u}(0) = \boldsymbol{u}_0$, satisfying the quasi-variational inequality

$$\int_\Omega \partial_t \boldsymbol{u}(t)\cdot(\boldsymbol{w} - \boldsymbol{u}(t))) + \delta\int_\Omega D\boldsymbol{u}(t)\cdot D(\boldsymbol{w} - \boldsymbol{u}(t)) \geq \int_\Omega \boldsymbol{f}(t)\cdot(\boldsymbol{w} - \boldsymbol{u}(t)),$$

for all $\boldsymbol{w} \in \mathbb{K}_{G[\boldsymbol{u}](t)}$ and a.e. $t \in (0,T)$.

This result can be generalized to the Navier-Stokes flows, i.e. with convection (see [80]). □

References

1. Ambrosio, L.: Lecture notes on optimal transport problems. In: mathematical aspects of evolving interfaces, Funchal 2000 Lecture Notes in Math., vol.1812, pp. 1–52. Springer, Berlin (2003)
2. Amrouche, C., Seloula, N.: L^p-theory for vector potentials and Sobolev's inequalities for vector fields: application to the Stokes equations with pressure boundary conditions. Math. Models Methods Appl. Sci. **23**, 37–92 (2013)
3. Andersson, J., Shahgholian, H. and Weiss, G.: Double obstacle problems with obstacles given by non-C^2 Hamilton-Jacobi equations. Arch. Ration. Mech. Anal. **206**, 779–819 (2012)
4. Antil, H., Rautenberg, C. Fractional elliptic quasi-variational inequalities: theory and numerics. Interfaces Free Bound. **20** no. 1, 1–24 (2018)
5. Aronsson, G., Evans, L.C., Wu, Y.: Fast/slow diffusion and growing sandpiles. J. Differ. Equations **131**, 304–335 (1996)
6. Azevedo, A., Miranda, F., Santos, L.: Variational and quasivariational inequalities with first order constraints. J. Math. Anal. Appl. **397**, 738–756 (2013)

7. Azevedo, A., Santos, L.: Convergence of convex sets with gradient constraint. Journal of Convex Analysis **11**, 285–301 (2004)
8. Azevedo, A., Santos, L.: Lagrange multipliers and transport densities. J. Math. Pures Appl. **108**, 592–611 (2017)
9. Baiocchi, C., Capelo, A.: Variational and quasivariational inequalities: applications to free boundary problems. John Wiley and Sons, New York, 1984 (translation of the 1978 Italian edition).
10. Barrett, J., Prigozhin, L.: Dual formulation in critical state problems. Interfaces Free Bound. **8**, 349–370 (2006)
11. Barrett, J., Prigozhin, L.: A quasi-variational inequality problem in superconductivity. Math. Models Methods Appl. Sci. **20** no. 5, 679–706 (2010)
12. Barrett, J., Prigozhin, L.: A quasi-variational inequality problem arising in the modeling of growing sandpiles. ESAIM Math. Model. Numer. Anal. **47**, 1133–1165 (2013)
13. Barrett, J., Prigozhin, L.: Sandpiles and superconductors: nonconforming linear finite element approximations for mixed formulations of quasi-variational inequalities. IMA J. of Numerical Analysis, **35**, 1–38 (2015)
14. Barrett, J., Prigozhin, L.: Lakes and rivers in the landscape: a quasi-variational inequality approach. Interfaces Free Bound. **16** no. 2, 269–296 (2014)
15. Bhattacharya, T., Di Benedetto, E., Manfredi, J.: Limits as $p \to \infty$ of $\Delta_p u_p = f$ and related extremal problems. Rend. Sem. Mat. Univ. Politec. Torino 1989, Special Issue, 15–68 (1991)
16. Bensoussan, A. and Lions, J.-L.: Contrôle impulsionnel et inéquations quasi-variationnelles. Gauthier-Villars, Paris, 19822
17. Brézis, H.: Equations et inéquations non linéaires dans les espaces vectoriels en dualité. Ann. Inst. Fourier, **18** 115–175 (1968)
18. Brézis, H.: Multiplicateur de Lagrange en torsion elasto-plastique. Arch. Ration. Mech. Anal. **49**, 32–40 (1972)
19. Brézis, H.: Opérateurs maximaux monotones et semigroups de contractions dans les espaces de Hilbert. North-Holland Math.Stud.,vol.5, North-Holland Publishing Co./American Elsevier Publishing Co., Inc. (1973)
20. Brézis, H., Sibony, M.: Equivalence de deux inéquations variationnelles et applications. Arch. Ration. Mech. Anal. **41**, 254–265 (1971)
21. Caffarelli, L. A., Friedman, A.: Reinforcement problems in elasto-plasticity. Roc. Mount. J. Math. **10**, 155–184 (1980)
22. Caffarelli, L.A., McCann, R.J.: Free boundaries in optimal transport and Monge-Ampère obstacle problems. Ann. of Math. **171**, 673–730 (2010)
23. Caffarelli, L. A., Rivière, N. M.: The smoothness of the elastic-plastic free boundary of a twisted bar. Proc. Am. Math. Soc. **63**, 56–58 (1977)
24. Cannarsa, P., Cardaliaguet, P., Representation of equilibrium solutions to the table problem for growing sandpiles. J. Eur. Math. Soc. **6**, 435–464 (2004)
25. Cannarsa, P., Cardaliaguet, P., Sinestrari, C.: On a differential model for growing sandpiles with non-regular sources. Comm. Partial Differential Equations **34**, 656–675 (2009)
26. Chiadò Piat, V., Percivale, D.: Generalized Lagrange multipliers in elastoplastic torsion. J. Differential Equations **114**, 570–579 (1994)
27. Choe, H. J., Souksomvang, P.: Elliptic gradient constraint problem. Comm. Partial Differential Equation **41**, 1918–1933 (2016)
28. De los Reyes, J. C., Stadler, G.: A non smooth model for discontinuous shear thickening fluids: analysis and numerical solution. Interfaces Free Bound. **16** (4), 575–602 (2014)
29. De Pascale, L.; Evans, L. C., Pratelli, A.: Integral estimates for transport densities, Bull. London Math. Soc. **36**, 383–395 (2004)
30. Dumont, S., Igbida, N.; On a dual formulation for the growing sandpile problem, Euro. J. of Appl. Math. **20**, 169–185 (2009)
31. Duvaut, G., Lions, J.-L.: Les inéquations en mécanique et en physique. Dunod, Paris (1972), English transl. Springer, Berlin (1976)

32. Evans, L. C.: A second order elliptic equation with gradient constraint. Comm. Part. Diff. Eq. **4**, 555–572 (1979)
33. Evans, L. C.: Partial differential equations and Monge-Kantorovich mass transfer. In: Bott, Raoul et al. (ed.), International Press, 65–126 (1999)
34. Facchinei, F., Kanzow, C., Sagratella, S.: Solving quasi-variational inequalities via their KKT conditions. Math. Program. **144** no. 1-2, Ser. A, 369–412 (2014)
35. Fichera, G.: Problemi elastotatici con vincoli unilaterali: il problema di Signorini con ambigue condizioni al contorno. Atti Accad. Naz. Lincei Mem. Cl. Sci. Fis. Mat. Net. Scz. la. **7** (8 ,) 91–140 (1963)
36. Friedman, A.: Variational principles and free-boundary problems. Pure and Applied Mathematics. John Wiley & Sons, Inc., New York (1982)
37. Figalli, A.: The optimal partial transport problem. Arch. Ration. Mech. Anal. **195** , 533–560 (2010)
38. Fukao, T.,Kenmochi, N.: Parabolic variational inequalities with weakly time-dependent constraints. Adv. Math. Sci. Appl. **23** (2), 365–395 (2013)
39. Gerhardt, C.: On the existence and uniqueness of a warpening function in the elastic-plastic torsion of a cylindrical bar with multiply connected cross-section. In: applications of methods of functional analysis to problems in mechanics (Joint Sympos., IUTAM/IMU, Marseille, 1975), Springer, 1976, Lecture Notes in Math. **503** 328–342 (1976)
40. Hintermüller, M., Rautenberg, C.: A sequential minimization thecnique for elliptic quasi-variational inequalities with gradient constraints. SIAM J. Optim. **22**, 1224–1257 (2013)
41. Hintermüller, M., Rautenberg, C.: Parabolic quasi-variational inequalities with gradient-type constraints. SIAM J. Optim., **23** (4), 2090–2123 (2013)
42. Hintermüller, M., Rautenberg, C.: On the uniqueness and numerical approximation of solutions to certain parabolic quasi-variational inequalities. Port. Math. **74**, 1–35 (2017)
43. Hintermüller, M., Rautenberg, C., Strogies, N.: Dissipative and non-dissipative evolutionary quasi-variational inequalities with gradient constraints. Set-Valued Var. Anal, https://doi.org/10.1007/s11228-018-0489-0, (2018)
44. Igbida, N.: Equivalent formulations for Monge-Kantorovich equation. Nonlinear Anal. **71**, 3805–3813 (2009)
45. Igbida, N., Nguyen, V. T.: Optimal partial mass transportation and obstacle Monge-Kantorovich equation. J. DifferentialEquations **264**, 6380–6417 (2018)
46. Juutinen, P., Parviainen, M., Rossi, J.D.: Discontinuous gradient constraints and the infinity Laplacian. Int. Math. Res. Not. IMRN **8**, 2451–2492 (2016)
47. Kenmochi, N.: Monotonicity and compactness methods for nonlinear variational inequalities in "Handbook of Differential Equations: Stationary Partial Differential Equations, Vol. IV" (ed. M. Chipot), Elsevier/North Holland, Amsterdam, (2007)
48. Kenmochi, N.: Solvability of nonlinear evolution equations with time-dependent constraints and applications. Bull. Fac. Educ., Chiba Univ., Part II **30** 1–87 (1981)
49. Kenmochi, N.: Parabolic quasi-variational diffusion problems with gradient constraints. Discrete Contin. Dyn. Syst Ser. S **6**, 423–438 (2013)
50. Murase, Y., Kano, R., Kenmochi, N.: Elliptic quasi-variational inequalities and applications. Discrete Contin. Dyn. Syst., Dynamical systems, differential equations and applications. 7th AIMS Conference, suppl., 583–591 (2009)
51. Kenmochi, N. and Niezgódka, M.: Weak Solvability for Parabolic Variational Inclusions and Application to Quasivariational Problems. Adv. Math. Sci. Appl. **25**, 63–98 (2016)
52. Kinderlehrer, D., Stamppachia, G.: An introduction to variational inequalities and their applications. Academic Press, New York, (1980)
53. Kubo, M.: Quasi-variational analysis. Sugaku Expositions, **30**, 17–34 (2017)
54. Kubo, M., Yamazaki, N.: Global strong solutions to abstract quasi-variational evolution equations. J. Differential Equations **265**, 4158–4180 (2018)
55. Kubo, M.; Murase, Y.. Quasi-subdifferential operator approach to elliptic variational and quasi-variational inequalities. Math. Methods Appl. Sci. **39** no. 18, 5626–5635 (2016)

56. Kunze, M., Rodrigues, J.-F.: An elliptic quasi-variational inequality with gradient constraints and some of its applications. Math. Methods Appl. Sci. **23**, 897–908 (2000)
57. Lions, J.-L.: Contrôle optimal de systèmes gouvernés par des équations aux dérivées partielles. Dunod, Paris; Gauthier-Villars, Paris (1968)
58. Lions, J.-L.: Quelques méthodes de résolution des problèmes aux limites non linéaires. Dunod (1969)
59. Lions, P.-L.: Generalized solutions of Hamilton-Jacobi equations. Pitman (Advanced Publishing Program) (1982)
60. Lions, P. -L., Perthame, B.: Une remarque sur les opérateurs non linéaires intervenant dans les inéquations quasi-variationelles. Annales de la Faculté des Sciences de Toulouse, 5e serie 259–263 (1983)
61. Lions, J.-L., Stampacchia, G.: Variational inequalities. Comm. Pure Appl. Math. **20**, 493–519 (1967)
62. Mignot, F.and Puel, J.-P.: Inéquations d'évolution paraboliques avec convexes dépendant du temps. Applications aux inéquations quasi variationnelles d'évolution. Arch. Ration. Mech. Anal. **64** (1) 59–91 (1977)
63. Miranda, F., Rodrigues, J.-F.: On a variational inequality for incompressible non-Newtonian thick flows. Recent advances in partial differential equations and applications, Contemp. Math., **666**, Amer. Math. Soc., Providence, RI, 305–316, (2016)
64. Miranda, F., Rodrigues, J.-F. and Santos, L.: A class of stationary nonlinear Maxwell systems. Math. Models Methods Appl. Sci. **19**, 1883–1905 (2009)
65. Miranda, F., Rodrigues, J.-F., Santos, L.: On a p-curl system arising in electromagnetism. Discrete Contin. Dyn. Syst Ser. S **5**, 605–629 (2012)
66. Miranda, F., Rodrigues, J.-F., Santos, L.: Evolutionary quasi-variational and variational inequalities with constraints on the derivatives. Advances in Nonlinear Analysis, **9**, (2020) 25–277
67. Mosco, U.: Convergence of convex sets and of solutions of variational inequalities. Adv. Math. **3**, 610–585 (1969)
68. Prager, W.: On ideal locking materials. Transactions of the Society of Rheology, Vol.1, 169–175 (1957)
69. Prigozhin, L.: Quasivariational inequality in a poured pile shape problem. (in Russian) Zh. Vychisl. Mat. i Mat. Fiz. **26** no. 7, 1072–1080, 1119 (1986)
70. Prigozhin, L.: Sandpiles and river networks: extended systems with nonlocal interactions. Phys. Rev. E (3) **49**, 1161–1167 (1994)
71. Prigozhin, L.: Variational model of sandpile growth. European J. Appl. Math. **7**, 225–235 (1996)
72. Prigozhin, L.: On the Bean critical state model in superconductivity. European J. Appl. Math. **7**, 237–247 (1996)
73. Prigozhin, L., Zaltzman, B.: On the approximation of the dynamics of sandpile surfaces. Port. Math. (N.S.) **60**, 127–137 (2003)
74. Rodrigues, J.-F.: Obstacle problems in mathematical physics. North-Holland Mathematics Studies 134, (1987)
75. Rodrigues, J.-F.: On some nonlocal elliptic unilateral problems and applications. In: Nonlinear analysis and applications (Warsaw, 1994), GAKUTO Internat. Ser. Math. Sci. Appl., pp. 343–360 Tokyo (1996)
76. Rodrigues, J.-F. : On the Mathematical Analysis of Thick Fluids. J. Math. Sci. (N.Y.) **210**, 835–848 (2015) (also published in Zapiski Nauchnykh Seminarov POMI **425**, 117–136 (2014)).
77. Rodrigues, J.-F., Santos, L.: A parabolic quasi-variational inequality arising in a superconductivity model. Ann. Scuola Norm. Sup. Pisa Cl. Sci. (4) **29**, 153–169 (2000)
78. Rodrigues, J.-F., Santos, L., Quasivariational solutions for first order quasilinear equations with gradient constraint. Arch. Ration. Mech. Anal. **205**, 493–514 (2012)
79. Rodrigues, J.-F., Santos, L.: Solutions for linear conservation laws with gradient constraint. Port. Math. (N.S.) **72**, 161–192 (2015)
80. Rodrigues, J.-F., Santos, L.: Quasi-variational solutions to thick flows. (To appear)

81. Roubíček, T.: Nonlinear partial differential equations with applications. 2nd ed., International Series of Numerical Mathematics 153, Basel, Birkhäuser, 2013.
82. Safdari, M., On the shape of the free boundary of variational inequalities with gradient constraints. Interfaces Free Bound. **19**, 183–200 (2017)
83. Santos, L.: A diffusion problem with gradient constraint and evolutive Dirichlet condition, Port. Math. **48**, 441–468 (1991)
84. Santos, L.: Variational problems with non-constant gradient constraints. Port. Math. **59**, 205–248 (2002)
85. Showalter, R. E.: Monotone operators in Banach space and nonlinear partial differential equations. Mathematical Surveys and Monographs, **49**, American Mathematical Society, Providence, RI, (1997)
86. Stampacchia, G.: Formes bilinéaires coercitives sur les ensembles convexes. C. R. Acad. Sci. Paris **258** 4413–4416 (1964)
87. Treu, G., Vornicescu, M.: On the equivalence of two variational problems. Calc. Var. Partial Differential Equations **11** 307–319 (2000)
88. Zhu, H.: Characterization of variational inequalities in singular stochastic control. Ph.D. thesis, Brown University, U.S.A. (1992)

Models of Dynamic Damage and Phase-field Fracture, and their Various Time Discretisations

Tomáš Roubíček

Abstract Several variants of models of damage in viscoelastic continua under small strains in the Kelvin-Voigt rheology are presented and analyzed by using the Galerkin method. The particular case, known as a phase-field fracture approximation of cracks, is discussed in detail. All these models are dynamic (i.e. involve inertia to model vibrations or waves possibly emitted during fast damage/fracture or induced by fast varying forcing) and consider viscosity which is also damageable. Then various options for time discretisation are devised. Eventually, extensions to more complex rheologies or a modification for large strains are briefly exposed, too.

1 Introduction

Damage in continuum mechanics of solids is an important part of engineering modelling (and also experimental research), focusing on the attribute of various degradation of materials. During past few decades, some of the engineering models had been also under rigorous mathematical scrutiny.

Phenomenological damage models structurally represent the simplest example of the concept of internal variables, where only one scalar-valued variable (here denoted by α) is considered. Cf. G.A. Maugin [46] for a thorough historical survey of this concept. This scalar-phenomenological-damage concept was invented by L.M. Kachanov [35] and Yu.N. Rabotnov [59], the damage variable ranging the interval $[0,1]$ and having an intuitive microscopical interpretation as a density of microcracks or microvoids. There are two conventions: damaging means α increasing and $\alpha = 1$ means maximal damage (which is used in engineering or e.g. also in geophysics) or, conversely, damaging means α decreasing and $\alpha = 0$ means

Mathematical Institute, Charles University, Sokolovská 83, CZ-186 75 Praha 8, Czech Republic, and
Institute of Thermomechanics, Czech Academy of Sciences, Dolejškova 5, CZ-182 00 Praha 8, Czech Republic e-mail: `tomas.roubicek@mff.cuni.cz`

M. Hintermüller, J. F. Rodrigues (eds.), *Topics in Applied Analysis and Optimisation*, CIM Series in Mathematical Sciences, https://doi.org/10.1007/978-3-030-33116-0_14

maximal damage (which is used in mathematical literature and also here), cf. e.g. the monographs [21, Ch.12] and [22, Ch.6]. Let us still note that, although damage as a single variable is most often used in applications, some models with more variables are sometimes considered in engineering, too. Most generally, one may think about 8th-order tensor as a damage variable, transforming 4th-order elastic-moduli tensor $\mathbb{C}$, cf. e.g. [55].

Damage can be (and typically is) a very fast process, usually much faster than the time scale of external loading. This is reflected by an (often accepted) idealization to model it as a rate-independent process which can have arbitrary speed. No matter whether the model is rate-independent or involves some sort of damage viscosity, the fast damage may generate elastic waves in the continuum. Conversely, waves can trigger damage. This combination of damage at usually localized areas and inertial effects in the whole bulk needs a bit special methods both for rigorous analysis and for numerical approximation, some of them being suitable rather for vibration (where transfer of kinetic energy is not dominant) than waves. On top of it, in some applications even the loading itself can vary fast in time during various impacts or explosions. Such *dynamic damage* or dynamic *fracture mechanics* [23] is the main focus of this chapter, the (often considered) quasistatic variants being thus intentionally avoided here. It is also important that inertia suppresses artificial global long-range interactions which otherwise make various unphysical effects and causes a need of some rather artificial quasistatic models, cf. Remark 1 below.

Most of this exposition will be formulated at small strains. Plain damage will be presented in several variants in Section 2. Its usage for fracture mechanics exploiting the so-called phase-field approximation will then be in Section 3, outlining a wide menagerie of models towards distinguishing crack initiation and crack propagation, possibly sensitive to modes (i.e. opening versus shearing), some of these models being likely new. Various discretisations of these models in time may exhibit various useful properties, which will be presented in Section 4. Eventually, the basic scenario of small-strain models with just one scalar-valued damage variable can be enriched in many ways, by involving some other internal variables like plastic strain or a diffusant content and also temperature, which is certainly motivated by specific applications. One can make it either in the framework of small strains considered in the previous sections, or even at large strains. Some of these enhancements will be briefly outlined in Section 5.

2 Models of damage at small strains

Beside the already mentioned alternative of damage being rate-dependent versus rate-independent, there are many variants. Basic alternatives are unidirectional damage (i.e. no healing is allowed, relevant in most engineering materials) versus reversible damage (i.e. a certain reconstruction of the material is possible, relevant e.g. in rock mechanics in the time scales of thousands years or more). And, of course, damage models can be incorporated into various viscoelastic models, and damage can influence not only the stored energy but also the dissipation potential. The damage can be complete (which is mathematically much more difficult, cf. [52] at least for some partial results) or incomplete. In addition to the simplest mode-insensitive damage, many applications need a mode-sensitive damage (damage by tension/opening easier than by shearing). On top of all this, there is a conceptual discussion whether rather stress or energy (or a combination of both) causes damage.

In addition to these options, some nonlocal theories are typically used. This concern the damage variable and sometimes also the strain. Here we have in mind so-called weakly nonlocal concepts which involve usual local gradients. The former case thus involves damage gradient into the stored energy and allows us to introduce length-scale into the damage, while the latter option allows for weaker assumptions on lower-order terms and for involving dispersion into elastic waves, as discussed in [34].

There are many options of damage models outlined above, some of them complying with rigorous analysis while some others which making troubles. Most mathematical models at small strains consider the specific stored energy $\varphi = \varphi(e,\alpha)$ quadratic in terms of the small-strain variable $e \in \mathbb{R}^{d\times d}_{\rm sym}$.

From an abstract viewpoint, the evolution is governed by *Hamilton's variational principle* generalized for dissipative systems [6], which says that, among all admissible motions $q = q(t)$ on a fixed time interval $[0,T]$, the actual motion makes

$$\int_0^T \mathscr{L}\big(t,q,\dot{q}\big)\,\mathrm{d}t \quad \text{stationary (i.e. } q \text{ is its } \textit{critical point}\text{)}, \tag{2.1}$$

where $\dot{q} = \frac{\partial}{\partial t}q$ and $\mathscr{L}(t,q,\dot{q})$ is the *Lagrangian* defined by

$$\mathscr{L}\big(t,q,\dot{q}\big) := \mathscr{T}\big(\dot{q}\big) - \mathscr{E}(t,q) + \langle F(t),q\rangle\,, \tag{2.2}$$

where $F = -\partial_{\dot{q}}\mathscr{R}(q,\dot{q})$ is a nonconservative force assumed for a moment fixed, with $\mathscr{R}(q,\cdot)$ denoting the (Rayleigh's *pseudo*)*potential* of the *dissipative force*. Then (2.1) leads after by-part integration in time to

$$\partial_q\mathscr{L}\big(t,q,\dot{q}\big) - \frac{\mathrm{d}}{\mathrm{d}t}\partial_{\dot{q}}\mathscr{L}\big(t,q,\dot{q}\big) = 0. \tag{2.3}$$

This gives the abstract 2nd-order evolution equation

$$\mathscr{T}'\,\ddot{q} + \partial_{\dot{q}}\mathscr{R}(q,\dot{q}) + \partial_q\mathscr{E}(t,q) = 0 \tag{2.4}$$

where the apostrophe (or ∂) indicates the (partial) Gâteaux differential.

In the context of this section, the state $q = (u,\alpha)$ consists from the displacement $\Omega \to \mathbb{R}^d$ and the damage profile $\Omega \to [0,1]$ with $\Omega \subset \mathbb{R}^d$ a bounded domain with a Lipschitz boundary Γ, and we specify the overall kinetic energy, stored energy (including external loading), and dissipation-potential as

$$\mathscr{T}(\dot{q}) = \mathscr{T}(\dot{u}) := \int_\Omega \frac{\varrho}{2}|\dot{u}|^2\,\mathrm{d}x, \tag{2.5a}$$

$$\mathscr{E}(t,q) = \mathscr{E}(t,u,\alpha) := \int_\Omega \varphi(e(u),\alpha) + \frac{\kappa}{2}|\nabla\alpha|^2 - f(t)\cdot u + \delta_{[0,1]}(\alpha)\,\mathrm{d}x + \int_\Gamma g(t)\cdot u\,\mathrm{d}S, \tag{2.5b}$$

$$\mathscr{R}(q,\dot{q}) = \mathscr{R}(\alpha,\dot{u},\dot{\alpha}) = \int_\Omega \frac{1}{2}\mathbb{D}(\alpha)e(\dot{u})\colon e(\dot{u}) + \zeta(\dot{\alpha})\,\mathrm{d}x \tag{2.5c}$$

with the small-strain tensor $e(u) = \frac{1}{2}(\nabla u^\top + \nabla u)$ and with some specific damage dissipation-force potential $\zeta : \mathbb{R} \to [0,+\infty]$ convex with $\zeta(0) = 0$, with a 4th order tensor $\mathbb{D} : [0,1] \to \mathbb{R}^{d\times d\times d\times d}$ smoothly dependent on α, $\kappa > 0$ a phenomenological coefficient determining a length-scale of damage (which is a usual engineering concept, cf. e.g. [4], useful also from analytical reasons), and with $\delta_{[0,1]}(\cdot) : \mathbb{R} \to \{0,+\infty\}$ denoting the indicator function of the interval $[0,1]$ where the damage variable is assumed to take its values.

This general framework gives a relatively simple model of damage in the linear Kelvin-Voigt viscoelastic solids where the only internal variable is the damage. Feeding (2.4) by the functionals (2.5), we arrive at the system of partial differential equation and inclusions

$$\varrho\ddot{u} - \mathrm{div}\big(\mathbb{D}(\alpha)e(\dot{u}) + \partial_e\varphi(e(u),\alpha)\big) = f \qquad \text{in } Q, \tag{2.6a}$$

$$\partial\zeta(\dot{\alpha}) + \partial_\alpha\varphi(e(u),\alpha) - \mathrm{div}\big(\kappa\nabla\alpha\big) + r_\mathrm{c} \ni 0 \qquad \text{in } Q, \tag{2.6b}$$

$$r_\mathrm{c} \in N_{[0,1]}(\alpha) \qquad \text{in } Q, \tag{2.6c}$$

where $N_{[0,1]} = \partial\delta_{[0,1]}$ is the normal cone to the interval $[0,1]$ where α is supposed to be valued, together with the boundary conditions

$$\big(\mathbb{D}(\alpha)e(\dot{u}) + \partial_e\varphi(e(u),\alpha)\big)\mathbf{n} = g \quad \text{and} \quad \nabla\alpha\cdot\mathbf{n} = 0 \qquad \text{on } \Sigma, \tag{2.6d}$$

where $Q = \Omega\times I$ and $\Sigma = \Gamma\times I$ with $I = [0,T]$ for a fixed time horizon $T > 0$, and where $\mathbf{n}$ is the outward unit normal to Γ. In fact, (2.6b,c) can be understood as one doubly-nonlinear inclusion if the "reaction pressure" r_c would be substituted from (2.6b) into (2.6c). We will consider an initial-value problem and thus complete (2.6) by the initial conditions

$$u|_{t=0} = u_0, \qquad \dot{u}|_{t=0} = v_0, \qquad \alpha|_{t=0} = \alpha_0 \qquad \text{in } \Omega. \tag{2.7}$$

The energetics can be obtained by testing (2.6a) by $\dot{u}$ and (2.6b) by $\dot{\alpha}$. After integration over Ω with using Green's formula and by-part integration over a time interval $[0,t]$, this test yields, at least formally,[1]

[1] This means that (2.8) can rigorously be proved only for sufficiently smooth solutions, e.g. $\dot{\alpha}$ is to be in duality with $\mathrm{div}(\kappa\nabla\alpha)$, as e.g. in Proposition 2.3 below.

$$\int_\Omega \underbrace{\frac{\varrho}{2}|\dot u(t)|^2}_{\substack{\text{kinetic}\\ \text{energy at time } t}} + \underbrace{\varphi(e(u(t)),\alpha(t)) + \frac{\kappa}{2}|\nabla\alpha(t)|^2}_{\text{stored energy at time } t}\,\mathrm{d}x$$

$$+\int_0^t\!\!\int_\Omega \underbrace{\mathbb{D}(\alpha)e(\dot u)\colon e(\dot u) + \dot\alpha\partial\zeta(\dot\alpha)}_{\text{dissipation rate}}\,\mathrm{d}x\mathrm{d}t$$

$$= \int_\Omega \underbrace{\frac{\varrho}{2}|v_0|^2}_{\substack{\text{initial}\\ \text{kinetic energy}}} + \underbrace{\varphi(e(u_0),\alpha_0) + \frac{\kappa}{2}|\nabla\alpha_0|^2}_{\text{initial stored energy}}\,\mathrm{d}x$$

$$+\int_0^t\!\!\int_\Omega \underbrace{f\cdot\dot u}_{\substack{\text{power of}\\ \text{bulk load}}}\,\mathrm{d}x\mathrm{d}t + \int_0^t\!\!\int_\Gamma \underbrace{g\cdot\dot u}_{\substack{\text{power of}\\ \text{surface load}}}\,\mathrm{d}x\mathrm{d}t. \tag{2.8}$$

In fact, the model (2.6) may simplify in some particular situations when $r_\mathrm{c} = 0$ and (2.6c) can be omitted, in particular when

$$\partial_\alpha\varphi(e,0) \le 0 \quad\text{and}\quad \begin{cases}\partial_\alpha\varphi(e,1) \ge 0, & \text{or}\\ \zeta(\dot\alpha) = +\infty & \text{for } \dot\alpha > 0\,.\end{cases} \tag{2.9}$$

The first option allows for healing if ζ is finite also for $\dot\alpha > 0$, while the second option is called unidirectional damage. The condition $\partial_\alpha\varphi(e,0) = 0$ needs infinitely large driving force to achieve $\alpha = 0$, i.e. some sort of large hardening when damaging evolves.

The weak formulation of (2.6a) with the initial/boundary conditions from (2.6d)-(2.7) is quite standard, using usually one Green formula in space and one or two by-part integrations in time. The weak formulation of (2.6b,c) consists in two variational inequalities. Writing the convex subdifferential in (2.6b), one see the term $\int_Q \dot\alpha\partial_\alpha\varphi(e(u),\alpha)$ which is not a-priori integrable and we substitute it by using the calculus

$$\int_Q \dot\alpha\partial_\alpha\varphi(e(u),\alpha)\,\mathrm{d}x\mathrm{d}t = \int_\Omega \varphi\big(e(u(T)),\alpha(T)\big) - \varphi\big(e(u_0),\alpha_0\big)\,\mathrm{d}x$$
$$- \int_Q \partial_e\varphi(e(u),\alpha)\cdot e(\dot u)\,\mathrm{d}x\mathrm{d}t. \tag{2.10}$$

Thus, using the standard notation L^p, $W^{k,p}$, and $L^p(I;\cdot)$ or $W^{1,p}(I;\cdot)$ for Lebesgue, Sobolev, and Bochner or Bochner-Sobolev spaces using also the convention $H^k := W^{k,2}$, we arrive at:

Definition 1 (Weak formulation) A triple $(u,\alpha,r_\mathrm{c}){\in}H^1(I;H^1(\Omega;\mathbb{R}^d)){\times}H^1(Q){\times}$ ${\times}L^2(Q)$ is called a weak solution to the initial-boundary-value problem (2.6)–(2.7) if $u|_{t=0} = u_0$ and $0 \le \alpha \le 1$ hold a.e. together with

$$\int_Q \big(\mathbb{D}(\alpha)e(\dot{u}) + \partial_e\varphi(e(u),\alpha)\big) : e(v) - \varrho\dot{u}\cdot\dot{v}\,\mathrm{d}x\mathrm{d}t$$
$$= \int_\Omega v_0\cdot v(0,\cdot)\,\mathrm{d}x + \int_Q f\cdot v\,\mathrm{d}x\mathrm{d}t + \int_\Sigma g\cdot v\,\mathrm{d}S\mathrm{d}t \qquad (2.11a)$$

for all $v \in L^2(I;H^1(\Omega;\mathbb{R}^d)) \cap H^1(I;L^2(\Omega;\mathbb{R}^d))$ with $v|_{t=T} = \dot{v}|_{t=T} = 0$, and

$$\int_Q \partial_\alpha\varphi(e(u),\alpha)z + r_\mathrm{c}(z-\dot{\alpha}) + \kappa\nabla\alpha\cdot\nabla z + \zeta(z)\,\mathrm{d}x\mathrm{d}t$$
$$+ \int_\Omega \varphi\big(e(u_0),\alpha_0\big) + \frac{\kappa}{2}|\nabla\alpha_0|^2\,\mathrm{d}x \ge \int_Q \zeta(\dot{\alpha}) + \partial_e\varphi(e(u),\alpha) : e(\dot{u})\,\mathrm{d}x\mathrm{d}t$$
$$+ \int_\Omega \varphi\big(e(u(T)),\alpha(T)\big) + \frac{\kappa}{2}|\nabla\alpha(T)|^2\,\mathrm{d}x \qquad (2.11b)$$

to be valid for all $z \in C^1(\bar{Q})$ and with r_c satisfying (2.6c) a.e. on Q.

Let us now analyze the model with the (partly) damageable viscosity in the special situation that $\mathbb{D}(\alpha) = \mathbb{D}_0 + \chi\partial_e\varphi(\cdot,\alpha)$ with a relaxation time $\chi > 0$ possibly dependent on $x \in \Omega$, cf. [39] or also [49, Sect. 5.1.1 and 5.2.5] for the rate-independent unidirectional damage. This means that $\varphi(\cdot,\alpha)$ is quadratic and we thus specify the stored-energy density $\varphi : \mathbb{R}^{d\times d}_\mathrm{sym} \times [0,1] \to \mathbb{R}$ as

$$\varphi(e,\alpha) = \frac{1}{2}\mathbb{C}(\alpha)e : e - \phi(\alpha) \qquad (2.12)$$

with a 4th order elastic-moduli tensor $\mathbb{C} : [0,1] \to \mathbb{R}^{d\times d\times d\times d}$ continuously dependent on α and with ϕ standing for the specific energy of damage which (if ϕ is increasing) gives rise to a driving force for healing.

This specifies the system (2.6a-c) as

$$\dot{u} = v, \qquad \varrho\dot{v} - \mathrm{div}\big(\mathbb{D}(\alpha)e(v) + \mathbb{C}(\alpha)e(u)\big) = f \qquad \text{in } Q, \qquad (2.13a)$$
$$\partial\zeta(\dot{\alpha}) + \frac{1}{2}\mathbb{C}'(\alpha)e(u) : e(u) - \mathrm{div}\big(\kappa\nabla\alpha\big) \ni \phi'(\alpha) \qquad \text{in } Q, \qquad (2.13b)$$

when we confine ourselves to (2.9). Let us note that we introduce the auxiliary variable v standing for velocity and write, rather for later purposes in Sect. 4 the 1st-order system instead of the 2nd-order one. The mathematical treatment relies on the linearity of (2.13a) in terms of u but, on the other hand, (2.13b) is nonlinear in terms of $e = e(u)$.

Rather for simplicity, we consider the scenarios (2.9), which now means that $\mathbb{C}'(0) = 0$ and possibly (in the first option in (2.9)) also $\mathbb{C}'(1) = 0$. We consider a nested sequence of finite-dimensional subspaces of $H^1(\Omega;\mathbb{R}^d)$ and $H^1(\Omega)$ indexed by $k \in \mathbb{N}$ whose union is dense in these Banach spaces, and then an H^1-conformal Galerkin approximation, denoting the approximate solution thus obtained by (u_k,α_k). For simplicity, we assume that $u_0, v_0 \in V_1 \subset V_k \subset H^1(\Omega;\mathbb{R}^d)$ as used for the Galerkin approximation; in fact, a natural qualification $v_0 \in L^2(\Omega;\mathbb{R}^d)$ would in general need an approximation $v_{0,k} \in V_k$ such that $v_{0,k} \to v_0$ strongly in $L^2(\Omega;\mathbb{R}^d)$.

We allow for a *complete damage* in the elastic response, although a resting Stokes-type viscosity due to D_0 is needed for the following assertion relying on the linearity of $\partial_e\varphi(\cdot,\alpha)$, i.e. on that $\varphi(\cdot,\alpha)$ is quadratic:

Proposition 2.1 (Existence in the linear model) *Let the ansatz* (2.5) *be considered, let also* $\varrho,\kappa\in L^\infty(\Omega)$ *with* $\operatorname{ess\,inf}\varrho>0$ *and* $\operatorname{ess\,inf}\kappa>0$, $f\in L^1(I;L^2(\Omega;\mathbb{R}^d))$, $g\in L^2(\Sigma;\mathbb{R}^d))$, $u_0\in H^1(\Omega;\mathbb{R}^d)$, $v_0\in L^2(\Omega;\mathbb{R}^d)$, $\alpha_0\in H^1(\Omega)$ *with* $0\le\alpha_0\le 1$ *a.e. on* Ω *be supposed,* $\zeta:\mathbb{R}\to\mathbb{R}^+$ *be convex and lower semicontinuous with* $\zeta(\cdot)\ge\epsilon|\cdot|^2$ *for some* $\epsilon>0$, *and let* (2.9) *hold, and let also* (2.12) *be considered with*

$$\mathbb{C}\in C^1([0,1];\mathbb{R}^{(d\times d)^2})\ \text{be symmetric positive-semidefinite valued,} \tag{2.14a}$$

$$\mathbb{D}(\cdot)=\mathbb{D}_0+\chi\mathbb{C}(\cdot)\ \text{with}\ \chi\ge 0\ \text{and}\ \mathbb{D}_0\ \text{symmetric positive-definite.} \tag{2.14b}$$

Then the Galerkin approximation (u_k,α_k) *exists and, for selected subsequences, we have*

$$u_k\to u\quad \text{weakly* in}\ H^1(I;H^1(\Omega;\mathbb{R}^d))\cap W^{1,\infty}(I;L^2(\Omega;\mathbb{R}^d))\ \text{and} \tag{2.15a}$$

$$\text{strongly in}\ L^2(I;H^1(\Omega;\mathbb{R}^d)),\ \text{and} \tag{2.15b}$$

$$\alpha_k\to\alpha\quad \text{weakly* in}\ H^1(I;L^2(\Omega))\cap L^\infty(I;H^1(\Omega))\cap L^\infty(Q), \tag{2.15c}$$

and every such a limit (u,α) *is a weak solution in the sense of Definition 1 with* $r_c=0$. *Moreover, even* $e(\dot u_k)\to e(\dot u)$ *strongly in* $L^2(Q;\mathbb{R}^{d\times d}_{\mathrm{sym}})$.

Proof. The apriori estimates in the spaces occurring in (2.15a,c) can be obtained by standard energetic test by $\dot u_k$ and $\dot\alpha_k$, which leads to (2.8) written for the Galerkin approximation, and using Hölder's, Young's, and Gronwall's inequalities.

After selecting a subsequence weakly* converging in the sense (2.15a,c) and using the Aubin-Lions theorem for the damage and then continuity of the superposition operator induced by $\mathbb{C}(\cdot)$, we can pass to the limit first in the semilinear force-equilibrium equation. We put $w:=u+\chi\dot u$ and write the limit equation (2.6a) as

$$\frac{\varrho}{\chi}\dot w-\operatorname{div}\big(\mathbb{D}_0e(\dot u)+\mathbb{C}(\alpha)e(w)\big)=f+\frac{\varrho}{\chi}\dot u \tag{2.16}$$

accompanied with the corresponding initial/boundary conditions from (2.6d)–(2.7).

For the damage flow rule, we need the strong convergence of $\{e(u_k)\}_{k\in\mathbb{N}}$, however. Furthermore, we denote $w_k:=u_k+\chi\dot u_k$ and, using the linearity of $\partial_e\varphi(\cdot,\alpha)$, write the Galerkin approximation of the force equilibrium as[2]

$$\frac{\varrho}{\chi}\dot w_k-\operatorname{div}\big(\mathbb{D}_0e(\dot u_k)+\mathbb{C}(\alpha_k)e(w_k)\big)=f+\frac{\varrho}{\chi}\dot u_k\,. \tag{2.17}$$

Then we subtract (2.16) and (2.17), and test it by w_k-w, and integrate over the time interval $[0,t]$. This gives

[2] More precisely, (2.17) is to be understood valued in V_k^*.

$$\int_\Omega \frac{\varrho}{2\chi}|w_k(T){-}w(T)|^2 + \frac{1}{2}\mathbb{D}_0 e(u_k(T){-}u(T)){:}e(u_k(T){-}u(T))\,\mathrm{d}x \\ + \int_Q \mathbb{D}_0\chi e(\dot{u}_k{-}\dot{u}){:}\ e(\dot{u}_k{-}\dot{u}) + \mathbb{C}(\alpha_k)e(w_k{-}w){:}\ e(w_k{-}w)\,\mathrm{d}x\mathrm{d}t \\ = \int_Q \big(\mathbb{C}(\alpha_k){-}\mathbb{C}(\alpha)\big)e(w){:}\ e(w_k{-}w) + \frac{\varrho}{\chi}(\dot{u}_k{-}\dot{u}){\cdot}(w_k{-}w)\,\mathrm{d}x\mathrm{d}t \to 0\,. \tag{2.18}$$

Here we used that $\dot{u}_k{-}\dot{u} \to 0$ strongly in $L^2(Q;\mathbb{R}^d)$ by the Aubin-Lions theorem and also that $(\mathbb{C}(\alpha_k){-}\mathbb{C}(\alpha))e(w) \to 0$ strongly in $L^2(Q;\mathbb{R}^{d\times d}_{\rm sym})$. This gives (2.15b). In fact, (2.18) is again a rather conceptual strategy and still a strong approximation of (u,w) is needed to facilitate usage of the Galerkin identity and convergence-to-zero of the additional terms thus arising.

The limit passage in the damage variational inequality towards (2.11b) is then easy by (semi)continuity. □

In some applications a non-quadratic $\varphi(\cdot,\alpha)$ is a reasonable ansatz in particular because damage may act very differently on compression than on tension, cf. (2.28a) below. Examples are concrete- or masonry-, or rock-type materials where mere compression practically does not cause damage while tension (as well as shear) may cause damage relatively easily. Unfortunately, Proposition 2.1 does not cover such models. Two options allowing for α-dependent $\mathbb{D}$ are doable: a unidirectional damage with hardening-like effect and bi-directional (i.e. with possible healing) damage. Note that (2.14) is not needed. In the first option, the constraint $\alpha \geq 0$ is never active and $\alpha \leq 1$ is only "semi-active", both leading to zero Lagrange multiplier $r_{\rm c}$.

Proposition 2.2 (Unidirectional damage in nonlinear models) *Let the data ϱ, $\zeta(\cdot)$, κ, f, g, u_0, v_0, and α_0 be as in Proposition 2.1, and let also $\varrho \in W^{1,r}(\Omega)$ with $r = 3$ for $d = 3$ or $r > 1$ for $d = 2$,*

$$|\partial_e\varphi(e,\alpha)| \leq C\big(1 + |e|\big) \quad \text{and} \quad |\partial_\alpha\varphi(e,\alpha)| \leq C\big(1 + |e|^2\big). \tag{2.19}$$

Let moreover $\mathbb{D} : [0,1] \to \mathbb{R}^{(d\times d)^2}$ be symmetric-valued, continuous, monotone (nondecreasing) with respect to the Löwner ordering (i.e. ordering of $\mathbb{R}^{d\times d}_{\rm sym}$ by the cone of positive semidefinite matrices), and with $\mathbb{D}(0)$ positive definite. Let moreover the second option in (2.9) hold. Then the Galerkin approximate solutions do exist with $r_{{\rm c},k} = 0$. The sequence $\{(u_k,\alpha_k)\}_{k\in\mathbb{N}}$ possesses subsequences such that again (2.15) hold. The limit of each such a subsequence solves the initial-boundary-value problem (2.6)–(2.7) weakly in the sense of Definition 1 with $r_{\rm c} = 0$.

Proof. We perform the test of (2.6a,b) in the Galerkin approximation by $(\dot{u}_k,\dot{\alpha}_k)$. By using the data qualification and Hölder and Gronwall inequalities, this gives the estimates in the spaces occurring in (2.15a,c). By comparison from (2.6a), we obtain also the bound[3] for $\ddot{u}_k$ in $L^2(I;H^1(\Omega;\mathbb{R}^d)^*)$ by estimating

[3] More precisely, this bound is valid only in Galerkin-induced seminorms or for the Hahn-Banach extension, cf. [61, Sect. 8.4].

$$\int_Q \ddot{u}_k v\,\mathrm{d}x\mathrm{d}t = \int_Q (f + \mathrm{div}\,\sigma_k)\cdot\frac{v}{\varrho}\,\mathrm{d}x\mathrm{d}t = \int_Q \frac{f\cdot v}{\varrho} - \sigma_k : \nabla\frac{v}{\varrho}\,\mathrm{d}x\mathrm{d}t$$
$$= \int_Q \frac{f\cdot v}{\varrho} - \frac{\sigma_k : \nabla v}{\varrho} + \frac{\sigma_k : \nabla\varrho}{\varrho^2}\,\mathrm{d}x\mathrm{d}t + \int_\Sigma \frac{g\cdot v}{\varrho}\,\mathrm{d}S\mathrm{d}t \le C\|v\|_{L^2(I;H^1(\Omega;\mathbb{R}^d))}, \tag{2.20}$$

where C is dependent on the already obtained estimates (2.15a,c); note also that we need a certain smoothness of ϱ, as supposed.

After selection of weakly* convergent subsequences, we prove the strong convergence (2.15b). We use a slightly different estimation comparing to (2.18) based on a test by $\dot{u}_k - \dot{u}$, namely now we employ the test by $u_k - u$ to estimate

$$\int_\Omega \frac{1}{2}\mathbb{D}(\alpha_k(t))e(u_k(t)-u(t)) : e(u_k(t)-u(t))\,\mathrm{d}x$$
$$+ \int_0^t\int_\Omega \big(\partial_e\varphi(e(u_k),\alpha_k) - \partial_e\varphi(e(u),\alpha_k)\big) : e(u_k-u)\,\mathrm{d}x\mathrm{d}t$$
$$= \int_0^t\int_\Omega \frac{\partial}{\partial t}\frac{1}{2}\mathbb{D}(\alpha_k)e(u_k-u) : e(u_k-u)$$
$$+ \big(\partial_e\varphi(e(u_k),\alpha_k) - \partial_e\varphi(e(u),\alpha_k)\big) : e(u_k-u)\,\mathrm{d}x\mathrm{d}t$$
$$= \int_0^t\int_\Omega \big(\mathbb{D}(\alpha_k)e(\dot{u}_k-\dot{u}) + \partial_e\varphi(e(u_k),\alpha_k) - \partial_e\varphi(e(u),\alpha_k)\big) : e(u_k-u)$$
$$+ \frac{1}{2}\dot{\alpha}_k\mathbb{D}'(\alpha_k)e(u_k-u) : e(u_k-u)\,\mathrm{d}x\mathrm{d}t$$
$$\le \int_0^t\int_\Omega (f - \varrho\ddot{u}_k)\cdot(u_k-u) - \big(\mathbb{D}(\alpha_k)e(\dot{u}) + \partial_e\varphi(e(u),\alpha_k)\big) : e(u_k-u)\,\mathrm{d}x\mathrm{d}t$$
$$= \int_0^t\int_\Omega f\cdot(u_k-u) + \varrho\dot{u}_k\cdot(\dot{u}_k-\dot{u}) - \big(\mathbb{D}(\alpha_k)e(\dot{u}) + \partial_e\varphi(e(u),\alpha)\big) : e(u_k-u)\,\mathrm{d}x\mathrm{d}t$$
$$+ \int_0^t\int_\Omega \big(\partial_e\varphi(e(u),\alpha) - \partial_e\varphi(e(u),\alpha_k)\big) : e(u_k-u)\,\mathrm{d}x\mathrm{d}t$$
$$- \int_\Omega \varrho\dot{u}_k(t)\cdot(u_k(t)-u(t))\,\mathrm{d}x \to 0. \tag{2.21}$$

because $\dot{\alpha}_k\mathbb{D}'(\alpha_k)e(u_k-u) : e(u_k-u) \le 0$ a.e. on Q since $\dot{\alpha}_k \le 0$ due to the assumption that $\zeta_1(\dot{\alpha}) = +\infty$ for $\dot{\alpha} > 0$ and $\mathbb{D}'(\cdot)$ is positive semidefinite due to the assumption of monotone dependence of $\mathbb{D}(\cdot)$.

Then, having this strong convergence, we can easily pass to the limit by (semi)continuity both towards the identity (2.11a) and towards variational inequality (2.11b).□

It is interesting that the usual "limsup-argument" relying on the energy conservation to prove the strong convergence (2.15b) could not be used while (2.21) worked, relying on the unidirectionality of damage evolution. Let us further illustrate the opposite situation when (2.21) does not work while the energy conservation holds and facilitates the mentioned limsup-argument:

Proposition 2.3 (Damage with healing in nonlinear models) *Let the data* ϱ, $\mathbb{D}(\cdot)$, κ, f, g, u_0, v_0, *and* α_0 *be as in Proposition 2.2, and again* (2.19) *hold. Let moreover*

$$\exists 0 < \varepsilon \le C\ \forall (e,z) \in \mathbb{R}^{d\times d}_{\rm sym} \times \mathbb{R}: \quad \begin{array}{r}\varepsilon|z|^2 \le \zeta(z) \le C(1+|z|^2),\\ |\partial_\alpha\varphi(e,z)| \le C(1+|e|).\end{array} \tag{2.22}$$

and the Galerkin approximation is H^2*-conformal so that* $\mathrm{div}(\kappa\nabla\alpha_k)$ *is well defined. Then the mentioned Galerkin approximate solutions do exist. The sequence* $\{(u_k,\alpha_k,r_{\mathrm{c},k})\}_{k\in\mathbb{N}}$ *possesses subsequences such that again* (2.15) *hold together with*

$$r_{\mathrm{c},k} \to r_\mathrm{c} \quad \textit{weakly in } L^2(Q). \tag{2.23}$$

The limit of each such a subsequence solves the initial-boundary-value problem (2.6)–(2.7) *weakly in the sense of Definition 1. Moreover,* $\mathrm{div}(\kappa\nabla\alpha) \in L^2(Q)$, *the damage flow rule (2.6b,c) holds a.e. on* Q, *and the energy conservation holds.*

Proof. Let us outline only the differences from the proof of Proposition2.3. Beside the a-priori estimates there, we further test the approximated damage flow-rule by $\mathrm{div}(\kappa\nabla\alpha_k)$ We thus obtain a bound for $\mathrm{div}(\kappa\nabla\alpha_k)$ in $L^2(Q)$, and eventually also for $r_{\mathrm{c},k} \in \mathrm{div}(\kappa\nabla\alpha_k) + \phi'(\alpha_k) - \partial\zeta(\dot\alpha_k) - \partial_\alpha\varphi(e(u_k),\alpha_k)$ in $L^2(Q)$.[4] Here we used also the growth conditions (**??**) guaranteeing that both $\partial\zeta(\dot\alpha_k)$ and $\partial_\alpha\varphi(e(u_k),\alpha_k)$ are bounded in $L^2(Q)$.

We now can pass to the limit in the force equilibrium just by the weak convergence and monotonicity of $\partial_e\varphi(\cdot,\alpha_k)$ and the Aubin-Lions compactness theorem used for α_k. Having the limit equation (2.6a) at disposal in the weak sense (2.11a), we can test it by $v = \dot u$ and show energy conservation in this part of the system. To this goal, it is important that both $\dot\alpha$ and $\partial_\alpha\varphi(e(u),\alpha)$ are in $L^2(Q)$ so that the chain rule (2.10) rigorously holds and that $\sqrt{\varrho}\ddot u \in L^2(I;H^1(\Omega;\mathbb{R}^d)^*)$ is in duality with $\sqrt{\varrho}\dot u \in L^2(I;H^1(\Omega;\mathbb{R}^d))$ so that also the chain rule $\int_Q \varrho\dot u\colon \ddot u\,\mathrm{d}x\mathrm{d}t = \int_\Omega \frac12\varrho|\dot u(T)|^2 - \frac12\varrho|\dot u(0)|^2\,\mathrm{d}x$; the information about $\sqrt{\varrho}\ddot u$ can be obtained by a simple modification of (2.20). By this test, we obtain

$$\int_\Omega \frac12\varrho|\dot u(T)|^2 + \varphi(e(u(T)),\alpha(T))\,\mathrm{d}x + \int_Q \mathbb{D}(\alpha)e(\dot u)\colon e(\dot u)\,\mathrm{d}x\mathrm{d}t = \int_\Gamma g\cdot\dot u\,\mathrm{d}S\mathrm{d}t$$
$$+ \int_\Omega \frac12\varrho|v_0|^2 + \varphi(e(u_0),\alpha_0)\,\mathrm{d}x + \int_Q f\cdot\dot u - \dot\alpha\partial_\alpha\varphi(e(u),\alpha)\,\mathrm{d}x\mathrm{d}t \tag{2.24}$$

Instead of (2.21), we now estimate by weak semicontinuity

[4] Here we have employed also the calculus $\int_\Omega r_{\mathrm{c},k}\mathrm{div}(\kappa\nabla\alpha_k)\,\mathrm{d}x = \int_\Omega \partial\delta_{[0,1]}(\alpha_k)\mathrm{div}(\kappa\nabla\alpha_k)\,\mathrm{d}x = -\int_\Omega \kappa\nabla\big[\partial\delta_{[0,1]}(\alpha_k)\big]\cdot\nabla\alpha_k\,\mathrm{d}x = -\int_\Omega \kappa\partial^2\delta_{[0,1]}(\alpha_k)\nabla\alpha_k\cdot\nabla\alpha_k\,\mathrm{d}x \le 0$.

$$
\begin{aligned}
\int_Q \mathbb{D}(\alpha)e(\dot u)\colon e(\dot u)\,\mathrm{d}x\mathrm{d}t &\le \liminf_{k\to\infty}\int_Q \mathbb{D}(\alpha_k)e(\dot u_k)\colon e(\dot u_k)\,\mathrm{d}x\mathrm{d}t\\
&= \int_\Omega \frac12\varrho|v_0|^2 + \varphi(e(u_0),\alpha_0)\,\mathrm{d}x + \lim_{k\to\infty}\int_Q f\cdot\dot u_k - \dot\alpha_k\partial_\alpha\varphi(e(u_k),\alpha_k)\,\mathrm{d}x\mathrm{d}t\\
&\qquad - \liminf_{k\to\infty}\int_\Omega \frac12\varrho|\dot u_k(T)|^2 + \varphi(e(u_k(T)),\alpha_k(T))\,\mathrm{d}x\\
&\le \int_\Omega \frac12\varrho|v_0|^2 + \varphi(e(u_0),\alpha_0)\,\mathrm{d}x - \int_\Omega \frac12\varrho|\dot u(T)|^2 + \varphi(e(u(T)),\alpha(T))\,\mathrm{d}x\\
&\qquad + \int_Q f\cdot\dot u - \dot\alpha\partial_\alpha\varphi(e(u),\alpha)\,\mathrm{d}x\mathrm{d}t = \int_Q \mathbb{D}(\alpha)e(\dot u)\colon e(\dot u)\,\mathrm{d}x\mathrm{d}t \qquad (2.25)
\end{aligned}
$$

where the last equality is due to (2.24). Altogether, we have proved that $\liminf_{k\to\infty}\int_Q\mathbb{D}(\alpha_k)e(\dot u_k)\colon e(\dot u_k)\,\mathrm{d}x\mathrm{d}t = \int_Q \mathbb{D}(\alpha)e(\dot u)\colon e(\dot u)\,\mathrm{d}x\mathrm{d}t$. From this, we obtain even $e(\dot u_k) \to e(\dot u)$ strongly in $L^2(Q;\mathbb{R}^{d\times d}_{\rm sym})$. More in detail, using uniform positive definiteness of $\mathbb{D}(\cdot)$, we perform the estimate

$$
\begin{aligned}
\min_{\alpha\in[0,1]}|\mathbb{D}^{-1}(\alpha)|^{-1}\|e(\dot u_k-\dot u)\|^2_{L^2(Q;\mathbb{R}^{d\times d})} &\le \int_Q \mathbb{D}(\alpha_k)e(\dot u_k-\dot u)\colon e(\dot u_k-\dot u)\,\mathrm{d}x\mathrm{d}t\\
= \int_Q \mathbb{D}(\alpha_k)e(\dot u_k)\colon e(\dot u_k) - 2\mathbb{D}(\alpha_k)e(\dot u_k)&\colon e(\dot u) + \mathbb{D}(\alpha_k)e(\dot u)\colon e(\dot u)\,\mathrm{d}x\mathrm{d}t \to 0.
\end{aligned} \qquad (2.26)
$$

Hence, we obtained even more that the desired strong convergence $e(u_k) \to e(u)$.

Now, beside the limit passage as in the proof of Proposition 2.3, also the limit passage towards the inclusion (2.6c) using that $N_{[0,1]}(\cdot)$ has a closed monotone graph is easy since $\alpha_k \to \alpha$ strongly in $L^2(Q)$ due to Aubin-Lions theorem while $r_{{\rm c},k} \to r_{\rm c}$ weakly in $L^2(Q)$. In particular, we have the chain rule $\int_Q r_{\rm c}\dot\alpha\,\mathrm{d}x\mathrm{d}t = 0$ at disposal.

Eventually, as both $\dot\alpha$ and $\mathrm{div}(\kappa\nabla\alpha)$ are in $L^2(Q)$, we have also the chain rule $\int_Q \dot\alpha\mathrm{div}(\kappa\nabla\alpha)\,\mathrm{d}x\mathrm{d}t = \int_\Omega \frac12\kappa|\nabla\alpha_0|^2 - \frac12\kappa|\nabla\alpha(T)|^2\,\mathrm{d}x$ and we can test the damage flow rule by $\dot\alpha$, and then sum it with (2.24) to obtain the energy balance (2.8). □

Let us notice that a unidirectional damage eveolution (i.e. $\zeta(\dot\alpha) = +\infty$ for $\dot\alpha > 0$) together with the damage, where the indicator function $\delta_{[0,1]}$ in (2.5b) and (2.6c) must be considered, is not covered by Propositions 2.1–2.3. A particular case when $\zeta(\cdot)$ is positively homogeneous (i.e. the damage-process itself is rate-independent) allows a particular treatment by using a so-called energetic formulation, invented for rate-independent systems by A. Mielke at al. [48, 49, 53], and later adapted for dynamical systems containing rate-independent sub-systems in [60]. We need a space of functions $I \to L^1(\Omega)$ of bounded variations, denoted by $\mathrm{BV}(I;L^1(\Omega))$, i.e. the Banach space of functions $\alpha: I \to L^1(\Omega)$ with $\sup_{0\le t_0<t_1<\ldots t_N\le T,\ N\in\mathbb{N}}\sum_{i=1}^N \|\alpha(t_i)-\alpha(t_{i-1})\|_{L^1(\Omega)}$ finite.

Definition 2 (Energetic formulation) A pair $(u,\alpha) \in H^1(I;H^1(\Omega;\mathbb{R}^d)) \times \mathrm{BV}(I;L^1(\Omega)) \cap L^\infty(I;H^1(\Omega))$ is called an energetic solution to the initial-boundary-value problem (2.6)–(2.7) if again (2.11a) holds for all $v \in L^2(I;H^1(\Omega;\mathbb{R}^d)) \cap$

$H^2(I; L^2(\Omega;\mathbb{R}^d))$ with $v|_{t=T} = \dot{v}|_{t=T} = 0$, if also the energy balance (2.8) holds, and if the so-called semi-stability

$$\int_\Omega \varphi\big(e(u(t)),\alpha(t)\big) + \frac{\kappa}{2}|\nabla\alpha(t)|^2 \le \int_\Omega \varphi\big(e(u(t)),z\big) + \frac{\kappa}{2}|\nabla z|^2 + \zeta\big(z-\alpha(t)\big)\mathrm{d}x$$

holds for all $t \in I$ and all $z \in H^1(\Omega)$ with $0 \le z \le 1$ a.e. on Ω, and also $\alpha|_{t=0} = \alpha_0$.

Under the semistability of the initial damage profile α_0, existence of energetic solutions can be shown by the implicit or semi-implicit time discretisation and by an explicit construction of a so-called mutual recovery sequence [50].[5] For even an anisothermally enhanced model, we also refer to [40].

An important special case consist in isotropic materials, where one can easily distinguish response under volumetric and shear load which might be very diverse. To this goal, we use the decomposition of the strain to its compression/tension spherical and its deviatoric parts:

$$e = \mathrm{sph}^+ e + \mathrm{sph}^- e + \mathrm{dev}\, e \quad \text{with} \quad \mathrm{sph}^\pm e = \frac{(\mathrm{tr}\, e)^\pm}{d}\mathbb{I}, \tag{2.27}$$

where $(\mathrm{tr}\, e)^+ = \max(\sum_{i=1}^d e_{ii}, 0)$ and $(\mathrm{tr}\, e)^- = \min(\sum_{i=1}^d e_{ii}, 0)$. Note that the deviatoric and the spherical strains from (2.27) are orthogonal to each other and the deviatoric strain is trace-free, i.e. $\mathrm{dev}\, e : \mathrm{sph}^\pm e = 0$ and $\mathrm{tr}(\mathrm{dev}\, e) = 0$. This decomposition allows for distinguishing the response under compression (usually not causing damage), tension (so-called Mode I damage), or shearing (Mode II damage). A combination is called a mixed mode, and altogether we speak about a mode sensitive damage.

The latter growth restriction in (2.22) excludes quadratic energies of the type (2.12). An example of a model with the mode-sensitive isotropic stored energy satisfying this restriction and the damage dissipation potential with quadratic coercivity might be

$$\varphi(e,\alpha) = \frac{d}{2}\bigg(K(1)|\mathrm{sph}^- e|^2 + \frac{K(\alpha)|\mathrm{sph}^+ e|^2}{\sqrt{1+\epsilon(\mathrm{tr}\, e)^2}}\bigg) + \frac{G(\alpha)|\mathrm{dev}\, e|^2}{\sqrt{1+\epsilon|\mathrm{dev}\, e|^2}}, \tag{2.28a}$$

$$\zeta(e;\dot\alpha) = \begin{cases} -\mathfrak{g}_\mathrm{c}\dot\alpha + \nu\dot\alpha^2 & \text{if } \dot\alpha \le 0, \\ +\infty & \text{if } \dot\alpha > 0, \end{cases} \tag{2.28b}$$

---

[5] This specifically here means that, having $u_k \to u$ strongly in $H^1(\Omega;\mathbb{R}^d)$ and $\alpha_k \to \alpha$ weakly in $H^1(\Omega)$ with $0 \le \alpha_k \le 1$, we need to find a sequence $\{z_k\}_{k\in\mathbb{N}}$ such that

$$\limsup_{k\to\infty} \int_\Omega \varphi\big(e(u_k), z_k\big) + \frac{\kappa}{2}|\nabla z_k|^2 + \zeta(z_k-\alpha_k) - \varphi\big(e(u_k),\alpha_k\big) - \frac{\kappa}{2}|\nabla\alpha_k|^2\mathrm{d}x$$
$$\le \int_\Omega \varphi\big(e(u),z\big) + \frac{\kappa}{2}|\nabla z|^2 + \zeta(z-\alpha) - \varphi\big(e(u),\alpha(t)\big) - \frac{\kappa}{2}|\nabla\alpha|^2\mathrm{d}x\,.$$

This needs quite sophisticated construction devised in [74]. When the quadratic gradient term in (2.5b) would be replaced by the p-power with $p > d$, a simpler construction would apply, cf. [74].

where $K = K(\alpha)$ is the bulk modulus and $G = G(\alpha)$ is the shear modulus; recall that $K = \lambda + 2G/d$ with λ and G the so-called Lamé constants. The coefficient $\mathfrak{g}_c > 0$ in (2.28b) is called a fracture toughness while the coefficient $\nu > 0$ makes fast damage more dissipative (more heat producing) than slower damage, which might be sometimes relevant and which makes mathematics sometimes easier, as in Propositions 2.1 and 2.2 above. The (small) regularizing parameter $\epsilon > 0$ makes the tension and shear stress bounded if $|e|$ is (very) large and makes the growth restriction on $\partial_\alpha\varphi$ in (2.22) satisfied, while $\varepsilon = 0$ is admitted in the case of the second option in (2.9).

Let us illustrate heuristically how the flow-rule $\zeta(\dot\alpha) + \partial_\alpha\varphi(e,\alpha) \ni 0$ with the initial condition $0 < \alpha(0) = \alpha_0 \le 1$ operates when the loading gradually increases. For the example (2.28a) with $\epsilon = 0$ and (2.28b) with $\nu = 0$, the stress is $\sigma = \partial_e\varphi(e,\alpha) = \sigma^-_{\rm sph} + \sigma^+_{\rm sph} + \sigma_{\rm dev}$ with $\sigma^-_{\rm sph} = dK(1){\rm sph}^- e$, $\sigma^+_{\rm sph} = dK(\alpha){\rm sph}^+ e$, and $\sigma_{\rm dev} = 2G(\alpha){\rm dev}\, e$. The driving force for damage evolution expressed in therm of the actual stress is

$$\partial_\alpha\varphi(e,\alpha) = \frac{1}{2}K'(\alpha)|{\rm sph}^+ e|^2 + G'(\alpha)|{\rm dev}\, e|^2 = \frac{K'(\alpha)}{2dK(\alpha)^2}|\sigma^+_{\rm sph}|^2 + \frac{G'(\alpha)}{4G(\alpha)^2}|\sigma_{\rm dev}|^2.$$

Then the criterion $\partial_\alpha\varphi(e,\alpha_0) = \mathfrak{g}_c$ reveals the stress needed to start damaging the material. In the pure shear or pure tension, this critical stress is

$$|\sigma_{\rm dev}| = G(\alpha_0)\sqrt{\frac{4\mathfrak{g}_c}{G'(\alpha_0)}} = \text{“effective fracture stress” in Mode II.} \qquad (2.29)$$

$$|\sigma^+_{\rm sph}| = K(\alpha_0)\sqrt{\frac{2d\mathfrak{g}_c}{K'(\alpha_0)}} = \text{“effective fracture stress” in Mode I.} \qquad (2.30)$$

respectively. If $G(\cdot)G'(\cdot)^{-1/2}$ and $K(\cdot)K'(\cdot)^{-1/2}$ are increasing (in particular if $G(\cdot)$ and $K(\cdot)$ are concave), and the loading is via stress rather than displacement, damage then accelerates when started so that the rupture happens immediately (if any rate and spatial-gradient effects are neglected).

3 Phase-field concept towards fracture

The concept of bulk damage can (asymptotically) imitate the philosophy of fracture along surfaces (*cracks*) provided the damage stored energy ϕ is big. A popular ansatz takes the basic model (2.5) with (2.12) for

$$\mathbb{C}(\alpha) := \Big(\frac{\varepsilon^2}{\varepsilon_0^2}+\alpha^2\Big)\mathbb{C}_1, \quad \phi(\alpha) := -\mathfrak{g}_c\frac{(1-\alpha)^2}{2\varepsilon}, \quad \text{and} \quad \kappa := \varepsilon\mathfrak{g}_c \qquad (3.1)$$

with $\mathfrak{g}_c$ denoting the energy of fracture and with ε controlling a “characteristic” width of the *phase-field fracture* zone(s). This width is supposed to be small with respect to the size of the whole body. Then, (2.5b) looks (up to the forcing f and g) as

$$\mathscr{E}(u,\alpha) := \int_\Omega \gamma(\alpha)\mathbb{C}e(u){:}e(u) + \underbrace{\mathfrak{g}_\mathrm{c}\Big(\frac{1}{2\varepsilon}(1-\alpha)^2+\frac{\varepsilon}{2}|\nabla\alpha|^2\Big)}_{\text{crack surface density}}\,\mathrm{d}x \quad\text{with}\quad \underbrace{\gamma(\alpha) = \frac{(\varepsilon/\varepsilon_0)^2+\alpha^2}{2}}_{\text{degradation function}} \tag{3.2}$$

and with $\varepsilon_0 > 0$. The physical dimension of ε_0 as well as of ε is m (meters) while the physical dimension of $\mathfrak{g}_\mathrm{c}$ is J/m^2. This is known as the so-called *Ambrosio-Tortorelli functional.* [6] The fracture toughness $\mathfrak{g}_\mathrm{c}$ is now involved in (3.2) instead of the dissipation potential (2.28b), i.e. ζ in (2.28b) is now considered with $\mathfrak{g}_\mathrm{c} = 0$.

It should be emphasized that, in the "crack limit" for $\varepsilon \to 0$, the phase-field fracture model (3.2) approximates (at least in the static and quasistatic cases) the true infinitesimally thin cracks in Griffith's [27] variant (i.e. competition of energies), which works realistically for crack propagation but might have unrealistic difficulties with crack initiation,[7] while scaling of the fracture energy to 0 if $\varepsilon \to 0$ might lead to opposite effects, cf. also the discussion e.g. in Remark 2 below. This is partly reflected by the fact that, in its rate-independent variant, the damage and phase-field fracture models admit many various solutions of very different characters, as presented in [49]. In the dynamic variant, the influence of overall stored energy during fast rupture (which may be taken into account during quasistatic evolution) seems eliminated because of finite speed of propagation of information about it.[8]

Various modifications have been devised. For example, Bourdin at al. [11] used

$$\phi(\alpha) = \frac{3\mathfrak{g}_\mathrm{c}\alpha}{8\varepsilon} \quad\text{and}\quad \kappa = \frac{3\mathfrak{g}_\mathrm{c}\varepsilon}{4} \tag{3.3}$$

in (2.5) and $\mathfrak{g}_\mathrm{c}$ the energy of fracture in J/m^2 and with ε controlling a "characteristic" width of the phase-field fracture zone. Although it activates damage process only when stress achieves some threshold, it exhibits a similar undesired behaviour as (3.2) when $\varepsilon \to 0$ and leads to consider $\varepsilon > 0$ as another parameter (without intention to put it 0) in addition to $\mathfrak{g}_\mathrm{c}$ to tune the model. Various modifications of the degradation function from (3.2) therefore appeared in literature. E.g. a cubic degradation function γ has been used e.g. in [7, 75]. Inspired by (2.29), keeping

---

[6] In the static case, this approximation was proposed by Ambrosio and Tortorelli [2, 3] originally for the scalar Mumford-Shah functional [54] and the asymptotic analysis for $\varepsilon \to 0$ was rigorously executed. A generalization (in some sense in the spirit of finite-fracture mechanics) is in [14]. The generalization for the vectorial case is in [19, 20, 33]. Later, it was extended for evolution situation, namely for a rate-independent damage, in [25], see also also [9, 10, 12, 39, 49] where also inertial forces are sometimes considered.

[7] In fact, as $\phi'(1) = 0$, the initiation of damage has zero threshold and is happening even on very low stress but then, if $\varepsilon > 0$ is very small, stops and high stress is needed to continue damaging.

[8] Yet, in this dynamic case, the analysis for $\varepsilon \to 0$ remains open and, even worse, in the limit crack problem one should care about non-interpenetration, which is likely very difficult; cf. the analysis for the damage-to-delamination problem [51].

still the original motivation $\varepsilon \to 0$, one can think about some γ convex increasing with $\gamma'(1) = \mathscr{O}(1/\varepsilon)$. The mentioned cubic ansatz is not compatible with these requirement. Some more sophisticated γ's have been devised in [69, 79]. We have thus more independent parameters than only $\mathfrak{g}_c$ and ε in (3.2) to specify the lengh-scale of damage zone, fracture propagation and fracture initiation.

Remark 1 (Finite fracture mechanics (FFM)) In contrast to the Griffith model relevant rather for infinitesimally short increments of cracks[9], the finite (large) increments needs rather the concept of energetic solution. As already mentioned, it does not seem much realistic to count with the overall strain energy in very distant spots (particularly in dynamical problems with finite speed of propagation of information), so rather only energy around a current point $x \in \Omega$ is to be considered and cracks can propagate only by finite distance during incremental stepping. This is concept is commonly called a finite fracture mechanics (FFM); this term has been suggested by Z. Hashin [28], but being developed rather gradually by several authors, see e.g. [73]. It occurs useful in particular in quasistatic problems which neglect inertia to compensate (rather phenomenologically) this simplification.

Remark 2 (Coupled stress-energy criterion) In addition to FFM, in fracture (or in general damage) mechanics, there is a disputation whether only sufficiently big stress can lead to rupture or (in reminiscence to Grifith's concept) whether (also or only) some sufficiently big energy in the specimen or around the crack process zone is needed for it. A certain standpoint is that both criteria should be taken into account. This concept is nowadays referred as coupled stress-energy criterion.[10] Here, this coupled-criterion concept can be reflected by making ζ dependent on the strain energy. Having in mind FFM, one can think to let $\zeta = \zeta(\widetilde{\varepsilon}; \dot{\alpha})$ with $\widetilde{\varepsilon}(x) = \int_\Omega k(x-\widetilde{x})\varphi(e(u(\widetilde{x})), \alpha(\widetilde{x}))\,\mathrm{d}\widetilde{x}$ for some kernel $k : \mathbb{R}^d \to \mathbb{R}^+$, making $-\partial_{\dot{\alpha}}\zeta(\widetilde{\varepsilon}; \dot{\alpha})$ larger if $\widetilde{\varepsilon}$ is small.

Remark 3 (Mixity-mode sensitive cracks) Combining the mixity-mode sensitive model (2.28a) with the crack surface density from (3.2), one can distinguish the fracture by tension while mere compression does not lead to fracture, cf. [47, 70], and one can also distinguish the *Mode I* (*fracture* by opening) from *Mode II* (fracture by shear), cf. [38]. More in detail, like in (2.28a), we can use different degradation functions γ's for the deviatoric part and the spherical compressive part and the spherical tension part.

Remark 4 (Various other models) An alternative option how to distinguish Mode I from Mode II is in the dissipation potential, reflecting the experimental observation that Mode II needs (dissipates) more energy than Mode I. Thus, one can take a state-dependent $\zeta = \zeta(e; \dot{\alpha})$ e.g. as (2.28b) with $\mathfrak{g}_c = \mathfrak{g}_c(\mathrm{sph}\, e, \mathrm{dev}\, e) > 0$ to be

---

[9] Cf. e.g. the analysis and discussion in [71] in the quasistatic situations.

[10] Cf. the survey [78], and has been devised and implemented in many variants in engineering literature, cf. e.g. [15,24,41,43,44], always without any analysis of numerical stability and convergence and thus computational simulations based on these models, whatever practical applications they have, stay in the position of rather speculative playing with computers.

rather small if $\mathrm{tr}\,e \gg |\mathrm{dev}\,e|$ (which indicates Mode I) and bigger if $|\mathrm{tr}e| \ll |\mathrm{dev}\,e|$ (i.e. Mode II), or very large if $\mathrm{tr}e \ll -|\mathrm{dev}\,e|$ (compression leading to no fracture). Moreover, in the spirit of FFM from Remark 2, one can consider energy in a finite neighbourhood of a current point, here split into the spherical and the shear parts to make ζ mode sensitive. Of course, combination of both alternatives (i.e. also from Remark 3) is possible, too. On top of it, one can also consider a combination with other dissipative processes triggered only in Mode II, a prominent example being isochoric plasticity with hardening, cf. Sect. 5.1. Altogether, there are many parameters with clear physical interpretation in the model to fit the model with many possible experiments in concrete situations.

4 Various time discretisations

In principle, the 2nd-order time derivative $\ddot{u}$ can be discretised by 2nd-order time differences, either as an explicit (as in (4.19) below) or an implicit scheme. This typically requires a fixed time step, and in the implicit variant exhibits unacceptably spurious numerical dissipation. Therefore, we avoid such discretisation here and work rather with the 1st-order system (2.13a) so that variable time-step is easily possible. Anyhow, for notational simplicity, we consider an equidistant partition of the time interval $I = [0,T]$ with a fixed time step $\tau > 0$ with $T/\tau \in \mathbb{N}$.

Considering some approximate values $\{u_\tau^k\}_{k=0,\dots,K}$ of the displacement u with $K = T/\tau$, we define the piecewise-constant and the piecewise affine interpolants respectively by

$$\overline{u}_\tau(t) = u_\tau^k, \qquad \underline{u}_\tau(t) = u_\tau^{k-1}, \qquad \overline{\underline{u}}_\tau(t) = \frac{1}{2}u_\tau^k + \frac{1}{2}u_\tau^{k-1}, \quad \text{and} \tag{4.1a}$$

$$u_\tau(t) = \frac{t-(k-1)\tau}{\tau}u_\tau^k + \frac{k\tau - t}{\tau}u_\tau^{k-1} \qquad \text{for } (k-1)\tau < t \le k\tau. \tag{4.1b}$$

Similar meaning will have also v_τ, $\overline{v}_\tau$, etc.

4.1 Implicit "monolithic" discretisation in time

Some applications need to reflect the coupled character of the problem in the truly coupled discrete fully-implicit scheme, in contrast to the decoupled scheme considered in Sect. 4.2 below. This is indeed often solved in engineering, but only an approximate solution can be expected by some iterative procedures.[11] Such schemes are known in engineering literature under the adjective "monolithic" and the mentioned iterative solution is e.g. by the Newton-Raphson (or here equivalently

[11] Some models are even formulated only in quasistatic time-discrete variants without having much chance to converge to some time-continuous problem; an example might be models with sharp interface between undamaged and partly damaged regions, as in [1, 80].

SQP = sequential quadratic programming) method without any guaranteed convergence, however, or alternating-minimization algorithm (AMA)[12]. In general, such schemes even do not seem numerically stable because the a-priori estimates are not available. The semiconvexity here with respect to the $(H^1{\times}L^2)$-norm can be exploited provided the Kelvin-Voigt viscosity is used, as it is indeed considered in Sect. 2 and 3.

In contrast to the usual fully implicit scheme discretising the inertial term by the second-difference formula $\varrho(u_\tau^k{-}2u_\tau^{k-1}{+}u_\tau^{k-2})/\tau^2$ as e.g. in serving satisfactorily for analytical purpoces but causing an unacceptably large spurious numerical dissipation, cf. e.g. [10,39], we discretise the inertial part by the mid-point (Crank-Nicolson) formula rather than the backward Euler one in order to reduce unwanted numerical attenuation, and we use a semi-implicit (but not the fully implicit backward-Euler) formula for the visco-elastic stress while α is taken in an explicit way for the viscous part in order to keep the variational structure of the incremental problems, cf. (4.6) below, and to guarantee existence of the discrete solutions. The resulted recursive coupled boundary-value problems here are:

$$\frac{u_\tau^k{-}u_\tau^{k-1}}{\tau} = v_\tau^{k-1/2} \qquad \text{with} \ \ v_\tau^{k-1/2} := \frac{v_\tau^k{+}v_\tau^{k-1}}{2}, \tag{4.2a}$$

$$\varrho\frac{v_\tau^k{-}v_\tau^{k-1}}{\tau} - \operatorname{div}\Big(\mathbb{D}(\alpha_\tau^{k-1})e(v_\tau^{k-1/2}) + \mathbb{C}(\alpha_\tau^k)e(u_\tau^k)\Big) = f_\tau^k \tag{4.2b}$$

$$\text{with} \ \ f_\tau^k := \frac{1}{\tau}\int_{(k-1)\tau}^{k\tau} f(t)\,\mathrm{d}t, \ \ \text{and} \tag{4.2c}$$

$$\partial\zeta\Big(\frac{\alpha_\tau^k{-}\alpha_\tau^{k-1}}{\tau}\Big) + \frac{1}{2}\mathbb{C}'(\alpha_\tau^k)e(u_\tau^k){:}e(u_\tau^k) - \operatorname{div}(\kappa|\nabla\alpha_\tau^k|^{p-2}\nabla\alpha_\tau^k) \ni \phi'(\alpha_\tau^k) \tag{4.2d}$$

considered on Ω while completed with the corresponding boundary conditions

$$\Big(\mathbb{C}(\alpha_\tau^k)e(u_\tau^k) + \mathbb{D}(\alpha_\tau^{k-1})e(v_\tau^{k-1/2})\Big)\mathbf{n} = g_\tau^k \ \ \text{and} \tag{4.3a}$$

$$\kappa\nabla\alpha_\tau^k{\cdot}\mathbf{n} = 0, \ \ \text{where} \ \ g_\tau^k := \frac{1}{\tau}\int_{(k-1)\tau}^{k\tau} g(t)\,\mathrm{d}t. \tag{4.3b}$$

It is to be solved recursively for $k = 1,...,T/\tau$, starting for $k = 1$ with

$$u_\tau^0 = u_0, \qquad v_\tau^0 = v_0, \qquad \alpha_\tau^0 = \alpha_0. \tag{4.4}$$

In terms of the interpolants, see (4.1), one can write the scheme (4.2) more "compactly" as

[12] In the rate-independent quasistatic variant, AMA is similar the splitting scheme as in Sect. 4.2 if the loading is modified as piecewise-constant in (rescaled) time except that the irreversibility constraint on the the damage profiles is up-dated differently. It was scrutinized e.g. in [36,45,56] and used e.g. in [38].

$$\dot{u}_\tau = \underline{v}_\tau \quad \text{and} \quad \varrho\dot{v}_\tau - \operatorname{div}\big(\mathbb{D}(\underline{\alpha}_\tau)e(\underline{v}_\tau) + \mathbb{C}(\overline{\alpha}_\tau)e(\overline{u}_\tau)\big) = \overline{f}_\tau\,, \tag{4.5a}$$

$$\partial\zeta\big(\dot{\alpha}_\tau\big) + \frac{1}{2}\mathbb{C}'(\overline{\alpha}_\tau)e(\overline{u}_\tau)\colon e(\overline{u}_\tau) - \operatorname{div}(\kappa|\nabla\overline{\alpha}_\tau|^{p-2}\nabla\overline{\alpha}_\tau) \ni \phi'(\overline{\alpha}_\tau)\,. \tag{4.5b}$$

The boundary conditions (4.3) can be written analogously.

Actually, we slightly modified the model used in Sections 2 and Section 3 by considering a p-Laplacian. For $p = 2$, we obtain the previous ansatz but for the convergence analysis we will need $p > d$.[13] Using the ansatz (3.2) with γ smooth, positive, and strictly convex, then $\frac{1}{2}\mathbb{C}(\alpha)e{:}e + \frac{1}{2}K|e|^2$ is convex for all K large enough. These underlying potentials are strongly convex[14] for the time-step $\tau > 0$ small enough and, assuming also a conformal space discretisation, the iterative solvers have guaranteed convergence towards a unique (globally minimizing) solution of the implicit scheme (4.2). This is satisfied for φ from (2.28a) with $\varepsilon = 0$ or from (3.2). The mentioned potential of the boundary-value problem (4.2)–(4.3) is

$$(u,\alpha) \mapsto \int_\Omega \frac{\varrho}{2\tau}\Big|\frac{u-u_\tau^{k-1}}{\tau} - v_\tau^{k-1}\Big|^2 + \frac{1}{2}\mathbb{C}(\alpha)e(u){:}e(u) - \phi(\alpha) + \tau\zeta\Big(\frac{\alpha-\alpha_\tau^{k-1}}{\tau}\Big)$$
$$+ \frac{1}{2\tau}\mathbb{D}(\alpha_\tau^{k-1})e(u-u_\tau^{k-1}){:}e(u-u_\tau^{k-1}) + \frac{\kappa}{p}|\nabla\alpha(t)|^p - f_\tau^k\cdot u\,\mathrm{d}x - \int_\Gamma g_\tau^k\cdot u\,\mathrm{d}S\,. \tag{4.6}$$

It is weakly lower semicontinuous on $H^1(\Omega;\mathbb{R}^d) \times H^1(\Omega)$ and coercive, so it serves also for proving existence of a weak solution to (4.2)–(4.3). For any $(u_\tau^k, v_\tau^k, \alpha_\tau^k) \in H^1(\Omega;\mathbb{R}^d) \times L^2(\Omega;\mathbb{R}^d) \times W^{1,p}(\Omega)$ solving (in the usual weak sense) the boundary value problem (4.2)–(4.3), the couple $(u_\tau^k, \alpha_\tau^k)$ is a critical point of this functional. Also, conversely, any critical point (u,α) of (4.6) gives a weak solution $(u_\tau^k, v_\tau^k, \alpha_\tau^k)$ to (4.2)–(4.3) when putting $u_\tau^k = u$, $v_\tau^k = 2(u_\tau^k - u_\tau^{k-1})/\tau - v_\tau^{k-1}$, and $\alpha_\tau^k = \alpha$. For $\tau > 0$ small enough, the mentioned convexity even ensures uniqueness to this solution which is simultaneously a global minimizer of (4.6).

The strategy (2.18) now uses the the piecewise affine and the piecewise constant interpolants respectively as

$$w_\tau = u_\tau + \chi v_\tau \quad \text{and} \quad \overline{w}_\tau = \overline{u}_\tau + \chi\dot{u}_\tau = \overline{u}_\tau + \chi\underline{v}_\tau\,. \tag{4.7}$$

Then we can write the time-discrete approximation of the force equilibrium (4.5a) as

$$\frac{\varrho}{\chi}\dot{w}_\tau - \operatorname{div}\big(\mathbb{D}_0 e(\underline{v}_\tau) + \mathbb{C}(\overline{\alpha}_\tau)e(\overline{w}_\tau)\big) = \overline{f}_\tau + \frac{\varrho}{\chi}\dot{u}_\tau + \operatorname{div}\big((\mathbb{C}(\overline{\alpha}_\tau) - \mathbb{C}(\underline{\alpha}_\tau))e(\dot{u}_\tau)\big).$$

We can test it by $\overline{w}_\tau$. By using in particular

[13] See (4.10) below. In fact, the presence of $\mathbb{D}(\alpha_\tau^{k-1})$ instead of $\mathbb{D}(\alpha_\tau^k)$ brings difficulties in proving the strong convergence of rates, because the analog of the argumentation used later in Sect. 4.2 does not work. The mentioned non-quadratic modification of the gradient term is here algorithmically tolerable because the strain energy $(e,\alpha) \mapsto \frac{1}{2}\mathbb{C}(\alpha)e\cdot e$ is not quadratic anyhow.

[14] To see it, one should analyze the Hessian on the the functional (4.6), which is a bit technical; cf. [64] for more details.

$$
\begin{aligned}
\int_Q \dot{w}_\tau\cdot\overline{w}_\tau \,\mathrm{d}x\mathrm{d}t &= \int_Q (\dot{u}_\tau + \chi\dot{v}_\tau)\cdot(\overline{u}_\tau + \chi\underline{\overline{v}}_\tau)\,\mathrm{d}x\mathrm{d}t \\
&= \int_Q (\dot{u}_\tau + \chi\dot{v}_\tau)\cdot(\underline{u}_\tau + \chi\underline{v}_\tau) + (\dot{u}_\tau + \chi\dot{v}_\tau)\cdot(\overline{u}_\tau - \underline{u}_\tau)\,\mathrm{d}x\mathrm{d}t \\
&= \int_\Omega \frac{1}{2}|u_\tau(T) + \chi v_\tau(T)|^2 - \frac{1}{2}|u_0 + \chi v_0|^2\,\mathrm{d}x + \underbrace{\frac{\tau}{2}\int_Q (\dot{u}_\tau + \chi\dot{v}_\tau)\cdot\dot{u}_\tau\,\mathrm{d}x\mathrm{d}t}_{=\mathscr{O}(\tau)}
\end{aligned}
\tag{4.8}
$$

we obtain the estimate

$$
\begin{aligned}
&\limsup_{\tau\to 0}\int_Q \chi\mathbb{D}_0 e(\underline{\overline{v}}_\tau){:}e(\underline{\overline{v}}_\tau)\,\mathrm{d}x\mathrm{d}t \le \int_\Omega \frac{\varrho}{2\chi}|u_0+\chi v_0|^2 + \frac{1}{2}\mathbb{D}_0 e(u_0){:}e(u_0)\,\mathrm{d}x \\
&\quad - \liminf_{\tau\to 0}\bigg(\int_Q \mathbb{C}(\overline{\alpha}_\tau)e(\overline{w}_\tau){:}e(\overline{w}_\tau)\,\mathrm{d}x\mathrm{d}t \\
&\qquad + \int_\Omega \frac{\varrho}{2\chi}|u_\tau(T)+\chi v_\tau(T)|^2 + \frac{1}{2}\mathbb{D}_0 e(u_\tau(T)){:}e(u_\tau(T))\,\mathrm{d}x\bigg) \\
&\quad + \lim_{\tau\to 0}\bigg(\int_Q \overline{f}_\tau\cdot\overline{w}_\tau + (\mathbb{C}(\underline{\alpha}_\tau) - \mathbb{C}(\overline{\alpha}_\tau)e(\dot{u}_\tau){:}\,e(\overline{w}_\tau)\,\mathrm{d}x\mathrm{d}t + \int_\Sigma \overline{g}_\tau\cdot\overline{w}_\tau\,\mathrm{d}S\mathrm{d}t + \mathscr{O}(\tau)\bigg) \\
&\le \int_\Omega \frac{\varrho}{2\chi}|u_0+\chi v_0|^2 + \frac{1}{2}\mathbb{D}_0 e(u_0){:}e(u_0)\,\mathrm{d}x + \int_Q f\cdot w - \mathbb{C}(\alpha)e(w){:}e(w)\,\mathrm{d}x\mathrm{d}t \\
&\qquad + \int_\Omega \frac{\varrho}{2\chi}|u(T)+\chi v(T)|^2 + \frac{1}{2}\mathbb{D}_0 e(u(T)){:}e(u(T))\,\mathrm{d}x + \int_\Sigma g\cdot w\,\mathrm{d}S\mathrm{d}t \\
&= \int_Q \chi\mathbb{D}_0 e(v){:}e(v)\,\mathrm{d}x\mathrm{d}t,
\end{aligned}
\tag{4.9}
$$

where $\mathscr{O}(\tau)$ if from (4.8). The (last) equality in (4.9) is due to the energy conservation in the limit equation (2.17). In (4.9), we used also

$$
\begin{aligned}
&\bigg|\int_Q (\mathbb{C}(\underline{\alpha}_\tau) - \mathbb{C}(\overline{\alpha}_\tau))e(\dot{u}_\tau){:}\,e(\overline{w}_\tau)\,\mathrm{d}x\mathrm{d}t\bigg| \\
&\quad \le \big\|\mathbb{C}(\underline{\alpha}_\tau) - \mathbb{C}(\overline{\alpha}_\tau)\big\|_{L^\infty(Q;\mathbb{R}^{d^4})}\big\|e(\dot{u}_\tau)\big\|_{L^2(Q;\mathbb{R}^{d\times d})}\big\|e(\overline{w}_\tau)\big\|_{L^2(Q;\mathbb{R}^{d\times d})} \to 0
\end{aligned}
\tag{4.10}
$$

Here we used the compact embedding of $L^\infty(I;W^{1,p}(\Omega))\cap H^1(I;L^2(\Omega))$ into $C(\overline{Q})$ for $p>d$. This is actually one of the spot where $p>d$ is needed.

As we already know $e(\underline{\overline{v}}_\tau)\to e(v)$ weakly in $L^2(Q;\mathbb{R}^{d\times d})$, from (4.9) we can see even the strong convergence. Since $e(\dot{u}_\tau) = e(\underline{\overline{v}}_\tau)$, it also says $e(\dot{u}_\tau)\to e(\dot{u})$ strongly, from which the desired strong convergence $e(\overline{u}_\tau)\to e(u)$ needed for the limit passage in the damage flow rule follows.

4.2 Fractional-step (staggered) discretisation

The damage problem typically involves the stored energies $\varphi = \varphi(e,\alpha)$ which are separately convex (or even separately quadratic). This encourages for an illustration of the *fractional-step method*, also called *staggered scheme*. In addition, to suppress a unwanted numerical attenuation within vibration, the time discretisation of the inertial term by the Crank-Nicholson scheme can also be considered, leading to an energy-conserving discrete scheme.

This falls into a broader class of the so-called HHT numerical integration methods devised by Hilber, Hughes, and Taylor [30], generalizing the class of Newmark's methods [57], widely used in engineering and computational physics. In fact, for a special choice of parameters,[15] the latter method gives the classical Crank-Nicolson scheme [17] here applied to a transformed system of three 1st-order equations/inclusions (2.13). Actually, the Crank-Nicolson scheme was originally devised for heat equation and later used for 2nd-order problems in the form (2.6), see e.g. [26, Ch.6, Sect.9]. It is different if applied to the dynamical equations transformed into the form (2.13); then it is sometimes called just a central-difference scheme or generalized midpoint scheme, cf. e.g. [77, Sect. 12.2] or [72, Sect. 1.6], respectively. For usage of Nemark's method in dynamical damage see e.g. [8,31,42,70].

To allow for damage acting nonlinearly, we assume $\mathbb{C}(\cdot)$ and $\phi(\cdot)$ smooth and introduce the notation

$$\mathbb{C}^\circ_{ijkl}(\alpha,\tilde\alpha) = \begin{cases} \dfrac{\mathbb{C}_{ijkl}(\alpha)-\mathbb{C}_{ijkl}(\tilde\alpha)}{\alpha-\tilde\alpha}, \\ \mathbb{C}'_{ijkl}(\alpha)=\mathbb{C}'_{ijkl}(\tilde\alpha), \end{cases} \qquad \phi^\circ(\alpha,\tilde\alpha) = \begin{cases} \dfrac{\phi(\alpha)-\phi(\tilde\alpha)}{\alpha-\tilde\alpha} & \text{if } \alpha\ne\tilde\alpha, \\ \phi'(\alpha)=\phi'(\tilde\alpha) & \text{if } \alpha=\tilde\alpha, \end{cases}$$

cf. e.g. [13, 66]. Let us note that $\mathbb{C}^\circ(\alpha_\tau^k,\alpha_\tau^{k-1}) = \mathbb{C}'$ or $\phi^\circ(\alpha_\tau^k,\alpha_\tau^{k-1}) = \phi'$ if $\mathbb{C}(\cdot)$ or $\phi(\cdot)$ are affine. It leads to the recursive boundary-value decoupled problems:

$$\frac{u_\tau^k-u_\tau^{k-1}}{\tau} = v_\tau^{k-1/2} := \frac{v_\tau^k+v_\tau^{k-1}}{2}, \tag{4.11a}$$

$$\varrho\frac{v_\tau^k-v_\tau^{k-1}}{\tau} - \mathrm{div}\,\big(\mathbb{D}(\alpha_\tau^{k-1})e(v_\tau^{k-1/2}) + \mathbb{C}(\alpha_\tau^{k-1})e(u_\tau^{k-1/2})\big) = f_\tau^k, \tag{4.11b}$$

$$\partial\zeta\Big(\frac{\alpha_\tau^k-\alpha_\tau^{k-1}}{\tau}\Big) + \frac12\mathbb{C}^\circ(\alpha_\tau^k,\alpha_\tau^{k-1})e(u_\tau^k){:}\, e(u_\tau^k) - \mathrm{div}(\kappa\nabla\alpha_\tau^{k-1/2}) \ni \phi^\circ(\alpha_\tau^k,\alpha_\tau^{k-1}) \tag{4.11c}$$

with $u_\tau^{k-1/2} := \frac12 u_\tau^k + \frac12 u_\tau^{k-1}$ and $\alpha_\tau^{k-1/2} := \frac12\alpha_\tau^k + \frac12\alpha_\tau^{k-1}$, considered on Ω while completed with the corresponding boundary conditions discretized analogously. It is to be solved recursively for $k = 1,...,T/\tau$, starting with

$$u_\tau^0 = u_0, \qquad v_\tau^0 = v_0, \qquad \alpha_\tau^0 = \alpha_0, \tag{4.12}$$

---

[15] In the standard notation used for the HHT-formula which uses three parameters, this special choice is $\alpha = \beta = 1/2$ and $\gamma = 1$.

and solving alternately (4.11a,b) and (4.11c). Both these boundary-value problems have their own potentials.

In terms of the interpolants, see (4.1), one can write the scheme (4.11) more "compactly" as

$$\dot{u}_\tau = \overline{v}_\tau \quad \text{and} \quad \varrho\dot{v}_\tau - \operatorname{div}\big(\mathbb{D}(\underline{\alpha}_\tau)e(\overline{v}_\tau) + \mathbb{C}(\underline{\alpha}_\tau)e(\overline{u}_\tau)\big) = \overline{f}_\tau, \tag{4.13a}$$

$$\partial\zeta(\dot{\alpha}_\tau) + \frac{1}{2}\mathbb{C}^\circ(\overline{\alpha}_\tau,\underline{\alpha}_\tau)e(\overline{u}_\tau): e(\overline{u}_\tau) - \operatorname{div}(\kappa\nabla\underline{\alpha}_\tau) \ni \phi^\circ(\overline{\alpha}_\tau,\underline{\alpha}_\tau). \tag{4.13b}$$

The boundary conditions can be written analogously. The basic energetic test of (4.13a) is to be done by $\dot{u}_\tau = \overline{v}_\tau$ and of (4.13b) by $\dot{\alpha}_\tau$. We can use a binomial formula several times, in particular for

$$\varrho\frac{v_\tau^k - v_\tau^{k-1}}{\tau}\cdot\frac{v_\tau^k + v_\tau^{k-1}}{2} = \frac{\frac{1}{2}\varrho|v_\tau^k|^2 - \frac{1}{2}\varrho|v_\tau^{k-1}|^2}{\tau}, \tag{4.14a}$$

$$\kappa\nabla\frac{\alpha_\tau^k + \alpha_\tau^{k-1}}{2}\cdot\nabla\frac{\alpha_\tau^k - \alpha_\tau^{k-1}}{\tau} = \frac{\frac{1}{2}\kappa|\nabla\alpha_\tau^k|^2 - \frac{1}{2}\kappa|\nabla\alpha_\tau^{k-1}|^2}{\tau}, \quad \text{and} \tag{4.14b}$$

$$\begin{aligned}&\underbrace{\mathbb{C}(\alpha_\tau^{k-1})e(u_\tau^{k-1/2}): e(v_\tau^{k-1/2})}_{=\frac{e(u_\tau^k): e(u_\tau^k) - e(u_\tau^{k-1}): e(u_\tau^{k-1})}{2\tau}} + \underbrace{\frac{1}{2}\frac{\alpha_\tau^k - \alpha_\tau^{k-1}}{\tau}\mathbb{C}^\circ(\alpha_\tau^k,\alpha_\tau^{k-1})\,e(u_\tau^k): e(u_\tau^k)}_{=(\mathbb{C}(\alpha_\tau^k) - \mathbb{C}(\alpha_\tau^{k-1}))/\tau}\\ &\qquad = \Big(\frac{1}{2}\mathbb{C}(\alpha_\tau^k)e(u_\tau^k): e(u_\tau^k) - \frac{1}{2}\mathbb{C}(\alpha_\tau^{k-1})e(u_\tau^{k-1}): e(u_\tau^{k-1})\Big)\Big/\tau;\end{aligned} \tag{4.14c}$$

note that we have enjoyed the cancellation of the terms $\pm\frac{1}{2}\mathbb{C}(\alpha_\tau^{k-1})e(u_\tau^k): e(u_\tau^k)$, cf. also [63]. Thus we obtain the discrete analog of energy equality (2.8):

$$\begin{aligned}&\int_\Omega \frac{\varrho}{2}|\dot{u}_\tau(t)|^2 + \frac{1}{2}\mathbb{C}(\alpha_\tau(t))e(u_\tau(t)): e(u_\tau(t)) - \phi(\alpha(t)) + \frac{\kappa}{2}|\nabla\alpha(t)|^2\,\mathrm{d}x\\ &\quad + \int_0^t\!\!\int_\Omega \mathbb{D}(\underline{\alpha}_\tau)e(\dot{u}_\tau): e(\dot{u}_\tau) + \dot{\alpha}_\tau\partial\zeta(\dot{\alpha}_\tau)\,\mathrm{d}x\mathrm{d}t\\ &\qquad = \int_\Omega \frac{\varrho}{2}|v_0|^2 + \frac{1}{2}\mathbb{C}(\alpha_0)e(u_0): e(u_0) - \phi(\alpha_0) + \frac{\kappa}{2}|\nabla\alpha_0|^2\,\mathrm{d}x\\ &\qquad\qquad + \int_0^t\!\!\int_\Omega \overline{f}_\tau\cdot\dot{u}_\tau\,\mathrm{d}x\mathrm{d}t + \int_0^t\!\!\int_\Gamma \overline{g}_\tau\cdot\dot{u}_\tau\,\mathrm{d}S\mathrm{d}t\end{aligned} \tag{4.15}$$

at each mesh point $t = k\tau$ with $k \in \{0,...,T/\tau\}$. Let us note that this is indeed an equality, not only an estimate. This discrete energy conservation can advantageously be used to check a-posteriori correctness of a computational code.

We introduce the variables $w_\tau^k = u_\tau^k + \chi v_\tau^k$ for $k \in \{0,...,T/\tau\}$ and the corresponding interpolants $w_\tau = u_\tau + \chi v_\tau$ and $\overline{w}_\tau = \overline{u}_\tau + \chi\overline{v}_\tau$. Likewise (2.17), we can rewrite (4.13a) as

$$\frac{\varrho}{\chi}\dot{w}_\tau - \operatorname{div}\big(\mathbb{D}_0 e(\dot{u}_\tau) + \mathbb{C}(\underline{\alpha}_\tau)e(\overline{w}_\tau)\big) = \overline{f}_\tau + \frac{\varrho}{\chi}\dot{u}_\tau. \tag{4.16}$$

To replicate the strategy (2.18), we use a test of (4.16) by $\overline{w}_\tau - w$ and the calculus

$$
\begin{aligned}
&\int_Q \varrho(\dot{w}_\tau - \dot{w})\cdot(\overline{w}_\tau - w)\,\mathrm{d}x\mathrm{d}t\\
&\quad= \int_\Omega \frac{\varrho}{2}|w_\tau(T) - w(T)|^2\,\mathrm{d}x + \int_Q \varrho(\dot{w}_\tau - \dot{w})\cdot(\overline{w}_\tau - w_\tau)\,\mathrm{d}x\mathrm{d}t\\
&\quad= \int_\Omega \frac{\varrho}{2}|w_\tau(T) - w(T)|^2\,\mathrm{d}x - \int_Q \underbrace{\varrho\dot{w}\cdot(\overline{w}_\tau - w_\tau)}_{\to 0 \text{ in } L^1(Q) \text{ weakly}}\mathrm{d}x\mathrm{d}t
\end{aligned}
\tag{4.17}
$$

because $\int_0^T \dot{w}_\tau\cdot(\overline{w}_\tau - w_\tau)\,\mathrm{d}t = 0$ a.e. on Ω. Similarly, still we use the calculus

$$
\begin{aligned}
&\int_Q \mathbb{D}_0 e(\overline{v}_\tau - v) : e(\overline{w}_\tau - w)\,\mathrm{d}x\mathrm{d}t = \int_Q \mathbb{D}_0 e(\dot{u}_\tau - \dot{u}) : e(\overline{u}_\tau - u)\\
&\quad + \chi\mathbb{D}_0 e(\overline{v}_\tau - v) : e(\overline{v}_\tau - v)\,\mathrm{d}x\mathrm{d}t = \int_\Omega \mathbb{D}_0 e(u_\tau(T) - u(T)) : e(u_\tau(T) - u(T))\,\mathrm{d}x\\
&\qquad + \int_Q \chi\mathbb{D}_0 e(\overline{v}_\tau - v) : e(\overline{v}_\tau - v) - \underbrace{\mathbb{D}_0 e(\dot{u}) : e(\overline{u}_\tau - u)}_{\to 0 \text{ in } L^1(Q) \text{ weakly}}\,\mathrm{d}x\mathrm{d}t\,.
\end{aligned}
\tag{4.18}
$$

Thus we obtain the strong convergence $e(\overline{v}_\tau) = e(\dot{u}_\tau) \to e(\dot{u})$ in $L^2(Q;\mathbb{R}^{d\times d}_{\mathrm{sym}})$, and thus also $e(\overline{u}_\tau) \to e(u)$ needed to pass to the limit in (4.13b).

When using the separately quadratic ansatz (3.2) and when combined the time discretisation (4.11) with P1 finite-element space discretisation, it gives an alternating linear-quadratic programming problems and thus very efficient numerical algorithms; in fact, it can be implemented without any iterative procedure needed, and the energy balance (4.15) is satisfied exactly up to only round-off errors.

Let us briefly illustrate this algorithm on a 2-dimensional computational experiment considering an isotropic material occupying a rectangular domain Ω. The left side of this rectangular vertically stretched specimen is left free while the right-hand side is allowed to slide. This asymmetry also causes a slight asymmetry of the solution and not completely straight fracture line, cf. Figure 1. Although the discretisation scheme is unconditionally convergent, to see reasonable numerical results, one should respect the maximal wave speed by choosing reasonably small time step, cf. the CFL-condition in the following Sect. 4.3. For details about the implementation and data and more complete presentation of an overall experiments we refer to [67].

4.3 Explicit time discretisation outlined

Implicit schemes from Sect. 4.1 and 4.2 are not causal and not much efficient for real wave propagation calculations usually containing higher frequencies in comparison with mere vibrations. For this, more often, explicit schemes are used for real wave calculations, at least in linear elastodynamic models. These time-discretisation schemes work only if combined with space discretisation.

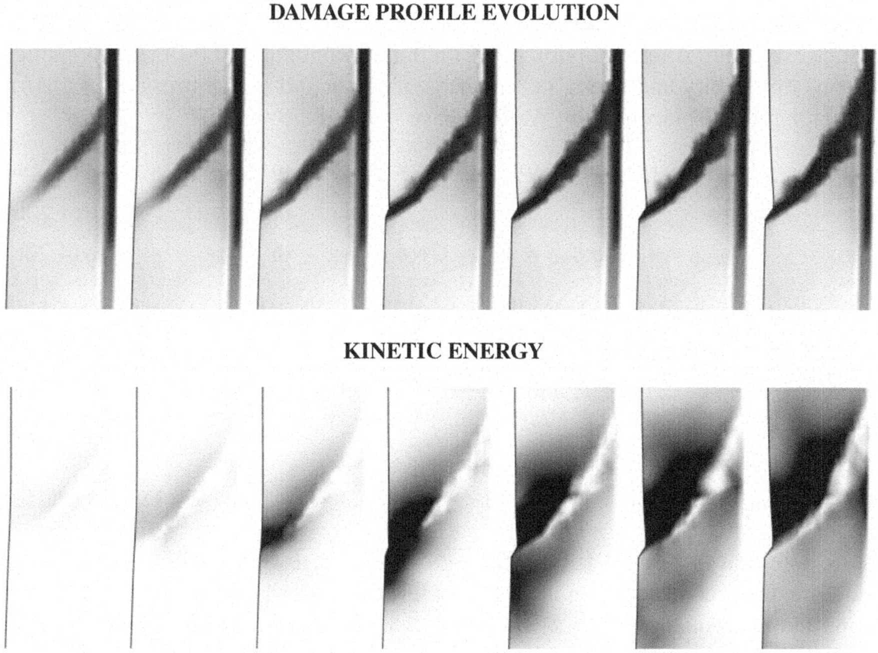

Fig. 1: *Simulations of a rupture in a two-dimensional specimen loaded by tension in a vertical direction, modelled by the phase-field crack approximation, and subsequent emission of an elastic wave. Seven selected snapshots are depicted. The decoupled energy-preserving time discretisation and* $P1$*-finite elements have been used.*
Courtesy of Roman Vodička (Technical University Košice, Slovakia)

Disregarding damage and the viscous rheology, one efficient option often considered for waves in purely elastic materials is a so-called leapfrog scheme (also known as Verlet's integration), i.e. central differences for the kinetic term

$$\mathscr{T}'\frac{v_\tau^{k+1/2}-v_\tau^{k-1/2}}{\tau}+\partial_u\mathscr{E}_h(u_\tau^k)=\mathscr{F}_\tau^k \quad \text{with} \quad v_\tau^{k+1/2}=\frac{u_\tau^{k+1}-u_\tau^k}{\tau}. \tag{4.19}$$

The test by $v_\tau^{k+1/2}$ leads to a slightly twisted energy (im)balance:

$$\mathcal{T}\big(v_\tau^{k+1/2}\big)+\frac{1}{2}\big\langle\partial_u\mathscr{E}_h(u_\tau^k),u_\tau^{k+1}\big\rangle=\mathcal{T}\big(v_\tau^{k-1/2}\big)+\frac{1}{2}\big\langle\partial_u\mathscr{E}_h(u_\tau^{k-1}),u_\tau^k\big\rangle+\big\langle\mathscr{F}_\tau^k,v_\tau^{k+1/2}\big\rangle.$$

This gives a correct kinetic energy but the stored energy is correct only asymptotically under the Courant-Friedrichs-Lewy (so-called CFL) condition [16], which needs also a space discretisation (here indicated by the abstract "mesh parameter" $h > 0$) and the time step sufficiently small with respect to h; typically $\tau < Vh$ with V the maximal speed of arising waves if h has the meaning of a size of the largest element in a finite-element discretisation.

The option (4.19) does not seem directly amenable for being merged with the damage evolution. Another option relies on the reformulation of the elastodynamics in terms of velocity and stress, i.e. in terms of $v = \dot{u}$ and of the stress $\sigma := \mathbb{C}e(u)$, eliminating the displacement u. We thus have in mind the system

$$\dot{\sigma} = \mathbb{C}e(v) \quad \text{and} \quad \varrho\dot{v} - \operatorname{div}\sigma = f \qquad \text{in } Q, \tag{4.20a}$$

$$\dot{\sigma}\mathbf{n} = \dot{g} \qquad \text{on } \Sigma, \tag{4.20b}$$

$$v|_{t=0} = v_0, \quad \sigma|_{t=0} = \sigma_0 := \mathbb{C}e(u_0) \qquad \text{in } \Omega. \tag{4.20c}$$

The explicit staggered (called also "leap-frog") time-discretisation can now be done as

$$\frac{\sigma_\tau^k - \sigma_\tau^{k-1}}{\tau} = \mathbb{C}e(v_\tau^{k-1}) \text{ and } \varrho\frac{v_\tau^k - v_\tau^{k-1}}{\tau} - \operatorname{div}\sigma_\tau^k = f_\tau^k \quad \text{in } Q, \tag{4.21a}$$

$$\frac{\sigma_\tau^k - \sigma_\tau^{k-1}}{\tau}\mathbf{n} = \frac{g_\tau^k - g_\tau^{k-1}}{\tau} \qquad \text{on } \Sigma, \tag{4.21b}$$

$$v_\tau^0 = v_0, \quad \sigma_\tau^0 = \sigma_0 := \mathbb{C}e(u_0) \qquad \text{in } \Omega. \tag{4.21c}$$

Let us note that (4.21a) is decoupled, i.e. one is first to compute σ_τ^k and then v_τ^k. Averaging the second equation in (4.21a) at level k and $k-1$ and testing it v_τ^{k-1} while testing the first equation in (4.21a) by $(\sigma_{\tau h}^k + \sigma_{\tau h}^{k-1})/2$, we obtain the approximate energy balance as

$$\frac{1}{2}\big\langle \mathcal{T}'v_\tau^k, v_\tau^{k-1}\big\rangle + \Phi_h(\sigma_\tau^k) = \frac{1}{2}\big\langle \mathcal{T}'v_\tau^{k-1}, v_\tau^{k-2}\big\rangle + \Phi_h(\sigma_\tau^{k-1}) + \big\langle \mathscr{F}_\tau^k, v_\tau^{k-1}\big\rangle, \tag{4.22}$$

with Φ the stored energy expressed in terms of stress. Now the stored energy is correct while the kinetic energy needs the CFL-condition, cf. [5]. In contrast with (4.19), this option is more compatible with possible enhancement of the stored energy by internal parameters as e.g. damage.

Assuming $\mathbb{C}(\alpha) = \gamma(\alpha)\mathbb{C}_1$ as in (3.2), we consider the energy $\Phi = \Phi(\varsigma, \alpha)$ with a "proto-stress" $\varsigma = \mathbb{C}_1 e(u)$ and with $\Phi(\varsigma, \alpha) = \int_\Omega \frac{1}{2}\gamma(\alpha)\mathbb{C}_1^{-1}\varsigma{:}\varsigma - \phi(\alpha) + \frac{\kappa}{2}|\nabla\alpha|^2\,\mathrm{d}x$; for a general concept see [65] although, in damage mechanics, this proto-stress is also called an effective stress, having a specific mechanical meaning [59]. An important trick is that the proto-stress does not explicitly involve α and its time derivate does not lead to $\dot{\alpha}$. The system (4.20) enhnaced by damage like (2.13b) then looks as

$$\dot{\varsigma} = \mathbb{C}_1 e(v) \quad \text{and} \quad \varrho\dot{v} - \operatorname{div}\sigma = f \quad \text{with} \quad \sigma = \gamma(\alpha)\varsigma \qquad \text{in } Q, \tag{4.23a}$$

$$\partial\zeta(\dot{\alpha}) + \frac{1}{2}\gamma'(\alpha)\mathbb{C}_1^{-1}\varsigma{:}\varsigma - \operatorname{div}(\kappa\nabla\alpha) \ni \phi'(\alpha) \qquad \text{in } Q, \tag{4.23b}$$

$$\dot{\sigma}\mathbf{n} = \dot{g} \quad \text{and} \quad \kappa\nabla\alpha\cdot\mathbf{n} = 0 \qquad \text{on } \Sigma, \tag{4.23c}$$

$$v|_{t=0} = v_0, \quad \sigma|_{t=0} = \sigma_0 := \mathbb{C}e(u_0), \quad \alpha|_{t=0} = \alpha_0 \qquad \text{in } \Omega. \tag{4.23d}$$

Applying the staggered discretisation like in Sect. 4.2, we obtain a 3-step scheme:

$$\frac{\varsigma_\tau^k - \varsigma_\tau^{k-1}}{\tau} = \mathbb{C}_1 e(v_\tau^{k-1}) \qquad \text{in } Q, \tag{4.24a}$$

$$\partial\zeta\Big(\frac{\alpha_\tau^k - \alpha_\tau^{k-1}}{\tau}\Big) + \frac{1}{2}\gamma^\circ(\alpha_\tau^k, \alpha_\tau^{k-1})\mathbb{C}_1^{-1}\varsigma_\tau^k : \varsigma_\tau^k \\ - \operatorname{div}(\kappa\nabla\alpha_\tau^{k-1/2}) \ni \phi^\circ(\alpha_\tau^k, \alpha_\tau^{k-1}) \qquad \text{in } Q, \tag{4.24b}$$

$$\varrho\frac{v_\tau^k - v_\tau^{k-1}}{\tau} - \operatorname{div}\sigma_\tau^k = f_\tau^k \quad \text{with} \quad \sigma_\tau^k = \gamma(\alpha_\tau^k)\varsigma_\tau^k \qquad \text{in } Q, \tag{4.24c}$$

to be completed by the respective boundary conditions.

The analysis of (4.24) is however rather nontrivial and the analog of (4.22) with the corresponding damage terms like in (4.15) contains still some other term vanishing in the limit under the CFL condition, cf. [65] for details. Even more, as there is no Kelvin-Voigt viscosity which would be troublesome for such explicit discretisation, one needs still some higher-order gradient term not subject to damage and acting on ς to guarantee convergence of such a scheme; cf. also [37, Sect.7.5.3].

To conclude, it should be mentioned that a really efficient (i.e. explicit) numerical scheme with granted stability and convergence for the simple inviscid or viscous material undergoing damage does not seem to be devised so far.

5 Concluding remarks – some modifications

Many other phenomena can be combined with the plain damage in the Kelvin-Voigt vicoelastic model considered so far. Typically one can think about more complicated viscoelastic rheologies, involving possibly some inelastic processes as plasticity, which will be in a simple variant in Sect. 5.1.

Also, some diffusant (like water in poroelastic rocks or hydrogen in metals or some solvent in polymers) can propagate through the bulk by a Fick/Darcy law, interacting with mechanical properties including fracture toughness. Of course, full thermodynamical context should involve heat production and transfer through the Fourier law. Here we only refer to [63] where a staggered energy-conserving time discretisation like in Sect. 4.2 is devised. Damage with plasticity accompanied by heat production and heat transfer allows for fitting to the popular rate-and-state-dependent friction model [62].

Moreover, the plain models from Sections 2–3 together with all these extensions can be considered within the large strains, too. We will outline it Sect. 5.2.

5.1 Combination with creep or plasticity

The Kelvin-Voigt rheology is mathematically the most basic viscoelastic rheology of parabolic type. In particular from the wave-propagation viewpoint, physically more natural is that Maxwell rheology but it is rather hyperbolic and mathematically troublesome if accompanied with inelastic processes like damage. A certain reasonable compromise it the Jeffreys' rheology combining the Norton-Hoff (also called Stokes) and Kelvin-Voigt rheology in series (or alternatively Maxwell's and Norton-Hoff's rheology in parallel). It can capture creep effects, which have sense in the shear part rather than the spherical part.

Instead (or in addition) to the linear Norton-Hoff dumper in the shear part, one can consider also the activated plastic element. The schematic rheological model is depicted in Figure 2, distinguishing also the compression and the tension in the spherical part.

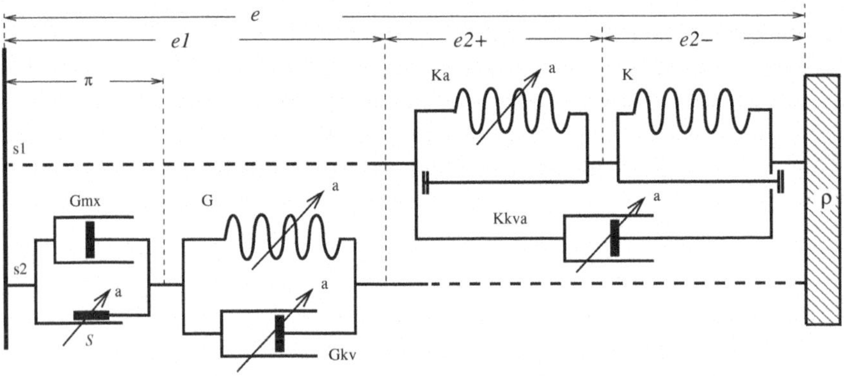

Fig. 2: *Schematic diagram for the viscoelastic Jeffreys rheology (if* $\sigma_{\rm YLD} = 0$*) which is subjected to damage* α *in the deviatoric part except undamageable creep (the* $G_{\rm NH}$*-dashpot) while the Kelvin-Voigt rheology in the spherical (volumetric) part is subjected to damage only under tension but not compression. For* $\sigma_{\rm YLD} > 0$*, it models (visco)plasticity. Evolution of damage is not depicted.*

The additional dissipation due to isochoric plastification is then achieved when damage is performed in a shear mode (i.e. Mode II) comparing to damage by opening (i.e. Mode I) where plastification is not triggered. When considering the isotropic stored energy (2.28a) with damage without any hardening-like effects (i.e. linearly-depending $K_{\rm E}(\alpha) = \alpha K$ and $G_{\rm E}(\alpha) = \alpha G$) and with $\phi(\alpha)$ also linear and the elastic strain $e - \pi$ in place of the total strain e combined with the isotropic hardening with $\sigma_{\rm DAM} = 0$, we altogether arrive at the model governed by

$$\varphi_{\rm E}(e,\alpha,\pi,\nabla\alpha,\nabla\pi) = \frac{d}{2}\Big(K_{\rm E}|{\rm sph}^- e|^2 + K_{\rm E}(\alpha)|{\rm sph}^+ e|^2\Big) + G_{\rm E}(\alpha)|{\rm dev}e{-}\pi|^2 \\ - \phi'(\alpha) + \frac{1}{2}H|\pi|^2 + \frac{\kappa_1}{2}|\nabla\pi|^2 + \frac{\kappa_2}{2}|\nabla\alpha|^2, \qquad (5.1a)$$

$$\zeta(\alpha;\dot e,\dot\alpha,\dot\pi) = \frac{d}{2}K_{\rm KV}(\alpha)|{\rm sph}\,\dot e|^2 + G_{\rm KV}(\alpha)|{\rm dev}\dot e{-}\dot\pi|^2 + \frac{1}{2}G_{\rm NH}|\dot\pi|^2 \\ + \sigma_{\rm YLD}(\alpha)|\dot\pi| + \delta^*_{[0,+\infty)}(\dot\alpha) + \frac{\nu}{2}\dot\alpha^2, \qquad (5.1b)$$

where the specific dissipation potential now contains another damper $G_{\rm NH}$ which facilitates to the Jeffrey's model in the shear part and a yield stress $\sigma_{\rm YLD} \geq 0$ possibly depending on damage, which can model activated inelastic plastic response. Starting from undamaged material, the energy needed (dissipated) by damaging in opening without plastification is just the toughness $\mathfrak{g}_{\rm c} := \phi'(1)$, while in shearing mode it is larger, namely $\mathfrak{g}_{\rm c} + \sigma_{\rm YLD}(\sqrt{2G\mathfrak{g}_{\rm c}} - \sigma_{\rm YLD})/H$ provided the parameters are tuned in a way to satisfy $\sqrt{G\mathfrak{g}_{\rm c}/2} < \sigma_{\rm YLD} \leq \sqrt{2G\mathfrak{g}_{\rm c}}$. This was first devised for an interfacial delamination model [68], being inspired just by such bulk plasticity.

Let us illustrate the staggered scheme (4.13) in the case of a linearly responding material, i.e. $K_{\rm E}|{\rm sph}^-e|^2 + K_{\rm E}(\alpha)|{\rm sph}^+e|^2$ is simplied to $K_{\rm E}(\alpha)|{\rm sph}\,e|^2$ in (5.1a), denoting $\mathbb{C}_{ijkl}(\alpha) = K_{\rm E}(\alpha)\delta_{ij}\delta_{kl} + G_{\rm E}(\alpha)(\delta_{ik}\delta_{jl} + \delta_{il}\delta_{jk} - \frac{2}{d}\delta_{ij}\delta_{kl})$ and $\mathbb{D}_{ijkl}(\alpha) = K_{\rm KV}(\alpha)\delta_{ij}\delta_{kl} + G_{\rm KV}(\alpha)(\delta_{ik}\delta_{jl} + \delta_{il}\delta_{jk} - \frac{2}{d}\delta_{ij}\delta_{kl})$ with δ standing for the Kronecker symbol. More specifically, introducing a notation for the elastic strain $e_{\rm el} = e(u){-}\pi$ and its discretisation $e_{{\rm el},\tau} = e(u_\tau){-}\pi_\tau$ and $\overline{e}_{{\rm el},\tau} = e(\overline{u}_\tau){-}\overline{\pi}_\tau$, the system (4.13) can be expanded as

$$\dot{u}_\tau = \underline{\overline{v}}_\tau \qquad \text{and} \qquad \varrho\dot{v}_\tau - {\rm div}\big(\mathbb{D}(\underline{\alpha}_\tau)\dot{e}_{{\rm el},\tau} + \mathbb{C}(\underline{\alpha}_\tau)\overline{e}_{{\rm el},\tau}\big) = \overline{f}_\tau, \tag{5.2a}$$

$$\sigma_{\rm YLD}(\underline{\alpha}_\tau){\rm Dir}(\dot{\pi}_\tau) + H\underline{\overline{\pi}}_\tau - {\rm div}(\kappa_1\nabla\underline{\overline{\pi}}_\tau) \ni {\rm dev}\big(\mathbb{D}(\underline{\alpha}_\tau)\dot{e}_{{\rm el},\tau} + \mathbb{C}(\underline{\alpha}_\tau)\underline{\overline{e}}_{{\rm el},\tau}\big), \tag{5.2b}$$

$$\partial\zeta\big(\dot{\alpha}_\tau\big) + \frac{1}{2}\mathbb{C}^\circ(\overline{\alpha}_\tau,\underline{\alpha}_\tau)\overline{e}_{{\rm el},\tau} : \overline{e}_{{\rm el},\tau} - {\rm div}(\kappa_2\nabla\underline{\overline{\alpha}}_\tau) \ni \phi^\circ(\overline{\alpha}_\tau,\underline{\alpha}_\tau)\,. \tag{5.2c}$$

The boundary conditions can be written analogously. Now, the splitting during the recursive time-stepping procedure concerns separately (5.2a,b) and (5.2c), both these boundary-value problems at particular time levels having a potential. The basic energy estimates can be obtained by testing the particular equations/inclusions in (5.2) subsequently by v_τ, $\dot{\pi}_\tau$, and $\dot{\alpha}_\tau$, using the quadratic trick several times (e.g. for $\varrho\dot{v}_\tau \cdot \underline{\overline{v}}_\tau = \frac{\partial}{\partial t}\frac{1}{2}|\underline{\overline{v}}_\tau|^2$ a.e. o Q) and the cancellation of the terms $\pm\mathbb{C}(\underline{\alpha}_\tau)\overline{e}_{{\rm el},\tau} : \overline{e}_{{\rm el},\tau}$ arising by these tests.

For the strong convergence of $\overline{e}_{{\rm el},\tau}$, instead of the strategies (2.18) or (2.21), we now rely rather on the test of (5.2a,b) respectively by $\dot{u}_\tau{-}\dot{u}$ and $\dot{\pi}_\tau{-}\dot{\pi}$. We need $\mathbb{C}$ monotone (nondecreasing) with respect to the Löwner ordering, and we use the unidirectionality of the damage evolution, i.e. $\dot{\alpha}_\tau \leq 0$. We first approximate the limit u and π, defining $\widetilde{u}_\tau^k := \frac{1}{\tau}\int_{(k-1)\tau}^{k\tau} u(t)\,{\rm d}t$ and $\widetilde{\pi}_\tau^k := \frac{1}{\tau}\int_{(k-1)\tau}^{k\tau} \pi(t)\,{\rm d}t$, and then put $\varepsilon_{{\rm el},\tau}^k := e(\widetilde{u}_\tau^k) - \widetilde{\pi}_\tau^k$. Then also the interpolants $\varepsilon_{{\rm el},\tau}$ and $\overline{\varepsilon}_{{\rm el},\tau}$ which both converges to $e_{\rm el}$ strongly. This approximation allows us to estimate[16]

[16] Here, abbreviating $E_\tau^k = e_{{\rm el},\tau}^k - \varepsilon_{{\rm el},\tau}^k$, we used the algebra

$$\begin{aligned}
&\frac{1}{2}\mathbb{C}(\alpha_\tau^k)E_\tau^k : E_\tau^k - \frac{1}{2}\mathbb{C}(\alpha_\tau^{k-1})E_\tau^{k-1} : E_\tau^{k-1}\\
&\quad= \frac{1}{2}(\mathbb{C}(\alpha_\tau^k) - \mathbb{C}(\alpha_\tau^{k-1}))E_\tau^k : E_\tau^k + \frac{1}{2}\mathbb{C}(\alpha_\tau^{k-1})(E_\tau^k : E_\tau^k - E_\tau^{k-1} : E_\tau^{k-1})\\
&\quad= \frac{1}{2}(\alpha_\tau^k - \alpha_\tau^{k-1})\mathbb{C}^\circ(\alpha_\tau^k, \alpha_\tau^{k-1})E_\tau^k : E_\tau^k + \mathbb{C}(\alpha_\tau^{k-1})\Big(\frac{E_\tau^k + E_\tau^{k-1}}{2}\Big) : (E_\tau^k - E_\tau^{k-1})\,.
\end{aligned}$$

$$
\begin{aligned}
&\frac{1}{2}\mathbb{C}(\alpha_\tau(T))(e_{\mathrm{el},\tau}(T)-\varepsilon_{\mathrm{el},\tau}(T))\colon(e_{\mathrm{el},\tau}(T)-\varepsilon_{\mathrm{el},\tau}(T))\\
&=\int_0^T\Big(\mathbb{C}(\underline{\alpha}_\tau)(\underline{e}_{\mathrm{el},\tau}-\underline{\varepsilon}_{\mathrm{el},\tau})\colon(\dot{e}_{\mathrm{el},\tau}-\dot{\varepsilon}_{\mathrm{el},\tau})-\frac{1}{2}\dot{\alpha}_\tau\mathbb{C}^\circ(\overline{\alpha}_\tau,\underline{\alpha}_\tau)(\underline{e}_{\mathrm{el},\tau}-\underline{\varepsilon}_{\mathrm{el},\tau})\colon\\
&\qquad\colon(\underline{e}_{\mathrm{el},\tau}-\underline{\varepsilon}_{\mathrm{el},\tau})\Big)\,\mathrm{d}t\geq\int_0^T\mathbb{C}(\underline{\alpha}_\tau)(\underline{e}_{\mathrm{el},\tau}-\underline{\varepsilon}_{\mathrm{el},\tau})\colon(\dot{e}_{\mathrm{el},\tau}-\dot{\varepsilon}_{\mathrm{el},\tau})\,\mathrm{d}t
\end{aligned}
$$

a.e. on Ω. When integrate over Ω, this allows us to estimate:

$$
\begin{aligned}
&\int_\Omega\frac{1}{2}\mathbb{C}(\alpha_\tau(T))(e_{\mathrm{el},\tau}(T)-\varepsilon_{\mathrm{el},\tau}(T))\colon(e_{\mathrm{el},\tau}(T)-\varepsilon_{\mathrm{el},\tau}(T))\,\mathrm{d}x\\
&\qquad+\int_Q\mathbb{D}(\underline{\alpha}_\tau)(\dot{e}_{\mathrm{el},\tau}-\dot{\varepsilon}_{\mathrm{el},\tau})\colon(\dot{e}_{\mathrm{el},\tau}-\dot{\varepsilon}_{\mathrm{el},\tau})\,\mathrm{d}x\mathrm{d}t\\
&\leq\int_Q\Big(\mathbb{D}(\underline{\alpha}_\tau)(\dot{e}_{\mathrm{el},\tau}-\dot{\varepsilon}_{\mathrm{el},\tau})+\mathbb{C}(\underline{\alpha}_\tau)(\underline{e}_{\mathrm{el},\tau}-\underline{\varepsilon}_{\mathrm{el},\tau})\Big)\colon(\dot{e}_{\mathrm{el},\tau}-\dot{\varepsilon}_{\mathrm{el},\tau})\,\mathrm{d}x\mathrm{d}t\\
&=\int_Q(\overline{f}_\tau-\varrho\dot{v}_\tau)\cdot(\underline{v}_\tau-\dot{\tilde{u}}_\tau)-\Big(\mathbb{D}(\underline{\alpha}_\tau)\dot{\varepsilon}_{\mathrm{el},\tau}+\mathbb{C}(\underline{\alpha}_\tau)\underline{\varepsilon}_{\mathrm{el},\tau}\Big)\colon(\dot{e}_{\mathrm{el},\tau}-\dot{\varepsilon}_{\mathrm{el},\tau})\\
&\qquad+\sigma_{\rm YLD}(\alpha_\tau)(|\dot{\pi}|-|\dot{\pi}_\tau|)+H\underline{\pi}_\tau\colon(\dot{\pi}-\dot{\pi}_\tau)+\kappa_1\nabla\underline{\pi}_\tau\vdots\nabla(\dot{\pi}-\dot{\pi}_\tau)\to 0.
\end{aligned}
\tag{5.3}
$$

From this $\dot{e}_{\mathrm{el},\tau}-\dot{\varepsilon}_{\mathrm{el},\tau}\to 0$ strongly in $L^2(Q;\mathbb{R}^{d\times d}_{\mathrm{sym}})$. Since $\dot{\varepsilon}_{\mathrm{el},\tau}\to\dot{e}_{\mathrm{el}}$, we obtain $\dot{e}_{\mathrm{el},\tau}\to\dot{e}_{\mathrm{el}}$ and hence also $\overline{e}_{\mathrm{el},\tau}\to e_{\mathrm{el}}$ strongly in $L^2(Q;\mathbb{R}^{d\times d}_{\mathrm{sym}})$ needed for the limit passage in (5.2c). The convergence in the other terms in (5.2) is then simple.

When built into the *phase-field fracture* model of the type (3.2), we obtain the *mode-sensitive fracture*. Since the fracture toughness $\mathfrak{g}_{\rm c}$ is scaled as $\mathscr{O}(1/\varepsilon)$ in (3.2), $\sigma_{\rm YLD}$ in (5.1b) is to be scaled as $\mathscr{O}(1/\sqrt{\varepsilon})$. A combination of damage in its *phase-field fracture* or the crack approximation with plasticity is referred to (an approximation of) *ductile cracks*, in contrast to *brittle cracks* without possibility of plastification on the crack tips. The idea to involve plastification processes into fracture mechanics is due to G. Irwin [32].

A combination of damage with the *perfect plasticity* (i.e. $G_{\rm NH}=H=0$ and $\kappa_1=0$) has been analysed in [18] by using the strategy (5.3) except that $\ddot{u}$ was approximated by the backward second time difference.

5.2 Damage models at large strains

In some applications, the small-strain approximation is not appropriate and one must take into account large strains. In solid mechanics, mathematical analysis is to be performed in the fixed reference configuration $\Omega\subset\mathbb{R}^d$, the deformation being $y:\Omega\to\mathbb{R}^d$. The stored energy then depends on the deformation gradient ∇y, and can be considered enhanced as

$$\varphi_{\rm E}(F,\nabla F,\alpha,\nabla\alpha):=\varphi(F,\alpha)+\mathscr{H}(\nabla F)+\delta_{[0,1]}(\alpha)+\mathscr{G}(F,\nabla\alpha)$$
$$\text{with}\quad \mathscr{H}(G):=\frac{1}{4}\int_{\Omega\times\Omega}(G(x){-}G(\widetilde x)):\mathbb{K}(x{-}\widetilde x):(G(x){-}G(\widetilde x))\,{\rm d}x{\rm d}\widetilde x$$
$$\text{and with}\quad \mathscr{G}(F,\nabla\alpha)=\frac{\kappa}{2}|\nabla\alpha|^2\quad\text{or}\quad\frac{\kappa}{2}|F^{-\top}\nabla\alpha|^2. \tag{5.4}$$

The former option in (5.4) is the gradient-damage theory in the material (reference) configuration, and is mathematically simpler and even a local nonsimple-material concept $\mathscr{H}(G)=\frac{1}{2}\int_\Omega G:\mathbb{K}:G\,{\rm d}x$ might be used.

The latter, mathematically more difficult option is in the actual (deformed) configuration, the factor $F^{-\top}:=(F^{-1})^\top$ being the push-forward transformation of the vector $\nabla\alpha$ from the reference configuration into the actual one. Both options are relevant in particular situations, the latter one being mathematically more difficult and, except [37, Sect. 9.5.1], has been so far rather devised without any rigorous proofs, cf. e.g. [58, 76]. In this latter option, the system resulted via the extended Hamilton variational principle (2.1)–(2.2) then reads as

$$\varrho\ddot y-{\rm div}\big(\partial_F\varphi(\nabla y,\alpha)+\sigma_{\rm K}(\nabla y,\nabla\alpha)-{\rm div}\,\mathfrak{H}(\nabla^2y)\big)=f$$
$$\text{with}\quad\big[\mathfrak{H}(G)\big](x)=\int_\Omega\mathbb{K}(x{-}\widetilde x)(G(x)-G(\widetilde x))\,{\rm d}\widetilde x\quad\text{and}$$
$$\text{with}\quad\sigma_{\rm K}(F,\nabla\alpha)=\kappa F^{-\top}{:}(F^{-\top})'{:}(\nabla\alpha\otimes\nabla\alpha)\qquad\text{in }Q, \tag{5.5a}$$
$$\partial\zeta(\dot\alpha)+\partial_\alpha\varphi(\nabla y,\alpha)\ni{\rm div}\big(\kappa(\nabla y)^{-1}(\nabla y)^{-\top}\nabla\alpha\big)\qquad\text{in }Q, \tag{5.5b}$$
$$(\partial_F\varphi(\nabla y,\alpha)+\sigma_{\rm K}(\nabla y,\nabla\alpha))\mathbf{n}-{\rm div}_{\rm S}((\mathfrak{H}\nabla^2y){\cdot}\mathbf{n})=g$$
$$\text{and}\quad(\mathfrak{H}\nabla^2y){:}\,(\mathbf{n}\otimes\mathbf{n})=0\qquad\text{on }\Sigma, \tag{5.5c}$$
$$\kappa(\nabla y)^{-1}(\nabla y)^{-\top}\nabla\alpha{\cdot}\mathbf{n}=0\qquad\text{on }\Sigma, \tag{5.5d}$$

where ${\rm div}_{\rm S}$ is the surface divergence. The analysis now needs the strong convergence of $\nabla\alpha$ which now occurs nonlinearly in the Korteweg-like stress $\sigma_{\rm K}=\sigma_{\rm K}(F,\nabla\alpha)$ in (5.5a). An important aspect is that we need to have a control over $(\nabla y)^{-1}$, i.e. $\det(\nabla y)$ should be kept surely away 0. As also physically desirable, this can be ensured by by preventing local self-interpenetration by assuming a singulariy in the stored energy when $\det F\to 0+$. More specifically, the potential φ is to be qualified as

$$\varphi:{\rm GL}^+(d)\times[0,1]\to\mathbb{R}\quad\text{continuously differentiable and} \tag{5.6a}$$
$$\exists\epsilon>0\ \forall F\in{\rm GL}^+(d),\ \alpha\in[0,1]:$$
$$\varphi(F,\alpha)\ge\frac{\epsilon}{(\det F)^q}\quad\text{with}\quad q>\frac{2d}{2\gamma{+}2{-}d}\quad\text{for some}\quad\gamma>\frac{d}{2}-1 \tag{5.6b}$$

while $\varphi(F,\alpha)=+\infty$ if $\det F\le 0$, where γ related to the qualification of the kernel $\mathbb{K}$ in (5.4) as

$$\exists\varepsilon>0\,\forall x\in\Omega,\ F\in\mathbb{R}^{d\times d}:\quad\Big(\frac{\varepsilon|F|^2}{|x|^{d+2\gamma}}-\frac{1}{\varepsilon}\Big)^+\le F{:}\,\mathbb{K}(x){:}\,F\le\frac{|F|^2}{\varepsilon|x|^{d+2\gamma}}. \tag{5.6c}$$

Together with the intertial term, (5.6c) grants coercivity in $H^{2+\gamma}(\Omega;\mathbb{R}^d)$ which is embedded, by (5.6b), into $W^{2,p}(\Omega;\mathbb{R}^d)$ with $p > d$.

Again we use Galerkin approximation and denote the approximate solution by (y_k, α_k). Testing (5.5a,b) in its Galerkin approximation by $(\dot{y}_k, \dot{\alpha}_k)$, we obtain the estimates

$$\|y_k\|_{L^\infty(I;H^{2+\gamma}(\Omega;\mathbb{R}^d))\cap W^{1,\infty}(I;L^2(\Omega;\mathbb{R}^d))} \le K, \qquad \Big\|\frac{1}{\det(\nabla y_k)}\Big\|_{L^\infty(Q)} \le K, \tag{5.7a}$$

$$\|\alpha_k\|_{L^\infty(Q)} \le K \quad \text{and} \quad \|(\nabla y_k)^{-\top}\nabla\alpha_k\|_{L^\infty(I;L^2(\Omega;\mathbb{R}^d))} \le K\,. \tag{5.7b}$$

For the latter estimate in (5.7a), we use the result by Healey and Krömer [29], which also excludes the Lavrentiev phenomenon.

Based on the weak convergence $y_k \to y$ and $\alpha_k \to \alpha$ and the Aubin-Lions compactness arguments, we prove the convergence in the damage flow rule.

To prove the mentioned strong convergence $\nabla\alpha_k$, we use the uniform (with respect to y) strong monotonicity of the mapping $\alpha \mapsto -\mathrm{div}\big((\nabla y)^{-1}\kappa(\nabla y)^{-\top}\nabla\alpha\big)$. Taking $\widetilde{\alpha}_k$ an approximation of α valued in the respective finite-dimensional spaces used for the Galerkin approximation and converging to α strongly, we can test (5.5b) in its Galerkin approximation by $\alpha_k-\widetilde{\alpha}_k$ and use it in the estimate

$$\begin{aligned}\limsup_{k\to\infty}\int_Q &(\nabla y_k)^{-1}\kappa(\nabla y_k)^{-\top}\nabla(\alpha_k-\widetilde{\alpha}_k)\cdot\nabla(\alpha_k-\widetilde{\alpha}_k)\,\mathrm{d}x\mathrm{d}t\\ &= \lim_{k\to\infty}\int_Q \big(\partial_\alpha\varphi(\nabla y_k,\alpha_k) + \partial\zeta(\dot{\alpha}_k)\big)(\widetilde{\alpha}_k-\alpha_k)\\ &\qquad - (\nabla y_k)^{-1}\kappa(\nabla y_k)^{-\top}\nabla\widetilde{\alpha}_k\cdot\nabla(\alpha_k-\widetilde{\alpha}_k)\,\mathrm{d}x\mathrm{d}t = 0\end{aligned}$$

because $\partial_\alpha\varphi(\nabla y_k,\alpha_k) + \partial\zeta(\dot{\alpha}_k)$ is bounded in $L^2(Q)$ while $\widetilde{\alpha}_k-\alpha_k \to 0$ strongly in $L^2(Q)$ by the Aubin-Lions compactness theorem and because $(\nabla y_k)^{-1}\kappa(\nabla y_k)^{-\top}\nabla\widetilde{\alpha}_k$ converges strongly in $L^2(Q;\mathbb{R}^d)$ while $\nabla(\alpha_k-\widetilde{\alpha}_k) \to 0$ weakly in $L^2(Q;\mathbb{R}^d)$. As $(\nabla y_k)^{-1}\kappa(\nabla y_k)^{-\top}$ is uniformly positive definite, we thus obtain that $\nabla(\alpha_k-\widetilde{\alpha}_k) \to 0$ strongly in $L^2(Q;\mathbb{R}^d)$, and thus $\nabla\alpha_k \to \nabla\alpha$ strongly in $L^2(Q;\mathbb{R}^d)$. Then we have the convergence in the Korteweg-like stress $\sigma_{\mathrm{K}}(\nabla y_k,\nabla\alpha_k) \to \sigma_{\mathrm{K}}(\nabla y,\nabla\alpha)$ even strongly in $L^p(I;L^1(\Omega;\mathbb{R}^{d\times d}))$ for any $1 \le p < +\infty$. The limit passage in the force equilibrium towards (5.5a) formulated weakly is then straightforward.

Acknowledgements Special thanks are to Roman Vodička for providing sample snapshots from numerical simulations presented in Fig. 1. Also discussion with Vladislav Mantič about finite fracture mechanics and coupled criterion has been extremely useful, as well as discussions with Martin Kružík about the actual gradient of damage at large strains. Careful reading of the manuscript and many comments of Elisa Davoli and Roman Vodička, as well as of an anonymous referee are also appreciated very much. A partial support from the Czech Science Foundation projects 17-04301S and 19-04956S, the institutional support RVO: 61388998 (ČR), and also by the Austrian-Czech project 16-34894L (FWF/CSF) are acknowledged, too.

References

1. Allaire, G., Jouve, F., Goethem, N.V.: A level set method for the numerical simulation of damage evolution. In: R. Jeltsch and G. Wanner (ed.) Proc. ICIAM 2007, pp. 3–22. EMS, Zürich (2009)
2. Ambrosio, L., Tortorelli, V.M.: Approximation of functional depending on jumps via by elliptic functionals via Γ-convergence. Comm. Pure Appl. Math. **43**, 999–1036 (1990)
3. Ambrosio, L., Tortorelli, V.M.: On the approximation of free discontinuity problems. Bollettino Unione Mat. Italiana **7**, 105–123 (1992)
4. Bažant, Z., Jirásek, M.: Nonlocal integral formulations of plasticity and damage: Survey of progress. J. Engr. Mech. ASCE **128**, 1119–1149 (2002)
5. Bécache, E., Joly, P., Tsogka, C.: A new family of mixed finite elements for the linear elastodynamic problem. SIAM J. Numer. Anal. **39**, 2109–2132 (2002)
6. Bedford, A.: Hamilton's Principle in Continuum Mechanics. Pitman, Boston (1985)
7. Borden, M.: Isogeometric analysis of phase-field models for dynamic brittle and ductile fracture. Ph.D. thesis, Univ. of Texas, Austin (2012)
8. Borden, M., Verhoosel, C., Scott, M., Hughes, T., Landis, C.: A phase-field description of dynamic brittle fracture. Comput. Meth. Appl. Mech. Engr. **217–220**, 77–95 (2012)
9. Bourdin, B., Francfort, G.A., Marigo, J.J.: The variational approach to fracture. J. Elasticity **91**, 5–148 (2008)
10. Bourdin, B., Larsen, C.J., Richardson, C.L.: A time-discrete model for dynamic fracture based on crack regularization. Int. J. of Fracture **10**, 133–143 (2011)
11. Bourdin, B., Marigo, J.J., Maurini, C., Sicsic, P.: Morphogenesis and propagation of complex cracks induced by thermal shocks. Phys. Rev. Lett. **112**, 014301 (2014)
12. Caponi, M.: Existence of solutions to a phase-field model of dynamic fracture with a crack-dependent dissipation. Preprint SISSA 06/2018/MATE
13. Condette, N., Melcher, C., Süli, E.: Spectral approximation of pattern-forming nonlinear evolution equations with double-well potentials of quadratic growth. Math. Comp. **80**, 205–223 (2011)
14. Conti, S., Focardi, M., Iurlano, F.: Phase field approximation of cohesive fracture models. Ann. Inst. H. Poincaré Anal. Non Linéaire **33**(4), 1033–1067 (2016).
15. Cornetti, P., Pugno, N., Carpinteri, A., Taylor, D.: Finite fracture mechanics: a coupled stress and energy failure criterion. Engineering Fracture Mechanics **73**, 2021–2033 (2006)
16. Courant, R., Friedrichs, K., Lewy, H.: Über die partiellen Differenzengleichungen der mathematischen Physik. Mathematische Annalen **100**, 32–74 (1928)
17. Crank, J., Nicolson, P.: A practical method for numerical evaluation of solutions of partial differential equations of the heat conduction type. Proc. Camb. Phil. Soc. **43**, 50–67 (1947)
18. Davoli, E., Roubíček, T., Stefanelli, U.: Dynamic perfect plasticity and damage in visco-elastic solids. In preparation
19. Focardi, M.: On the variational approximation of free-discontinuity problems in the vectorial case. Math. Models Methods Appl. Sci. **11**, 663–684 (2001)
20. Focardi, M., Iurlano, F.: Asymptotic analysis of Ambrosio-Tortorelli energies in linearized elasticity. SIAM J. Math. Anal. **46**(4), 2936–2955 (2014).
21. Frémond, M.: Non-Smooth Thermomechanics. Springer, Berlin (2002)
22. Frémond, M.: Phase Change in Mechanics. Springer, Berlin (2012)
23. Freund, L.B.: Dynamic Fracture Mechanics. Cambridge Univ. Press (1998)
24. García, I.G., Carter, B.J., Ingraffea, A.R., Mantič, V.: A numerical study of transverse cracking in cross-ply laminates by 3D finite fracture mechanics. Composites Part B **95**, 475–487 (2016)
25. Giacomini, A.: Ambrosio-Tortorelli approximation of quasi-static evolution of brittle fractures. Calc. Var. Partial Diff. Eqs. **22**, 129–172 (2005)
26. Glowinski, R., Lions, J.L., Trémolières, R.: Numerical analysis of variational inequalities. North-Holland, Amsterdam (1981). (French original Dunod, Paris, 1976)

27. Griffith, A.A.: The phenomena of rupture and flow in solids. Phil. Trans. R. Soc. London A **221**, 163–198 (1921)
28. Hashin, Z.: Finite thermoelastic fracture criterion with application to laminate cracking analysis. J. Mech. Phys. Solids **44**, 1129–1145 (1996)
29. Healey, T., Krömer, S.: Injective weak solutions in second-gradient nonlinear elasticity. ESAIM: Control, Optim. & Cal. Var. **15**, 863–871 (2009)
30. Hilber, H.M., Hughes, T.J.R., Taylor, R.L.: Improved numerical dissipation for time integration algorithms in structural dynamics. Earthquake Eng. Struct. Dyn. **5**, 283–292 (1977)
31. Hofacker, M., Miehe, C.: Continuum phase field modeling of dynamic fracture: variational principles and staggered FE implementation. Int J Fract **178**, 113–129 (2012)
32. Irwin, G.: Analysis of stresses and strains near the end of a crack traversing a plate. J. Appl. Mech. **24**, 361–364 (1957)
33. Iurlano, F.: A density result for GSBD and its application to the approximation of brittle fracture energies. Calc. Var. Partial Diff. Eqs. **51**(1-2), 315–342 (2014).
34. Jirásek, M.: Nonlocal theories in continuum mechanics. Acta Polytechnica **44**, 16–34 (2004)
35. Kachanov, L.: Time of rupture process under creep conditions. Izv. Akad. Nauk SSSR **8**, 26 (1958)
36. Knees, D., Negri, M.: Convergence of alternate minimization schemes for phase field fracture and damage. Math. Models Methods Appl. Sci. **27**, 1743–1794 (2017)
37. Kružík, M., Roubíček, T.: Mathematical Methods in Continuum Mechanics of Solids. Springer, Switzeland (2019)
38. Lancioni, G., Royer-Carfagni, G.: The Variational Approach to Fracture Mechanics. A Practical Application to the French Panthéon in Paris. J. Elasticity **95**, 1–30 (2009)
39. Larsen, C.J., Ortner, C., Süli, E.: Existence of solution to a regularized model of dynamic fracture. Math. Models Meth. Appl. Sci. **20**, 1021–1048 (2010)
40. Lazzaroni, G., Rossi, R., Thomas, M., Toader, R.: Rate-independent damage in thermo-viscoelastic materials with inertia. J. Dynam. Diff. Eqs. **30**, 1311–1364 (2018)
41. Leguillon, D.: Strength or toughness? A criterion for crack onset at a notch. European J. of Mechanics A/Solids **21**, 61–72 (2002)
42. Li, T., Marigo, J.J., Guilbaud, D., Potapov, S.: Numerical investigation of dynamic brittle fracture via gradient damage models. Adv. Model. and Simul. in Eng. Sci. **3**, 26 (2016)
43. Mantič, V.: Interface crack onset at a circular cylindrical inclusion under a remote transverse tension. Application of a coupled stress and energy criterion. Intl. J. Solids Structures **46**, 1287–1304 (2009)
44. Mantič, V.: Prediction of initiation and growth of cracks in composites. Coupled stress and energy criterion of the finite fracture mechanics. In: ECCM-16th Europ. Conf. on Composite Mater. 2014, pp. 1–16. Europ. Soc. Composite Mater. (ESCM), http://www.escm.eu.org/eccm16/assets/1252.pdf (2014)
45. Marigo, J.J., Maurini, C., Pham, K.: An overview of the modelling of fracture by gradient damage models. Meccanica **51**, 3107–3128 (2016)
46. Maugin, G.A.: The saga of internal variables of state in continuum thermo-mechanics (1893-2013). Mechanics Research Communications **69**, 79–86 (2015)
47. Miehe, C., Welschinger, F., Hofacker, M.: Thermodynamically consistent phase-field models of fracture: Variational principles and multi-field FE implementations. Intl. J. Numer. Meth. Engr. **83**, 1273–1311 (2010)
48. Mielke, A.: Evolution in rate-independent systems (Ch. 6). In: C. Dafermos, E. Feireisl (eds.) Handbook of Differential Equations, Evolutionary Equations, vol. 2, pp. 461–559. Elsevier B.V., Amsterdam (2005)
49. Mielke, A., Roubíček, T.: Rate-Independent Systems – Theory and Application. Springer, New York (2015)
50. Mielke, A., Roubíček, T., Stefanelli, U.: Γ-limits and relaxations for rate-independent evolutionary problems. Calc. Var. Part. Diff. Eqns. **31**, 387–416 (2008)
51. Mielke, A., Roubíček, T., Thomas, M.: From damage to delamination in nonlinearly elastic materials at small strains. J. Elasticity **109**, 235–273 (2012)

52. Mielke, A., Roubíček, T., Zeman, J.: Complete damage in elastic and viscoelastic media and its energetics. Comput. Methods Appl. Mech. Engrg. **199**, 1242–1253 (2010)
53. Mielke, A., Theil, F.: On rate-independent hysteresis models. Nonl. Diff. Eqns. Appl. **11**, 151–189 (2004)
54. Mumford, D., Shah, J.: Optimal approximations by piecewise smooth functions and associated variational problems. Comm. Pure Appl. Math. **42**, 577–685 (1989)
55. Murakami, S.: Continuum Damage Mechanics. Springer, Dordrecht (2012)
56. Negri, M.: Quasi-static evolutions in brittle fracture generated by gradient flows: sharp crack and phase-field approaches. In: K. Weinberg, A. Pandolfi (eds.) Innovative Numerical Approaches for Multi-Field and Multi-Scale Problems, pp. 197–216. Springer (2016)
57. Newmark, N.: A method of computation for structural dynamics. J. Eng. Mech. Div. **85**, 67–94 (1959)
58. Pawłow, I.: Thermodynamically consistent Cahn-Hilliard and Allen-Cahn model in elastic solids. Disc. Cont. Dynam. Syst. **15**, 1169–1191 (2006)
59. Rabotnov, Y.: Creep problems in structural members. North-Holland, Amsterdam (1969)
60. Roubíček, T.: Rate independent processes in viscous solids at small strains. Math. Methods Appl. Sci. **32**, 825–862 (2009). Erratum Vol. 32(16) p. 2176
61. Roubíček, T.: Nonlinear Partial Differential Equations with Applications, 2nd edn. Birkhäuser, Basel (2013)
62. Roubíček, T.: A note about the rate-and-state-dependent friction model in a thermodynamical framework of the Biot-type equation. Geophys. J. Intl., **199**, 286–295 (2014)
63. Roubíček, T.: An energy-conserving time-discretisation scheme for poroelastic media with phase-field fracture emitting waves and heat. Disc. Cont. Dynam. Syst. S **10**, 867–893 (2017)
64. Roubíček, T.: Coupled time discretisation of dynamic damage models at small strains. IMA J. Numer. Anal. (2019, on line: DOI 10.1093/imanum/drz014)
65. Roubíček, T., Panagiotopoulos, C., Tsogka, C.: Explicit time-discretisation of elastodynamics with some inelastic processes at small strains. (Preprint arXiv no.1903.11654. Submitted.)
66. Roubíček, T., Panagiotopoulos, C.G.: Energy-conserving time discretization of abstract dynamic problems with applications in continuum mechanics of solids. Numer. Funct. Anal. Optim. **38**, 1143–1172 (2017)
67. Roubíček, T., Vodička, R.: A monolithic model for seismic sources and seismic waves. Intl. J. Fracture, submitted
68. Roubíček, T., Mantič, V., Panagiotopoulos, C.G.: Quasistatic mixed-mode delamination model. Disc. Cont. Dynam. Syst. - S **6**, 591–610 (2013)
69. Sargado, J., Keilegavlen, E., Berre, I., Nordbotten, J.: High-accuracy phase-field models for brittle fracture based on a new family of degradation functions. J. Mech. Phys. Solids **111**, 458–489 (2018)
70. Schlüter, A., Willenbücher, A., Kuhn, C., Müller, R.: Phase Field Approximation of Dynamic Brittle Fracture. Comput Mech **54**, 1141–1161 (2014)
71. Sicsic, P., Marigo, J.J.: From gradient damage laws to Griffith's theory of crack propagation. J. Elasticity **113**, 55–74 (2013)
72. Simo, J.C., Hughes, J.R.: Computational Inelasticity. Springer, Berlin (1998)
73. Taylor, D., Cornetti, P., Pugno, N.: The fracture mechanics of finite crack extension. Engineering Fracture Mechanics **72**, 1021–1038 (2005)
74. Thomas, M., Mielke, A.: Damage of nonlinearly elastic materials at small strain. Existence and regularity results. Zeitschrift angew. Math. Mech. **90**, 88–112 (2010)
75. Vignollet, J., May, S., de Borst, R., Verhoosel, C.: Phase-field models for brittle and cohesive fracture. Meccanica **49**, 2587–2601 (2014)
76. Waffenschmidt, T., Polindara, C., Menzel, A., Blanco, S.: A gradient-enhanced large-deformation continuum damage model for fibre-reinforced materials. Comput. Methods Appl. Mech. Engrg. **268**, 801–842 (2014)
77. Wang, L.: Foundations of Stress Waves. Elsevier, Amsterdam (2007)
78. Weißgraeber, P., Leguillon, D., Becker, W.: A review of Finite Fracture Mechanics: crack initiation at singular and non-singular stress raisers. Arch. Appl. Mech. **86**, 375–401 (2016)

79. Wu, J.Y., Nguyen, V.: A length scale insensitive phase-field damage model for brittle fracture. J. Mech. Phys. Solids **119**, 20–42 (2018)
80. Xavier, M., E.Fancello, Farias, J., Goethem, N.V., Novotny, A.: Topological Derivative-Based Fracture Modelling in Brittle Materials: A Phenomenological Approach. Eng. Frac. Mech. **179**, 13–27 (2017)

The manufacturer's authorised representative in the EU is Springer Nature Customer Service Centre GmbH, Europaplatz 3, 69115 Heidelberg, Germany. If you have any concerns regarding our products, please contact ProductSafety@springernature.com

Printed and bound by CPI Group (UK) Ltd, Croydon, CR0 4YY
30/09/2026
02225652-0015